T0387164

INFLAMMATION AND NATURAL PRODUCTS

INFLAMMATION AND NATURAL PRODUCTS

Edited by

SREERAJ GOPI

Aurea Biolabs (P) Ltd, Kolenchery, Cochin, Kerala, India

AUGUSTINE AMALRAJ

Aurea Biolabs (P) Ltd, Kolenchery, Cochin, Kerala, India

AJAIKUMAR KUNNUMAKKARA

Indian Institute of Technology Guwahati, Guwahati, Assam, India

SABU THOMAS

Mahatma Gandhi University, Kottayam, Kerala, India

Academic Press is an imprint of Elsevier
125 London Wall, London EC2Y 5AS, United Kingdom
525 B Street, Suite 1650, San Diego, CA 92101, United States
50 Hampshire Street, 5th Floor, Cambridge, MA 02139, United States
The Boulevard, Langford Lane, Kidlington, Oxford OX5 1GB, United Kingdom

© 2021 Elsevier Inc. All rights reserved.

No part of this publication may be reproduced or transmitted in any form or by any means, electronic or mechanical, including photocopying, recording, or any information storage and retrieval system, without permission in writing from the publisher. Details on how to seek permission, further information about the Publisher's permissions policies and our arrangements with organizations such as the Copyright Clearance Center and the Copyright Licensing Agency, can be found at our website: www.elsevier.com/permissions.

This book and the individual contributions contained in it are protected under copyright by the Publisher (other than as may be noted herein).

Notices
Knowledge and best practice in this field are constantly changing. As new research and experience broaden our understanding, changes in research methods, professional practices, or medical treatment may become necessary.

Practitioners and researchers must always rely on their own experience and knowledge in evaluating and using any information, methods, compounds, or experiments described herein. In using such information or methods they should be mindful of their own safety and the safety of others, including parties for whom they have a professional responsibility.

To the fullest extent of the law, neither the Publisher nor the authors, contributors, or editors, assume any liability for any injury and/or damage to persons or property as a matter of products liability, negligence or otherwise, or from any use or operation of any methods, products, instructions, or ideas contained in the material herein.

Library of Congress Cataloging-in-Publication Data
A catalog record for this book is available from the Library of Congress

British Library Cataloguing-in-Publication Data
A catalogue record for this book is available from the British Library

ISBN: 978-0-12-819218-4

For information on all Academic Press publications
visit our website at https://www.elsevier.com/books-and-journals

Publisher: Andre Gerhard Wolff
Acquisitions Editor: Erin Hill-Parks
Editorial Project Manager: Billie Jean Fernandez
Production Project Manager: Omer Mukthar
Cover Designer: Christian J. Bilbow

Typeset by SPi Global, India

Contents

Contributors

Mohammad H. Abukhalil Department of Biology, Faculty of Science, Al-Hussein Bin Talal University, Ma'an, Jordan

Augustine Amalraj R&D Centre, Aurea Biolabs (P) Ltd, Kolenchery, Cochin, Kerala, India

Thahira Banu Azeez School of Sciences, Department of Home Science, The Gandhigram Rural Institute-Deemed to be University, Gandhigram, Dindigul, Tamil Nadu, India

A. Thahira Banu School of Sciences, Department of Home Science, The Gandhigram Rural Institute—Deemed to be University, Gandhigram, Dindigul, Tamil Nadu, India

May Bin-Jumah Department of Biology, College of Science, Princess Nourah Bint Abdulrahman University, Riyadh, Saudi Arabia

Fernão C. Braga Department of Pharmaceutical Products, Faculty of Pharmacy, Federal University of Minas Gerais (UFMG), Belo Horizonte, Brazil

Ana Laura Tironi de Castilho Department of Structural and Functional Biology, São Paulo State University (UNESP), Botucatu, SP, Brazil

Muhammad Daniyal TCM and Ethnomedicine Innovation and Development International Laboratory, School of Pharmacy, Hunan University of Chinese Medicine, Changsha, China

Sreeraj Gopi Department of Polymer Technology, Gdansk University of Technology, Gdańsk, Poland; R&D Centre, Aurea Biolabs (P) Ltd, Kolenchery, Cochin, Kerala, India

N.S.K. Gowthaman Materials Synthesis and Characterization Laboratory, Institute of Advanced Technology, Universiti Putra Malaysia, Serdang, Selangor, Malaysia

Józef T. Haponiuk Chemical Faculty, Gdansk University of Technology, Gdańsk, Poland

Joby Jacob R&D Centre, Aurea Biolabs (P) Ltd, Kolenchery, Cochin, Kerala, India

Shintu Jude R&D Centre, Aurea Biolabs (P) Ltd, Kolenchery, Cochin, Kerala, India

H.N. Lim Department of Chemistry, Faculty of Science, Universiti Putra Malaysia, Serdang, Selangor, Malaysia

Janeline Lunghar School of Sciences, Department of Home Science, The Gandhigram Rural Institute-Deemed to be University, Gandhigram, Dindigul, Tamil Nadu, India

Tooba Mahboob Department of Medical Microbiology, Faculty of Medicine, University of Malaya, Kuala Lumpur, Malaysia

Ayman M. Mahmoud Physiology Division, Department of Zoology, Faculty of Science; Biotechnology Department, Research Institute of Medicinal and Aromatic Plants, Beni-Suef University, Beni-Suef, Egypt

Akhila Nair Department of Polymer Technology, Gdansk University of Technology, Gdańsk, Poland; R&D Centre, Aurea Biolabs (P) Ltd, Kolenchery, Cochin, Kerala, India

Anjana S. Nair R&D Centre, Aurea Biolabs (P) Ltd, Kolenchery, Cochin, Kerala, India

Veeranoot Nissapatorn School of Allied Health Sciences, Southeast Asia Water Team (SEA Water Team) and World Union for Herbal Drug Discovery (WUHeDD), Walailak University, Nakhon Si Thammarat, Thailand

Diego P. de Oliveira Department of Pharmaceutical Products, Faculty of Pharmacy, Federal University of Minas Gerais (UFMG), Belo Horizonte, Brazil

Anupam Paliwal R&D Centre, Aurea Biolabs (P) Ltd, Kolenchery, Cochin, Kerala, India

Jithin Raj R&D Centre, Aurea Biolabs (P) Ltd, Kolenchery, Cochin, Kerala, India

Chandramathi Samudi Raju Department of Medical Microbiology, Faculty of Medicine, University of Malaya, Kuala Lumpur, Malaysia

Ariane Leite Rozza Department of Structural and Functional Biology, São Paulo State University (UNESP), Botucatu, SP, Brazil

Cristina C. Salibay College of Science and Computer Studies, De La Salle University-Dasmariñas, Dasmariñas, Cavite, Philippines

Jonnacar S. San Sebastian College of Science and Computer Studies, De La Salle University-Dasmariñas, Dasmariñas, Cavite, Philippines

Matheus Chiaradia de Souza Department of Structural and Functional Biology, São Paulo State University (UNESP), Botucatu, SP, Brazil

Hazel Anne Tabo College of Science and Computer Studies, De La Salle University-Dasmariñas, Dasmariñas, Cavite, Philippines

Carolina Mendes Tarran Department of Structural and Functional Biology, São Paulo State University (UNESP), Botucatu, SP, Brazil

Leonardo de Liori Teixeira Department of Structural and Functional Biology, São Paulo State University (UNESP), Botucatu, SP, Brazil

Mauro M. Teixeira Department of Biochemistry and Immunology, Institute of Biological Sciences, Federal University of Minas Gerais (UFMG), Belo Horizonte, Brazil

Roshin U. Thankachen Department of Polymer Technology, Gdansk University of Technology, Gdańsk, Poland

Bincicil Annie Varghese R&D Centre, Aurea Biolabs (P) Ltd, Kolenchery, Cochin, Kerala, India

Karthik Varma R&D Centre, Aurea Biolabs (P) Ltd, Kolenchery, Cochin, Kerala, India

Ajoy Kumar Verma National Institute of Tuberculosis and Respiratory Diseases (NITRD), New Delhi, India

Wei Wang TCM and Ethnomedicine Innovation and Development International Laboratory, School of Pharmacy, Hunan University of Chinese Medicine, Changsha, China

Mateus Souza Zabeu Department of Structural and Functional Biology, São Paulo State University (UNESP), Botucatu, SP, Brazil

1

Inflammation, symptoms, benefits, reaction, and biochemistry

Akhila Nair[a,b], Roshin U. Thankachen[a], Jithin Raj[b], and Sreeraj Gopi[a]

[a]Department of Polymer Technology, Gdansk University of Technology, Gdańsk, Poland, [b]R&D Centre, Aurea Biolabs (P) Ltd, Kolenchery, Cochin, Kerala, India

1.1 Introduction

The term inflammation is known from the Old Testament biblical era when Moses mentions that "If the bright spot stay in his place, and spread not in the skin, but it be somewhat dark; it is a rising of the burning, and the priest shall pronounce him clean: for it is an inflammation of the burning" (Translation of Latin term inflammationem) [1]. Since then, this primeval term has undergone explications. Cornelius Celsus, a Roman encyclopedist, explained it as "redness and swelling with heat and pain," which was later refined by Rudolf Virchow by adding "loss of function." In 2007, Ferrero Miliani clarified that inflammation is a nonspecific immune response that develops as an answer to any type of injury, and indicates accelerated blood flow, vasodilation, extravasation of fluids, increased cellular metabolism, soluble mediator response, cellular influx, and extravasation of fluids [2]. Currently, the medical lexicon states that inflammation is a local or systemic reaction in tissue generated due to internal or external stimuli in order to remove an injurious agent to prevent further progression and repair tissue damage. Injury in tissue and exposure to irritants or pathogens are assumed to be the main reasons for this acute tissue or cellular process. Although under normal conditions its reactions are circumscribed, it converts into a chronic state upon prolonged exposure to inflammatory stimuli [2].

1.2 Causes and symptoms of inflammation

The skin is considered an immunological and mechanical barrier that safeguards one's body from the external environment.

Inflammation and Natural Products. https://doi.org/10.1016/B978-0-12-819218-4.00003-1
© 2021 Elsevier Inc. All rights reserved.

However, any sort of damage to this shield opens a gateway for the inflammation-causing agents to invade the body. These causative agents are multifarious such as viral or bacterial pathogens; matter such as metal parts, sharp objects, or foreign particles that enter tissue; chemical agents such as radiation, alcohol, and autoimmunity; and local tissue injury. These agents stimulate inflammation but are self-limiting in the acute phase and turn chronic through the perpetual exposure of these causative agents [2]. Generally, inflammation gives rise to redness, heat, and swelling. However, there may be other noninflammatory causes for these symptoms. To illustrate, myositis and tendinitis are often misunderstood with inflammation. Therefore, at the cellular level, inflammation that arises due to the delay in the onset of muscular sores, which consequently cause mild discomfort or tenderness upon palpation, could be considered the cardinal signs of inflammation [3].

1.3 Types of inflammation

Inflammation developed in response to tissue injury or pathogens can be subdivided into acute and chronic inflammation. The major differences in acute and chronic inflammation are shown in Table 1.1. Acute inflammation does not persist long and can be controlled without even forming lesions. Chronic inflammation lasts for a longer duration and is formed when the subject with acute inflammation is continuously exposed to causative agents. Further, chronic inflammation is subdivided into primary chronic inflammation and secondary chronic inflammation [2, 4].

Table 1.1 Difference between acute and chronic inflammation.

S. No.	Inflammation	Acute inflammation	Chronic inflammation
1.	Causative agents	Injured tissue, pathogen	Prolonged acute inflammation: autoimmune reactions, nondegradable pathogen, prolonged invasion of foreign bodies
2.	Duration	Couple of days	Months or years
3.	Onset	Immediate	Delayed
4.	Major cells involved	Mononuclear cells (macrophages, monocytes), neutrophils	Mononuclear cells (fibroblast, macrophages, monocytes, plasma cells, lymphocytes)
5.	Primary mediators	Eicosanoids, vasoactive amines	Growth factors, hydrolytic enzymes, reactive oxygen species, IFN-γ, cytokines
6.	Outcomes	Resolution, chronic inflammation, abscess formation	Fibrosis, tissue destruction

1.3.1 Acute inflammation

The invasion by inflammation-causing agents or a nonself-antigen stimulates the innate immune system and thereby the immune responses. These are actuated mechanisms such as serotonin and histamine release; escalation of vascular penetrability; chemotactic factor secretion; and adhesion molecule hyperexpression on endothelial cells [4]. It is identified as the expulsion of plasma proteins and fluids with the simultaneous relocation of leukocytes, especially neutrophils, into the affected area [5]. Besides, the production of antibodies facilitates the release of mediators to accelerate the local reaction along with the continual intake of cells such as granulocytes, monocytes, lymphocytes, and plasma proteins from the peripheral blood [4]. An acute inflammatory response is a defense mechanism developed against the causative agents such as viruses, bacteria, and parasites to facilitate wound repair. The chemical mediators produced commonly in acute inflammation are leukotrienes, bradykinin, prostaglandin, histamine, anaphylotoxin, complement system, and nitric oxide. To cease inflammation, the cyclooxygenase (Cox) enzyme must be inhibited. It is the prime responsible factor that converts arachidonic acid to prostaglandin H2, where prostaglandin H2 radically increases during inflammation [5]. Thus, this process is temporary and exists until the inflammation-causing agents are debarred [4].

1.3.2 Chronic inflammation

Chronic inflammation is also called nonresolving inflammation or inflammaging, which is a dysregulatory, prolonged, and maladaptive response that produces constant active inflammation, followed by tissue destruction and unsuccessful tissue repair. Moreover, age-related inflammation occurs in a low and continuous way where the escalated levels of proinflammatory cytokines and C-reactive proteins (CRP) are activated and antiinflammatory cytokines are reduced but asymptomatic with the level variation of pathophysiological modification. Although the mechanism of chronic inflammation is unknown, mitochondrial dysfunction, chronic inflection, hormonal changes, redox stress, epigenetic damage, immunosenescence, and glycation are suspected modes of action of this type of inflammation [6]. This inexorable inflammation is an age-associated disease that has a pernicious effect on the host cells as it fraternizes with numerous pathogenic diseases such as cancer, rheumatoid arthritis, coronary heart disease, obesity, inflammatory bowel disease, atherosclerosis, Crohn's disease, autoimmune diseases, diabetes, and so on [5, 7]. It is further divided into primary and secondary chronic inflammation.

1.3.2.1 Primary chronic inflammation

In this category, the onset of inflammation projects a clear reaction marked with increased permeability and vascularity as well as no or minimal neutrophil infiltration. In addition, cell-mediated immune responses are generated against the body cells that become prey to the immune system. Primary chronic inflammation is associated with autoimmune diseases such as thyroiditis and certain tumors (exhibiting lymphocytic infiltration) as well as rheumatoid arthritis (exhibiting T and B mixed cells, neutrophils, and plasma cells).

1.3.2.2 Secondary chronic inflammation

This type of chronic inflammation occurs when acute inflammation persists due to the continuous exposure to causative agents that converts the inflammatory lesions into chronic inflammation to expel polymorphonuclear cells and normalize endothelial activation, vascular permeability, and vasodilation. The progression of inflammation is suggestive of the infiltration of cells that are mainly mononuclear in nature such as lymphocytes and monocyte-macrophage series cells. Examples include a chronic infection such as tuberculosis that forms sarcoidosis, chronic granulomas, and contact dermatitis; human immunodeficiency virus (HIV); and cytomegalovirus (CMV) [4]. Further, tissue immunity is the major local source as well as outlying inflammation that could be considered responsible for chronic inflammation. Certain cases are related to the development of this type of inflammation, even in the absence of pathogens. Helicobacter pylori infection exemplifies such cases where the unwavering inflammation eventually leads to cancer [8].

1.4 Benefits of inflammation

Inflammation is regarded as a necessary evil that makes the surrounding immune cells aware of infection existing at any area. This involvement of cellular pathways plays a vital role in regulating normal cellular activities [5].

1.4.1 Inflammation as a necessary evil

The immune cells such as dendritic cells (DCs) and macrophages liberate proinflammatory cytokines such as tumor necrosis factor-α (TNF-α) and interferon gamma (IFN-γ) by combining the pattern recognition receptors (PRRs) and pathogen-associated molecular patterns (PAMPs) that exist on the surface of bacteria or on the DNA/RNA of the viruses. Subsequently, the vasodilation of blood vessels occurs due to the release of other chemicals that accelerate the intake of innate immune cells (monocytes, neutrophils) to the site of inflammation. This process continues until the existence of inflammation, which is a positive indication of the active mechanism that manages various activities such as managing pathogens, controlling the resolution of the collateral damage linked with either the pathogen or injury, and directing the outflow of damaged tissue. Further, the loss of control, low-grade sustained inflammation, low amplitude in patients with immune suppression are other processes during inflammation. Hence, these concepts explain the analogy presented by Dr. Jekyll and Mr. Hyde, which represents inflammation, immune response, and dysregulation, respectively, to project that inflammation is necessary because it requires strict control over the immune system and its regulatory functions. Moreover, it provides significant help in controlling infection as well as the intensity of inflammation and subsides the commencement of diseases without causing much damage to the immune response [8]. In addition, persistent chronic inflammation is a source of development of any chronic disease such as cancer, rheumatoid arthritis, heart disease, diabetes, gout, neurodegenerative disease, Alzheimer's disease, inflammatory bowel disease, infections (fungi, parasite, bacteria), and so on. In inflammation, the T cells play an important role in charging the cell-mediated immunity. The activated T-cells such as CD4 + and CD8 + produce cytokines and chemokines that charge other inflammatory cells such as mast cells, neutrophils, and macrophages. The mast cells introduce cytokines, namely interleukin (IL-,3,4,5,6), tumor necrosis factor (TNF-α), interferon (IFN-γ), and other mediators, that produce inflammatory responses [9]. These macrophages, chemokines, and cytokines promoting inflammation, if controlled, could help overcome these diseases [10].

1.4.2 Indicative of chronic diseases

Persistent inflammation is indicative of major inflammatory diseases such as cancer, cardiovascular diseases, neurodegenerative and Alzheimer's diseases, autoimmune diseases, and inflammatory bowel diseases (IBD) (Fig. 1.1). The knowledge of inflammatory pathways that causes these diseases is beneficial in designing any particular treatment regimen or methodology; see Table 1.2.

1.4.2.1 Cancer

The hypothesis that inflammation has a strong connection with carcinogenesis led researchers to comprehensively explore the mechanism of inflammation that plausibly leads to cancer. Therefore, Maeda et al. reviewed the nuclear factor kappa (NF-κB), a cardinal pathway, to conclude that it had to be the major targeting candidate [11]. Recently, investigations suggest that the nuclear factor kappa

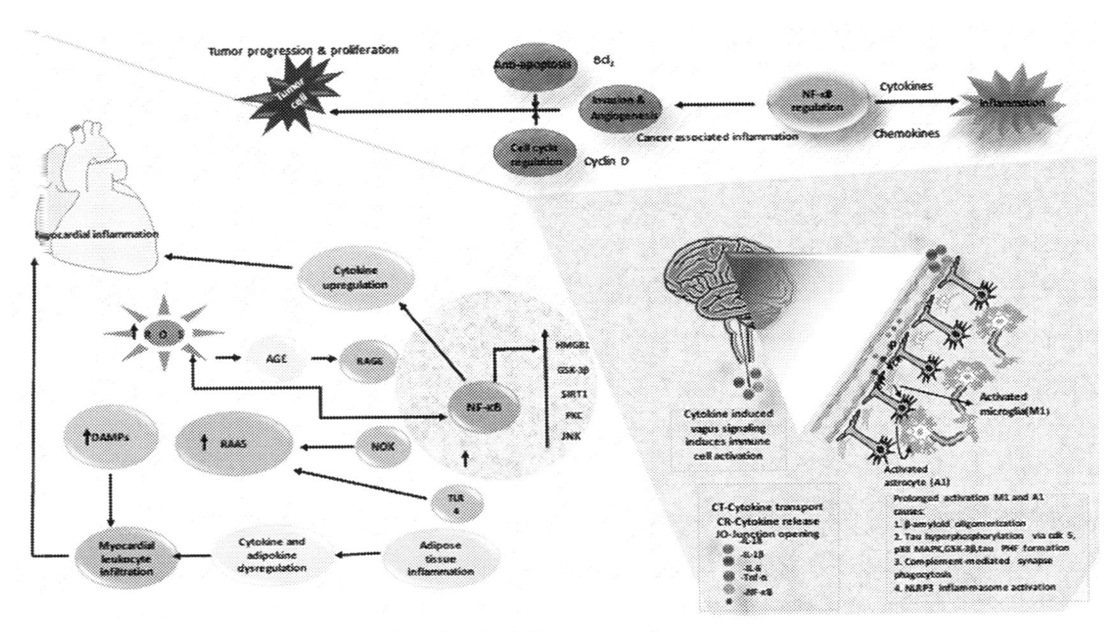

Fig. 1 1 Persistent inflammation is indicative of major inflammatory diseases.

Table 1.2 Inflammatory pathways indicative of inflammatory diseases.

S. No.	Disease	Inflammatory mediators	Reference
1.	Cancer	NF-κB, TNF-α, IL4, 6, 11, 13, 17, MIF	[11–13]
2.	Cardiovascular diseases	ROS, NF-κB, RAAS, AGE, DAMP, TLRS	[14]
3.	Alzheimer's disease	Proinflammatory cytokines	[15]
4.	Autoimmune uveortinitis	Cytokines: IFN-c, IL-6, IL-17, TNF-α, T-cells, ROS, NCF1	[10, 16]
5.	Autoimmune myocarditis	Th1/Th 2, ratio of CD4 +/CD8 +	[17–20]
6.	Inflammatory bowel diseases	MicroRNA (miR)-219a-5p, Th1/Th17, STAT3, STAT 4, DAMPs	[21, 22]

light chain enhancer of the activated B cell (NF-κB) is a vital candidate in bridging inflammation to cancer. Therefore, focusing on the retardation of NF-κB could be beneficial in the management of cancer. Hence, this link between inflammation and cancer could open a wide area of opportunity in the form of new therapies and combinations in different types of cancer that could help eradicate this disease to a larger extent [5]. Besides, inflammation and cancer have been shown to have similar modes of action such as angiogenesis or the gravity of cell proliferation. The existence of inflammatory cells for a longer duration and tumor microenvironment factors increase their growth subsequently, constraining the apoptosis of the affected cells [12]. Treatment with antiinflammatory drugs is very effective in cancer patients as it alleviates the tumor incidence. Any malignant disease at any stage, whether the beginning, progression, dissemination, mobility, or morbidity, could be treated by targeting the different modes related to inflammation. The probability of tumor occurrence in obese patients is due to the energy metabolism and adipose tissue inflammation [13].

1.4.2.2 Cardiovascular diseases

It is well documented that the signaling pathways and protein regulators leading to inflammation also influence chronic diseases. New therapeutic options are frequently investigated for therapeutic benevolence in case of inflammation related to chronic cardiovascular conditions. Nuclear medicines with hybrid imaging such as single photon emission computed tomography or computed tomography imaging devices and hybrid positron emission tomography or computed tomography have become an important treatment regimen to overcome the severity of the inflammatory processes involved in cardiovascular infections [23]. The risk factors of cardiovascular diseases such as malnutrition and chronic inflammation were investigated in 27 patients on hemodialysis. Various markers of inflammation were studied, including albumin, prealbumin, ferritin, transferrin, C-reactive protein (CRP), and fibrinogen. It was observed that CRP levels had a negative correlation with prealbumin, albumin, HDL, apoprotein A1, and hemoglobin and a positive association with Htc ratios and erythropoietin. Also, the ferritin, erythrocyte, and CRP levels were higher and the transferrin levels were lower when compared to the control in selected hemodialysis patients. This reflected that markers related to chronic inflammation, especially CRP levels, could reduce the risk factors such as dyslipidemia, anemia, and malnutrition related to cardiovascular diseases, and necessary therapeutic measures could control these risk factors [24]. In the case of a high glucose level or dyslipidemia, the chemokines, cytokines, and adhesion molecules are upregulated to activate NF-κB signaling. Apart from this, accumulated advanced glycation end products (AGE), renin-angiotensin-aldosterone system (RAAS), and damage-associated molecular pattern (DAMP) provoke inflammation through TLRs. Thereafter, myocardium infiltration by leucocytes and initiate inflammation through ROS production, secretion of cytokines as well as pro-fibrotic factors, which convert to signaling mode to cause mitochondrial dysfunction, cardiomyocyte hypertrophy, endoplasmic reticulum stress (ER) are indicative of diabetic cardiac myopathy [25]. In addition, reactive oxygen species (ROS) badly affect myocardial calcium that leads to arrhythmia and causes cardiac remodeling to provoke hypertrophic signaling, necrosis, and apoptosis [14]. Hence, the elevated inflammation targets are indicative of the existence of any particular disease.

1.4.2.3 Neurodegenerative and Alzheimer's diseases

Inflammation that occurs around the central nervous system (CNS) encourages neurodegeneration, cognitive decline, and Alzheimer's disease. This inflammation is indicative of increased blood levels of proinflammatory chemokines and cytokines. The proinflammatory cytokines that cross the blood-brain barrier are proficient in creating a proinflammatory ground in the CNS by circumventricular organs or endothelial cell signaling as well as stimulating the vagus cell, which detects the inflammatory proteins by connecting directly to the brain stem. Inflammation proceeds to induce proinflammatory and reactive microglia as well as astrocytic phenotypes to encourage β-amyloid oligomerization, hyperphosphorylation, complement activation, and neurotransmitter breakdown to dangerous metabolites. These modifications commence or aggravate and reveal neurodegenerative processes that cause dementia or cognitive decline [15].

1.4.2.4 Autoimmune diseases

Targeting retinal inflammation could be effective in autoimmune uveortinitis. The production of factors, namely complement factor B (CFB) and complement factor H (CFH), of retinal pigment epithelial cells is regulated by inflammatory cytokines. This reflects that targeting or hindering the alternative pathways of complement activation in autoimmune uveortinitis by the complement receptor of the Ig superfamily protein (CRIg-Fc) remarkably decreased the C3d deposition and CFB expression, along with a reduction in nitric acid production in BM-derived macrophages and T-cell proliferation as well as their production of IFN-c, IL-6, IL-17, and TNF-α cytokine [10]. Another salient inflammatory pathway is the ROS, which is investigated for treating chronic diseases. The autoimmune inflammatory disorders could also be treated by induced ROS and regulating neutrophil cystolic factor1 (NCF1). The NADPH oxidase 2 (NOX2) and NCF1 complex channeled ROS are vital parameters to modulate chronic inflammatory disorders such as gout, psoriasis arthritis, psoriasis, lupus, multiple sclerosis, and rheumatoid arthritis. Therefore, ROS regulation is a promising inflammatory pathway that could help in the prevention of inflammation-related chronic diseases [16]. Another autoimmune disease that leads to heart failure is autoimmune myocarditis. Chen et al. reported that the upregulation of Th1 or Th2 is responsible for myocardial inflammation [17, 18]. The higher the CD4 +/CD8 + ratio, the higher the chances of autoimmune diseases [19]. In addition, innate and CD1d restricted $V\gamma4^+$ T cell response encourages the adaptive $CD4^+$ $\gamma\delta$ T cell response initially that aids the $CD8^+$ $\alpha\beta$ TCR^+ T cell that causes cardiac damage. The α myosin specific T cell translocates myocarditis from virus-infected mice to SCID mice devoid of T and B cells. Thus, three distinct T cells are responsible for viral myocarditis and these mechanisms are clinically proven to contribute to the pathogenesis of autoimmune myocarditis [20].

1.4.2.5 Inflammatory bowel diseases (IBD)

The chronic condition when inflammation turns severe along with mucosal destruction in the intestine is characterized by inflammatory bowel disease (IBD), which is of two types: Crohn's disease and ulcerative colitis. MicroRNA (miR)-219a-5p expression is vital in triggering autoimmune diseases, carcinoma, and IBD. Other proinflammatory cytokines such as TNF-α, IL-6, IL-12, and IL-23 were observed to inhibit microRNA (miR)-219a-5p in CD4 + T cells. The luciferase assays confirmed that the ETS variant 5 (ETV5), a functional target of miR-219a-5p, is accelerated drastically when inflammation occurs in intestinal mucosa and PB-CD4 + T cells, increasing the immune response (Th1/Th17) and facilitating the phosphorylation of STAT3 and STAT 4. Therefore, by targeting this expression, the Th1/Th17-mediated immune responses are retarded with the help of proinflammatory cytokines to suppress the intestinal inflammation favoring IBD [21]. The recognition of the various receptors and inflammatory pathways associated with inflammation could help in the construction of a suitable treatment regimen for pathogenic diseases. The damage-associated molecular patterns (DAMPs), an endogenous host-derived molecule that is released or produced by damaged or dying cells, encourage inflammation and related inflammatory diseases such as neurodegenerative diseases, metabolic disorders, cancer, and autoimmune diseases. Hence, discovering the role of these types of receptors could overcome the severity of such diseases, as a suitable drug treatment therapy could be designed [22].

1.5 Reactions and biochemistry

The inflammation process ignites various modifications such as the release of signals, the hemodynamic effector molecules, and leucocyte and platelet intake, which are time-regulated and depend upon the severity of the incidence. It gets converted to chronic or

Table 1.3 Reaction of different signaling expressions during inflammation.

S. No.	Different signaling pathways	Function	Reference
1	Nuclear factor kappa-light-chain-enhancer of activated B cells (NF-κB)	**a.** IκB-α phosphorylate **b.** Transfer of p50, p65 from cytoplasm to nucleus	[5]
2	Cytokines	**a.** IL-17 produces IL-6 and IL-8, IL-1β **b.** Produce chemokines and cytokines when IL-1 receptor antagonist is absent	[27]
3	Tumor necrosis factor alpha (TNF-α)	**a.** M1 macrophage and insulin resistance	[28]
4	Protein kinase	**a.** Help signaling of immune receptors: T-cell receptor (TCR), natural killer (NK) cell receptors, E-cell receptors, Fc receptors	[29]
5	P-38 Mitogen-activated protein kinase (MAPK)	**a.** Preserve S_TKc sector **b.** Phosphorylation of upstream MAPK kinase MKK3/6 **c.** Phosphorylate downstream transcription factors: transcription factor 2 (ATF-2) activation, NF-κB, and activator protein-1 (AP-1), regulate proinflammatory factors such as TNF-α. IL-1β, and IL-6	[30]
6	CD8 + T cells	**a.** Transfer of antigen-nonspecific activated T cells to affected site **b.** Vital synergism between mesenchymal cells (synovial fibroblasts) and activated T cells	[31, 32]
7	Regulatory T cells (Treg cells)	**a.** Reduce immune response **b.** Inhibit the proliferation of T cells, cytokine production	[33]
8	Toll-like receptor ligand (TLR)	**a.** Provoke NF-κB activation, immune and non-immune cells follow TLR-dependent signaling pathways to produce inflammatory mediators	[34]
9	G protein-coupled receptors (GPCRs)	**a.** NOD-like receptor family, pyrin domain 3 (NLRP3) inflammasome activation	[35]

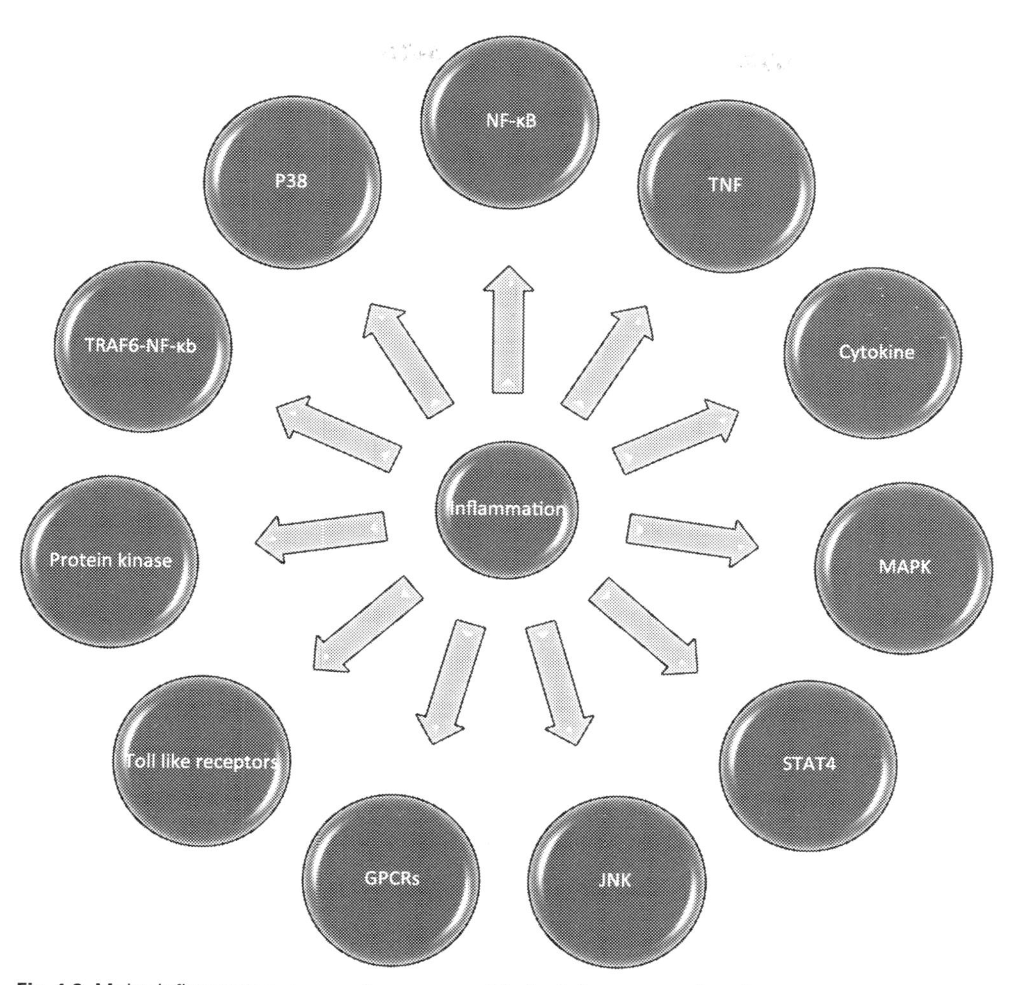

Fig. 1.2 Major inflammatory expressions responsible for inflammatory disorders.

complex with the simultaneous modulation of various functional elements, especially by controlling the intake of numerous immune cells and managing gene expression and signaling pathways. The cellular level protein configuration is a vital parameter that coordinates transcriptional regulation by encoding gene to channel inflammation and its processes [26]. The major inflammatory expressions responsible for inflammatory disorders are listed in Table 1.3 and shown in Fig. 1.2.

1.5.1 Nuclear factor kappa-light-chain-enhancer of activated B cells (NF-κB)

The nuclear factor of the kappa light polypeptide gene enhancer in the B-cells inhibitor, alpha (IκBα), a family member of the cellular protein, is responsible for blocking NF-κB. The family of NF-κB constitutes five members—p50, p52, p65, c-Rel, and RelB—that form various homodimers and heterodimers. NF-κB is categorized as a classical and alternative activation pathway where the classical pathway is considered prime and is activated by numerous stimuli such as bacteria, products, viral products, and proinflammatory cytokines as well as stress-influencing stimuli such as ROS, ultraviolet light (UV), and γ-radiation. When inflammation is kindled due to external stimuli, the IκB-α is phosphorylated to transfer the units of p50 and p65 from the cytoplasm to the nucleus and activate the gene expression that consequently activates the regulators promoting apoptosis (Bcl2 and Fas), cytokines (IL-2 and IL-6), chemokines (human monocyte chemo attractant protein-1, IL-8), and enzymes such as cyclooxygenase-2 (Cox-2), prostaglandins, and receptors to bring about inflammation [5].

1.5.2 Cytokines

Cytokines are a signaling expression regulating the inflammatory responses and TNF-α has been studied abundantly. Certain recent studies have been highlighted. Another cytokine, interleukin (IL-17) produced by T-helper cells (Th17), carries out a host defense mechanism against chronic inflammation and infection. Human IL-17 facilitates the production of IL-6, which is an important cytokine responsible for host defense and inflammation as well as IL-8, which is a chemokine ligand that recruits neutrophils in the synoviocytes of rheumatoid arthritis. In the case of osteoblast and chondrocytes, it is considered to inhibit their matrix production. IL-17 is suggested to be the prime suspect in IBD, cancer, multiple sclerosis, and joint damage. However, clinical trials have shown limited outcomes in IBD and rheumatoid arthritis, but positive outcomes for psoriasis and autoimmunity. Moreover, the polymorphisms of IL-17 such as rs 2,275,913 and rs 763,780 are reported to have strong links with cancer [27]. IL-6 plays a multifarious role by modulating autoimmune as well as inflammation-mediated diseases. Their polymorphisms in IL-6 are linked with various diseases such as rs1800795 (idiopathic arthritis), rs1800797 (rheumatic heart disease), and G174C (cardiovascular disease, myocardial infarction in type 2 diabetes). This cytokine has been confirmed as playing an active role in cardiovascular diseases by exacerbating left ventricular hypertrophy, which induces elevated blood pressure. Though it is capable of either preventing or promoting inflammation, various biological therapies have been designed to inhibit this expression. IL-1 (IL-1β isoform) is also considered vital in local and systemic inflammation. In the inadequacy of the IL-1 receptor antagonist, the IL-1β enormously produces proinflammatory cytokines and chemokines. Their polymorphism rs1143634 causes periodontitis.

1.5.3 Tumor necrosis factor alpha (TNF-α)

TNF-α is regarded as a protypical multifunctional cytokine that regulates the endocrine, cardiovascular, and metabolic systems as well as being functional in inflammation and immunity. The elevated levels of TNF are reported in respiratory distress syndrome and are responsible for high mortality rate. It is also responsible for a prolonged proinflammatory condition, especially in the synovial tissue of rheumatoid arthritis. Therefore, targeting TNF alone is investigated to be adequate for managing inflammation [36]. Abdollahzade et al. studied that among the proinflammatory cytokines, including TNF-α, IL-1, and IL-6. TNF-α-238G/A and TNF-α-308 G/A single nucleotide polymorphisms (Snps) in association with other inflammatory mediators are crucial in invertebral disc degeneration pathogenesis (IVDD) [37]. All inflammatory diseases such as cancer, rheumatoid arthritis, and obesity are mediated by cytokines in one or another way, of which TNF-α is studied as the key mediator especially in obesity, which involves M1 macrophage and insulin resistance [28].

1.5.4 Protein kinase

Protein kinase is defined as the enzymes that are capable of transferring a phosphate group on an acceptor amino acid in a protein substrate. This process is defined as phosphorylation. According to the structural preference, especially the amino acid substrate, they could be classified as tyrosine kinases, serine/threonine kinases, and dual kinases. Other CNS-acting protein kinases are the protein kinases A, B, and C (PKA, PKB, and PKC, respectively) [38]. PKC is reported useful in tumor promotion and cell division (spindle orientation) [39]. In eukaryotic cells, reversible phosphorylation manages protein activity and other cellular activities such as cell shape, growth, movement, metabolism, differentiation, cell cycle, and apoptosis. The dysregulation and mutation of protein kinase is conducted through protein phosphorylation. The immune receptors such as the T-cell receptor (TCR), natural killer (NK) cell receptors, B-cell receptors, and Fc receptors manage signaling through protein phosphorylation. The initial signaling by multichain immune recognition receptors is the tyrosine phosphorylation of adaptor molecules such as the linker of activated T cells (LAT) and the receptor itself. These are controlled by the Src family protein tyrosine kinase (PTK), and consequently the uptake of PTK members Zap70 and spleen tyrosine kinase (Syk), which leads to adapter phosphorylation that includes the SH2 domain containing leukocyte phosphoprotein 76 kDa (SLP-76) and Tec family PTK activation, followed by serine-threonine kinases, namely protein kinase C and MAPKs. Initially, the phosphorylation activation leads to numerous cytokine receptor signalings. Receptor tyrosine kinases (RTKs) are responsible for growth factor cytokines, namely platelet-derived growth factor (PDGF) and stem cell factor. Serine-threonine kinase receptors are responsible for transforming growth factor family cytokines; IL-1 and TNF also initiate kinase-dependent signaling. Protein phosphorylation us important in the inflammatory and immune mechanisms. Therefore, targeting the protein kinase proves promising to act against various inflammatory diseases [29]. The vascular smooth muscles (VSM) such as coronary artery disease, diabetic vasculopathy, hypertension, and ischemia-reperfusion injury are channeled and accelerated by PKC. Hence, PKC inhibitors were developed to test PKC such as ruboxistaurin. These inhibitors are isoform-specific and are investigated to be clinically safe and efficient in vascular diseases [40].

1.5.5 P38 mitogen-activated protein kinase (MAPK)

P38 mitogen-activated protein kinase (MAPK) is preserved serine/threonine protein kinase with varied functions at different stages such as immune responses, cell differentiation, proliferation, and apoptosis. It functions to safeguard the S_TKc sector, which has both an ATRW substrate binding site and a Thr-Gly-Tyr (TGY) motif that interacts with the linear kinase interaction motif (KIM). The phosphorylation of upstream MAPK kinase MKK3/6 facilitated by P38 MAPKs and a combination of P38 MAPKs along with Tyr182 and Thr180 in the TGY motif proceeds to phosphorylate downstream transcription factors such as activating transcription factor 2 (ATF-2), NF-κB, and activator protein-1 (AP-1) that modulate the target gene expression. P38 MAPKs channel the production of various proinflammatory cytokines and hence become vital in multiple immune responses. This could be illustrated by the increased production of IL-1β in the microglial cells by the activation of P38-induced lipopolysaccharide (LPS). In intervertebral mast and disc cells, the P38 MAPK pathway could channel proinflammatory factors such as TNF-α, IL-1β, and IL-6. Besides, in human intestinal epithelial cells (IECs), iron chelators persuade the phosphorylation of P38 that leads to the activation of AP-1 and thereafter facilitates the generation of IL-8 and regulates this expression in reaction to the *Vibrio cholerea*'s outer membrane protein U. They also participate in the NF-κB signaling pathway with TNF-α to induce the production of IL-8 in human hepatocellular carcinoma cells [41]. In the case of acute lung injury (ALI), the associated immune response brings about changes in microRNA (miRNA) expression by targeting mitogen-activated protein kinase (MAPK14) to suppress the activation of the MAPK signaling pathway. This mechanism downregulated the proinflammatory cytokine activities to facilitate cell proliferation and apoptosis, which was monitored by TUNEL staining and immunohistochemistry [30]. Moreover, p38γ MAPK, a subclass of MAPK, is considered responsive to cellular stress including LPS, UV light, osmotic shock, growth factors, and inflammatory cytokines. Recent studies investigated that p38γ MAPK expression greatly influenced the aggressiveness of cancer and tumorigenesis, and hence, it is an important signaling pathway activated by inflammatory cytokines to promote p38γ MAPK-mediated tumors [42].

1.5.6 CD8 + T cells

There are numerous recently investigated pathways that form the fundamentals of any inflammatory process. CD8 + T cells are investigated to play an important role in chronic inflammation as antigen-nonspecific activated T cells are translocated to the affected site by changing the metabolism in the cardiovascular system by inducing the innate immunity and macrophages [31]. In joints, the synergistic activity between mesenchymal cells (synovial fibroblasts) and activated T cells is vital in establishing the development of chronic inflammation [32].

1.5.7 Regulatory T cells (Treg cells)

Regulatory T cells (Treg cells) are an identifiable fraction of the T cells that reduce immune response and are capable of inhibiting the proliferation of T cells as well as cytokine production to regulate autoimmunity. The chronic inflammatory response and autoantibody production exerted by auto antibodies are suppressed by Treg cells to modulate autoimmune inflammation [33].

1.5.8 Toll-like receptor ligand (TLR)

The toll-like receptor ligand (TLR) belongs to the pattern recognition receptor family (PRRs) and is investigated to trigger acute, chronic, or postischemic inflammation. There are 13 TLRs in mice and 11 in humans, among which TLR 10 function as a TLR2 coactivator. These TLRs provoke NF-κB activation and are type I single spanning membrane glycoproteins with a leucine-rich repeat of the extracellular domain that facilitates recognition of the ligand as well as a TIR intracellular domain to mediate the intake of adaptors and activate downstream signaling [34]. These are sensory receptors produced by microbial components such as lipoproteins, nucleotides, and lipopolysaccharides (LPSs). They play a vital role in the recognition of pathogen-associated molecular patterns (PAMPs) and thereafter activate the immune system. Both immune and nonimmune cells follow TLR-dependent signaling pathways to produce inflammatory mediators [43].

1.5.9 G protein-coupled receptors (GPCRs)

The NOD-like receptor family and the pyrin domain 3 (NLRP3) inflammasome, which is an intracellular multimeric protein complex present in stimulated cytosolic immune cells such as dentritic cells, macrophages, and monocytes, are cardinal in the pathogenesis of inflammatory diseases such as Alzheimer's, diabetes [35], and atherosclerosis. These protein complexes are activated and regulated by numerous G protein-coupled receptors (GPCRs) by metabolites, neurotransmitters, and sensing multiple ions [44]. Therefore, GPCRs are an important expression because the protein-ligand interaction is considered vital in any biological process and the identification of protein binding sites for ligands is crucial in comprehending both drug molecules and endogenous ligand functions [45]. GPCRs are diverse extracellular signal molecules accessible to drug sites that possess cell specific expression and are capable of transferring signals across the membrane via G-protein interactions; this makes them attractive candidates for drug targets [46]. The GPCR expression data are used with functional and signaling activities to provide remedies for therapeutics and disease-relevant GPCR targets [47].

1.6 Conclusion

Understanding inflammation and its symptoms is beneficial, as the inflammatory pathway involved during inflammation triggers chronic diseases. Inflammation is referred as a biological process developed in response to any external stimuli. It can be either a short-term process known as acute inflammation or a prolonged inflammatory response termed chronic inflammation. Acute inflammation develops due to tissue injury and exists from a few minutes to a few hours. It is characterized by certain cardinal signs such as immobility, heat, pain, redness, and swelling. Chronic inflammation involves a progressive change in the cells at the inflammation site and is indicative of the tissue repair and destruction caused during this process. Chronic inflammation leads to numerous inflammation-related diseases such as cancer, neurodegenerative, cardiovascular, IBD, autoimmune diseases, and so on. The inflammatory pathways involved in these disease states are primarily

NF-κB, TLR-4, MAPK, STAT, and GPCRs. Exhaustive knowledge of these inflammatory expressions is beneficial as active targeting of these expressions could control the severity of numerous contagious diseases.

References

[1] C.R.P. George, From Fahrenheit to cytokines: fever, inflammation and the kidney, J. Nephrol. 19 (2006) S88–S97.
[2] R. Sreedhar, K. Watanabe, S. Arumugam, General mechanisms of immunity and inflammation, in: Japanese Kampo Medicines for the Treatment of Common Diseases: Focus on Inflammation, Elsevier, 2017, pp. 23–29, ISBN: 9780128094440.
[3] A. Scott, K.M. Khan, J.L. Cook, V. Duronio, What is "inflammation"? Are we ready to move beyond Celsus? Br. J. Sports Med. 38 (2014) 248–249.
[4] Signore, S.J. Mather, G. Piaggio, G. Malviya, R.A. Dierck, Molecular imaging of inflammation/infection: nuclear medicine and optical imaging agents and methods, Chem. Rev. 110 (2010) 3112–3145.
[5] K.N.C. Murthy, G.K. Jayaprakasha, B.S. Patil, Anti-inflammatory mediated applications of monoterpenes found in fruits, in: Tropical and Subtropical Fruits: Flavors, Color, and Health Benefits, ACS Symposium Series, American Chemical Society, Washington, DC, 2013, pp. 121–131.
[6] J.K. Chhetri, P.d.S. Barreto, B. Fougère, Y. Rolland, B. Vellas, M. Cesari, Chronic inflammation and sarcopenia: a regenerative cell therapy perspective, Exp. Gerontol. 103 (2018) 115–123.
[7] K.C. Wu, C.J. Lin, The regulation of drug-metabolizing enzymes and membrane transporters by inflammation: evidences in inflammatory diseases and age-related disorders, J. Food Drug Anal. 27 (2019) 48–59, https://doi.org/10.1016/j.jfda.2018.11.005.
[8] W. Xu, A. Larbi, Immunity and inflammation: from Jekyll to Hyde, Exp. Gerontol. 107 (2018) 98–101.
[9] B. Javadi, A. Sahebkar, Natural products with anti-inflammatory and immunomodulatory activities against autoimmune myocarditis, Pharmacol. Res. 124 (2017) 34–42.
[10] M. Chen, E. Muckersie, C. Luo, J.V. Forrester, H. Xu, Inhibition of the alternative pathway of complement activation reduces inflammation in experimental autoimmune uveoretinitis, Eur. J. Immunol. 40 (2010) 2870–2881.
[11] S. Maeda, M. Omata, Inflammation and cancer: role of nuclear factor-kappaB activation, Cancer Sci. 99 (2008) 836–842.
[12] A. Korniluk, O. Koper, H. Kemona, V.D. Piekarska, From inflammation to cancer, Ir. J. Med. Sci. 186 (2017) 57–62.
[13] G. Trinchieri, Cancer and inflammation: an old intuition with rapidly evolving new concepts, Annu. Rev. Immunol. 30 (2012) 677–706.
[14] T. Münzel, G.G. Camici, C. Maack, N.R. Bonetti, V. Fuster, J.C. Kovacic, Impact of oxidative stress on the heart and vasculature, J. Am. Coll. Cardiol. 70 (2017) 212–229.
[15] K.A. Walker, B.N. Ficek, R. Westbrook, Understanding the role of systemic inflammation in Alzheimer's disease, ACS Chem. Nerosci. 10 (2019) 3340–3342.
[16] R. Holmdahl, O. Sareila, L.M. Olsson, L. Backdahl, K. Wing, Ncf1 polymorphism reveals oxidative regulation of autoimmune chronic inflammation, Immunol. Rev. 269 (2016) 228–247.
[17] X. Cheng, Y.-H. Liao, H. Ge, B. Li, J. Zhang, J. Yuan, M. Wang, Y. Liu, Z.Q. Guo, J. Chen, J. Zhang, L. Zhang, Th1/Th2 functional imbalance after acute myocardial

infarction: coronary arterial inflammation or myocardial inflammation, J. Clin. Immunol. 25 (2005) 246–253.
[18] M.D. Daniels, K.V. Hyland, K. Wang, D.M. Eengman, Recombinant cardiac myosin fragment induces experimental autoimmune myocarditis via activation of Th1 and Th17 immunity, J. Autoimmun. 41 (2008) 490–499.
[19] L. Fengqin, W. Yulin, Z. Xiaoxin, J. Youpeng, C. Yan, W. Qing-qin, C. Hong, S. Jia, H. Lei, The heart-protective mechanism of Qishaowuwei formula on murine viral myocarditis induced by CVB3, J. Ethnopharmacol. 127 (2010) 221–228.
[20] U. Eriksson, J.M. Penninger, Autoimmune heart failure: new understandings of pathogenesis, Int. J. Biochem. Cell Biol. 37 (2005) 27–32.
[21] Y. Shi, S. Dai, C. Qiu, T. Wang, Y. Zhou, C. Xue, J. Yao, Y. Xu, MicroRNA-219a-5p suppresses intestinal inflammation through inhibiting Th1/Th17-mediated immune responses in inflammatory bowel disease, Mucosal Immunol. 13 (2020) 303–312.
[22] T. Gong, L. Liu, W. Jiang, R. Zhou, DAMP-sensing receptors in sterile inflammation and inflammatory diseases, Nat. Rev. Immunol. 20 (2020) 95–112.
[23] S. Ben-Haim, S. Gacinovic, O. Israel, Cardiovascular infection and inflammation, Semin. Nucl. Med. 39 (2009) 103–114.
[24] N. Cengiz, E. Baskin, P.I. Agras, N. Sezgin, U. Saatci, Relationship between chronic inflammation and cardiovascular risk factors in children on maintenance hemodialysis, Transplant. Proc. 37 (2005) 2915–2917.
[25] K. Taniguchi, M. Karin, NFκB, inflammation, immunity and cancer: coming of age, Nat. Rev. Immunol. 18 (2018) 309–324.
[26] A. Kielland, H. Carlsen, Molecular imaging of transcriptional regulation during inflammation, J. Inflamm. (Lond.) 7 (2010) 20.
[27] S. Suman, P.K. Sharma, G. Rai, S. Mishra, D. Arora, P. Gupta, Y. Shukla, Current perspectives of molecular pathways involved in chronic inflammation mediated breast cancer, Biochem. Biophys. Res. Commun. 472 (2016) 401–409.
[28] I. Peluso, M. Palmery, The relationship between body weight and inflammation: lesson from anti-TNF-a antibody therapy, Hum. Immunol. 77 (2016) 47–53.
[29] A. Laurence, M. Gadina, J.J. O'Shea, Protein kinase antagonists in therapy of immunological and inflammatory diseases, in: R.R. Rich, T.A. Fleisher, W.T. Shearer, H.W. Schroeder, A.J. Frew, C.M. Weyand (Eds.), Part-10, Prevention and Therapy of Immunological Diseases, Elsevier, 2016, pp. 1185–1196.
[30] W. Pan, N. Wei, W. Xu, G. Wang, F. Gong, N. Li, MicroRNA-124 alleviates the lung injury in mice with septic shock through inhibiting the activation of the MAPK signaling pathway by downregulating MAPK14, Int. Immunopharmacol. 76 (2019), 105835.
[31] C. Mauro, F.M. Marelli-Berg, T cell immunity and cardio vascular metabolic disorders: does metabolism fuel inflammation, Front. Immunol. 3 (2012) 173.
[32] N. Komatsu, H. Takayanagi, Inflammation and bonedestruction in arthritis: synergistic activity of immune and mesenchymal cells in joints, Front. Immunol. 3 (2012) 77.
[33] K. Fujio, T. Okamura, S. Sumitomo, K. Yamamoto, Regulatory T cell-mediated control of autoantibody-induced inflammation, Front. Immunol. 3 (2012) 28.
[34] Y. Feng, W. Chao, Toll-like receptors and myocardial inflammation, Int. J. Inflamm. 2011 (2011), 170352.
[35] E. Gendaszewska-Darmach, A. Drzazga, M. Koziołkiewicz, Targeting GPCRs activated by fatty acid-derived lipids in type 2 diabetes, Trends Mol. Med. 25 (2019) 915–929.
[36] D. Mulleman, M. Ohresser, H. Watier, The case of anti-TNF agents, in: Introduction to Drug Research and Development, Theory and Case Studies, Elsevier, 2013, pp. 385–397.

[37] S. Abdollahzade, S. Hanaeia, M. Sadra, M.H. Mirbolouk, E. Fattahi, N. Rezaeid, A. Khoshnevisan, Significant association of TNF-α, but not other pro-inflammatory cytokines, single nucleotide polymorphisms with intervertebral disc degeneration in Iranian population, Clin. Neurol. Neurosurg. 173 (2018) 77–83.
[38] A.P. Raval, M.A. Perez-Pinzon, K.R. Dave, Protein Kinases in Cerebral Ischemia, Primer on Cerebrovascular Diseases, second ed., Academic Press, 2017, pp. 246–250.
[39] I. Kramer, Protein Kinase C in Oncogenic Transformation and Cell Polarity, Signal Transduction, third ed., Elsevier, 2016, pp. 529–588.
[40] H.C. Ringvold, R.A. Khalil, Protein kinase C as regulator of vascular smooth muscle function and potential target in vascular disorders, in: Protein Kinase C in Vascular Smooth Muscle, Advances in Pharmacology, Elsevier, 2017, pp. 203–301.
[41] J. Sun, L. Wang, Z. Wu, S. Han, L. Wang, M. Li, Z. Liu, L. Song, P38 is involved in immune response by regulating inflammatory cytokine expressions in the Pacific oyster Crassostrea gigas, developmental and comparative, J. Immunol. (2019) 108–114.
[42] M. Xu, S. Wang, Y. Wang, H. Wu, J.A. Frank, Z. Zhang, J. Luo, Role of p38γ MAPK in regulation of EMT and cancer stem cells, Biochim. Biophys. Acta (BBA): Mol. Basis Dis. 1864 (2018) (2018) 3605–3617.
[43] Y. Nihashi, T. Ono, H. Kagami, T. Takaya, Toll-like receptor ligand-dependent inflammatory responses in chick skeletal muscle myoblasts, Dev. Comp. Immunol. 91 (2019) 115–122.
[44] T. Tang, T. Gong, W. Jiang, R. Zhou, GPCRs in NLRP3, inflammasome activation, regulation, and therapeutics, Trends Pharmacol. Sci. 39 (2018) 798–811.
[45] H.C.S. Chan, Y. Li, T. Dahoun, H. Vogel, S. Yuan, New binding sites, new opportunities for GPCR drug discovery, Trends Biochem. Sci. 44 (2019) 312–330.
[46] M. Seyedabadi, M.H. Ghahremani, P.R. Albert, Biased signaling of G protein coupled receptors (GPCRs): molecular determinants of GPCR/transducer selectivity and therapeutic potential, Pharmacol. Ther. (2019) 148–178.
[47] P.A. Insel, K. Sriram, M.W. Gorr, S.Z. Wiley, A. Michkov, C. Salmerón, A.M. Chinn, GPCRomics: an approach to discover GPCR drug targets, Trends Pharmacol. Sci. 40 (2019) 378–387.

2

Molecular pharmacology of inflammation: Medicinal plants as antiinflammatory agents

Muhammad Daniyal and Wei Wang
TCM and Ethnomedicine Innovation and Development International Laboratory, School of Pharmacy, Hunan University of Chinese Medicine, Changsha, China

2.1 Introduction

Inflammation is a highly dynamic process that can be characterized as the first protective response of the body's immune system. The immediate goal is protection against microbial invasions, entrance of antigens, and any injury or damage to cells and tissues. It involves complex interactions of soluble mediators, resident cells, and infiltrating cells and molecules that belong to the extracellular matrix. A successful and controlled inflammatory response is a useful process that leads to the clearance of injurious stimuli and restores normal physiology that is regulated accurately by a complex molecular cascade. Any imperfection of the inflammatory response may cause morbidity and shorten the life span. The extent of the inflammatory response is critically important, as if acute inflammation fails to regulate the proinflammatory stimulus, this leads to chronic inflammation, autoimmunity, and excessive tissue damage [1]. Meanwhile, excess responses are morbid and fatal in diseases such as osteoarthritis, rheumatoid arthritis, inflammatory bowel disease, Crohn's disease, metabolic syndrome-associated disorders, retinal neovascularization, and cancers [2–4]. In addition to this, atherosclerosis, myocardial infarction, chronic heart failure, Parkinson's disease, Alzheimer's disease, asthma, diabetes mellitus, psoriasis, osteoporosis, angiotensin II-derived hypertension, tumor progression, and DNA damage are also associated with chronic and persistent inflammation and autoimmune response [5, 6].

The initiation and resolution of inflammation is mediated by signals and controlled by a number of soluble mediators that work through a mechanism that results in decreasing the inflammation and returning the

Inflammation and Natural Products. https://doi.org/10.1016/B978-0-12-819218-4.00005-5
© 2021 Elsevier Inc. All rights reserved.

inflamed tissues to their normal physiology. The process of inflammation involves different types of cells and mediators that can regulate cell chemotaxis, migration, and proliferation in a highly coordinated manner.

Throughout human civilization, medicinal plants have been used to cure a variety of ailments. Medicinal plants and their synthetic and semisynthetic derivatives contribute to most of the clinically used drugs to treat infectious diseases as well as cancers of different origins. However, hundreds of molecules of therapeutic significance are awaiting discovery as the worldwide prevalence of cardiovascular diseases, microbial infections, and cancers continues to grow. Even though traditional medicines have been used along with the fast development of successful treatments against inflammation, the undesirable side effects limit the use of antiinflammatory drugs. The study of natural compounds in consort with pharmacological and ethnobotanical information is a significant contribution toward further improving these traditional compounds [7]. Natural compounds have long been used for antiinflammatory purposes. The contribution of phytochemical and ethnopharmacological studies played a key role that led to the identification, isolation, characterization, and research on the mechanisms of action of a variety of natural active compounds. Furthermore, the potential manipulations of antiinflammatory effects utilizing the knowledge from molecular pharmacology enhanced our knowledge about the natural extracts as well as made their clinical uses healthier.

Molecular pharmacology was initiated during the last decade to take a prevailing role in the discipline of pharmacology. A molecular understanding of pharmacology not only provides a venue for studies that discovered the mechanism of drug action through advances in receptor signaling, molecular probes, drug disposition and metabolism, and their biological targets, but also has spawned innovative maneuvers to influence biological systems, thus providing new prospects in drug design and suggesting important new implications for existing clinical medicine. Nowadays, it is essential to study the molecular aspect not only in pharmacology but also in all the studies involving biological processes.

Herein, we review the molecular pharmacology of inflammation, chemical components, and biological activities of a few medicinal plants as well as their mechanism of action during inflammation at the molecular levels.

2.2 Antiinflammatory agent molecular targets

As a fundamental biological process that is regarded as the most frequent sign of disease, inflammation is an expression of the body's local and protective response against microbial invasion, antigen entrance, and damage to cells and tissues. A successful and controlled inflammatory response is a useful process that leads to the clearance of injurious stimuli and restores normal physiology. The resolution of inflammation and the diffusion of the inflammatory stimulus is controlled by a number of soluble mediators. This complicated process involves different types of cells and pro- and antiinflammatory mediators to regulate cell migration, chemotaxis, and proliferation in a highly coordinated manner. If acute inflammation fails to control the proinflammatory stimulus, this leads to chronic inflammation, autoimmunity, and excessive tissue damage [1]. Meanwhile, excess responses are morbid and fatal in diseases such as osteoarthritis, rheumatoid arthritis, gastric ulcers, Crohn's disease, metabolic syndrome-associated disorders, retinal neovascularization, and cancers. Chronic and persistent inflammation and autoimmune response are also associated with atherosclerosis, myocardial infarction, chronic heart failure, Parkinson's disease, Alzheimer's disease, asthma, diabetes mellitus, psoriasis, osteoporosis, angiotensin II-derived hypertension, tumor progression, and DNA damage [5, 6, 8, 9].

2.3 Inflammatory mediators, receptors, and major signaling pathways

2.3.1 Cytokines

Cytokines are major signaling proteins in the host response against inflammation and the immune system. They are classified as interleukins, chemokines, interferons, tumor necrosis factors, growth factors, and colony-stimulating factors. They are further subdivided into proinflammatory cytokines (IL-1, IL-6, IL-15, IL-17, IL-23, tumor necrosis factor α (TNFα)) and antiinflammatory cytokines (IL-4, IL-10, IL-13, transforming growth factor β (TGFβ), interferon γ (IFNγ)) [10]. Each participates in several functions and is secreted by stromal cells and immune cells in response to inflammatory or injurious stimuli (Fig. 2.1). The following are the major inflammatory mediators that belong to the cytokine class.

2.3.1.1 TNFα

Tumor necrosis factor α (TNFα) is a major mediator of inflammation among cytokines with numerous effects on several types of cells such as inflammatory cells, endothelial cells, and fibroblasts [10]. Multiple inflammatory stimuli lead to the activation of macrophages and T cells, which further secrete TNFα. The secreted TNFα acts via a positive feedback mechanism, thus leading to more TNFα secretion as well as other cytokines such as IL-8 [11]. However, an uncontrolled

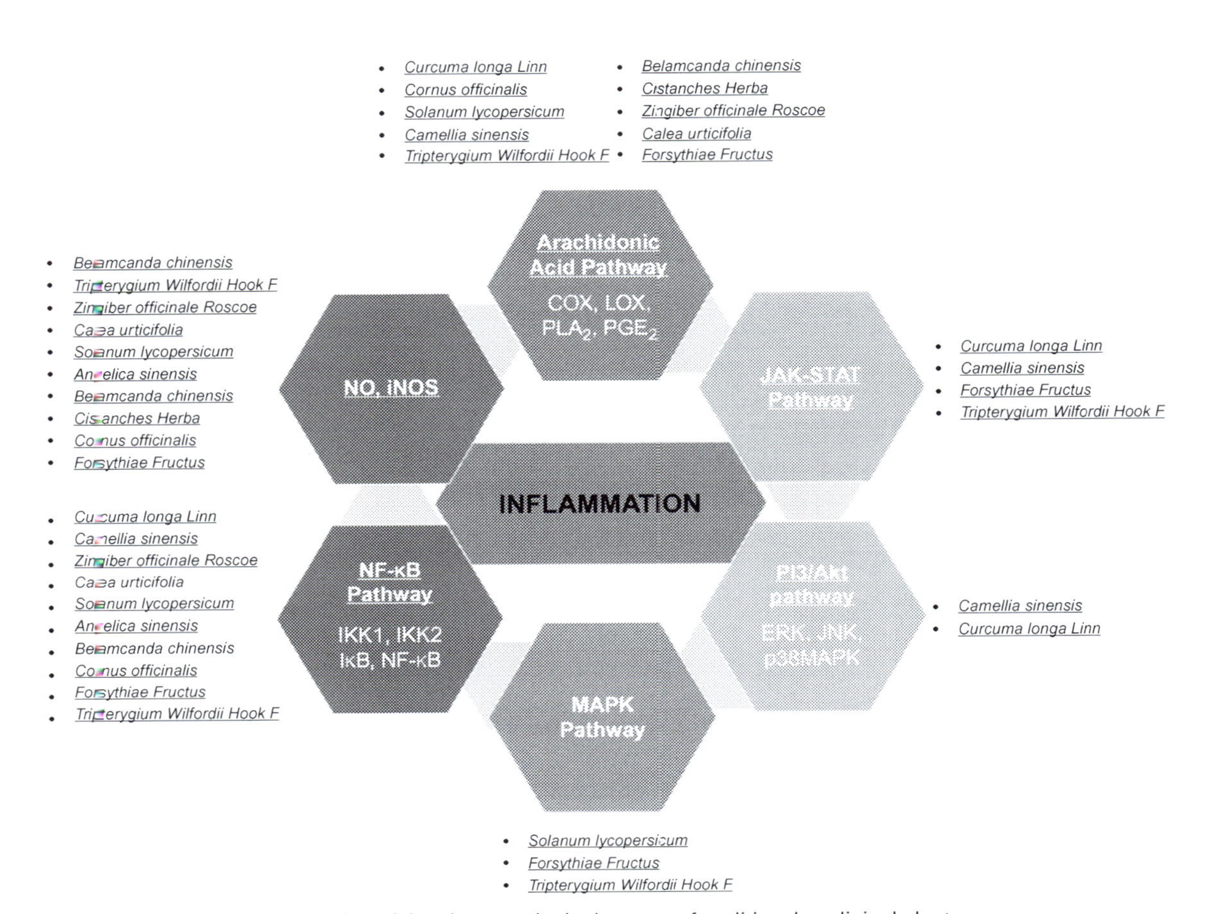

Fig. 2.1 Schematic representation of the pharmacological targets of traditional medicinal plants.

or chronic secretion of TNFα may mediate various diseases, including chronic inflammatory diseases, cancer, and autoimmune diseases, through the secretion of additional inflammatory mediators and proteases [12]. This can further cause direct DNA damage and play an apoptotic or antiapoptotic role depending upon the downstream signaling promoters [13]. TNFα levels are found to be raised in the blood and intestinal mucosa of Crohn's disease, ulcerative colitis, and inflammatory bowel disease (IBD) patients [14].

TNF works by two receptors: TNFRI, presented in most cells of the body, and TNFRII, primarily expressed in hematopoietic cells. Upon activation, the intracellular adapter proteins of TNF receptors undergo recruitment and trigger multiple signaling pathways. The activation of TNFRI stimulates the FAS-associated signal via the death domain (FAS-associated death domain, FADD)/caspase3/caspase8, mitogen-activated kinase (MAPK), Jun kinase (JNK)/activation protein-1 (AP-1), receptor-interacting protein (RIP) 3, and IκB kinase (IKK)- NF-κB pathways (Fig. 2.1). The MAPK, JNK/AP-1, and NF-κB pathways induce the gene expression of COX-2, IL-1, IL-6, MMP, chemokines, and adhesion molecules. The activation of FADD/caspase3/caspase8 can induce apoptosis and receptor-interacting protein (RIP) 3 activation, leading to necrosis [15]. TNFRII activation induces a range of inflammatory mediators and growth factors via the activation of the transcription factors AP-1 or NF-κB. This cascade induces the activation of negative regulators of apoptosis such as c-FLIP (c-FLICE-like inhibitory protein), Bcl-2 (B-cell lymphoma 2), and superoxide dismutase.

2.3.1.2 Interleukins

Interleukins are another major class under cytokines that play a pivotal role in immune modulation. Among them, IL-1 is involved in the transformation of phagocyte infiltrates during inflammatory states or cancers for the production of ROS and RNS as well as the synthesis of inflammatory molecules such as chemokines, integrins, and MMPs [16]. There are two IL-1 agonist forms, IL-1α and IL-1β, and one antagonist form, the IL-1 receptor antagonist (IL-1Ra) [17]. IL-1α is localized in the cytosol or cell membrane and works in the intracellular environment. IL-1β is secreted extracellularly after conversion to its active form by the interleukin-1β-converting enzyme (ICE). The binding of IL-1 and IL-6 to their receptors activates the IKKβ kinase, which encourages the degradation of IkBα. Further, proteins are secreted to the nucleus, where they facilitate gene transcription alone or accompanied with STAT proteins. Phosphorylated STAT binding to p65 induces histone acetyltransferase p300 and retains active p65 in the nucleus, thus working in a secondary way (Fig. 2.1). IL-1 is involved in the transformation of phagocyte infiltrates during inflammatory states or cancers for the production of ROS and RNS as well as the production of inflammatory mediators such as chemokines, integrins, and MMPs [16].

IL-6 is another chief molecule during acute inflammation and its uncontrolled production leads to many inflammatory diseases [13]. Monocytes, macrophages, and T cells at the site of inflammation are the major sources for its production. Meanwhile, the production and release of IL-6 depends on several transcription factors, particularly NF-κB and AP-1 [18]. Binding of IL-6 to its receptor IL-6R induces the dimerization of gp130 chains and activation of the linked Janus kinases (JAKs). JAKs further phosphorylate the gp130, which leads to the downstream recruitment and activation of STAT3 and STAT1 transcription factors (Fig. 2.1) and a few other molecules as well (phosphoinositide-3-kinase PI3K, Ras-MAPK, and SHP2) [18].

2.3.1.3 Chemokines

Chemokines, chemo-attractant cytokines, act in a coordinated manner and are involved in many biological processes such as cell invasion, cell motility, cell survival, and interactions with the extracellular matrix during immunological and inflammatory reactions. More than 50 chemokines and at a minimum 18 human chemokine receptors have been discovered so far [19]. Chemokine receptors are coupled to G-proteins, which contain seven trans-membrane domains. IL-8 is a major proinflammatory chemokine whose biosynthesis is controlled by TNFα, IL-1β, hypoxia, and steroidal hormones (estrogens, androgens). IL-8 binds to its widely distributed receptors with high affinity. IL-8 promotes chemotaxis and regulates cell proliferation and apoptosis by utilizing the PI3K pathway and downstream signaling with AKT. AKT phosphorylates the proapoptotic protein BAD and inhibits its interaction with the anti-apoptotic protein Bcl-xl [20]. Moreover, AKT activates IKK, and induces degradation of the NF-κB inhibitor, IκB [21]. Moreover, IL-8 triggers the signaling cascade Raf-1/MAP/ERK kinase 1/ERK cascade [22].

2.3.1.4 TGFβ

TGFβ (transforming growth factor β) is a regulatory cytokine that plays a part in the regulation of inflammation. Alteration of its activity can lead to chronic inflammation [23].

TGFβ is a dimer that is activated after cleavage from its precursor by different activators. The activated TGFβ dimer signals by pairing of its receptors, type I and type II, which are serine/threonine kinases. Upon pairing, the phosphorylation of the type I receptor happens by the type II receptors, which further transduce the signal by the downstream phosphorylation of the Smad transcription factors (Fig. 2.1) [24]. The activation of Smad complexes cause the recruitment of transcriptional coactivators and suppressors that can activate or suppress hundreds of target genes at once [24]. Furthermore, TGFβ also serves as a cytostatic regulator by suppressing c-Myc, a key transcriptional inducer of cell growth and division, and inhibiting the kinase inhibitors p15 and p21 [25].

2.3.2 Transcription factors

Transcription factors play a pivotal role by regulating the expression of proinflammatory mediators during inflammation involving several transcription factors (TFs), such as NF-κB, STAT1/STAT3, AP-1, Nrf2, and HIFs.

2.3.2.1 NF-κB

NF-κB is the most focal heterodimeric transcription factor. This protein chiefly comprises two subunits, p50 and p65, also known as the NF-κB/Rel complex, and remains inactive cytoplasmically with inhibitory-κB (IκB) as an inactive complex form [26]. Oxidative and proinflammatory stimuli such as cytokines, viruses, or LPS can activate NF-κB through proteasomal degradation and the phosphorylation of IkBα, resulting in the translocation and binding of NF-κB to the gene promoter region in the nucleus, encoding proinflammatory mediators such as COX-2, iNOS, and cytokines [27]. NF-κB is found to actively participate during chronic inflammatory conditions such as Crohn's disease, inflammatory bowel disease (IBD), and inflammatory lung and kidney diseases [28, 29]. Further, it plays a role during the transcriptional modulation of more than 150 genes [30], and the targeted genes are for inflammatory mediators (TNF-α, COX-2, IL-6, IL-8), inducers of cell proliferation (c-MYC, cyclin D1), effectors of invasion and metastasis (adhesion molecules, matrix metalloproteinases [MMPs]), antiapoptotic proteins (BCL-2), promoters of DNA damage (ROS,RNS), and angiogenic factors (VEGF, angiopoietin) [26].

The NF-κB signal transduction pathway is triggered by proinflammatory stimuli (IL-1, TNF-α), oncogenes in tumor cells, viruses, growth factors, the toll-like receptor (TLR)-MyD88 complex, genotoxic stress, and hypoxia (Fig. 2.1) [31, 32].

2.3.2.2 STATs

STATs are redox-sensitive transcription factors that are induced by the phosphorylation of tyrosine residues upon stimulation by several stimuli [33]. STATs are involved in downstream signaling initiated by cytokines such as IL-6 family, IL-23, IL-21, and VEGF, platelet-derived growth factor (PDGF), and epidermal growth factor (EGF). Among them, STATs are mostly activated via cytokine receptors, with downstream signaling through the phosphorylation of tyrosine kinases. The phosphorylation of the tyrosine residues of STAT stabilizes the Src-homology 2 (SH2) domain through phosphotyrosine-SH2 interactions, followed by STAT dimer formation that translocates and attaches to the promoter regions of genes in the nucleus, encoding inflammatory modulators and cell proliferation regulatory proteins (Fig. 2.1) [34]. The maintenance of NF-κB activation as well as its retention in the nucleus requires STAT3. Furthermore, STAT3 also interacts with the RelA/p65 subunit of NF-κB and recruits the p300 histone acetylase to the complex within the nucleus. These interactions are necessary for STAT3-dependent transcription and the acetylation of RelA that cumulatively result in an increase of nuclear retention and activity of NF-κB within the nucleus. In this manner, the transcriptional activity is prolonged.

2.3.2.3 Hypoxia-inducible factor HIF

Hypoxia-inducible factors (HIFs) are key elements in the control of the metabolic, functional, and vascular alteration in response to hypoxia [35]. Meanwhile, they are significant regulators in the controlling function and metabolism of a variety of immune cells [36]. They comprise the heterodimeric complex and have HIF-α isoforms: HIF-1α or HIF-2α and HIF-1β. The widely expressed HIF-1α is identified in all innate and adaptive immune cells such as macrophages, lymphocytes, neutrophils, and dendritic cells. HIF-α is controlled by abundant oxygen involving the iron-dependent enzymes prolylhydroxylases (PHDs), which hydroxylate HIF-α and make them more stable. During hypoxic states, PHDs become inactive, leading to the accumulation of HIF-α, which further propagates inflammation. Other than hypoxia, HIF-1α expression can also be induced by any stimuli that enhance the NF-κB activity, which further results in the induction of HIF-1α mRNA transcription [37]. TNFα, IL-6, TGFβ, oxidized LDL, and ROS can also activate HIF-1α expression [38–40]. Increased expression of HIF-1α was also found in the intestinal mucosa in IBD patients or in tumors of colorectal origin [41].

Upon hypoxia, HIF-1α initially translocates and dimerizes with HIF-1α in the nucleus, then further regulates the recruitment of coactivators CBP/P300. It finally binds to the HREs and regulates the transcription of genes worked during the hypoxic states [42]. In addition to NF-kB, the MAPK/ERK(p42/p44) pathway is also involved in HIF-1α induction induced by LPS and bacterial infections [43].

2.3.3 Complement activation pathways

Pathogen-associated molecular patterns (PAMPs) are diverse kinds of molecules in the exogenous bodies and pathogens that are recognized by immune cells. The presence of PAMPs leads to the activation of a complement system after being recognized by pattern recognition receptors (PRRs) such as mannose receptors, toll-like receptors (TLRs), and the nucleotide-binding oligomerization domain-like receptors are expressed on innate immune cells [44]. The complement system results in the activation of the central component C3 by C3 convertase via different pathways: the classical pathway, the alternative pathway, and the lectin pathway. The classical pathway is initiated with the formation of immune complexes that are formed with the release of the antibodies immunoglobulin M (IgM) or immunoglobulin G (IgG), which bind to pathogens or other foreign bodies. Multiple steps lead to the formation of the C3 convertase enzyme, which activates C3 and results in the synthesis of: (1) C3a, C4a, and C5a, (2) MAC (membrane attack complex) consisting of C5b, C6, C7, C8, and C9, and (3) C3b (opsonization molecule) [44, 45]. Secondly, binding of H-, M-, and L-ficolin or MBL to foreign bodies and recognition by the PRRs leads to the stimulation of the lectin pathway. These PRRs work in association with MBL-associated serine protease 1 (MASP1), MASP2, MASP3, and MAP19. Upon binding of the MBL-MASP complex with pathogens, that triggers the hydrolysis of C2 and C4 and activate the

C3 convertase, same as classical pathway [45]. On the other hand, the alternative pathway is activated by pathogenic carbohydrates, proteins, and lipids, which are bound and recognized by the receptors. This leads to the cleavage of C3 and its interaction with complement factors B and D further leads to the activation of the C3 convertase [45].

C3a, C4a, and C5a are strong inflammatory proteins with varied actions on various cells. They act as chemoattractants, induce oxidative bursts, are involved in the generation of proinflammatory cytokines, release histamine, and take part in various other functions. C3a and C5a work by binding to the G protein-coupled receptor (GPCR), namely, C3aR (C3a receptor), C5aR (C5a receptor), and C5L2 (C5a receptor-like 2). GPCRs are among the largest family of receptors and regulate extracellular signals to diverse physiological events. Upon binding to the complement, GPCR undergoes a conformational change that causes GDP displacement by GTP as well as Gα subunit dissociation from the Gβγ. This enables the interactions of free subunits with several effector molecules and propagates the downstream signaling [46]. C3a and C5a binding to C3aR and C5aR, respectively, produces a varied range of reactions such as inflammatory cell migration and the production of inflammatory mediators. When C3a binds to C3aR, the activation of phosphoinositol-3-kinase (PI3K) occurs, which then activates phospholipase C (PLC)β and PLCγ (Fig. 2.1). This further generates inositol triphosphate (IP3) and diacylglycerol (DAG), leading to Ca^{2+} mobilization and phosphokinase C (PKC) activation, respectively. PI3K also facilitates the MEK/ERK (Raf/mitogen-activated protein kinase (MAPK)/extracellular-signal regulated kinase kinase) cascade that leads to the enhanced expression of numerous inflammatory cytokines and chemokines. C5aR is associated with Gαi subunits, and involves downstream signaling by PI3K-γ, PLCβ, and phospholipase D (PLD) [47]. Additionally, C5aR prolongs the inflammatory response via delaying neutrophil apoptosis. This action involves the activation of the transcription factor, CREB (cAMP response element-binding protein) [48]. Furthermore, C5a activates p21-activated kinases (PAK) that are involved in the propagation of diverse MAPK signaling pathways, leading to NF-κB activation. Alternatively, upon stimulation of neutrophils by C5a, the ERK cascade activates and results in the phosphorylation of STAT3 [49]. The C5L2 receptor initially has been postulated as a nonsignaling recycling decoy receptor for C5a [50], although recent findings demonstrate its involvement in the regulation of pro- and antiinflammatory responses upon association with C5a [51]. Additionally we can say that C5L2 functions to regulate C5aR signaling such that at a higher concentration of C5a, it forms a heteromer and serves as an antiinflammatory while at a lower concentration of C5a, it serves as a proinflammatory [51].

2.3.3.1 Eicosanoids

Eicosanoids are locally acting bioactive signaling molecules synthesized by the oxidation of arachidonic acid or other polyunsaturated fatty acids (PUFAs) that produce prostaglandins, thromboxane, leukotrienes, endocannabinoids, and isoeicosanoids [52]. Eicosanoids regulate various physiological processes, especially during immunological responses. They play a dominant role that can be seen by the extensive consumption of NSAIDs (nonsteroidal antiinflammatory drugs) during inflammatory disorders. These are potent inhibitors of cyclooxygenase enzymes. Eicosanoid biosynthesis begins with the activation of the first enzyme, phospholipase A_2 (PLA_2), and with the release of arachidonic acid (AA) from membrane phospholipids, which is further metabolized by cyclooxygenase (COX), lipoxygenase (LOX), and cytochrome *p*450 enzymes [53]. Cyclooxygenases are mainly found in two isoforms: COX-1 and COX-2. COX-2 is shown to be the constitutive part during inflammation [52]. When arachidonic acid is metabolized by COX, the resultant moieties are prostaglandins, prostacyclin, and thromboxane. Lipoxygenase (LOX), found in different forms such as 5-LOX, 12-LOX, and 15-LOX, has two isoforms, 15-LOX 1 and 15-LOX 2 [52]. The immense group of biologically active moieties is generated during the arachidonic acid and associated biosynthetic pathways, and is the beginning of the vast signaling domains in cells through their respective receptors.

COX-induced signaling pathways collectively provoke inflammation, heat, swelling, redness, pain, and loss of function. The metabolism by COX produces prostaglandin H_2 (PGH_2) and downstream transformation by multiple isomerases to prostanoids, PGD_2 in the central nervous system, PGE_2 in the vascular system, $PGF_{2\alpha}$ in the smooth muscle, PGI_2 in the vascular endothelium and gastric mucosa, and thromboxane A_2 (TXA_2) in the platelets [54]. The binding of prostaglandin E_2 (PGE_2) to GPCR receptors activates the EP receptor in neurons, which starts the pain associated with inflammation. EP receptor binding in leukocytes can upregulate IL-10 production and downregulate TNF [55], and a net reduction in inflammatory signals can be observed. During inflammatory responses, the PGE2 and thromboxane A2 are predominant. The activity of COX/PGE_2 is by activation of the Ras-MAPK/ERK pathway. PGE_2 has also been reported to activate PI3/AKT, ERK signaling, EGFR signaling, and cAMP/protein kinase A signaling [54]. During hypoxia, the transcription factor hypoxia-inducible factor-1 (HIF-1) directly upregulates COX-2 expression and increases PGE_2 production. This PGE_2, in turn, will activate the Ras-MAPK pathway and the consequent positive feedback for COX-2 [56].

The 5-LOX pathway is more specifically operative during respiratory tract inflammation as well as to promote bronchoconstriction [57] and leukocyte recruitment to sites of tissue damage [58]. In this regard, the predominant functions are of leukotrienes and some may be ligands for peroxisome proliferator-activating receptor-α (PPARα) [59] and PPARγ [60], which induce antiinflammatory effects.

2.3.3.2 Reactive oxygen species (ROS) and reactive nitrogen species (RNS)

Reactive oxygen species (ROS) are moieties that are commonly presented in biological systems and are produced by the partial reduction of oxygen such as diatomic oxygen (O_2), superoxide ($O_2^{\bullet -}$), hydroxyl radicals ($^{\bullet}OH$), peroxide ions (O_2^{2-}), hyperchlorous acid (HOCl), and hydrogen peroxide (H_2O_2) [61]. In the physiological state, ROS production is counterbalanced by antioxidant protection systems including superoxide dismutase, catalase, glutathione peroxidase, thioredoxin, peroxiredoxins, etc. [62]. However, if ROS production exceeds the cellular antioxidant capacity, it leads to shear oxidative stress, which is injurious to DNA, proteins, and lipids. Oxidative stress plays a crucial role in the modulation of versatile immune responses against inflammatory stimuli. NADPH oxidase (NOX) is the family of enzymes that comprises seven isoforms that generate ROS as the primary function in phagocytic as well as nonphagocytic cells. The ROS generation process comprises multiple steps, beginning from the NADPH-dependent reduction of oxygen to superoxide, then superoxide conversion to hydrogen peroxide. Certain evidence suggests that the major source of ROS production during acute and chronic inflammation includes members of the NOX family [63]. NOX plays a significant part in the generation of ROS and defending the host against a variety of injurious stimuli [64]. Duox2 is a member of the NOX family and its expression is found to be increased prominently in IBD patients [65]. Numerous inflammatory mediators regulate the expression of NOX isoforms, such as cytokines and growth factors [66, 67]. Transcription factors such as STAT1, INF-γ IRF1, GATA, and NF-κB are involved in inducing NOX family member expression. Upregulation of NOX expression at sites of inflammation and DNA damage activates the NOX1-Rac1 pathway [67]. NOX-derived ROS exerts its action by increasing the growth factor-mediated tyrosine autophosphorylation, inactivates protein tyrosine phosphatases, hyperphosphorylation and induction of receptor tyrosine kinases. NOX-mediated DNA damage and senescence are also evident by the ROS-induced Ras oncogene genomic instability [68]. Numerous NLRs (NOD-like receptor) can form inflammasomes. The activation of NLRP3 inflammasome facilitates inflammation and apoptosis. However, various activators of NLRP3 can also initiate ROS generation and consequent NLRP3 inflammasome activation [69]. NF-κB has a major role during inflammation and the altered regulation of it caused by oxidative stress could lead to an imbalance between the mechanism of healing and apoptosis, serving to initiate chronic inflammation, autoimmune diseases, and cancer [70]. The mechanisms of redox regulation of NF-κB and other transcription factors are extremely complex and are thought to work by altering the extent of IKKβ activity [71]. ROS and RNS are active participants in innate and immune responses and they can affect several steps involved in these processes. Despite high-profile research in the field, the mechanisms of signaling and the nature of effects produced by ROS, either the activation or suppression of gene expression, remain the subject of intense scrutiny.

Reactive nitrogen species (RNS) include nitrogen dioxide, peroxynitrite, and other forms. They are synthesized with the interactions of nitric oxide with several ROS. Nitric oxide ($NO^{\bullet}$) is produced as a cellular or intracellular signaling molecule by the activation of nitric oxide synthases (NOSs). Three NOS isoforms have been identified. Two of them are nNOS and eNOS3 in neuronal and endothelial cell types, respectively, and they are highly dependent on increases of intracellular Ca^{2+} for enzyme activation [72]. The third isoform, iNOS, is primarily present in the cells of the inflammatory immune system, where it is induced during inflammatory conditions [73]. RNS can alter cell physiology by nitration or oxidation of various cellular moieties. The nitration of tyrosine residues is one of the significant mechanisms of protein regulation by RNS, which is potentially pathogenic during inflammatory diseases [74].

2.3.3.3 Toll-like receptors

Although the TLR family is very diverse, all members are involved in inflammatory response and the progression of certain inflammatory diseases such as atherosclerosis. The first discovery of the toll gene product was in the early 1980s [75] and the first toll-like receptor was discovered in 1996. In mammals, 13 identified TLRs are located at the cellular and intracellular spaces and 11 of those are present in humans. The TLRs present three structural features: (1) an extracellular region comprised of leucine, (2) a transmembrane region, and (3) a cytoplasmic region, or toll/interleukin-1-receptor (TIR), homologous to the IL-1 receptor and essential for initiating signaling pathways.

The first step after TLR activation is dimerization, or collaboration with other receptors, and a redistribution and aggregation at the cell surface [76]. Mostly the signaling cascade is dependent on MyD88 (myeloid differentiation primary response protein 88). The MyD88-dependent pathway is initiated in the cytoplasmic TIR region, which then promotes the combination of IRAK (IL-1RI-associated protein kinase) 4 and IRAK1 by recruiting the MyD88 adaptor molecule (Fig. 2.1). IRAK4 phosphorylates the IRAK1, which further interacts with TRAF 6 (TNF receptor-associated factor). This forms the IRAK1-TRAF6 complex that can interact with certain molecules and induce the activation of the inhibitory κB (IKB) kinase (IKK) complex. The IKK comprises IKKα and IKKβ, resulting in the phosphorylation of IκB, releasing the NF-κB dimer and allowing it to translocate to the nucleus. This is the most activated pathway working with the transcription factor NF-κB and utilizing the TLR during inflammatory response.

2.4 Cell signaling pathways involved in inflammation

2.4.1 The NF-κB pathway

Upon binding of ligands such as cytokines, growth factors, or microbial products to their respective cell surface receptors, the activation of the IKK complex occurs. This activation is by three different second messengers: (i) the activation of PIP2 and the downstream activation of DAG, (ii) the activation of PIP2 and the downstream activation of PIP3, and (iii) involving toll-like receptor signaling. The overall effect of all three mechanisms is the activation of the IKK complex, which comprises I kappa B kinase 1 (IKK1), I kappa B kinase 2 (IKK2), and the NF-κB essential modulator (Fig. 2.1). The activated IKK complex phosphorylates inhibitory IκB, a component of the inactive cytosolic NF-κB complex (NF-κB/RelA). NF-κB/RelA includes different proteins such as, p65, p50, p52, RelB, and C-Rel. The phosphorylated IκB undergoes proteasomal degradation, which in turn activates NF-κB that then translocates into the nucleus, binds to DNA, and induces the expression of specific target genes [77]. The natural products that regulate the NF-κB pathway include aspirin, curcumin, epigallocatechin gallate (EGCG), and triptolide.

2.4.2 Mitogen-activated protein kinase (MAPK) pathway

MAPKs are a family of serine/threonine protein kinases that include different kinases: (i) extracellular signal-regulated kinase (ERK), (ii) c-Jun N-terminal kinase (JNK), and (iii) p38 MAPK. The binding of various proinflammatory stimuli to G proteins or tyrosine kinase receptors allows the activation of Ras by GDP conversion to GTP, which activates MAP3K (e.g., Raf). RAF kinase phosphorylates and activates MEK (MEK1 and MEK2), which activates MAPK (ERK or p38 or JNK) [78] (Fig. 2.1). The activation of ERK promotes the activation of transcription factors such as cFos, c-Jun, and activating transcription factor 2 (ATF-2). Likewise, JNK activation further activates AP-1, c-Jun, and ATF-2. The p38 MAPK mediates both antiinflammatory and proinflammatory processes, and its action involves the transcription factors CREB (cAMP response element-binding protein) and ATF-2 [79]. Many traditional medicinal compounds have been shown to inhibit the MAPK pathway, including triptolide and compounds from *Forsythiae Fructus.*

2.4.3 The JAK-STAT pathway

The binding of mediators to the cytokine receptors leads to cytokine-cytokine receptor interaction at the cytoplasmic domain, which causes the phosphorylation of JAKs and STATs. Activated STATs form a dimer, and these dimers then translocate into the nucleus and modulate the expression of specific cytokine-responsive genes [80]. Several agents, including curcumin, triptolide, and EGCG, have been shown to modulate the JAK/STAT pathway.

2.4.4 The Phosphatidylinositol-3-kinase (PI3K)/Akt pathway

The PI3K/Akt pathway is another pathway that plays a role in the pathogenesis of immune-mediated disorders. Upon ligand binding to its receptor, the activation of PIP3 causes downstream signaling by activating PIP3, which further activates Akt and PKB (Fig. 2.1). Akt plays a crucial role in inflammation by activating the transcription factor NF-κB [81]. Epigallocatechin-3-gallate (EGCG) isolated from *Camellia sinensis* has been shown to inhibit the PI3/Akt pathway, and this inhibition contributes to its antiinflammatory activities.

2.4.5 The arachidonic acid pathway

Prostaglandins, thromboxanes, and leukotrienes are mediators that induce inflammation. Arachidonic acid (AA) is released from membrane phospholipids by the action of activated phospholipase A2 (PLA2). The released AA is then transformed into prostaglandins and thromboxanes by a series of enzymatic reactions following the action of cyclooxygenases (COX). Similarly, 5-lipoxygenase (5-LOX) activity results in the transformation of arachidonic acid into leukotrienes [82]. Compounds isolated from traditional medicinal plants such as *Camellia sinensis, Tripterygium Wilfordii Hook F,* and *Zingiber officinale* reduce the expression of COX and LOX and exert antiinflammatory effects.

2.4.6 Other pathways associated with inflammation

Inflammation alters multiple cellular functions and there is a cross-talk between different cellular pathways occurring simultaneously during inflammation. For example, NO is produced at higher rates by the activity of inducible NOS (iNOS) in response to proinflammatory signals produced by transcription factors and propagate inflammatory responses. Tectoridin induces the production of NO and iNOS [83]. Other inflammatory mediators such as nuclear factor erythroid 2-related factor 2 (Nrf2) [84] and cyclin-dependent kinases (CDKs) are targeted by EGCG and Forsythiaside A.

2.5 Antiinflammatory potential of medicinal plants and their active constituents

Acanthopanax senticosus Harms (AS). The dichloromethane soluble fraction showed the strongest antiinflammatory activities through the inhibition of expression of inducible NO synthase (iNOS), COX-2, TNF-α, IL-1β, and mRNAs and the generation of ROS in lipopolysaccharide-induced RAW 264.7 cells. AS decreased the level of NF-κB and the DNA-binding activity of NF-κB by inhibiting the NF-κB pathway [85].

Actinidia arguta. A chloroform layer of *Actinidia arguta* exerted antiinflammatory effects via the nuclear factor- (NF-) κB pathway. The chloroform extract inhibits NO production and iNOS mRNA expression in RAW 264.7 cells on LPS-stimulated macrophages. A reduction in the phosphorylation of mitogen-activated protein kinases, including extracellular signal-regulated kinase (ERK) 1/2, c-Jun N-terminal protein kinase, and p38, was accompanied. The chloroform extract of *Actinidia arguta* expresses antiinflammatory effect potential by the suppression of mitogen-activated protein kinase (MAPK) phosphorylation and the nuclear translocation of NF-κB in lipopolysaccharide- (LPS-) stimulated macrophages in RAW264.7 cells [86].

Ainsliaea fragrans Champ. The Ainsliaea fragrans Champ. (belongs to the Asteraceae family) is a well-known herbal medicine in China, and it has an extensive record of medicinal practice, particularly in China. It has two major active phenolic compounds: 3,5-dicaffeoylquinic acid and 4,5-dicaffeoylquinic acid (Fig. 2.2). The extract of medicinal plant reduces the expression of key inflammatory mediators through impact on the NF-κB signaling pathway. The active compound, 3,5-dicaffeoylquinic acid, only has an effect on inflammatory mediators but another compound, 4,5-dicaffeoylquinic acid, effectively inhibits the NF-κB-activated pathway. In this investigation, the extract and the compound 4,5-dicaffeoylquinic acid have significant antiinflammatory potential compared to another compound through suppression of signaling pathways [87].

Ampelopsis grossedentata. Ampelopsis grossedentata (belonging to the Vitaceae family) is an edible herb widely distributed in China. It is also an important traditional Chinese medicine for pharyngitis,

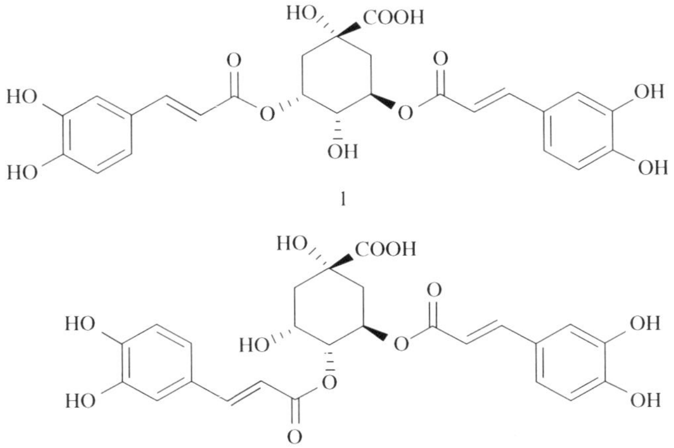

Fig. 2.2 Structures of compounds 1 and 2 isolated from *A. fragrans* (3,5-dicaffeoylquinic acid and 4,5-dicaffeoylquinic acid).

throat infection, fever, and allergenic skin disease. Houet al. [88] isolated the flavonoid bioactive compound, dihydromyricetin, and investigated the antiinflammatory potential using macrophages (Fig. 2.3). In this investigation, the authors demonstrated that this compound effectively inhibited proinflammatory cytokines and increased the production of the antiinflammatory cytokine, IL-10. A compound, dihydromyricetin, also downregulates the inflammatory protein expression, mainly iNOS and COX-2, in inflammation-induced macrophage cells. This effect was achieved through the suppression of the phosphorylation of NF-κB and IκBα in addition to the phosphorylation of p38 and JNK in inflammation-induced macrophages. Moreover, the antiinflammatory activity of this plant is proven to hit multiple mechanisms such as ROS/Akt/IκK/NF-κB signaling interdependently, which is believed to be a promising therapeutic agent [89].

Angelica keiskei. Angelica keiskei ethyl acetate-soluble fraction showed potent inhibitory action against the production of nitric oxide (NO) in LPS-activated RAW264.7 cells. It also inhibits the LPS-induced expression of inducible nitric oxide synthase (iNOS), cyclooxygenase-2 (COX-2) genes, and PGE2 production by inhibiting the degradation of IκBα and the nuclear translocation of NF-κB. The antiinflammatory effects by the HAK Angelica keiskei ethyl acetate-soluble fraction can be linked to interference with the signaling pathway of mitogen-activated protein kinases (MAPKs) and the activation pathway of NF-κB [90].

Artemisiae annuae Herba. The extract of *Artemisiae annuae herba* (AAH) inhibits NO production and various inflammatory mediators such as TNF-α, IL-6, and iNOS gene expression. Moreover, these extracts inhibited the nuclear translocation of p65 and IκBα degradation in the NF-κB pathway and decreased the extracellular signal-regulated kinase, p38, and

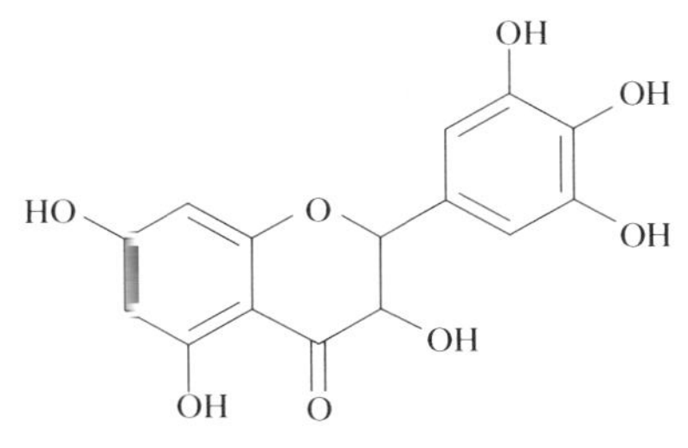

Fig. 2.3 Chemical structures of dihydromyricetin.

c-JunNH2-terminal kinase phosphorylation in the MAPK signaling pathway. These extracts possess antiinflammatory activities derived from the repression on the activation of the NF-κB and MAPK pathways [91].

Cassia occidentalis Roots. The bioactive compounds from *Cassia occidentalis* roots were identified and isolated from the ethyl acetate extract. They were found to suppress LPS-induced IL-1β, TNF-α, and NO production in a concentration-dependent manner in macrophages. From these active compounds, emodin and chrysophanol were also found active in inhibiting proinflammatory cytokines in an in vivo experimental model. This investigation has proven that the *C. occidentalis* root extract and the isolated active compounds are an effective natural therapy for the treatment and prevention of inflammation and associated ailments [92].

Cheilanthes albomarginata Clarke. Various fractions were isolated from the extract. The ethyl acetate fraction showed the strongest in vitro antioxidant properties, including phenolic content, DPPH radical scavenging, hydrogen peroxide scavenging, and nitrite scavenging activity. The in vitro antiinflammatory and antiadipogenic properties were measured in inflammation-stimulated cells, and the ethyl acetate fraction showed significant antiinflammatory and antiadipogenic activities [93].

Clerodendrum inerme. Clerodendrum inerme (L.) Gaertn. (belonging to the Verbenaceae family) appears commonly in the coastal mangrove forests of Thailand and South Asian countries and is used as a traditional medicine. The ethyl acetate fraction of the extract shows the most potent antiinflammatory effects among other extracts/fractions through the suppression of various inflammatory markers in LPS-induced macrophages. In addition, three known flavones, acacetin (1), hispidulin (2), and diosmetin (3), were isolated based on the suppression of inflammatory markers (Fig. 2.4). Among these three flavones, hispidulin is the most active antiinflammatory agent due to the suppression of NF-κB DNA-binding activity and the JNK signaling pathway, subsequently downregulating the key inflammatory targets such as iNOS and COX-2 expression [93].

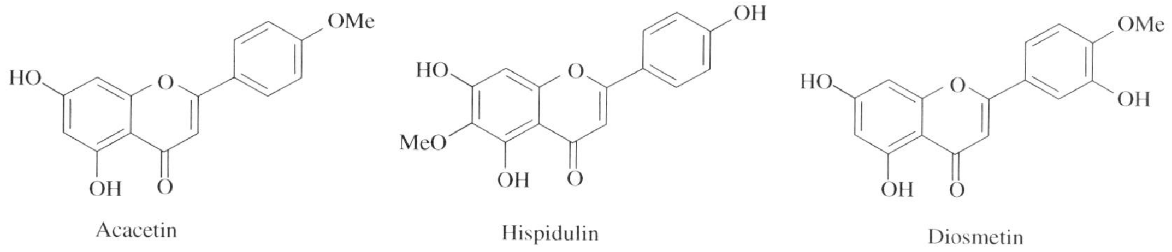

Fig. 2.4 Chemical structure of acacetin (1), hispidulin (2), and diosmetin (3).

Crataeva nurvala Buch. Ham. Crataeva nurvala Buch. Ham. (C. nurvala) has been traditionally measured as a medicinally important plant for treating immune function-related disorders and other metabolic disorders. It has various active components that are responsible for antiinflammatory and associated complications. These include lupeol, lupeol acetate, α-spinasterol acetate, Ψ-taraxasterol, 3-epilupeol, and β- sitosterol as its major components and lupenone and β-sitosterol acetate as its minor components [94]. The suppression of various inflammatory mediators' production by the extract of C. nurvala was followed by the downregulation of mitogen-activated protein kinases (MAPKs), specifically extracellular signal-regulated kinase (ERK) [95].

Cyperus rotundus. Cyperus rotundus L. (Cyperaceae) is a traditional Chinese medicine for various diseases/disorders. Recently, researchers have isolated the bioactive compounds, namely, fulgidic acid and pinellic acid, responsible for its antiinflammatory properties (Fig. 2.5). Fulgidic acid, an unsaturated trihydroxy C18 fatty acid, effectively suppressed the LPS-induced production of proinflammatory cytokines and various inflammatory mediators through the inactivation of the AP-1 transcription factor when compared to another active compound, pinellic acid [96].

Datura metel L. Nine new withanolides, named daturafolisides A-I (1–9), along with six known compounds, (22R) -27-hydroxy-7α-methoxy-1-oxowitha-3,5,24- trienolide-27-O-β-d-glucopyranoside, daturataturin A, daturametelin J, daturataurin B, baimantuoluoside B, and 12- deoxywithastramonolide, were identified and isolated from the leaves of *Datura metel* L. Various spectroscopic techniques including ^{1}D and ^{2}D NMR techniques, mass spectrometry, and circular dichroism (CD) were applied to elucidate the compounds. These isolated compounds were evaluated for in vitro antiinflammatory potential using LPS-stimulated murine macrophages, and these compounds have significant antiinflammatory properties [97].

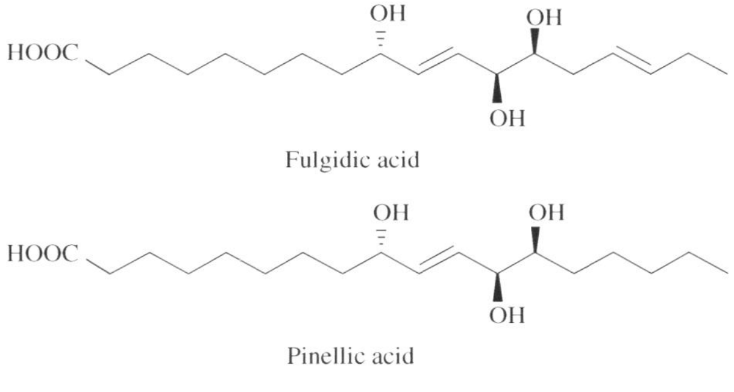

Fig. 2.5 Chemical structure of fulgidic acid and pinellic acid isolated from rhizomes of *C. rotundus.*

Dilodendron bipinnatum Radlk. These extracts significantly suppressed paw edema by carrageenan in the second hour at 20 mg/kg and by dextran in the first hour at 100 mg/kg, after induction with the phlogistic agents. It has also reduced total leukocyte and neutrophil migration at all different doses tested, producing a maximum effect at 20 mg/kg. It also suppressed the concentrations of proinflammatory cytokines (IL-1β and TNF-α) and increased the level of the antiinflammatory cytokine IL-10 in the peritonitis model [98].

Grateloupia lanceolata. Grateloupia lanceolata is a seaweed widely distributed in the seas of northeastern Asia and North America. Kim et al. [99] investigated the antiinflammatory potential of *Grateloupia lanceolate* extract in LPS-induced RAW 264.7 cells. The extract of this plant significantly inhibited the production of proinflammatory cytokines, particularly IL-1β expression, and is related to the blockade in extracellular signal-regulated kinases 1 and 2 (ERK1/2), c-Jun N-terminal kinases 1 and 2 (JNK1/2), and NF-κB signaling in LPS-stimulated RAW 264.7 cells. These findings suggest the antiinflammatory potential of the extract in RAW264.7 macrophages through the inhibition of LPS-stimulated p38MAPK/ERK/JNK inflammatory signaling pathways.

Hibiscus sabdariffa. Hibiscus sabdariffa L. (Malvaceae) is a well-known functional food and Chinese herbal medicine used to treat various inflammation-associated diseases. According to Sogo et al. [100], the active compound, delphinidin 3-sambubioside (Dp3-Sam), a Hibiscus anthocyanin from dried calices, was isolated and this compound was proven as an anticancer drug against human leukemia cells. In addition, Sogo et al. [100] discovered that anthocyanin aglycones, especially delphinidin (Dp) and cyanidin (Cy), and these compounds effectively inhibit the expression of inflammatory mediators, including COX-2 and PGE2 (Fig. 2.6). In the described mechanism of the active compound, the delphinidin 3- sambubioside (Dp3-Sam) compound significantly downregulated the NF-κB pathway and ERK1/2 signaling through the suppression of numerous inflammatory mediators.

Houttuynia cordata. The ethyl acetate fraction of the *H. cordata* extract downregulated the NO, PGE2, TNF-α, and IL-6 production in the inflammation-stimulated cells as well as iNOS and COX-2 expression. In addition, this fraction suppressed the nuclear translocation of the NF-κB p65 subunit, which was associated with an inhibitory effect on IκBα phosphorylation and also modulated the activation of MAPKs (p38 and JNK) [101].

Inula japonica Thunb. The *Inula japonica Thunb.* (*I. japonica*) flower extract showed significant antiinflammatory potential through preclinical studies. It has been used as a traditional Chinese medicine for the management of 1,10-secoeudesma-5(10),

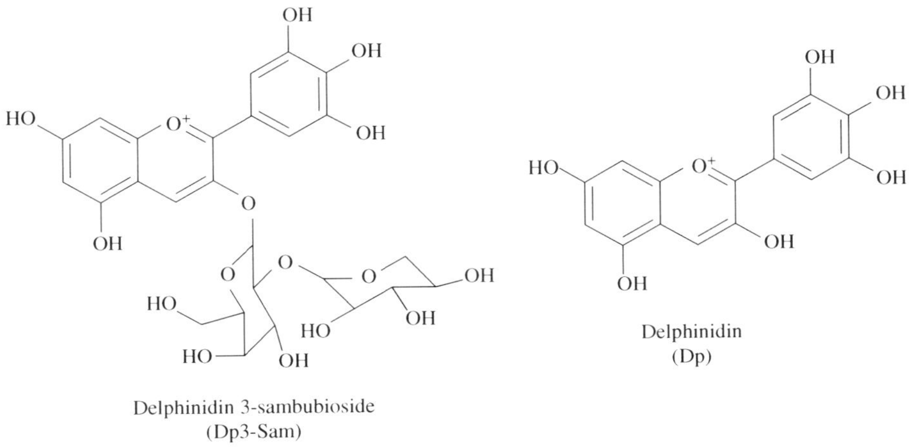

Delphinidin 3-sambubioside (Dp3-Sam)

Delphinidin (Dp)

Fig. 2.6 Chemical structures of delphinidin 3-sambubioside (Dp3-Sam) and delphinidin (Dp).

11(13)-dien-12,8β-olide (SE), 6α-isobutyryloxy- 1-hydroxy-4αH-1,10-secoeudesma-5(10),11(13)- dien-12,8β-olide (IBSE), and 6α-isovaleryloxy-1-hydroxy- 4αH-1,10-secoeudesma-5(10),11(13)-dien-12,8β-olide (IVSE)) from the flower extract (Fig. 2.7). These compounds effectively inhibited the production of NO and PGE2 in LPS-stimulated RAW264.7 macrophages [102]. Park et al. and Chen et al. [102, 103] indicated that the active compound IVSE has a potential inhibitory effect on LPS-stimulated NO production and iNOS protein expression in macrophages, and that these activities might be the suppression of two key signaling pathways, NF-κB and MAPK. Wang et al. [104] isolated the JEUD-38; 1-oxo-4aHeudesma-5(6),11(13)-dien-12,8β-olide as a new sesquiterpene lactone of the plant extract (Fig. 2.8). The JEUD-38 compound also significantly inhibited NO production and proinflammatory mediators in LPS-stimulated macrophages. In addition, JEUD-38 downregulates the NF-κB transcription factor through the inhibition of IκBα phosphorylation and degradation as well as suppressing MAPK activation.

Fig. 2.7 Chemical structure of SE, IBSE, and IVSE.

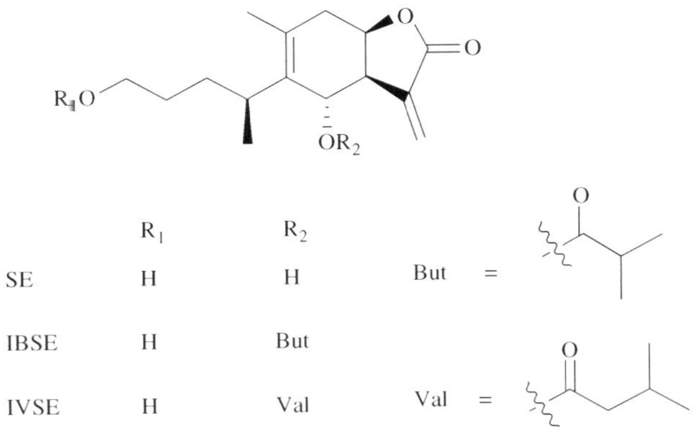

	R_1	R_2
SE	H	H
IBSE	H	But
IVSE	H	Val

Lignosus rhinocerotis. Various extracts including the cold water extract (CWE) and the hot water extract (HWE) of various herbs which associate with inflammatory diseases. Chen et al. [103] showed that the methanol extract (ME) of the sclerotial powder of L. rhinocerotis showed antiacute inflammatory activity by the carrageenan-induced paw edema test. In this method, CWE shows the most potent effect. The protein component of the high molecular weight fractions isolated from CWE contributes to the significant antiinflammatory potential through the inhibition of TNF-α production with an IC_{50} of 0.76 μg/mL [105].

Lycium barbarum (Lycii Radicis Cortex, LRC). The extract from plants showed antiinflammatory potential through a reduction in the LPS-induced production of inflammatory mediators such as NO and the proinflammatory cytokines IL-1β and IL-6 in the macrophages. Moreover, this extract inhibited the various LPS-induced inflammatory mediators such as inducible NO synthase (iNOS) and COX-2 mRNA as well as protein and inflammatory cytokine mRNA in the cells. The cellular mechanism of the extract inspires the suppression of LPS-mediated p38 and c-Jun N-terminal kinase (JNK), mitogen-activated protein kinases (MAPKs), and the nuclear factor- (NF-) κB signaling pathway in an inflammation-induced experimental model [106].

Matricaria recutita L. *Matricaria recutita* L. belongs to the Asteraceae family and has been used for centuries to treat numerous inflammatory-associated diseases. Chamomile preparations were approved for oral consumption in the treatment of inflammatory diseases, particularly of the gastrointestinal tract. The advantages of using chamomile preparations are connected to the presence of quite a few plant metabolites from various classes, including flavonoids, coumarins, proazulenes, their degradation product chamazulene, and essential oil. Among all other flavonoids, the apigenin and luteolin derivatives are the most abundant and have significant in vivo and in vitro antiinflammatory potential through the modulation of various associated pathways [107]. The most active component is guaianolide matricine (1), with significant antiinflammatory activities present in chamomile flower heads. Matricine (1) inhibits the ICAM-1 gene, NF-κβ signaling molecules, and protein expression that was induced by TNF-α and LPS in endothelial cells. Furthermore, these inhibitions were in a dose-dependent manner without causing cytotoxicity on the endothelial cells. Another degradation product of matricine (1) is chamazulene (2) and this product was inactive against NF-κB promotor activity [108] (Fig. 2.9).

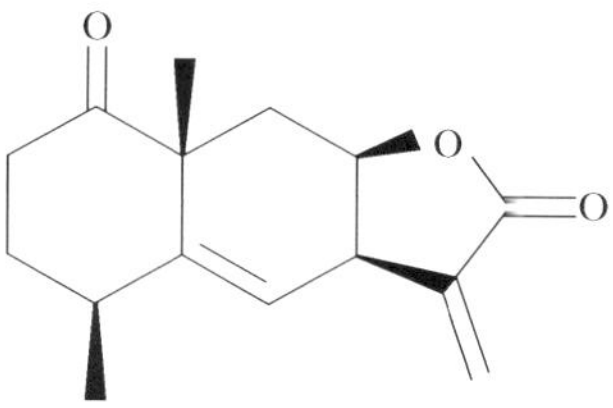

Fig. 2.8 Chemical structure of JEUD-38.

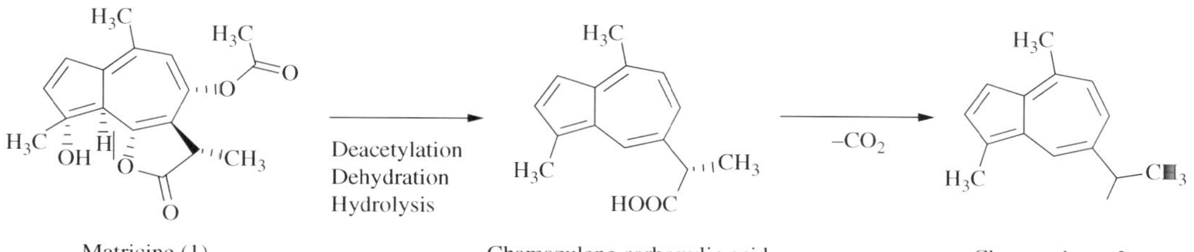

Fig. 2.9 Degradation of matricine (1) to chamazulene (2) via chamazulene carboxylic acid.

2.6 Inflammatory pathways as potential targets for natural bioactive compounds

Inflammation is a multifaceted process that engages molecular and cellular mechanisms, resulting in widespread physiological changes. The initial inflammation involves the recruitment of a wide range of immune cells to inflamed sites and the release of various proinflammatory cytokines and other agents [109]. This adaptive response evolved as a general reaction to a variety of stimuli and conditions, including infection and injuries [110], and the controlled acute inflammatory response is an essential part of the host's defense system. In sharp contrast with acute inflammation, systemic chronic inflammation, which is not apparently triggered by infection or injury, is strongly associated with a wide variety of diseases, including cancer, type 2 diabetes, and cardiovascular diseases. Chronic inflammation is linked to the malfunction of tissues and is believed to be functionally unrelated to tissue repair or to organism defense. It is well established that inflammatory processes contribute to the pathophysiology of cancer development and progression [111]. The intrinsic pathway and the extrinsic pathway connect inflammation and cancer [112]. The extrinsic pathway is driven by preexisting conditions in the specific anatomical location, for example, inflammation, infection, or injury, that increase the probability of neoplasia. In contrast, the intrinsic pathway is triggered by genetic events, including the activation of various types of oncogenes and the inactivation of tumor-suppressor genes. Both pathways converge, resulting in the activation of inflammatory signaling pathways [112]. The network of inflammatory pathways includes, besides cytokines and chemokines, a variety of transcription factors and enzymes that should be recognized for their critical regulatory functions during this complicated process (Fig. 2.1). Key molecular players linking cancer and inflammation include signal transducers and activators of transcriptions (STATs), nuclear factor-κB (NF-κB), the nuclear factor of activated T-cells (NFAT), activator protein-1 (AP-1), the CCAAT enhancer binding protein (C/EBP), the cAMP response element binding protein/p300 (CBP/p300), and the activator transcription factor (ATF) [113]. It has been recently recognized that NF-κB, the central coordinator of innate and adaptive immune responses, also plays a critical role in cancer development and progression [114]. Activated NF-κB often facilitates the transcription of numerous genes, including iNOS, COX-2, interleukin-6 (IL-6), IL-1β, tumor necrosis factor-R (TNF-R), 5-lipoxygenase (5-LOX), hypoxia inducible factor-1R (HIF-1R), and vascular endothelial growth factor (VEGF), resulting in inflammation and tumorigenesis. The activation of NF-κB is induced by a cascade of events leading to the activation of inhibitor κB (IκB) kinases (IKKs), which in turn phosphorylates IκB. The subsequent ubiquitination and proteasomal degradation of IκB leaves NF-κB free to translocate to the nucleus. These kinases can be activated through phosphorylation by upstream kinases, including NFκB-inducing kinase, mitogen-activated protein kinase (MAPK), and protein kinase C (PKC) [115]. In addition, many studies have confirmed the cytokine function in the induction of transcription activity of NFκB through the Janus kinase (JAK), the extracellular signal-regulated protein kinase 1/2 (Erk1/2) (p42/44), p38 MAPK, Ras, and phosphoinositide-3 kinase (PI3K)/Akt pathways [116]. NF-κB provides a mechanistic link between inflammation and cancer and is a major factor controlling the ability of both preneoplastic and malignant cells to resist apoptosis-based tumor surveillance mechanisms [114]. Recent studies revealed that the constitutive activation of STATs, particularly STAT3, is found in a number of primary human epithelial tumors and cancer cell lines. Persistently active STAT3 induces tumor angiogenesis by the upregulation of VEGF and its immune evasion. An understanding of molecular mechanisms linking inflammation and cancers is beneficial for the development of efficacious prevention and treatment of inflammation-associated tumorigenesis. Growing evidence clearly demonstrates that inflammatory pathways are critical targets in cancer treatment and prevention [111]. Many natural bioactive compounds have been reported to interfere with the initiation, promotion/progression, and invasion/metastasis of cancer through the control of intracellular signaling cascades of inflammation progress (Fig. 2.10). There is growing research on the effects of plant-derived compounds on the attenuation of proinflammatory gene expression [109, 117].

In response to tissue injury or inflammatory stimulation, the inflammatory and immune cells converge on the area and secrete large amounts of highly reactive chemicals and proinflammatory cytokines to eliminate foreign matters or recruit a wide range of immune cells. These inflammatory chemicals also attack normal tissue surrounding the infected tissue and result in oxidative DNA damage and genemutation, then finally promote cell proliferation. In tumor tissue, these inflammatory cells and tumor cells also create an inflammatory microenvironment. A network of signaling molecules not only promotes proliferation, angiogenesis, invasion, and metastasis, but also suppresses the ability of the host antitumor immune responses by the secretion of various immunosuppressive cytokines and chemokines [118].

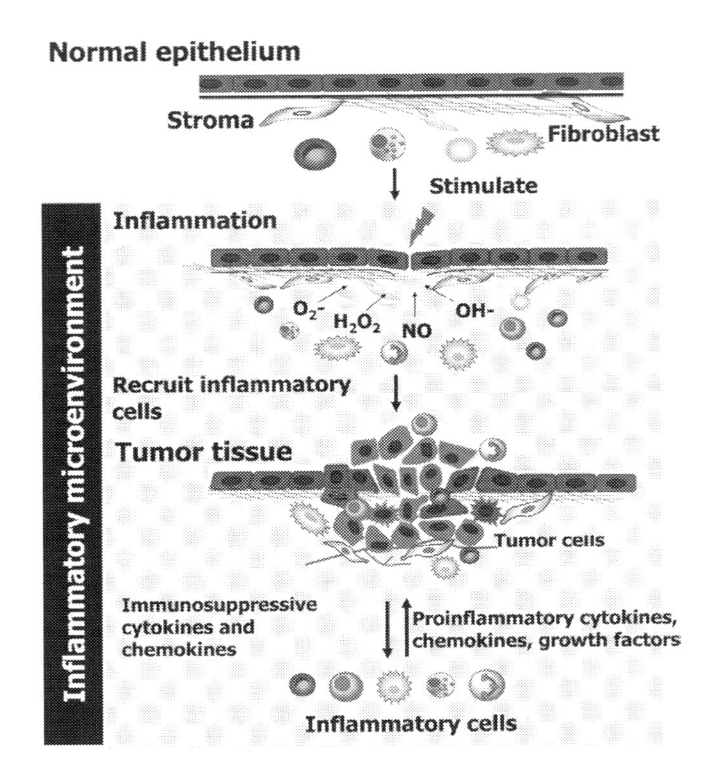

Fig. 2.10 Role of inflammation in tumorigenesis.

2.7 Antiinflammatory activity of natural bioactive compounds

The antiinflammatory activity of natural bioactive compounds is attracting growing interest from researchers and physicians. This interest is inspired both by broad recognition of the ethnomedical observation of how inflammation is treated with commonly cultivated and harvested wild plants [119, 120] and by advances in antiinflammatory screening methods [121, 122]. Many natural bioactive compounds in fruits and vegetables with potent antiinflammatory properties have been noted as plausible approaches for clinical cancer prevention trials. These compounds can be categorized into several classes. The molecular mechanisms underlying the modulating expression of inflammatory genes by selected natural dietary compounds (Figs. 2.11–2.13) are described below.

Flavonoids are ubiquitous in plants; almost all plant tissues are able to synthesize flavonoids. They can be classified into seven groups: flavones, flavonols, flavanones, flavanonols, flavanols, isoflavones, and anthocyanidins. In general, the leaves, flowers, fruits, or the plants themselves contain flavonoid glycosides, whereas the woody tissues

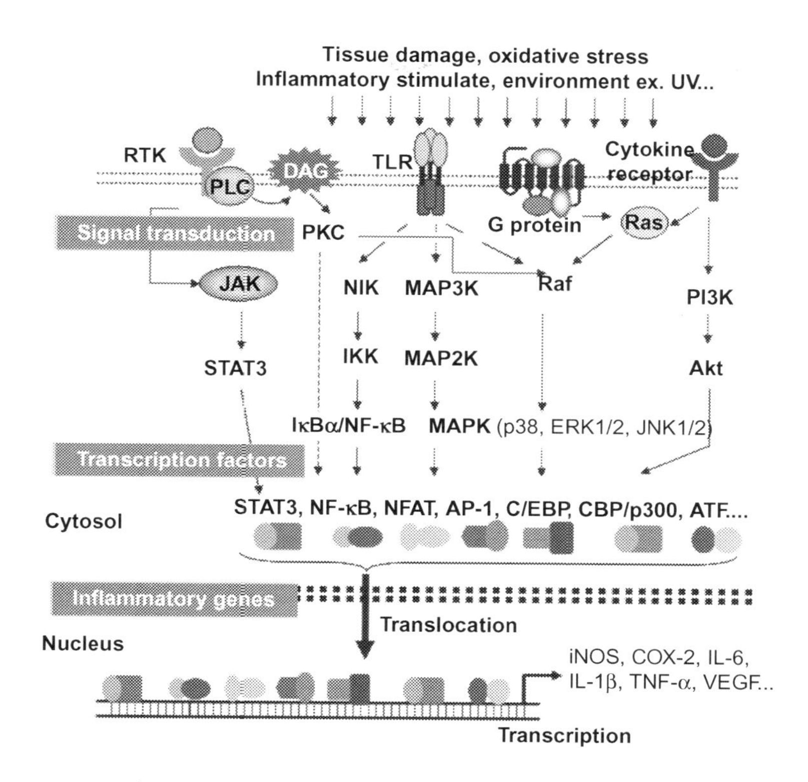

Fig. 2.11 Multiple signaling pathways in regulating inflammatory gene expression.

contain glycones and the seeds may contain both. Studies reported that isoflavone powders as well as isoflavone standard genistein were effective in inhibiting LPS-induced inflammation [123].

Tissue damage, oxidative stress, inflammatory stimulation, and environmental stimulation trigger the cellular signal transduction via ligand and receptor binding [receptor tyrosine kinase (RTK), the toll-like receptor (TLR), and the cytokine receptor], leading to receptor phosphorylation and conformational change in the protein structure. Receptor-mediated inflammatory signals recruit several adaptor proteins and result in the activation of downstream MAPK. These stimuli also induce the activation of phospholipase C (PLC) and activate the small G protein to yield downstream second messenger diacylglycerol (DAG), an activator of PKC. These intracellular molecules amplify the signaling transduction cascade through the phosphorylation of transcription factors, leading to activation, and translocate to the nucleus involved in the inflammatory response modulating the proinflammatory gene expression [118].

Apigenin, present in parsley and celery, has been found to block proinflammatory cytokines and HIFR, VEGF, and COX-2 expression through the inhibition of NF-κB, PI3K/Akt, and ATF/cyclic AMP responsive element (CRE) signaling pathways [124–126]. Numerous

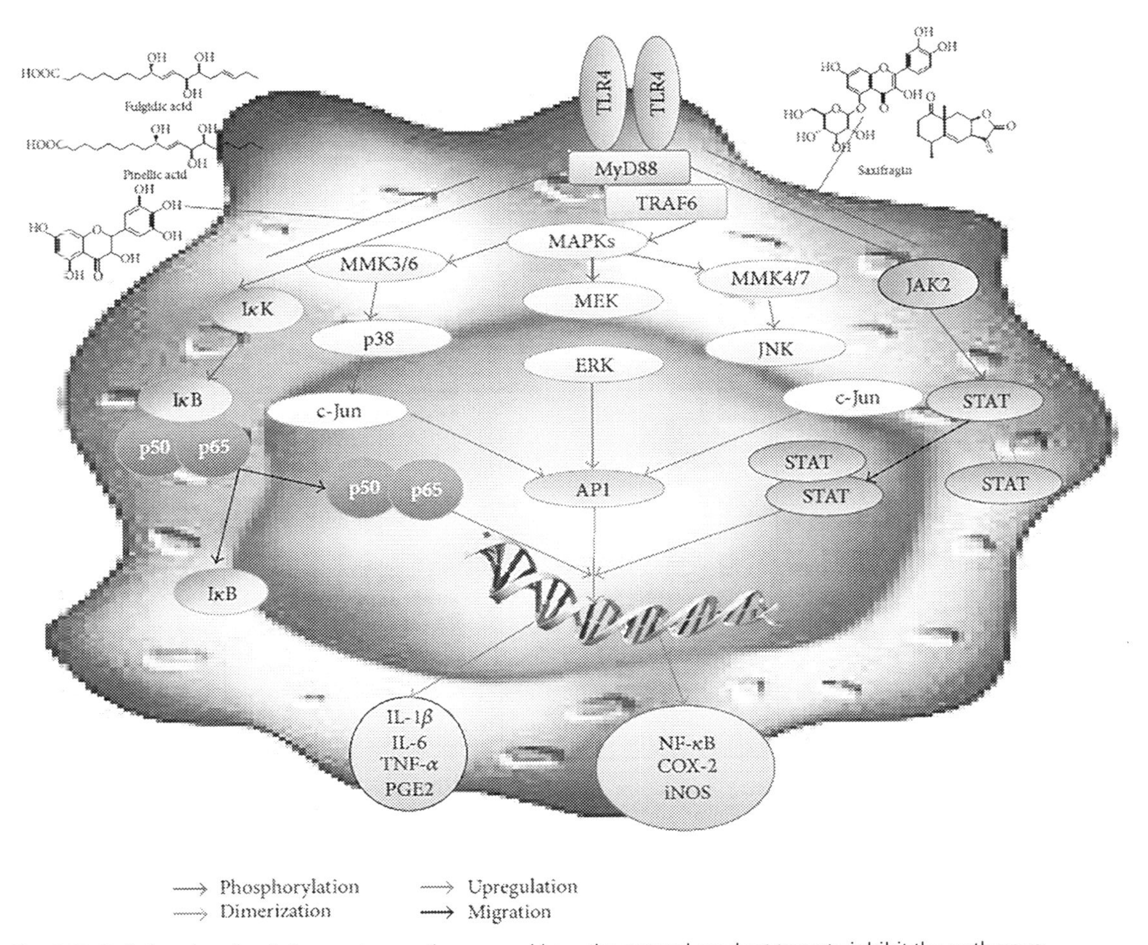

Fig. 2.12 Cellular signaling inflammatory pathways and how the natural product targets inhibit the pathways.

epidemiological and laboratory studies suggest that citrus flavonoids are of particular interest because many of these flavonoids exhibit a broad spectrum of biological activities, including anticancer and anti-inflammation [127–129]. Tangeretin belongs to the polymethoxylated flavones and is abundant in citrus peels. Tangeretin plays an important role in every stage of cancer development. It suppresses IL-1β-induced COX-2 expression through the inhibition of p38 MAPK, c-Jun N-terminal kinase (JNK), and Akt activation [130]. Lai et al. reported that 5-hydroxy-3,6,7,8,30,40-hexamethoxyflavone in citrus peels inhibits12-*O*-tetradecanoyl-phorbol-13-acetate (TPA)-induced skin inflammation and tumor promotion by suppressing the MAPK andPI3K/Akt signaling pathway [131]. Kaempferol, present in broccoli and tea, has been found to reduce the activity of inflammation-related genes such as iNOS and COX-2 by blocking the signaling of STAT-1,

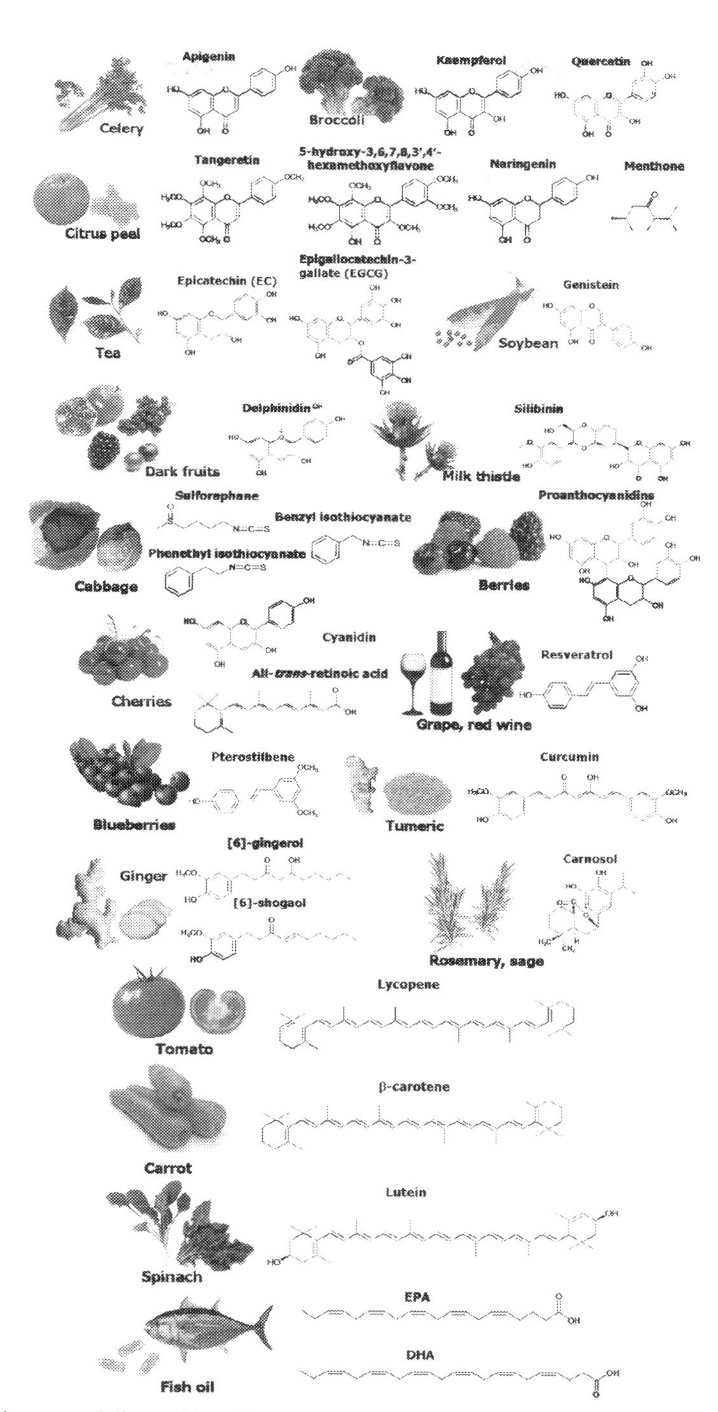

Fig. 2.13 Representative natural dietary bioactive compounds and their sources.

NF-κB, and AP-1 in activated macrophages [132] and human endothelial cells [133]. Quercetin, one of the ubiquitous plant secondary metabolites, is found typically in onions, broccoli, apples, grapes, wine, tea, and leafy green vegetables, and it has been shown to be a potent antioxidant and antiinflammatory agent. The antiinflammatory mechanism of quercetin is believed to inhibit the expression of proinflammatory cytokines in mast cells [134] and to suppress TNF-induced NF-κB and CBP/p300recruitment to proinflammatory gene promoters [135]. Naringenin, a flavanone present in oranges, is believed to inhibit LPS-induced IL-1β and TNF-R production [136] and inhibit iNOS protein and gene expression through blocking the activation of NF-κB (36). Tea is one of the most widely consumed beverages in the world. More than 300 different kinds of tea are produced from the leaves of *Camellia sinensis* by different manufacturing processes. Numerous health benefits have been attributed to the polyphenolic compounds in tea [137]. Catechins, the most abundant polyphenols in green tea, have been extensively studied in recent years. A typical cup of brewed green tea contains, by dry weight, 30%–40% catechins, including EGCG, epigallocatechin (EGC), epicatechin-3-gallate (ECG), and epicatechin (EC). In vivo, EC inhibited TPA-induced COX-2 expression through modulating NF-κB activation [138]. Recent studies indicated that EC reduced monocyte chemo-attractant protein-1 (MCP-1) production by suppressing the Akt phosphorylation TNFR in endothelial cells [139]. EGCG is the most abundant catechin in green tea. EGCG treatment suppressed IL-1β-dependent proinflammatory signal transduction in epithelial cells [140] and inhibited TPA-induced activation of NF-κB and CRP in mouse skin [141]. The aflavins from black tea and theaflavin-3,30-digallate in particular were demonstrated to be effective antiinflammatory agents. The beneficial antiinflammatory effect was linked to the aflavins' ability to inhibit the activation of NF-κB by inhibiting the phosphorylation, subsequent degradation of IκBR, and suppression of expression of proinflammatory genes, including interferon-γ (IFN-γ), IL-12, TNF-R, and iNOS [142]. Aflavin also affects brain injury by blocking inflammation-related events, including the overexpression of COX-2 and iNOS via downregulation of STAT-1 phosphorylation [143]. Recently, the aflavin monogallate and theaflavin-3,30-digallate were shown to modulate the regulators of G-protein signaling by selectively inducing the expression of the regulator of G-protein signaling (RGS)-10 [144]. The health benefits of soybeans and their products have been recognized in recent years [145]. Genistein, an isoflavone, is considered to be the main nutraceutical in soybeans. In the immune system, NF-κB is involved in the maturation of dendritic cells and is a critical mediator of inflammation progress, regulating the expression of a wide range of inflammation molecules. Genistein inhibited NF-κB-dependent gene expression in TLR

4-stimulated dendritic cells [146] and blocked the production of TNF-R and IL-1β in phytohemagglutinin-treated macrophages [147]. Cyanidin, an ananthocyanidin present in cherries and strawberries, exhibited a significant decrease in LPS-induced iNOS and COX-2 expression and platelet-derived growth factor ligand (PDGF)-induced VEGF expression in vascular smooth muscle cells [148, 149]. Cyanidin inhibited tumor promoter-induced carcinogenesis and tumor metastasis in vivo by modulating the expression of COX-2 and TNF-R [150]. Delphinidin, an anthocyanidin present in dark fruits, is believed to contribute to antiangiogenic activity by inhibiting the activation of the PDGF-BB/PDGF receptor (PDGFR)-β in smooth muscle cells [151]. Silymarin, a collective term for a mixture of seven flavonolignans from the milk thistle (*Silybum marianum* (L.) Gaertn.), is included in the pharmacopeia of many countries under the trademark Legalon or Hepatron and is often used as supportive therapy in food poisoning due to fungi and in chronic liver disorders [152]. Growing interest in using silymarin in cancer treatment and prevention is linked to its antiinflammatory properties [153]. Out of the mixture, silibinin is the best documented of the flavonolignans displaying beneficial effects. Silibinin was reported to inhibit UV-induced iNOS and COX-2 expression and to inhibit LPS-induced IL-12 expression by the downregulation of inflammatory-related signaling pathways [154, 155]. Isothiocyanates, proanthocyanidins, and terpenoids, compounds known as isothiocyanates, are formed during the mastication of some cruciferous vegetables, a process that promotes thioglucosidase (myrosinase) hydrolysis of the precursor conjugates known as glucosinolates. The vegetables belonging to the Brassica genus, which include cabbage, broccoli, kale, turnips, cauliflower, and Brussels sprouts, are the primary sources of glucosinolates and related breakdown products. Phenethyl isothiocyanate treatment suppressed LPS-induced inflammation in mouse macrophages and inhibited COX-2 expression through the modulation of multiple targets in the COX-2 gene promoter [156, 157]. The exposure of RAW 264.7 cells to phenethyl isothiocyanate suppressed the receptor for activation of NF-κB ligand (RANKL)-induced degradation of IκBR, the phosphorylation of p38 MAPK and ERK1/2, and the expression of NFAT [158]. Phenethyl isothiocyanate also displayed inhibition of NF-κB-mediated inflammatory transduction pathways in the colon and/or colon cancer [159]. Benzyl isothiocyanate was reported to decrease TPA-induced leukocyte infiltration and inhibit excessive superoxide generation in inflammatory leukocytes in mouse dermis [160, 161]. The topical application of benzyl isothiocyanate inhibited TPA-induced mouse skin inflammation [161]. Proanthocyanidins have been shown to inhibit COX-2 expression in LPS-activated mouse macrophages [162]. The proanthocyanindin A2 in the longan flower has potent antioxidative activity and delays LDL oxidation [163]. The terpenoids are a class of secondary metabolites from the common origin of mevalonate and isopentenyl pyrophosphate that are lipophilic in nature [164]. Menthone suppressed LPS-induced IL-1β and TNF-R production by suppressing the NF-κB signaling pathway in HaCat cells [165]. All-trans-retinoic acid treatment resulted in the inhibition of Th1- and Th2-related chemokine production via affecting the c-Raf-MAPK kinase (MKK)1/2-ERK/MAPK signaling pathway in monocytes [166]. Furthermore, all-trans-retinoic acid also suppressed IL-1-induced iNOS and COX-2 expression in human chondrocytes, which contributes to its antiinflammatory activity [167]. Besides flavonoids, many other polyphenolic compounds in nature also show health-promoting effects in food. One of the best researched is resveratrol, a compound found mainly in the skin of grapes, peanuts, mulberries, blueberries, and cranberries; most importantly, it is also found in red wines. Resveratrol was found to inhibit TPA-induced proinflammatory signaling pathways [168] and to inhibit COX-2 expression by blocking IKK activity in mouse skin [169]. Resveratrol has also been shown to suppress IL-6-induced intercellular cell adhesion molecule-1 (ICAM-1) gene expression in endothelial cells [170]. Pterostilbene, isolated from Vaccinium berries, has been shown to suppress LPS-induced iNOS and COX-2 expression in murine macrophages, which contributes to its antiinflammatory activity [171]. Pterostilbene was found to be as effective as resveratrol in inhibiting TPA-activated NFκB, AP-1, COX-2, and iNOS in mouse epidermis [172]. The dried rhizome of the turmeric plant (*Curcuma longa* Linn.) has been used for centuries as a naturally occurring medicine to treat topical inflammation and other diseases. The major pigment in the powdered rhizome, commonly known as turmeric spice, was identified as curcumin. Several laboratories have shown that curcumin and/or turmeric have potent antiinflammatory activity. Curcumin has also been reported to attenuate TNF-R-stimulated inflammatory cytokine production through modulating the phosphorylation of p38 and JNK and the activation of STAT-3 and NF-κB in human endothelial cells [173]. Curcumin also suppressed TPA-induced COX-2 expression by inhibiting NF-κB translocation and PKC activity [174]. Gingerols are the main pungent components of the rhizome ginger (*Zingiber officinale* Roscoe), which belongs to the ginger family Zingiberaceae. Common ginger has been used as a folk medicine for thousands of years. More recently, [6]-gingerol has been found to inhibit COX-2 expression by blocking the activation of p38MAPK and NF-κB in TPA-stimulated mouse skin [175]. Pan et al. demonstrated that [6]-shogaol suppressed the LPS-induced upexpression of iNOS and COX-2 by interfering with the activation of PI3K/Akt/IKK and MAPK in murine macrophages [176]. Rosemary and sage leaves are commonly used as spices and flavoring agents. Carnosol, an antioxidant in rosemary, suppressed LPS-induced iNOS expression through downregulating

NF-κB in macrophages [177]. Carnosol also inhibited COX-2 gene transcription by blocking PKC signaling and the binding of AP-1 to the CRE of the COX-2 promoter in human mammary epithelial cells [178]. Carotenoids are natural, fat-soluble pigments that provide bright coloration to plants and animals. Lycopene, a carotenoid found in tomatoes, watermelons, papaya, apricots, oranges, and pink grapefruit, has established antiinflammatory activities. Lycopene was reported to reduce the inflammatory response by lowering iNOS and COX-2 gene expression [179] and IL-12 production through blocking MAPK signaling and the activation of NF-κB in murine dendritic cells [180]. β-Carotene is primarily found in red palm oil, palm fruits, leafy green vegetables, carrots, sweet potatoes, mature squashes, pumpkins, mangoes, and papayas. β-Carotene is the most common carotenoid in food and the most potent of the provitamin A carotenoids. β-Carotene has been reported to inhibit LPS-induced iNOS, COX-2, and TNF-R expression by decreasing the phosphorylation and degradation of IκBR and the nuclear translocation of NF-κB in macrophages [181]. Lutein, found predominantly in dark green, leafy vegetables such as spinach and kale, is a yellow pigment that belongs to the class of nonprovitamin A carotenoids. Lutein treatment inhibited LPS- and H_2O_2-induced proinflammatory gene expression by decreasing the activity of PI3K and NF-κB inducing kinase (NIK) and the phosphorylation of Akt in RAW 264.7 cells [182]. Lutein has also been found to inhibit LPS-induced inflammatory cytokine signals and STAT3 phosphorylation in C57BL/6 mice [183]. A number of experimental and clinical studies have described potential health benefits of omega-3 polyunsaturated fatty acids (PUFA) abundant in marine oil and also present in some plant seed oils. These fatty acids are therapeutically useful in various diseases such as inflammatory disease and in prostate and colon cancers. Studies with fish oil, which contains eicosapentaenoicacid (EPA) and docosahexaenoic acid (DHA), showed that it has antiinflammatory and anticancer properties [184]. The pretreatment of EPA inhibited iNOS mRNA expression through the suppression of the DNA-binding activity of NF-κB [185]. EPA was also reported to inhibit the release of NO, PGE2, IL-1β, IL-6, and TNFR and to suppress NF-κB activation by blocking the MAPK and PI3K/Akt pathways in LPS-stimulated BV2 microglia [186]. DHA was reported to decrease RANKL-induced proinflammatory cytokine production by inhibiting the activation of c-fos, NFκB, and p38MAPK signaling pathways [187].

2.8 Conclusion

Inflammation plays a critical role during the pathology of multiple diseases and has become an imperative therapeutic target to develop novel approaches for pharmacological interventions. This review describes some of the recent work in the field that has elucidated the role of medicinal plants by regulating various inflammatory modulators and pathways. To regulate the level of inflammation in the system after an insult, a number of pharmacological targets have been identified and are currently being used therapeutically. The inhibition of inflammation can now be achieved through traditional interventional approaches, which has led to an increase in beneficial response during inflammatory diseases as well as a substantial reduction in associated complications and adverse effects. The current review has given an overview not only of the molecular mechanisms and approved targets for pharmacological intervention in inflammation, but additionally describes the medicinal plants and their pharmacological targets and how they are being developed as drugs to improve the quality of life while sparing the concomitant adverse effects during diseased states. The development of these drugs is essential to attain better control over the numerous acute and chronic inflammatory states. Taking into account the data analyzed in this study, we conclude that many plants have antiinflammatory activity, and clinical trials were conducted for many of them. The ancient knowledge of traditional Chinese medicine applied to the Western context has been discovered and increasingly gained greater importance. However, the exact chemical compound responsible for the observed pharmacological effects in most of the cases is still not known. The constituents of medicinal plants should therefore be investigated further to gain a better understanding of the individual components and pharmacological mechanisms. In addition, clinical studies in humans are necessary to confirm the claims of traditional phytotherapy. To achieve this, building a foundation for clinical trials of natural products is a greater approach. The side effects and potential interactions between medicinal plants and other synthetic drugs should also be further investigated. These may be important future challenges in drug discovery. Finally, pharmaceutical companies must aid the current knowledge by supporting relevant studies, even if their financial gain would be much lower compared to other kinds of treatment. Both international scientific societies and government organizations should take seriously the locally available opportunities of drug development by financially supporting relevant clinical studies.

Acknowledgments

This work was partially supported by the Natural Science Foundation of China (81673579 and 81874369) the Ministry of Science and Technology (2018YFC1707902 and 2018FY100703), the Hunan Department of Science and Technology (2018Wk2081 and 2018SK2110), the National Natural Science Foundation of China (No. 81803708), and the National Natural Science Foundation of Hunan Province (2018JJ3386).

Conflict of interest

The authors declare that they have no conflict of interest; these authors contributed to the work equally and should be regarded as co-first authors.

References

[1] R. Medzhitov, Inflammation 2010: new adventures of an old flame, Cell 140 (2010) 771–776.
[2] N. Eiró, F.J. Vizoso, Inflammation and cancer, World J. Gastrointest. Surg. 4 (2012) 62–72.
[3] S.Y. Kalinchenko, Y.A. Tishova, G.J. Mskhalaya, E.J. Giltay, L.J.G. Gooren, F. Saad, Effects of testosterone supplementation on markers of the metabolic syndrome and inflammation in hypogonadal men with the metabolic syndrome: the double-blinded placebo-controlled Moscow study, Clin. Endocrinol. (Oxf) 73 (2010) 602–612.
[4] E.H. Choy, G.S. Panayi, Cytokine pathways and joint inflammation in rheumatoid arthritis, N. Engl. J. Med. 344 (2001) 907–916.
[5] C.K. Glass, K. Saijo, B. Winner, M. Carolina, F.H. Gage, Mechanisms underlying inflammation in neurodegeneration, Cell 140 (2010) 918–934.
[6] B. Bozkurt, D.L. Mann, A. Deswal, Biomarkers of inflammation in heart failure, Heart Fail. Rev. 15 (2010) 331–341.
[7] C. Wiart, Medicinal Plants of Asia and the Pacific, CRC Press, 2006.
[8] S. Kawanishi, S. Ohnishi, N. Ma, Y. Hiraku, M. Murata, Crosstalk between DNA damage and inflammation in the multiple steps of carcinogenesis, Int. J. Mol. Sci. 18 (2017) 1808.
[9] S.C. Kneedler, L.E. Phillips, K.R. Hudson, K.M. Beckman, C.A. Lopez Gelston, J.M. Rutkowski, A.R. Parrish, P.A. Doris, B.M. Mitchell, Renal inflammation and injury are associated with lymphangiogenesis in hypertension, Am. J. Physiol. Renal Physiol. 312 (2017) F861–F869.
[10] I. Berczi, A. Szentivanyi, Cytokines and chemokines, in: I. Berczi, A. Szentivanyi (Eds.), NeuroImmune Biology, vol. 3, Elsevier, 2003, pp. 191–220.
[11] F. Balkwill, TNF-α in promotion and progression of cancer, Cancer Metastasis Rev. 25 (2006) 409–424.
[12] B.B. Aggarwal, S. Shishodia, K.S. Sandur, M.K. Pandey, G. Sethi, Inflammation and cancer: how hot is the link? Biochem. Pharmacol. 72 (2006) 1605–1621.
[13] F. Balkwill, A. Mantovani, Cancer and inflammation: implications for pharmacology and therapeutics, Clin. Pharmacol. Ther. 87 (2010) 401–406.
[14] S. Danese, What's hot in inflammatory bowel disease in 2011? World J. Gastroenterol. 17 (2011) 545–546.
[15] A. Kuraishy, M. Karin, S.I. Grivennikov, Tumor promotion via injury-and death-induced inflammation, Immunity 35 (2011) 467–477.
[16] R.N. Apte, Y. Krelin, X. Song, S. Dotan, E. Recih, M. Elkabets, Y. Carmi, T. Dvorkin, M.R. White, L. Gayvoronsky, S. Segal, E. Voronov, Effects of micro-environment- and malignant cell-derived interleukin-1 in carcinogenesis, tumour invasiveness and tumour-host interactions, Eur. J. Cancer 42 (2006) 751–759.
[17] W.P. Arend, M. Malyak, C.J. Guthridge, C. Gabay, Interleukin-1 receptor antagonist: role in biology, Annu. Rev. Immunol. 16 (1998) 27–55.
[18] P.C. Heinrich, I. Behrmann, S. Haan, H.M. Hermanns, G. Muller-Newen, F. Schaper, Principles of interleukin (IL)-6-type cytokine signalling and its regulation, Biochem. J. 374 (2003) 1–20.
[19] M. Locati, P.M. Murphy, Chemokines and chemokine receptors: biology and clinical relevance in inflammation and AIDS, Annu. Rev. Med. 50 (1999) 425–440.

[20] A. Li, S. Dubey, M.L. Varney, B.J. Dave, R.K. Singh, IL-8 directly enhanced endothelial cell survival, proliferation, and matrix metalloproteinases production and regulated angiogenesis, J. Immunol. 170 (2003) 3369–3376.
[21] X. Li, Y. Liu, L. Wang, Z. Li, X. Ma, Unfractionated heparin attenuates LPS-induced IL-8 secretion via PI3K/Akt/NF-κB signaling pathway in human endothelial cells, Immunobiology 220 (2015) 399–405.
[22] M. Profita, A. Bonanno, L. Siena, M. Ferraro, A.M. Montalbano, F. Pompeo, L. Riccobono, M.P. Pieper, M. Gjomarkaj, Acetylcholine mediates the release of IL-8 in human bronchial epithelial cells by a NFkB/ERK-dependent mechanism, Eur. J. Pharmacol. 582 (2008) 145–153.
[23] L. Yang, Y. Pang, H.L. Moses, TGF-β and immune cells: an important regulatory axis in the tumor microenvironment and progression, Trends Immunol. 31 (2010) 220–227.
[24] J. Massagué, D. Wotton, Transcriptional control by the TGF-β/Smad signaling system, EMBO J. 19 (2000) 1745–1754.
[25] J. Seoane, C. Pouponnot, P. Staller, M. Schader, M. Eilers, J. Massague, TGFβ influences Myc, Miz-1 and Smad to control the CDK inhibitor p15 INK4b, Nat. Cell Biol. 3 (2001) 400–408.
[26] T.D. Gilmore, Introduction to NF-κB: players, pathways, perspectives, Oncogene 25 (2006) 6680–6692.
[27] Y. Wu, S. Antony, J.L. Meitzler, J.H. Doroshow, Molecular mechanisms underlying chronic inflammation-associated cancers, Cancer Lett. 345 (2014) 164–173.
[28] Z.K. O'Brown, E.L.V. Nostrand, J.P. Higgnins, S.K. Kim, The inflammatory transcription factors NFκB, STAT1 and STAT3 drive age-associated transcriptional changes in the human kidney, PLoS Genet. 11 (2015) e1005734.
[29] A. Kaser, S. Zeissig, R.S. Blumberg, Inflammatory bowel disease, Annu. Rev. Immunol. 28 (2009) 573–621.
[30] H.L. Pahl, Activators and target genes of Rel/NF-κB transcription factors, Oncogene 18 (1999) 6853–6866.
[31] K.A. Fitzgerald, E.M.P. McDermott, A.G. Bowie, C.A. Jefferies, A.S. Mansell, G. Brady, E. Brint, A. Dunner, P. Gray, T.M. Harte, D. McMurray, E.D. Smith, E.J. Sims, T.A. Bird, L.A. O'Neill, Mal (MyD88-adapter-like) is required for toll-like receptor-4 signal transduction, Nature 413 (2001) 78–83.
[32] K. Takeda, S. Akira, TLR signaling pathways, Semin. Immunol. 16 (2004) 3–9.
[33] A.H. Brivanlou, J.E. Darnell, Signal transduction and the control of gene expression, Science 295 (2002) 813–818.
[34] H. Yu, D. Pardoll, R. Jove, STATs in cancer inflammation and immunity: a leading role for STAT3, Nat. Rev. Cancer 9 (2009) 798–809.
[35] G.L. Semenza, Oxygen sensing, homeostasis, and disease, N. Engl. J. Med. 365 (2011) 537–547.
[36] A. Palazon, A. Goldrath, V. Nizer, S.R. Johnson, HIF transcription factors, inflammation, and immunity, Immunity 41 (2014) 518–528.
[37] J. Rius, M. Guma, C. Schachtrup, K. Akassoglou, A.S. Zinkernagel, V. Nizer, R.S. Johnson, G.G. Haddad, M. Karin, NF-κB links innate immunity to the hypoxic response through transcriptional regulation of HIF-1α, Nature 453 (2008) 807–811.
[38] C.C. Blouin, E.L. Page, G.M. Soucy, D.E. Richard, Hypoxic gene activation by lipopolysaccharide in macrophages: implication of hypoxia-inducible factor 1α, Blood 103 (2004) 1124–1130.
[39] J.E. Albina, B. Mastrofrancesco, J.A. Vessella, C.A. Louis, W.L. Henry Jr., J.S. Reichner, HIF-1 expression in healing wounds: HIF-1α induction in primary inflammatory cells by TNF-α, Am. J. Physiol. Cell Physiol. 281 (2001) C1971–C1977.
[40] V.A. Shatrov, V.V. Sumbayev, J. Zhou, B. Brune, Oxidized low-density lipoprotein (oxLDL) triggers hypoxia-inducible factor-1α (HIF-1α) accumulation via redox-dependent mechanisms, Blood 101 (2003) 4847–4849.

[41] F. Mariani, P. Sena, L. Marzona, M. Riccio, R. Fano, P. Manni, C.D. Gregorio, A. Pezzi, M.P. Leon, S. Monni, A.D. Pol, L. Roncucci, Cyclooxygenase-2 and hypoxia-inducible factor-1α protein expression is related to inflammation, and up-regulated since the early steps of colorectal carcinogenesis, Cancer Lett. 279 (2009) 221–229.

[42] M. Ema, K. Hirota, J. Mimura, H. Abe, J. Yodoi, K. Sogawa, L. Poellinger, Y. Fujii-Kuriyama, Molecular mechanisms of transcription activation by HLF and HIF1α in response to hypoxia: their stabilization and redox signal-induced interaction with CBP/p300, EMBO J. 18 (1999) 1905–1914.

[43] S. Frede, C. Stockmann, P. Freitag, J. Fandrey, Bacterial lipopolysaccharide induces HIF-1 activation in human monocytes via p44/42 MAPK and NF-Kb, Biochem. J. 396 (2006) 517–527.

[44] M. Wills-Karp, Complement activation pathways: a bridge between innate and adaptive immune responses in asthma, Proc. Am. Thorac. Soc. 4 (2007) 247–251.

[45] E. Wagner, M.M. Frank, Therapeutic potential of complement modulation, Nat. Rev. Drug Discov. 9 (2010) 43–56.

[46] P. Pundir, M. Kulka, The role of G protein-coupled receptors in mast cell activation by antimicrobial peptides: is there a connection? Immunol. Cell Biol. 88 (2010) 632–640.

[47] M.J. Rabiet, E. Huet, F. Boulay, The N-formyl peptide receptors and the anaphylatoxin C5a receptors: an overview, Biochimie 89 (2007) 1089–1106.

[48] M. Perianayagam, V.S. Balakrishnan, B.J.G. Pereira, B.L. Jaber, C5a delays apoptosis of human neutrophils via an extracellular signal-regulated kinase and bad-mediated signalling pathway, Eur. J. Clin. Invest. 34 (2004) 50–56.

[49] M. Kuroki, J.T. O'Flaherty, Extracellular signal-regulated protein kinase (ERK)-dependent and ERK-independent pathways target STAT3 on serine-727 in human neutrophils stimulated by chemotactic factors and cytokines, Biochem. J. 341 (1999) 691–696.

[50] A.M. Scola, K.O. Johswich, B. Paul Morgan, A. Klos, P.N. Monk, The human complement fragment receptor, C5L2, is a recycling decoy receptor, Mol. Immunol. 46 (2009) 1149–1162.

[51] D.E. Croker, R. Jalai, D.P. Fairlie, M.A. Cooper, C5a, but not C5a-des Arg, induces upregulation of heteromer formation between complement C5a receptors C5aR and C5L2, Immunol. Cell Biol. 91 (2013) 625–633.

[52] P.K. Subhash, S.G. David, R.J. David, G. Letts, Eicosanoids in inflammation: biosynthesis, pharmacology, and therapeutic frontiers, Curr. Top. Med. Chem. 7 (2007) 311–340.

[53] N. Ellison, Goodman & Gilman's the pharmacological basis of therapeutics, tenth edition, Anesth. Analg. 94 (2002) 1377.

[54] E.A. Dennis, P.C. Norris, Eicosanoid storm in infection and inflammation, Nat. Rev. Immunol. 15 (2015) 511–523.

[55] S. Shinomiya, H. Naraba, A. Ueno, I. Utsunomiya, T. Maruyama, S. Ohuchida, F. Ushikubi, K. Yuki, S. Narumiya, Y. Sugimoto, A. Ichikawa, S. Oh-ishi, Regulation of TNFα and interleukin-10 production by prostaglandins I2 and E2: studies with prostaglandin receptor-deficient mice and prostaglandin E-receptor subtype-selective synthetic agonists, Biochem. Pharmacol. 61 (2001) 1153–1160.

[56] A. Greenhough, H.J.M. Smartt, A.E. Moore, H.R. Roberts, A.C. Williams, C. Paraskeva, A. Kaidi, The COX-2/PGE 2 pathway: key roles in the hallmarks of cancer and adaptation to the tumour microenvironment, Carcinogenesis 30 (2009) 377–386.

[57] B. Samuelsson, Leukotrienes: mediators of immediate hypersensitivity reactions and inflammation, Science 220 (1983) 568–575.

[58] T. Lämmermann, P.V. Afonso, B.R. Angermann, J.M. Wang, W. Kasternmuller, C.A. Parent, R.N. Germain, Neutrophil swarms require LTB4 and integrins at sites of cell death in vivo, Nature 498 (2013) 371–375.

[59] K. Murakami, T. Ide, M. Suzuki, T. Mochizuki, T. Kadowaki, Evidence for direct binding of fatty acids and eicosanoids to human peroxisome proliferators-activated receptor α, Biochem. Biophys. Res. Commun. 260 (1999) 609–613.
[60] J.T. Huang, J.S. Welch, M. Ricote, C.J. Binder, T.M. Willson, C. Kelly, J.L. Witztum, C.D. Funk, D. Conrad, C.K. Glass, Interleukin-4-dependent production of PPAR-γ ligands in macrophages by 12/15-lipoxygenase, Nature 400 (1999) 378–382.
[61] B. Halliwell, Reactive oxygen species in living systems: source, biochemistry, and role in human disease, Am. J. Med. 91 (1991) S14–S22.
[62] B. Halliwell, Antioxidants in human health and disease, Annu. Rev. Nutr. 16 (1996) 33–50.
[63] K. Block, Y. Gorin, Aiding and abetting roles of NOX oxidases in cellular transformation, Nat. Rev. Cancer 12 (2012) 627–637.
[64] T.L. Leto, M. Geiszt, Role of Nox family NADPH oxidases in host defense, Antioxid. Redox Signal. 8 (2006) 1549–1561.
[65] S. Lipinski, A. Till, C. Sina, A. Arlt, H. Grasberger, S. Schreiber, P. Rosenstiel, DUOX2-derived reactive oxygen species are effectors of NOD2-mediated antibacterial responses, J. Cell Sci. 122 (2009) 3522–3530.
[66] M. Katsuyama, NOX/NADPH oxidase, the superoxide-generating enzyme: its transcriptional regulation and physiological roles, J. Pharmacol. Sci. 114 (2010) 134–146.
[67] J.F. Woolley, A. Corcoran, G. Groeger, W.D. Landry, T.G. Cotter, Redox-regulated growth factor survival signaling, Antioxid. Redox Signal. 19 (2013) 1815–1827.
[68] R. Kodama, M. Kato, S. Furuta, S. Ueno, Y. Zhang, K. Matsuno, C.Y. Nishimura, E. Tanaka, T. Kamata, ROS-generating oxidases Nox1 and Nox4 contribute to oncogenic Ras-induced premature senescence, Genes Cells 18 (2013) 32–41.
[69] R. Zhou, A. Tardivel, B. Thorens, I. Choi, J. Tschopp, Thioredoxin-interacting protein links oxidative stress to inflammasome activation, Nat. Immunol. 11 (2010) 136–140.
[70] M. Khatami, 'Yin and Yang' in inflammation: duality in innate immune cell function and tumorigenesis, Expert Opin. Biol. Ther. 8 (2008) 1461–1472.
[71] K. Ckless, A. Lampert, J. Reiss, D. Kasahara, M.E. Poynter, C.G. Irvin, L.K.A. Lundblad, R. Norton, A. Vliet, Y.M.W.J. Heininger, Inhibition of arginase activity enhances inflammation in mice with allergic airway disease, in association with increases in protein S-nitrosylation and tyrosine nitration, J. Immunol. 181 (2008) 4255–4264.
[72] B. Gaston, J.M. Drazen, J. Loscalzo, J.S. Stamler, The biology of nitrogen oxides in the airways, Am. J. Respir. Crit. Care Med. 149 (1994) 538–551.
[73] W. Alderton, C. Cooper, R. Knowles, Nitric oxide synthases: structure, function and inhibition, Biochem. J. 357 (2001) 593–615.
[74] R. Radi, Nitric oxide, oxidants, and protein tyrosine nitration, Proc. Natl. Acad. Sci. 101 (2004) 4003–4008.
[75] K.V. Anderson, L. Bokla, C. Nüsslein-Volhard, Establishment of dorsal-ventral polarity in the Drosophila embryo: the induction of polarity by the toll gene product, Cell 42 (1985) 791–798.
[76] M. Triantafilou, F.G.J. Gamper, R.M. Haston, M. Angelos, S. Morath, T. Hartung, K. Triantafilou, Membrane sorting of toll-like receptor (TLR)-2/6 and TLR2/1 heterodimers at the cell surface determines heterotypic associations with CD36 and intracellular targeting, J. Biol. Chem. 281 (2006) 31002–31011.
[77] F.D. Herrington, R.J. Carmody, C.S. Goodyear, Modulation of NF-κB signaling as a therapeutic target in autoimmunity, J. Biomol. Screen. 21 (2016) 223–242.
[78] B. Kaminska, MAPK signalling pathways as molecular targets for anti-inflammatory therapy—from molecular mechanisms to therapeutic benefits, Biochim. Biophys. Acta 1754 (2005) 253–262.

[79] M.J. Thiel, C.J. Schaefer, M.E. Lesch, J.L. Mobley, D.T. Dudley, H. Tecle, S.D. Barrett, D.J. Schrier, C.M. Flory, Central role of the MEK/ERK MAP kinase pathway in a mouse model of rheumatoid arthritis: potential proinflammatory mechanisms, Arthritis Rheum. 56 (2007) 3347–3357.
[80] D. Boyle, K. Soma, J. Hodge, A. Kavanaugh, D. Mandel, P. Mease, R. Sjirmur, A.K. Singhal, N. Wei, S. Rosengren, I. Kaplan, S. Krishnaswami, Z. Luo, J. Bradley, G.S. Firestein, The JAK inhibitor tofacitinib suppresses synovial JAK1-STAT signalling in rheumatoid arthritis, Ann. Rheum. Dis. 74 (2015) 1311–1316.
[81] S.Y. Hwang, J.Y. Kim, K.W. Kim, M.K. Park, Y. Moon, W.U. Kim, H.Y. Kim, IL-17 induces production of IL-6 and IL-8 in rheumatoid arthritis synovial fibroblasts via NF-κB-and PI3-kinase/Akt-dependent pathways, Arthritis Res. Ther. 6 (2004) R120–R128.
[82] E. Ricciotti, G.A. FitzGerald, Prostaglandins and inflammation, Arterioscler. Thromb. Vasc. Biol. 31 (2011) 986–1000.
[83] J. Sharma, A. Al-Omran, S. Parvathy, Role of nitric oxide in inflammatory diseases, Inflammopharmacology 15 (2007) 252–259.
[84] H. Pan, H. Wang, L. Zhu, L. Mao, Depletion of Nrf2 enhances inflammation induced by oxyhemoglobin in cultured mice astrocytes, Neurochem. Res. 36 (2011) 2434–2441.
[85] Y. Jiang, M.H. Wang, Different solvent fractions of *Acanthopanax senticosus* harms exert antioxidant and anti-inflammatory activities and inhibit the human Kv1. 3 channel, J. Med. Food 18 (2015) 468–475.
[86] H.Y. Kim, K.W. Hwang, S.Y. Park, Extracts of *Actinidia arguta* stems inhibited LPS-induced inflammatory responses through nuclear factor-κB pathway in Raw 264.7 cells, Nutr. Res. 34 (2014) 1008–1016.
[87] X. Chen, J. MiO, H. Wang, F. Zhao, J. Hu, P. Gao, Y. Wang, L. Zhang, M. Yan, The anti-inflammatory activities of Ainsliaea fragrans Champ. extract and its components in lipopolysaccharide-stimulated RAW264. 7 macrophages through inhibition of NF-κB pathway, J. Ethnopharmacol. 170 (2015) 72–80.
[88] X. Hou, Q. Tong, W.Q. Wang, C.Y. Shi, W. Ziong, J. Chen, X. Liu, J.G. Fang, Suppression of inflammatory responses by dihydromyricetin, a flavonoid from Ampelopsis grossedentata, via inhibiting the activation of NF-κB and MAPK signaling pathways, J. Nat. Prod. 78 (2015) 1689–1696.
[89] S. Qi, Y. Xin, Y. Guo, Y. Diao, X. Kou, L. Luo, Z. Yin, Ampelopsin reduces endotoxic inflammation via repressing ROS-mediated activation of PI3K/Akt/NF-κB signaling pathways, Int. Immunopharmacol. 12 (2012) 278–287.
[90] H.R. Chang, H.J. Lee, J.H. Ryu, Cchalcones from *Angelica keiskei* attenuate the inflammatory responses by suppressing nuclear translocation of NF-Kb, J. Med. Food 17 (2014) 1306–1313.
[91] Y.C. Oh, Y.H. Jeong, T. Kim, W.K. Cho, Anti-inflammatory effect of *Artemisiae annuae* herba in lipopolysaccharide-stimulated RAW 264.7 Cells, Pharmacogn. Mag. 10 (2014) S588–S595.
[92] N.K. Patel, S. Pulipaka, S.P. Dubey, K.K. Bhutani, Pro-inflammatory cytokines and nitric oxide inhibitory constituents from *Cassia occidentalis* roots, Nat. Prod. Commun. 9 (2014) 661–664.
[93] R. Lamichhane, S.G. Kim, A. Poudel, D. Sharma, K.H. Lee, H.J. Jung, Evaluation of in vitro and in vivo biological activities of *Cheilanthes albomarginata* Clarke, BMC Complement. Altern. Med. 14 (2014) 342.
[94] T. Akihisa, S.G. Franzblau, M. Ukiya, H. Okuda, F. Zhang, K. Yasukawa, T. Suzuki, Y. Kimura, Antitubercular activity of triterpenoids from Asteraceae flowers, Biol. Pharm. Bull. 28 (2005) 158–160.
[95] Y.C. Cho, A. Ju, B.R. Kim, S. Cho, Anti-inflammatory effects of Crataeva nurvala Buch. Ham. are mediated via inactivation of ERK but not NF-Kb, J. Ethnopharmacol. 162 (2015) 140–147.

[96] J.S. Shin, Y. Hong, H.H. Lee, B. Ryu, Y.W. Cho, N.J. Kim, D.S. Jang, K.T. Lee, Fulgidic acid isolated from the rhizomes of *Cyperus rotundus* suppresses LPS-Induced iNOS, COX-2, TNF-α, and IL-6 expression by AP-1 inactivation in RAW264. 7 macrophages, Biol. Pharm. Bull. 38 (2015) 1081–1086.
[97] B.Y. Yang, R. Guo, T. Li, J.J. Wu, J. Zhang, Y. Liu, Q.H. Wang, H.X. Kuang, New anti-inflammatory withanolides from the leaves of *Datura metel* L, Steroids 87 (2014) 26–34.
[98] R.G. de Oliveira, C.P.A.N. Mahon, P.G.M. Ascencio, S.D. Scencio, S.O. Balogun, D.T. de Oliveira, Evaluation of anti-inflammatory activity of hydroethanolic extract of Dilodendron bipinnatum Radlk, J. Ethnopharmacol. 155 (2014) 387–395.
[99] D.H. Kim, M.E. Kim, J.S. Lee, Inhibitory effects of extract from *G. lanceolata* on LPS-induced production of nitric oxide and IL-1β via down-regulation of MAPK in macrophages, Appl. Biochem. Biotechnol. 175 (2015) 657–665.
[100] T. Sogo, N. Terahara, A. Hisanaga, T. Kumamoto, T. Yamashiro, S. Wu, K. Sakao, D.X. Hou, Anti-inflammatory activity and molecular mechanism of delphinidin 3-sambubioside, a Hibiscus anthocyanin, Biofactors 41 (2015) 58–65.
[101] J.M. Chun, K.J. Bho, H.S. Kim, A.Y. Lee, B.C. Moon, H.K. Kim, An ethyl acetate fraction derived from *Houttuynia cordata* extract inhibits the production of inflammatory markers by suppressing NF-κB and MAPK activation in lipopolysaccharide-stimulated RAW 264.7 macrophages, BMC Complement. Altern. Med. 14 (2014) 234.
[102] H.H. Park, M.J. Kim, Y. Li, Y.N. Park, J. Lee, Y.J. Lee, H.J. Park, J.K. Son, H.W. Chang, E. Lee, Britanin suppresses LPS-induced nitric oxide, PGE2 and cytokine production via NF-κB and MAPK inactivation in RAW 264.7 cells, Int. Immunopharmacol. 15 (2013) 296–302.
[103] X. Chen, S.A. Tang, E. Lee, Y. Qiu, IVSE, isolated from Inula japonica, suppresses LPS-induced NO production via NF-κB and MAPK inactivation in RAW264. 7 cells, Life Sci. 124 (2015) 8–15.
[104] X. Wang, Inhibitory effects of JEUD-38, a new sesquiterpene lactone from *Inula japonica* thunb, on LPS-induced iNOS expression in RAW264. 7 cells, Inflammation 38 (2015) 941–948.
[105] S.S. Lee, N.H. Tan, S.Y. Fung, S.M. Sim, C.S. Tan, S.T. Ng, Anti-inflammatory effect of the sclerotium of Lignosus rhinocerotis (Cooke) Ryvarden, the Tiger Milk mushroom, BMC Complement. Altern. Med. 14 (2014) 359.
[106] M.Y. Song, H.W. Jung, S.Y. Kang, K.H. Kim, Y.K. Park, Anti-inflammatory effect of Lycii radicis in LPS-stimulated RAW 264.7 macrophages, Am. J. Chin. Med. 42 (2014) 891–904.
[107] M. Flemming, B. Kraus, A. Rascle, G. Jurgenliemk, S. Fuchs, R. Furst, J. Heilmann, Revisited anti-inflammatory activity of matricine in vitro: comparison with chamazulene, Fitoterapia 106 (2015) 122–128.
[108] M.E. Gerritsen, W.W. Carley, G.E. Ranges, C.P. Shen, S.A. Phan, G.F. Lihon, C.A. Perry, Flavonoids inhibit cytokine-induced endothelial cell adhesion protein gene expression, Am. J. Pathol. 147 (1995) 278–292.
[109] M.H. Pan, C.T. Ho, Chemopreventive effects of natural dietary compounds on cancer development, Chem. Soc. Rev. 37 (2008) 2558–2574.
[110] R. Medzhitov, Origin and physiological roles of inflammation, Nature 454 (2008) 428–435.
[111] B.B. Aggarwal, R. Vijayalekshmi, B. Sung, Targeting inflammatory pathways for prevention and therapy of cancer: short-term friend, long-term foe, Clin. Cancer Res. 15 (2009) 425–430.
[112] A. Mantovani, P. Allavena, A. Sica, F. Balkwill, Cancer-related inflammation, Nature 454 (2008) 436–444.
[113] H. Lu, W. Ouyang, C. Huang, Inflammation, a key event in cancer development, Mol. Cancer Res. 4 (2006) 221–233.

[114] M. Karin, Nuclear factor-κB in cancer development and progression, Nature 441 (2006) 431–436.

[115] S. Maeda, M. Omata, Inflammation and cancer: role of nuclear factor-kappaB activation, Cancer Sci. 99 (2008) 836–842.

[116] A. Agarwal, K. Das, N. Lerner, S. Sathe, M. Cicek, G. Casey, N. Sizemore, The AKT/IκB kinase pathway promotes angiogenic/metastatic gene expression in colorectal cancer by activating nuclear factor-κB and β-catenin, Oncogene 24 (2005) 1021–1031.

[117] D.A. Evans, J.B. Hirsch, S. Dushenkov, Phenolics, inflammation and nutrigenomics, J. Sci. Food Agric. 86 (2006) 2503–2509.

[118] M.H. Pan, C.S. Lai, S. Dushenkov, C.T. Ho, Modulation of inflammatory genes by natural dietary bioactive compounds, J. Agric. Food Chem. 57 (2009) 4467–4477.

[119] E.P. Lansky, H.M. Paavilainen, A.D. Pawkysm, R.A. Bewnabm, Ficus spp.(fig): ethnobotany and potential as anticancer and anti-inflammatory agents, J. Ethnopharmacol. 119 (2008) 195–213.

[120] M. Balick, R. Lee, Inflammation and ethnomedicine: looking to our past, Exp. Dermatol. 1 (2005) 389–392.

[121] K. Subbaramaiah, P. Bulic, Y. Lin, A.J. Dannenberg, D.S. Pasco, Development and use of a gene promoter-based screen to identify novel inhibitors of cyclooxygenase-2 transcription, J. Biomol. Screen. 6 (2001) 101–110.

[122] M. Dey, C. Ripoll, R. Pouleva, R. Dorn, I. Aranovich, D. Zaurov, A. Kurmulov, M. Eliseyeva, I. Belolipov, A. Akimaliev, I. Sodombekov, D. Akimaliev, M.A. Lila, I. Raskin, Plant extracts from central Asia showing antiinflammatory activities in gene expression assays, Phytother. Res. 22 (2008) 929–934.

[123] T. Kao, W.M. Wu, C.F. Hung, W.B. Wu, B.H. Chen, Anti-inflammatory effects of isoflavone powder produced from soybean cake, J. Agric. Food Chem. 55 (2007) 11068–11079.

[124] J. Fang, C. Xia, Z. Cao, J.Z. Zheng, E. Reed, B.H. Jiang, Apigenin inhibits VEGF and HIF-1 expression via PI3K/AKT/p70S6K1 and HDM2/p53 pathways, FASEB J. 19 (2005) 342–353.

[125] C. Nicholas, S. Batra, M.A. Vargo, O.H. Voss, M.A. Gavrilin, M.D. Wewers, D.C. Guttridge, E. Grotewold, A.I. Doseff, Apigenin blocks lipopolysaccharide-induced lethality in vivo and proinflammatory cytokines expression by inactivating NF-κB through the suppression of p65 phosphorylation, J. Immunol. 179 (2007) 7121–7127.

[126] R.T. Van Dross, X. Hong, S. Essengue, S.M. Fischer, J.C. Pelling, Modulation of UVB-induced and basal cyclooxygenase-2 (COX-2) expression by apigenin in mouse keratinocytes: role of USF transcription factors, Mol. Carcinog. 46 (2007) 303–314.

[127] O. Benavente-Garcia, J. Castillo, Update on uses and properties of citrus flavonoids: new findings in anticancer, cardiovascular, and anti-inflammatory activity, J. Agric. Food Chem. 56 (2008) 6185–6205.

[128] J.A. Manthey, P. Bendele, Anti-inflammatory activity of an orange peel polymethoxylated flavone, 3′, 4′, 3, 5, 6, 7, 8-heptamethoxyflavone, in the rat carrageenan/paw edema and mouse lipopolysaccharide-challenge assays, J. Agric. Food Chem. 56 (2008) 9399–9403.

[129] S. Held, P. Schieberle, V. Somoza, Characterization of α-terpineol as an anti-inflammatory component of orange juice by in vitro studies using oral buccal cells, J. Agric. Food Chem. 55 (2007) 8040–8046.

[130] K.H. Chen, M.S. Weng, J.K. Lin, Tangeretin suppresses IL-1β-induced cyclooxygenase (COX)-2 expression through inhibition of p38 MAPK, JNK, and AKT activation in human lung carcinoma cells, Biochem. Pharmacol. 73 (2007) 215–227.

[131] C.S. Lai, S. Li, C.Y. Chai, C.Y. Lo, C.T. Ho, Y.J. Wang, M.H. Pan, Inhibitory effect of citrus 5-hydroxy-3, 6, 7, 8, 3′, 4′-hexamethoxyflavone on 12-O-tetradecanoylphorbol 13-acetate-induced skin inflammation and tumor promotion in mice, Carcinogenesis 28 (2007) 2581–2588.

[132] M. Hämäläinen, R. Nieminen, P. Vuorela, M. Heinonen, E. Moilanen, Anti-inflammatory effects of flavonoids: genistein, kaempferol, quercetin, and daidzein inhibit STAT-1 and NF-κB activations, whereas flavone, isorhamnetin, naringenin, and pelargonidin inhibit only NF-κB activation along with their inhibitory effect on iNOS expression and NO production in activated macrophages, Mediators Inflamm. 2007 (2007) 45673.

[133] I. Crespo, M.V. Mediavilla, B. Gurierrez, S. Campos, M.J. Tunon, J.G. Gallego, A comparison of the effects of kaempferol and quercetin on cytokine-induced pro-inflammatory status of cultured human endothelial cells, Br. J. Nutr. 100 (2008) 968–976.

[134] H.H. Park, S. Lee, H.Y. Son, S.B. Park, M.S. Kim, E.J. Choi, T.S.K. Singh, J.H. Ha, M.G. Lee, J.E. Kim, M.C. Hyun, T.K. Kwon, Y.H. Kim, S.H. Kim, Flavonoids inhibit histamine release and expression of proinflammatory cytokines in mast cells, Arch. Pharm. Res. 31 (2008) 1303.

[135] P.A. Ruiz, A. Braune, G. Holzlwimmer, L. Fend, D. Haller, Quercetin inhibits TNF-induced NF-κ B transcription factor recruitment to proinflammatory gene promoters in murine intestinal epithelial cells, J. Nutr. 137 (2007) 1208–1215.

[136] C. Bodet, V.D. La, F. Epifano, D. Grenier, Naringenin has anti-inflammatory properties in macrophage and ex vivo human whole-blood models, J. Periodontal Res. 43 (2008) 400–407.

[137] P.M. Coates, M.R. Blackman, G.M. Cragg, M. Levine, J. Moss, J.D. White, Encyclopedia of Dietary Supplements, CRC Press, NY, United States, 2004.

[138] K.W. Lee, J.K. Kundu, S.O. Kim, K.S. Chun, H.J. Lee, Y.J. Surh, Cocoa polyphenols inhibit phorbol ester-induced superoxide anion formation in cultured HL-60 cells and expression of cyclooxygenase-2 and activation of NF-κB and MAPKs in mouse skin in vivo, J. Nutr. 136 (2006) 1150–1155.

[139] H.Y. Ahn, Y. Xu, S.T. Davidge, Epigallocatechin-3-O-gallate inhibits TNFα-induced monocyte chemotactic protein-1 production from vascular endothelial cells, Life Sci. 82 (2008) 964–968.

[140] D.S. Wheeler, J.D. Catravas, K. Odoms, A. Denenberg, V. Malhotra, H.R. Wong, Epigallocatechin-3-gallate, a green tea–derived polyphenol, inhibits IL-1β-dependent proinflammatory signal transduction in cultured respiratory epithelial cells, J. Nutr. 134 (2004) 1039–1044.

[141] J.K. Kundu, Y.J. Surh, Epigallocatechin gallate inhibits phorbol ester-induced activation of NF-κB and CREB in mouse skin: role of p38 MAPK, Ann. N. Y. Acad. Sci. 1095 (2007) 504–512.

[142] A. Ukil, S. Maity, P. Das, Protection from experimental colitis by theaflavin-3, 3′-digallate correlates with inhibition of IKK and NF-κB activation, Br. J. Pharmacol. 149 (2006) 121–131.

[143] F. Cai, C.R. Li, J.L. Wu, J.G. Chen, C. Liu, Q. Min, C.H.M. Ouyang, J.H. Chen, Theaflavin ameliorates cerebral ischemia-reperfusion injury in rats through its anti-inflammatory effect and modulation of STAT-1, Mediators Inflamm. 2006 (2006) 30490.

[144] J. Lu, A. Gosslau, A.Y.C. Liu, K.Y. Chen, PCR differential display-based identification of regulator of G protein signaling 10 as the target gene in human colon cancer cells induced by black tea polyphenol theaflavin monogallate, Eur. J. Pharmacol. 601 (2008) 66–72.

[145] T. Cornwell, W. Cohick, I. Raskin, Dietary phytoestrogens and health, Phytochemistry 65 (2004) 995–1016.

[146] N. Dijsselbloem, S. Goriely, V. Albarani, S. Gerlo, S. Francoz, J.C. Marine, M. Goldman, G. Haegeman, W.V. Berghe, A critical role for p53 in the control of NF-κB-dependent gene expression in TLR4-stimulated dendritic cells exposed to genistein, J. Immunol. 178 (2007) 5048–5057.

[147] V. Kesherwani, A. Sodhi, Involvement of tyrosine kinases and MAP kinases in the production of TNF-α and IL-1 β by macrophages in vitro on treatment with phytohemagglutinin, J. Interferon Cytokine Res. 27 (2007) 497–506.
[148] M.H. Oak, J.E. Bedoui, S.V.F. Madeira, K. Chalupslu, V.B. Schini-Kerth, Delphinidin and cyanidin inhibit PDGFAB-induced VEGF release in vascular smooth muscle cells by preventing activation of p38 MAPK and JNK, Br. J. Pharmacol. 149 (2006) 283–290.
[149] Q. Wang, M. Xia, C. Liu, H. Guo, Cyanidin-3-O-β-glucoside inhibits iNOS and COX-2 expression by inducing liver X receptor alpha activation in THP-1 macrophages, Life Sci. 83 (2008) 176–184.
[150] M. Ding, R. Feng, S.Y. Wang, L. Bowman, Cyanidin-3-glucoside, a natural product derived from blackberry, exhibits chemopreventive and chemotherapeutic activity, J. Biol. Chem. 281 (2006) 17359–17368.
[151] S. Lamy, E. Beaulieu, D. Labbe, V. Bedard, A. Moghrabi, S. Barrette, D. Gingras, R. Beliveau, Delphinidin, a dietary anthocyanidin, inhibits platelet-derived growth factor ligand/receptor (PDGF/PDGFR) signaling, Carcinogenesis 29 (2008) 1033–1041.
[152] M. Parmar, T. Gandhi, Hepatoprotective herbal drug, silymarin from experimental pharmacology to clinical medicine—a review, Pharmacol. Rev. 2 (2008) 102–109.
[153] K. Ramasamy, R. Agarwal, Multitargeted therapy of cancer by silymarin, Cancer Lett. 269 (2008) 352–362.
[154] J.S. Lee, S.G. Kim, H.K. Kim, T.H. Lee, Y.I. Jeong, C.M. Lee, M.S. Yoon, Y.J. Na, D.S. Suh, N.C. Park, I.H. Choi, G.Y. Kim, Y.H. Choi, H.Y. Chung, Y.M. Park, Silibinin polarizes Th1/Th2 immune responses through the inhibition of immunostimulatory function of dendritic cells, J. Cell. Physiol. 210 (2007) 385–397.
[155] M. Gu, R.P. Singh, S. Dhanalakshimi, C. Agarwal, R. Agarwal, Silibinin inhibits inflammatory and angiogenic attributes in photocarcinogenesis in SKH-1 hairless mice, Cancer Res. 67 (2007) 3483–3491.
[156] W. Lin, R.T. Wu, T. Wu, T.O. Khorm, H. Wang, A.N. Kong, Sulforaphane suppressed LPS-induced inflammation in mouse peritoneal macrophages through Nrf2 dependent pathway, Biochem. Pharmacol. 76 (2008) 967–973.
[157] K.J. Woo, T.K. Kwon, Sulforaphane suppresses lipopolysaccharide-induced cyclooxygenase-2 (COX-2) expression through the modulation of multiple targets in COX-2 gene promoter, Int. Immunopharmacol. 7 (2007) 1776–1783.
[158] A. Murakami, M. Song, H. Ohigashi, Phenethyl isothiocyanate suppresses receptor activator of NF-kappaB ligand (RANKL)-induced osteoclastogenesis by blocking activation of ERK1/2 and p38 MAPK in RAW264. 7 macrophages, Biofactors 30 (2007) 1–11.
[159] W.S. Jeong, I.W. Kim, R. Hu, A.N.T. Kong, Modulatory properties of various natural chemopreventive agents on the activation of NF-κB signaling pathway, Pharm. Res. 21 (2004) 661–670.
[160] Y. Nakamura, Benzyl isothiocyanate inhibits oxidative stress in mouse skin: involvement of attenuation of leukocyte infiltration, Biofactors 21 (2004) 255–257.
[161] N. Miyoshi, S. Takabayashi, T. Osawa, Y. Nakamura, Benzyl isothiocyanate inhibits excessive superoxide generation in inflammatory leukocytes: implication for prevention against inflammation-related carcinogenesis, Carcinogenesis 25 (2004) 567–575.
[162] D.X. Hou, S. Masuzaki, F. Hashimoto, T. Uto, S. Tanigawa, M. Fujii, Y. Sakata, Green tea proanthocyanidins inhibit cyclooxygenase-2 expression in LPS-activated mouse macrophages: molecular mechanisms and structure–activity relationship, Arch. Biochem. Biophys. 460 (2007) 67–74.
[163] M.C. Hsieh, Y.J. Shen, Y.H. Kuo, L.S. Hwang, Antioxidative activity and active components of longan (*Dimocarpus longan* Lour.) flower extracts, J. Agric. Food Chem. 56 (2008) 7010–7016.

[164] T. Akihisa, K. Yasukawa, M. Yamaura, M. Ukiya, Y. Kimura, N. Shimizu, K. Arai, Triterpene alcohol and sterol ferulates from rice bran and their anti-inflammatory effects, J. Agric. Food Chem. 48 (2000) 2313–2319.
[165] L. Fong, T. Fong, A. Cooper, Inhibition of lipopolysaccharide-induced interleukin-1 beta mRNA expression in mouse macrophages by oxidized low density lipoprotein, J. Lipid Res. 32 (1991) 1899–1910.
[166] Y.C. Tsai, H.W. Chang, T.T. Chang, M.S. Lee, Y.T. Chu, C.H. Hung, Effects of all-trans retinoic acid on Th1- and Th2-related chemokines production in monocytes, Inflammation 31 (2008) 428–433.
[167] L.F. Hung, J.H. Lai, L.C. Lin, S.J. Wang, T.Y. Hou, D.M. Chang, C.C.T. Liang, L.J. Ho, Retinoid acid inhibits IL-1-induced iNOS, COX-2 and chemokine production in human chondrocytes, Immunol. Invest. 37 (2008) 675–693.
[168] J.K. Kundu, Y.K. Shin, Y.J. Surh, Resveratrol modulates phorbol ester-induced pro-inflammatory signal transduction pathways in mouse skin in vivo: NF-κB and AP-1 as prime targets, Biochem. Pharmacol. 72 (2006) 1506–1515.
[169] J.K. Kundu, Y.K. Shin, Y.J. Surh, Resveratrol inhibits phorbol ester-induced expression of COX-2 and activation of NF-κB in mouse skin by blocking IκB kinase activity, Carcinogenesis 27 (2006) 1465–1474.
[170] B. Wung, M.C. Hsu, C.W. Hsieh, Resveratrol suppresses IL-6-induced ICAM-1 gene expression in endothelial cells: effects on the inhibition of STAT3 phosphorylation, Life Sci. 78 (2005) 389–397.
[171] M.H. Pan, Y.H. Chang, M.L. Tsai, C.S. Lai, S.Y. Ho, V. Badmaev, C.T. Ho, Pterostilbene suppressed lipopolysaccharide-induced up-expression of iNOS and COX-2 in murine macrophages, J. Agric. Food Chem. 56 (2008) 7502–7509.
[172] M. Cichocki, J. Paluszczak, H. Szaefer, A. Piechowiak, A.M. Rimando, W. Dubowska, Pterostilbene is equally potent as resveratrol in inhibiting 12-O-tetradecanoylphorbol-13-acetate activated NFκB, AP-1, COX-2, and iNOS in mouse epidermis, Mol. Nutr. Food Res. 52 (2008) S62–S70.
[173] Y.S. Kim, Y. Ahn, M.H. Hong, S.Y. Joo, H.K. Kim, I.S. Sohn, H.W. Park, Y.J. Hong, J.H. Kim, W. Lim, M.H. Jeong, J.G. Cho, C.J. Parl, J.C. Kang, Curcumin attenuates inflammatory responses of TNF-α-stimulated human endothelial cells, J. Cardiovasc. Pharmacol. 50 (2007) 41–49.
[174] R. Garg, A.G. Ramchandani, G.B. Maru, Curcumin decreases 12-O-tetradecanoylphorbol-13-acetate-induced protein kinase C translocation to modulate downstream targets in mouse skin, Carcinogenesis 29 (2008) 1249–1257.
[175] S.O. Kim, K.J. Kundu, K.Y. Shin, J.H. Park, M.H. Cho, T.Y. Kim, Y.J. Surh, [6]-Gingerol inhibits COX-2 expression by blocking the activation of p38 MAP kinase and NF-κB in phorbol ester-stimulated mouse skin, Oncogene 24 (2005) 2558–2567.
[176] M.H. Pan, M.C. Hsieh, P.C. Hsu, S.Y. Ho, C.S. Lai, H. Wu, S. Sang, C.T. Ho, 6-Shogaol suppressed lipopolysaccharide-induced up-expression of iNOS and COX-2 in murine macrophages, Mol. Nutr. Food Res. 52 (2008) 1467–1477.
[177] A.H. Lo, Y.C. Liang, S.Y. Shiau, C.T. Ho, J.K. Lin, Carnosol, an antioxidant in rosemary, suppresses inducible nitric oxide synthase through down-regulating nuclear factor-κB in mouse macrophages, Carcinogenesis 23 (2002) 983–991.
[178] K. Subbaramaiah, P.A. Cole, A.J. Dannenberg, Retinoids and carnosol suppress cyclooxygenase-2 transcription by CREB-binding protein/p300-dependent and-independent mechanisms, Cancer Res. 62 (2002) 2522–2530.
[179] D. De Stefano, M.C. Maiuri, V. Simeon, G. Grassia, A. Soscia, M.P. Cinelli, R. Carnuccio, Lycopene, quercetin and tyrosol prevent macrophage activation induced by gliadin and IFN-γ, Eur. J. Pharmacol. 566 (2007) 192–199.
[180] G.Y. Kim, J.H. Kim, S.C. Ahn, H.J. Lee, D.O. Moon, C.M. Lee, Y.M. Park, Lycopene suppresses the lipopolysaccharide-induced phenotypic and functional matura-

tion of murine dendritic cells through inhibition of mitogen-activated protein kinases and nuclear factor-Kb, Immunology 113 (2004) 203–211.

[181] S.K. Bai, S.J. Lee, H.J. Na, K.S. Ha, J.A. Han, H. Lee, Y.G. Kwon, C.K. Chung, Y.M. Kim, β-Carotene inhibits inflammatory gene expression in lipopolysaccharide-stimulated macrophages by suppressing redox-based NF-κB activation, Exp. Mol. Med. 37 (2005) 323–334.

[182] J.H. Kim, H.J. Na, C.K. Kim, J.Y. Kim, K.S. Ha, H. Lee, H.T. Chung, H.J. Kwon, Y.G. Kwon, Y.M. Kim, The non-provitamin A carotenoid, lutein, inhibits NF-κB-dependent gene expression through redox-based regulation of the phosphatidylinositol 3-kinase/PTEN/Akt and NF-κB-inducing kinase pathways: role of H_2O_2 in NF-κB activation, Free Radic. Biol. Med. 45 (2008) 885–896.

[183] M. Sasaki, Y. Ozawa, T. Kurihara, K. Noda, Y. Imamura, S. Kobayashi, S. Ishida, T. Kazuo, Neuroprotective effect of an antioxidant, lutein, during retinal inflammation, Invest. Ophthalmol. Vis. Sci. 50 (2009) 1433–1439.

[184] M.N.A. Khan, J.Y. Cho, M.C. Lee, J.Y. Kang, N.G. Park, H. Fujii, Y.K. Hong, Isolation of two anti-inflammatory and one pro-inflammatory polyunsaturated fatty acids from the brown seaweed Undaria pinnatifida, J. Agric. Food Chem. 55 (2007) 6984–6988.

[185] Y. Xia, X. Feng, Y. Chen, Y. Su, Effects of EPA, DHA on the secretion of NO, expression of iNOS mRNA and DNA-binding activity of NFkappaB in human monocyte, Wei Sheng Yan Jiu 36 (2007) 445–448.

[186] D.O. Moon, K.C. Kim, C.Y. JIn, M.H. Han, C. Park, K.J. Lee, Y.M. Park, Y.H. Choi, G.Y. Kim, Inhibitory effects of eicosapentaenoic acid on lipopolysaccharide-induced activation in BV2 microglia, Int. Immunopharmacol. 7 (2007) 222–229.

[187] M.M. Rahman, A. Bhattacharya, G. Fernandes, Docosahexaenoic acid is more potent inhibitor of osteoclast differentiation in RAW 264.7 cells than eicosapentaenoic acid, J. Cell. Physiol. 214 (2008) 201–209.

3

Natural products with antiinflammatory activities against autoimmune myocarditis

Akhila Nair and Sreeraj Gopi
R&D Centre, Aurea Biolabs (P) Ltd, Kolenchery, Cochin, Kerala, India

3.1 Introduction

Nature meticulously embeds an illimitable source of compounds that scaffolds tremendous scope for new drug discovery. Although in the drug market synthetic products prevail because of their multitudinous benefits such as quick effects, quality, and cost, natural products due to their safety and efficacy profile have gathered surplus acclamation. It is estimated that about 80% of the world population is dependent upon green drugs or natural products derived from natural origin such as plants and animals [1]. Their primary and secondary metabolites such as alkaloids, glycoside, fats, oils, protein, dietary fiber, resins, gums, tannins, liver oil, and mucosal tissues (heparin) are mainly extracted through various extraction processes to obtain the concerned drug [2]. These natural products have been subjected to continuous investigation for the potency of their bioactive agents against diverse disease conditions from time immemorial [1]. Through well-documented literatures, it has been claimed that most of the green source possesses antiinflammatory and antioxidant activities, which are main cause of various diseases. Hence, exhaustive research on these pharmacological activities has succored in the treatment of numerous acute and chronic disease states.

In this disease series, myocarditis is apparently found the major cause of sudden death among different age groups, which is related to the damage of adjacent myocytes and mononuclear inflammation of the heart muscles. Therefore myocarditis is termed as the inflammation of the myocardium. It can also be redefined as mixed cellular or mononuclear infiltration that corresponds with degeneration and myocyte necrosis to develop dilated cardiomyopathy. This dilated myocarditis occurs in 21% cases that advances to cardiac arrest. Although great deal of investigation is required to understand the pathogenesis of myocarditis, it is supposed

Inflammation and Natural Products. https://doi.org/10.1016/B978-0-12-819218-4.00002-X
© 2021 Elsevier Inc. All rights reserved.

to possess an interrelation with viral infections. Meanwhile, patients with chronic myocarditis are prone to develop heart-reactive autoantibodies that depict an autoimmune mechanism revolving around this disease [3]. In addition, chemokines, cytokines, adhesion molecules, and other inflammatory responses that play a vital role through tissue damage in autoimmune myocarditis. Further, most of the nutraceuticals are considered to channel immune diseases through their antiinflammatory efficacy. Thus natural products considered for their ability to reduce inflammation with minimal side effects are highly demanded among patients with autoimmune myocarditis [4].

This chapter focuses on the basic understanding of myocarditis through their etiological and pathological aspects. Certain herbal drugs that possess antiinflammatory activity to circumscribe the supposed causes of autoimmune myocarditis are also discussed.

3.2 Myocarditis

In 19th century, literatures suggest that myocardial infraction and chronic ischemic heart conditions were known as myocarditis. Over the years of development in research, this term was further revised as acute or chronic inflammation of the myocardium stimulated by environmental or endogenous activators such as viruses and bacteria, fungi, and parasites and diagnosed by endomyocardial biopsy (EMB) using gene amplification techniques with established molecular criteria to identify infectious agents [5]. This definition has been greatly supported by the World Health Organization/International Society and Federation of Cardiology (WHO/ISFC), The American Heart Association (AHA), The European Society of Cardiology (ESC), and ESC expert consensus statements [6].

Myocarditis has reported for the development of dilated cardiomyopathy and the cause of sudden death [7]. In 2013 the death toll due to the incidence of myocarditis had reached 22% with about 1.5 million cases reported around the globe. A survey reported cardiac arrest among different aged group people and region varies about 0.4%–0.5%. It is suggested about 2% in neonate, about 5% in children, and about 12% in athletes, whereas Japan reported about 6% necropsy with sudden death [5]. The diagnosis of myocarditis is crucial, and it is diagnosed by standard procedures of endomyocardial biopsy that follows stringent guidelines in the selection of patients through histological, immunological, and immunohistochemical criteria [8]. However, this process of examination has its own shortcomings such as sampling errors and immune histochemical analysis of the biopsy material. Cardiac magnetic resonance is an alternative current diagnostic technique. However, both prove an inefficient tool in the assessment of the advancement toward dilated cardiomyopathy and to extract the underlying cause behind this disease [7]. Therefore cumulative evidences reflected that viruses such as coxsackievirus B3, cytomegalovirus, hepatitis A and C viruses, human immunodeficiency virus, Epstein-Barr virus, adenovirus, and parvovirus B19; bacteria such as Borrelia and Chlamydia; and autoimmune responses to cardiac antigens developed in chronic myocarditis patients play a vital role in permanent heart failure [9, 10]. Specifically, antigen specific autoantibody immune responses are developed, and these cardiac antibodies include cardiac myosin heavy chain, mitochondrial antigens (adenine nucleotide), cardiac troponin I, and branched-chain alpha-ketodehydrogenase [3].

3.3 Etiology and pathogenesis of autoimmune myocarditis

It is assumed that the most common cause of acute myocarditis is viruses. The most prominent viruses are human herpes virus 6 genomes, parvovirus B19, coxsackievirus, enterovirus, and so on. Table 3.1 presents the supposed causes of myocarditis. It is also seen that certain autoimmune syndromes, nonviral infections, allergenic carriers such as cardiac transplantation or medication, and toxic agents like amphetamine and cocaine are sources for myocarditis [7]. Autoimmune myocarditis is generally organ-specific disease that is either related to systemic immune-mediated diseases or constricted to heart. To elaborate, distinct autoreactive clones are widened and directed toward autoimmune myocarditis in response to the neoantigen and self antigen that the heart provides. Alternatively, through systemic hyperinflammatory reactions or localization as well as generalization of proliferation of immune-competent cells, the myocardium possibly gets infiltrated passively and consequently suffers damages. The diagnosis of autoimmune myocarditis involves certain stringent criteria to be fulfilled according to Rose-Witebsky, which include histocompatibility complex (HLA) association and familial aggregation, patients with abnormal serum cytokines and cardiac autoantibodies, and patients with nonavailability of infectious agents but detected with myocardium lymphomonocytic cellular infiltrates on tissue with molecular expression of adhesion on cardioendothelial cells and through immunomodulation or immunosuppression efficacy. These specifications enable detection of autoantibodies and clarified the pathogenic role of certain autoantibodies such as their functional effects and cardiomyopathy features [6].

Further, it is reported that the abnormality of class II category of major histocompatibility complex (MHC) expression, also known as human leucocyte antigen (HLA) complex, is the main components

Table 3.1 Etiology and pathology of autoimmune myocarditis.

S. No.	Type of etiology	Types	Examples
1	Infectious agents		
	Viral	RNA viruses	Coxsackievirus A and B, echovirus, influenza A and B virus, respiratory syncytial virus, human immunodeficiency virus-1 (rare)
		DNA viruses	Parvovirus B19, adenovirus, cytomegalovirus, herpes simplex virus-6, Epstein-Barr virus
	Protozoal	–	*Trypanosoma cruzi*
	Fungal	–	Uncommon
	Spirochetal	–	Borrelia, Leptospira
	Bacterial	–	*Haemophilus influenzae, Mycobacterium, Mycoplasma pneumoniae*
2	Immune mediated		
	Autoimmune	Organ specific	Lymphocytic, giant cell
		Autoimmune or immune-oriented disorders	Systemic lupus erythematosus, rheumatoid arthritis, Churg-Strauss syndrome, inflammatory bowel disease, insulin-dependent diabetes mellitus
3	Toxic drugs		
	Drugs	Anticancer	Fluorouracil, trastuzumab, cyclophosphamide
		Antipsychotic	Clozapine
		CNS stimulant	Amphetamines
		Recreational drug	Cocaine
	Heavy metals	–	Lead, iron, copper
	Hormones	–	Vitamins (Beri-Beri), pheochromocytoma
	Physical agents	–	Electric shock, radiation

for myocarditis. These complexes are capable of being recognized by T cells because the peptide-binding area forms stable complexes with numerous peptides. In addition, the presence of autoantibodies in low frequency and the week links with human leucocyte antigen (HLADR4) along with the decline of cardiac function in patients with antimyosin antibody-positive patients suggested that immune system is involved with certain haplotype-related inflammatory heart disease [11]. It is reported that humoral and cellular immune responses are involved in cardiac remodeling that affects the functioning such as degradation of extracellular matrix (ECM), cardiomyocyte hypertrophy, collagen deposition, sometimes apoptosis directing to vascular injuries, and cardiomyocyte ischemia. These functions are also operated by oxidative stress, mitochondrial dysfunction, nitric oxide (NO) production, autonomic dysfunction, disturbance of catecholaminergic stimulation and proinflammatory cytokines, metabolic alterations, and changes in micro-/macrocirculation. Moreover, Ca^{2+} homeostasis turmoil causes cardiomyocyte contractility to produce myocardial inflammation [8]. Still the pathogenesis linked to the channeling of immune mediated cardiac dysfunction requires exploration.

3.4 Antiinflammatory mechanism for autoimmune myocarditis

The infectious and noninfectious agents cause myocarditis through the deterioration of cardiac extracellular matrix or myocyte damage and death. Immune system identifies, eliminates, or repairs inflected cells [12]. The immune system in response develops self-antigen during the repair of cardiac tissue, which in return produces autoantibodies [13]. Although the mechanism of autoimmune myocarditis is not fully known, speculation through studies suggest apart from viral infection, autoimmune response is also blamed for the progression of myocarditis (Fig. 3.1) [14]. This reflects that the two classes of signals, namely, antigen-specific and antigen-nonspecific signals through which the activation of autoreactive lymphocytes by infectious agents takes place, determine autoimmune myocarditis. According to antigen-specific concept, the target antigen is identified by host immune system due to their structural similarity, to trick the immune response against host agent and to build autoreactive T cells that cause autoimmune destruction [15]. Another phenomenon is antigen-nonspecific concept where the host cells are destructed by the infection caused to release autoantigens, which direct autoimmune responses (bystander activation) [16]. Therefore, to strengthen the case, myosin-induced myocarditis animal model was considered to understand the mechanism and specific roles of CD + T cell, cytokines, and chemokines that facilitate autoimmune myocarditis and their control (Table 3.2).

3.4.1 Inhibition of regulatory T helper cells

The T-cell CD4 + are reported to be the coordinator of autoimmune response in myocarditis as they are not completely deleted from the thymus and continue to be regulated. Sometimes, these self-reactive T cells get vigorously stimulated to cause autoimmune diseases [3, 4]. These T cells are found in TH1 and TH2 cells, which are identified by the cytokines produced by them. Th1 cells (IFN-ϒ, TNF-α, IL12, and IL-2) regulate cellular immune response by intracellular viruses and bacteria, whereas Th2 cells (IL4, IL5, IL-10, and IL13) regulate a

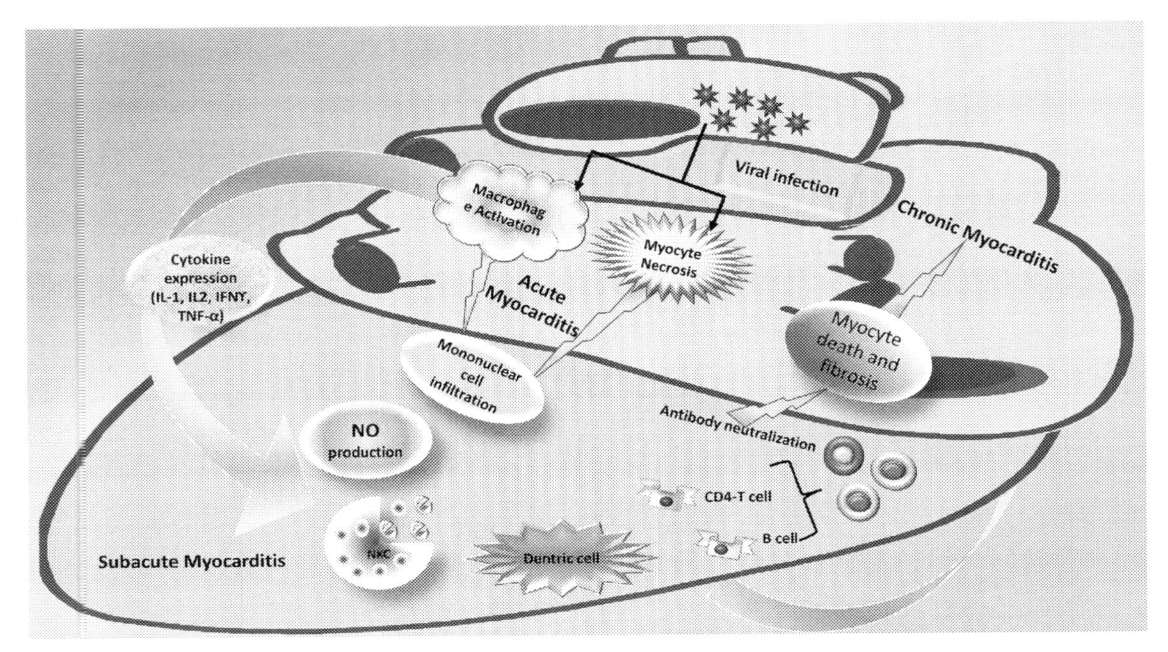

Fig. 3.1 Mechanism of action.

humoral immune response. The infectious or autoimmune diseases are caused by the imbalance between Th1 and Th2 [4].

3.4.2 Inhibition of cytokines and macrophage infiltration

It is reported that cytokines are major factors that channel and intensify autoimmunity [20]. Therefore experimental autoimmune myocarditis-induced animal model (EAM) is developed to study their mode of action, which is characteristic of Th17 cell activation, macrophage/monocyte infiltration, and inflammatory factors named high mobility group box1 (HMGB1), a nonhistone nuclear protein. It is believed that cytokines regulate the effector phase and activation of innate and specific immunity that control infection. Innate immunity is governed by IL-12 and IL-15, which modulate specific cell-mediated immune responses, while adhesion molecules, cytokines, and chemokines relocate and activate the responses of B cell, T cell, and macrophages in specific immunity [17]. In myocarditis the cytokine expression of IL1, IL6, and TNF-α plays a crucial role where the mechanism of TNF-α accelerates the enzyme activity in cardiac myocyte and nitric oxide synthase (NOS), which helps arginine to convert to NO (inhibition of myocardial contractility) and citrulline. Hence, EAM setup works to establish a relation between lower contractile performance and escalated iNOS mRNA, TNF-α, and IL-6 [21].

Table 3.2 Inflammatory behavior of the inflammatory mediators in autoimmune myocarditis.

S. No.	Mediators of inflammation	Role	Expression in autoimmune myocarditis	Behavior	Reference
1	Cytokines		Interleukins		
		IL-1, IL-6	Inflammatory response in acute phase	Increases	[17]
		IL-23	Inflammatory response in chronic phase	Increases	[18]
		IL-12	Activate natural killer cell (NKC)		[19]
		Interferon: IFNα	Inflammatory response in acute phase	Increases	[13]
		Tumor necrosis factor alpha (TNF-α)	Ameliorates the expression of other proinflammatory	Increases	[17]
			Nitric oxide synthase (NOS) activity	Increases	[17]
2	Chemokines	CCL2, CCL3, CCL4, CCL5, CCL11, CXCL8, CXCL10	Attract immune cells to the inflammatory site	Increase	[17]
3	Immune cells	Natural killer cell (NKC)	Development in acute phase	Decrease	[19]
		T cell	Development in acute phase (CD4-knockout myocarditis)	Decrease	[19]

Further, IL-6 regulates the IL-17-producing T helper cell-17 (Th17) in EAM model, which reflects that this cytokine is important therapeutic targets in reducing inflammation in autoimmune myocarditis [22]. Through EAM setup, Sonderegger et al. demonstrated that apart from IL-17, IL-23 is also an important candidate in autoimmune myocarditis and by active and passive vaccination strategies with IL-17, IL-23 is promising therapies for myocarditis and dilated cardiomyopathy [18]. In addition, HMGB1 blockage suppresses Th12 cells (CD_4T subset) to enhance the cardiac function in EAM [23]. Another characteristic of EAM is monocyte or macrophage infiltration that governs autoimmune myocarditis. It is speculated that inflammatory factor, namely, HMGB1, promotes macrophage reprogramming toward alternatively activated macrophage (M1) phenotype, which depends on TLR4-PI3Kϒ-Erk1/2 pathway and thus cardiac protection, especially by monitoring and modulating autoimmune myocarditis [24].

3.5 Natural products renowned for antiinflammatory activity

Certain natural products reported for their antiinflammatory activity, which would alleviate autoimmune myocarditis are enumerated in Table 3.3 (Fig. 3.2).

3.5.1 *Curcuma longa* L. (curcumin)

Curcumin is a pleiotropic polyphenol that is extracted and isolated from the dried rhizomes of *Curcuma longa* L. This compound is known for its diverse therapeutic activities such as antidiabetic, antifungal, antimalarial, antiinflammatory, cardiovascular, and antitumor. However, numerous studies have reflected that curcumin is a potentially active candidate against inflammation especially myocardial inflammation linked to ischemia [4]. There are various well-established mechanisms through which curcumin exhibits antiinflammatory activities. The inhibition of COX-2 regulates proliferation, and responsiveness of macrophages, DCs, NK cells, and lymphocytes is certain important mode of action reported for curcumin as antiinflammatory agent [19, 44]. The beneficial effects of curcumin to treat inflammation and oxidative stress through inactivation of NF-κB transcription factor and downregulation of proinflammatory cytokine and chemokine (Th-17 and IL-17) provide ample evidence that this herb is a promising for autoimmune diseases [25]. In addition, the severity of autoimmune diseases is well subsided by curcumin polyphenol as stated by Rahimi et al. The imbalance of different Th subsets that include Th17, a regulatory Th cells (Tregs), Th9, Th22, and follicular T helper cells are the main contributors to several autoimmune diseases. Targeting these subsets and their cytokine profiles by curcumin can efficiently inexpensively and safely treat inflammatory diseases [26].

Moreover, they regulate the myocardial protein level of IL-1β, GATA-4, TNF-α, NF-κB, and the area of the inflammatory lesions. This natural polyphenol exhibits protective effects against experimental autoimmune myocarditis rat model, and through echocardiographic and hemodynamic analysis, it was unveiled that both diastolic and systolic heart functions improved, and the HW/BW ratio was reduced [8].

3.5.2 *Cannabis sativa*/marijuana (cannabidiol)

Cannabidiol (CBD) is the major active compound of Marijuana or *Cannabis sativa* that is contemplated to be biologically inert. However, evidence-based studies suggest that they exhibit inflammatory, cytoprotective, antidiabetic, and antiinflammatory characteristics. They are characterized by nonpsychoactive cannabinoid cannabidiol

Table 3.3 Antiinflammatory green drugs for autoimmune myocarditis

S. No	Natural products	Mode of action	Reference
1	Curcumin	Regulate IL-1b, GATA-4, TNF-α, NF-κB, Th-17, IL-17, Treg cells, Th9, Th22	[4, 8, 19, 25, 26]
2	Cannabidiol	↓ Proinflammatory cytokines like interferon-γ (IFN-γ), IL-2 ↑ Antiinflammatory cytokines IL-6, IL-10, and transforming growth factor (TGF-β) Debilitated the inflammatory response mediated by CD3 + and CD4 +, fibrosis	[27–30]
3	Berberine	↓ Phosphorylated (p) signal transducer and activators of transcription STAT1, STAT3, STAT4 APC upregulation ↓ Upregulation of interleukin IL-17 and interferon (IFN-g)	[31–33]
4	Resveratrol	↓ Proinflammatory cytokines: interferon gamma (IFN-γ), tumor necrosis factor alpha (TNF-α) and ↑ antiinflammatory cytokines: interleukin-4 (IL-4) and interleukin-10 (IL-10) ↓ Inflammatory mediators such as interleukin-8 (IL-8), tumor necrosis factor alpha (TNF-α), monocyte chemoattractant protein-1 (MCP-1) ↓ Fibrosis, inflammatory cytokines, and cellular infiltration Regulation of silent mating type information regulation 2 homolog1 (STIRT)	[34–37]
5	Emodin	↓ Tumor necrosis factor (TNF-α), IL-1β, and transcription factor–nuclear factor-κBp65 response	[38]
6	Radix Astragali	↓ Th1 (IL, IFNϒ) and Th2 (IL4, IL10)	[39]
7	Catechin	↓ mRNA (messenger ribonucleic acid) ↑ Th 2 cytokines ↓ Nuclear factor kappa B (NF-κB) and intercellular adhesion molecule (ICAM)-1	[8, 40, 41]
8	Mulberry leaf	↓ Cardiac fibrosis, mitogen-activated protein kinase (MAPK) activation, endothelin-1, and vascular endothelial growth factor (VEGF) a fractional shortening ↑ Left ventricular ejection fraction, fractional shortening ↓ Brain natriuretic peptide levels, myosin-induced autoantibodies production, and the heart-to-body weight ratio	[42, 43]
9	Oleanolic acid	↑ Production of IL-10, IL-35, Treg cells ↓ Profibrotic and proinflammatory cytokines	[42]

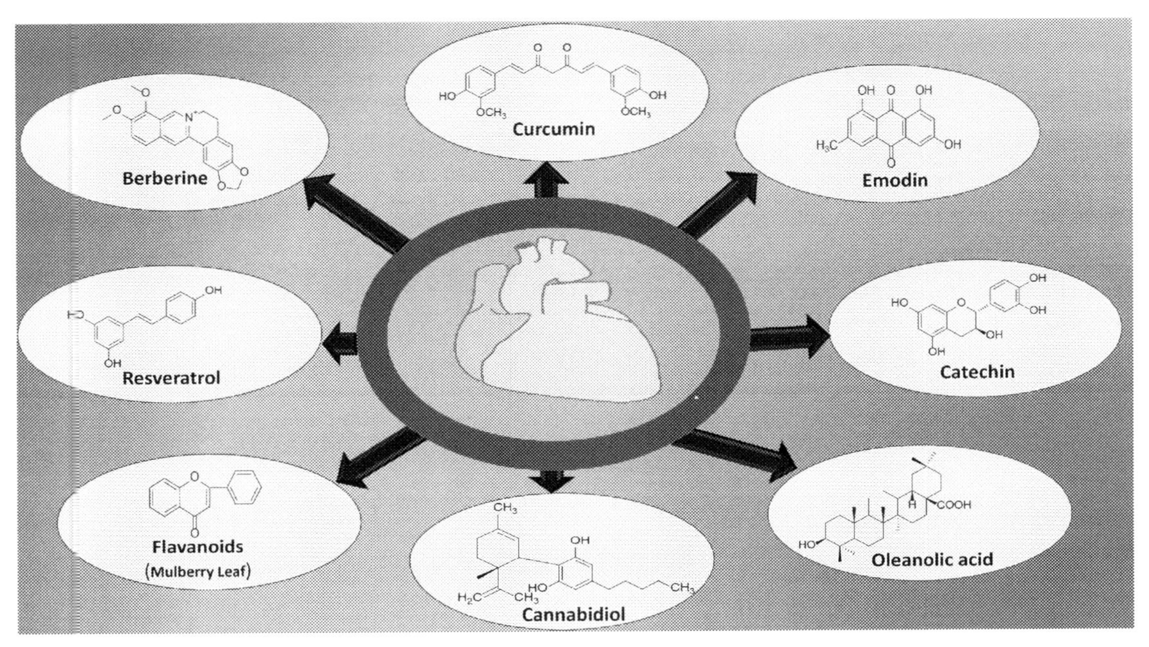

Fig. 3.2 Natural products possessing antiinflammatory activities.

(CBD) and psychoactive cannabinoid delta9-tetrahydrocannabinol (Δ 9-THC). Psychoactive and nonpsychoactive cannabinoids were first isolated from cannabis by group of Israel in 1964. Subsequently, many compounds were identified such as cannabidiolic acid, cannabidivarin, cannabivarin, cannabinol (CBN), and cannabinoid (CBD). Cannabinoid type 1 receptor (CB1) was identified by Matsuda and coworkers who were able to bind with Δ 9-THC in the early 1990s, and then cannabinoid type 2 receptor (CB2) was discovered, which proved to have immune-suppressive properties. On the one hand, Δ 9-THC is an important moiety for exerting antiinflammatory and immunomodulation as it is reported to alleviate the production of proinflammatory cytokines like interleukin-1 such as IL-1, IL-6, and IL-12; chemokines; and tumor necrosis factor alpha (TNF-α) and also regulate the proliferation, migration, and apoptosis of inflammatory cells. On the other hand, CBD are promising against inflammatory pain in rats treated with Freud's adjuvant intraplanar injection and help to reduce proinflammatory cytokines like interferon-γ (IFN-γ), IL-2 [27]. They are recognized by US Food and Drug Administration for the treatment of glioblastoma multiforme epilepsy in children. Cannabidiol is reported to effectively participate in the suppression of inflammation by decreasing the proinflammatory cytokines (IFN-γ and IL-17) and escalating antiinflammatory cytokines IL-6 and IL-10 and transforming growth factor-beta (TGF-β) analyzed by enzyme-linked immunosorbent assay (ELISA) to treat autoimmune diseases [28, 29].

Lee et al. studied the potency of cannabidiol through experimental autoimmune myocarditis-induced rat model through cardiac myosin immunization, consequently developing T cell-mediated inflammation, fibrosis, cardiomyocyte cell death, and myocardial dysfunction. Evaluation of theses function was based on quantitative analysis of histology, immunohistochemistry, real-time polymerase chain reaction to measure the inflammatory response, fibrosis, and infiltration of myocardial T cell. In addition, through volume conductance catheter method, the systolic and diastolic cardiac functions were measured. It was observed that treatment with CBD debilitated the inflammatory response mediated by CD3 + and CD4 +, fibrosis, and myocardial dysfunction. Hence, cannabidiol is efficient agent for the treatment of autoimmune myocarditis [30].

3.5.3 Berberine

Most commonly berberine is found in herbs *Coptis* and *Berberis*, which is a nonbasic and quaternary benzylisoquinoline alkaloid. They are extracted by applying percolation, maceration, Soxhlet, and cold and hot continuous extraction using various solvent such as ethanol, methanol, and chloroform. This extraction process is based on the principal of interconversion reaction between the base and photoberberine salt. It is utilized for the synthesis of numerous biologically active derivatives through substitution of functional group, condensation, and modification. The biological activities reported in them were antibacterial, antiinflammatory, antidiabetic, antitumor, and immunomodulatory effects against immune-mediated disorders. *Berberis asiatica* has occupied a special position in Yunani medicine for the treatment of jaundice, eye sores, asthma, skin pigmentation, toothache, swelling, and inflammation and for drying ulcers [31]. It is reported that berberine is efficient in treating the nonobese diabetic rat model by suppressing the inflammatory cytokines especially the Th1 cytokines (IL-2, IL-1β, IFN-γ, and TNF-α) and Th2 cytokines (IL-4, IL-5, IL-6, and IL-10), which were identified by ELIZA. 1.2 g berberine/day in combination with diuretics, digoxin, ACEI-/angiotensin-converting enzyme inhibitors were given to 51 patients who were diagnosed with III/IV cardiac failure and low left ventricular ejection fraction (LVEF) by New York Heart Association (NYHA). Berberine proved beneficial in escalating the LVEF and reduced the complexity and frequency of premature ventricular contractions. Moreover, berberine is found to be promising in ischemic conditions via regulation of various signaling pathways such as AKT (protein kinase B), Janus kinase or signal transducer (JAK/STAT) signaling and transcription activator, and AMP-activated kinase (AMPK) activity and inhibits serine/threonine protein-kinase enzyme named glycogen synthase kinase (GSK3β) and activates phosphoinoside 3-kinase (PI3K)/AKT pathway. PI3K/AKT activation restricts the proinflammatory functions and apoptotic processes and regulates toll-like receptors4 (TLR4) mediated signal transduction to alleviate ischemia [32]. Therefore it is well established that berberine is a potent cardioprotective candidate.

Furthermore, Liu et al. demonstrated the cardioprotective effect of berberine by treating them on porcine cardiac myosin–induced autoimmune myocarditis in rats. The underlying mechanism of action was studied by monitoring the changes in Th1 and Th17 and studying the signaling pathway of JAK-STAT (Janus kinase signal transducer and activator of transcription). For this, they kept JAK-specific inhibitor AG490 as positive control drug. Their findings suggested that berberine significantly elevated the phosphorylated (p)-STAT1, STAT3, and STAT4 in autoimmune myocarditis. Antigen-presenting cell (APC) was affected, and the Th 17 and Th1 cell responses were upregulated; however, their excessive responses were controlled. Upregulation of cytokines such as IL-17 and IFN-g and the function and expression of costimulatory molecules (involving sequential and reciprocal signals between cells) were suppressed [33].

3.5.4 Resveratrol

The active compound of *Polygonum capsidatum* or knotweed and red grape wine plant is resveratrol [34]. It is a natural polyphenol and stilbenoid with immunomodulatory, chemoprevention, antiproliferation, antiinflammatory, and cardioprotective effects, which are well documented. Recently, the study conducted by Wang et al. demonstrated that resveratrol when combined with mouse bone marrow mesenchymal stem cells (mBM-MSCs) augments the treatment of autoimmune encephalitis by inhibiting the proinflammatory cytokines: interferon gamma (IFN-γ), tumor necrosis factor alpha (TNF-α) and increased the anti-inflammatory cytokines: interleukin-4 (IL-4) and interleukin-10 (IL-10) which direct the T helper type 1 (Th1) and (Th2). In this experiment the presence of cytokines was detected by enzyme-linked immunosorbent assay (ELISA). This study established that resveratrol is a potent antiinflammatory candidate in treating autoimmune diseases [35]. Pinheiro et al. also displayed the potential of resveratrol as an antiinflammatory agent through transcriptomic analysis, which was performed on lipopolysaccharide (LPS) stimulation of monocyte cultures. This study monitored the inflammatory activity after LPS treatment and observed a decrease in inflammatory mediators such as interleukin-8 (IL-8), tumor necrosis factor alpha (TNF-α), monocyte chemoattractant protein-1(MCP-1), and devoid of cytotoxicity. The upregulation and downregulation of genes were related to the changes in the inflammatory responses through transcriptional regulation [36]. These literatures suggested that resveratrol is a promising candidate as an antiinflammatory agent against autoimmune diseases.

Their antiinflammatory activities paved way for a thorough investigation of this compound on autoimmune myocarditis. A study showcased that this polyphenol exhibited cardioprotective effect in rats immunized with cardiac myosin through echocardiographic, western blotting, histopathology, and immunohistochemical analysis. When 50 mg/kg per day of resveratrol was administered in rats, there was significant decrease in fibrosis, inflammatory cytokines, and cellular infiltration when 50 mg/kg per day of resveratrol was administered in rats. It also regulated silent mating type information regulation 2 homolog1 (STIRT) to circumscribe the lymphocyte proliferation. Hence, this drug retained cardiac function and beneficial in autoimmune myocarditis [8, 37].

3.5.5 Emodin

Emodin is an anthraquinone obtained from traditional herb Rhubarb or *Rheum officinale* and *Polygonum cuspidatum.* This has been evaluated for its immunosuppressive and antiinflammatory activities. Therefore, keeping the inflammatory effects in mind, Song et al. studied their antiinflammatory behavior for autoimmune myocarditis on Lewis male rats who were immunized for a week with porcine cardiac myosin to induce experimental autoimmune myocarditis, which resulted in inflammatory lesions. Through echocardiography, histopathology, western blot analysis, and measurement of TNF-α and IL-β, it was observed that there was significant improvement in left ventricular function. Emodin is capable in reducing the tumor necrosis factor-alpha (TNF-α) and interleukin (IL-1β), and the rapid response transcription factor–nuclear factor-κBp65 that acts as a potential tool in regulating proinflammatory cytokines was subdued, which exhibited that emodin was a potent weapon in experimental autoimmune myocarditis [38].

3.5.6 Radix Astragali

Radix Astragali is a traditional medicinal herb from China obtained from the *Astragalus membranaceus.* The main active compounds of Radix Astragali are flavonoids and polysaccharides. The predominant flavonoids were catycosin, kumatakenin, 7,3-dimercapto-4,1 methoxyisoflavone, 3-dimercapto-7,4,1-methoxyisoflavone, and formononetin, and polysaccharides that dominate are heteroglycans, namely, astragalus heteroglycan (AH)-1 and AH-2, and glucans, namely, astragalus glucan (AG)-1 and AG-2. Throughout Asian countries, it is used for the treatment of cardiovascular, liver, kidney diabetes, immune, and viral diseases. Their efficiency in autoimmune myocarditis was evaluated by Zhao et al. on porcine cardiac myosin-induced Lewis rats through transthoracic echocardiography. This study revealed that the myocardial contractile function impairment and effectively reduced Th1 (IL, IFNϒ) and Th2 (IL4, IL10) cytokines levels in experimental autoimmune myocarditis was resolved by Radix Astragali [39].

3.5.7 Catechins

Catechins are polyphenolic phytochemicals found in green tea and cocoa fruits. Other constituents found in green tea and cocoa fruits were (−)-epicatechin, (+)-gallocatechin, (+)-catechin, (+)-catechin, (−)-epigallocatechin, flavan-3-ols. Catechins are noted for their diverse biological activities such as antiobesity, antioxidative, antiinflammatory, antiatherosclerosis, antihyperglycemia, and antihypercholesterolemia. Suzuki et al. reported that catechin improves cardiac function in experimental autoimmune rats. The study showcased significant improvement in fibrosis and reduced myocardial cell infiltration. Further, it was observed through immunohistochemistry that messenger ribonucleic acid (mRNA) of tumor necrosis factor-alpha (TNF-α) were reduced and Th 2 cytokines were increased as well as nuclear factor kappa B (NF-κB) and ICAM-1 were reduced when compared with the group kept as control [40]. Similarly the study reported by Zempo et al. supported that cocoa polyphenol efficiently reduced the fibrotic area ratio and cardiac infiltration and increase in heart weight to body weight (HW/BW) (symbolizes myocardial hypertrophy) in experimental autoimmune myocarditis rats. Moreover, reduced cardiac H_2O_2concentration, cardiac myeloperoxidase (MPO), dihydroethidium staining (DHE) intensity, NF-κB P65 phosphorylation. Collagen type 1, vascular adhesion molecule (VCAM-1) and mRNA expression of IL-6, IL-1, E-selection as observed through reverse transcriptase-PCR [41]. To summarize, catechins play important role in inflammatory autoimmune myocarditis [8].

3.5.8 Mulberry leaf

Mulberry leaves (*Morus alba* L.) are from family Moraceae and are found in Korea, China, and Japan. These polyphenols are composed of active components such as moracin, flavonoids, quercetin, gallocatechin gallate, and naringenin. They are known for their antioxidant, hypolipidemic, hypoglycemic, antiinflammatory, and cardiovascular protective effects. Through echocardiography, western blotting, and histopathology, the therapeutic effects of mulberry leaves on experimental autoimmune myocarditis rat models were studied by Martin et al. [42] The intake of leaves reduced cardiac fibrosis, mitogen-activated protein kinase (MAPK) activation, endothelin-1, and vascular endothelial growth factor (VEGF) and enhanced left ventricular ejection fraction as well as fractional shortening when compared with control. The increase of endoplasmic stress markers was attenuated by mulberry leaves. These parameters suggested that these leaves hold great potential for cardiac remodeling and cardiac function improvement in postmyocarditis-dilated cardiomyopathy [43].

3.5.9 Oleanolic acid

Olea europaea (olive) are extracted for oleanolic acid from their root and leaves. Oleanolic acid has a plethora of biological activities such as immunomodulatory, antioxidant, antiinflammatory, and cardioprotective properties. Literatures suggest that it is promising against Th1 cell-channeled inflammatory diseases. Under this perspective, their ability to fight against autoimmune myocarditis was demonstrated by Martin et al. on experimental autoimmune myocarditis mouse model. They observed that oleanolic acid reduced the severity of the disease by reducing the brain natriuretic peptide levels, myosin-induced autoantibody production, and the heart-to-body weight ratio. Further, it increased the production of IL-10 and IL-35 along with Treg cells and reduced profibrotic and proinflammatory cytokines by generating cardiac-specific autoantibodies as well as exhibited protective effect on cardiac cells. This novel natural product mitigated fibrosis, cell penetration, and dystrophic calcifications and decreased cardiac fibroblast proliferation and weaken myocarditis-activated cytokine-induced calcium and collagen expulsion [42].

3.6 Conclusion

In summary the chapter deals with basic understanding of myocarditis and their most predicted cause such as viruses, autoimmune syndromes, allergic carriers, and nonviral infections. Besides, the most prominent mechanism, which reflect that myocyte/macrophage infiltration, Th17cell activation, and certain inflammatory factors namely HMGB1 modulate, experimental autoimmune animal model (EAM), and suppression of each, are helpful to build various treatment regimen. Apart from these, there is other mode of actions revolving around autoimmune myocarditis that needs exploration. Through EAM studies, it is clear that many of the phytochemicals such as cannabidiol, berberine, emodin, Radix Astragali, resveratrol, catechin, mulberry leaf, oleanolic acid, and curcumin are promising candidate for autoimmune myocarditis. Therefore exhaustive investigations on various disease-causing agents and their possible mechanism could facilitate the discovery of numerous natural products and utilize these agents for proper therapeutic strategies that could be built as platform to alleviate the intensity of this disease.

References

[1] C. Veeresham, Natural products derived from plants as a source of drugs, J. Adv. Pharm. Technol. Res. 3 (2012) 200–201.
[2] F. Chemat, M. Abert-Vian, A.S. Fabiano-Tixier, J. Strube, L. Uhlenbrock, V. Gunjevic, G. Cravotto, Green extraction of natural products. Origins, current status, and future challenges, Trends Anal. Chem. 118 (2019) 248–263.
[3] N.R. Rose, Learning from myocarditis: mimicry, chaos and black holes, F1000Prime Rep. 6 (2014) 25.
[4] J.J. Bright, Curcumin and autoimmune disease, Adv. Exp. Med. Biol. 595 (2007) 425–451.
[5] S. Heymans, U. Eriksson, J. Lehtonen, L.T. Cooper, The quest for new approaches in myocarditis and inflammatory cardiomyopathy, J. Am. Coll. Cardiol. 68 (2016) 2348–2364.
[6] A.L.P. Caforio, G. Malipiero, R. Marcolongo, S. Iliceto, Myocarditis, encyclopedia of cardiovascular research and medicine, Curr. Cardiol. Rep. 19 (2017) 63.
[7] P.S. Biesbroek, A.M. Beek, T. Germans, H.W.M. Niessen, A.C.V. Rossum, Diagnosis of myocarditis: current state and future perspectives, Int. J. Cardiol. 191 (2015) 211–219.
[8] B. Javadi, A. Sahebkar, Natural products with anti-inflammatory and immunomodulatory activities against autoimmune myocarditis, Pharmacol. Res. 124 (2017) 34–42.
[9] N. Makino, T. Toyofuku, N. Takegahar, H. Takamatsu, T. Okuno, Y. Nakagawa, S. Kang, S. Nojima, M. Horia, H. Kikutani, A. Kumanogoh, Involvement of Sema4A in the progression of experimental autoimmune myocarditis, FEBS Lett. 582 (2008) 3935–3940.
[10] V. Taneja, C.S. David, Spontaneous autoimmune myocarditis and cardiomyopathy in HLA-DQ8.NODAbo transgenic mice, J. Autoimmun. 33 (2009) 260–269.
[11] U. Eriksson, J.M. Penninger, Autoimmune heart failure: new understandings of pathogenesis, Int. J. Biochem. Cell Biol. 37 (2005) 27–32.
[12] R.R. Bernstein, D.L. Fairweather, Unresolved issues in theories of autoimmune disease using myocarditis as a framework, J. Theor. Biol. 375 (2015) 101–123.
[13] M.D. Daniel, K.V. Hyland, K. Wang, D.M. Engman, Recombinant cardiac myosin fragment induces experimental autoimmune myocarditis via activation of Th1 and Th17 immunity, Autoimmunity 41 (2008) 490–499.
[14] W. Yang, D. Lee, D.L. Lee, S. Hong, S. Lee, S. Kang, D. Choi, Y. Jang, S.H. Kim, S. Park, Blocking the receptor for advanced glycation end product activation attenuates autoimmune myocarditis, Circ. J. 78 (2014) 1197–1205.
[15] Å. Hjalmarson, M. Fu, R. Mobini, Who are the enemies? Inflammation and autoimmune mechanisms, Eur. Heart J. Suppl. 4 (Suppl. G) (2002) G27–G32.
[16] N. Girone, M. Fresno, Etiology of Chagas disease myocarditis: autoimmunity, parasite persistence, or both? Trends Parasitol. 19 (2003) 19–22.
[17] B. Maisch, I. Portig, A. Ristic, G. Hufnagel, S. Pankuweit, Definition of inflammatory cardiomyopathy (myocarditis): on the way to consensus, Herz 25 (2000) 200–209.

[18] I. Sonderegger, T.A. Rohn, M.O. Kurrer, G. Iezzi, Y. Zou, R.A. Kastelein, M.F. Bachmann, M. Kopf, Neutralization of IL-17 by active vaccination inhibits IL-23-dependent autoimmune myocarditis, Eur. J. Immunol. 36 (2006) 2849–2856.
[19] D.L. Fairweather, Z. Kaya, G.R. Shellam, C.M. Lawson, N.R. Rose, From infection to autoimmunity, J. Autoimmun. 16 (2001) 175–186.
[20] N. Kishore, P. Kumar, K. Shanker, A.K. Verma, Human disorders associated with inflammation and the evolving role of natural products to overcome, Eur. J. Med. Chem. 179 (2019) 272–309.
[21] T. Yamashita, T. Iwakura, K. Matsui, H. Kawaguchi, M. Obana, A. Hayama, M. Maeda, Y. Izumi, I. Komuro, Y. Ohsugi, M. Fujimoto, T. Naka, T. Kishimoto, H. Nakayama, Y. Fujio, IL-6-mediated Th17 differentiation through RORgt is essential for the initiation of experimental autoimmune myocarditis, Cardiovasc. Res. 91 (2011) 640–648.
[22] J. Qin, W. Wang, R. Zhang, Novel natural product therapeutics targeting both inflammation and cancer, Chin. J. Nat. Med. 15 (2017) 0401–0416.
[23] Z. Su, C. Sun, C. Zhou, Y. Liu, H. Zhu, S. Sandoghchian, D. Zheng, T. Peng, Y. Zhang, Z. Jiao, S. Wang, H. Xu, HMGB1 blockade attenuates experimental autoimmune myocarditis and suppresses Th17-cell expansion, Eur. J. Immunol. 41 (2011) 3586–3595.
[24] Z. Su, P. Zhang, Y. Yu, H. Lu, Y. Liu, P. Ni, X. Su, D. Wang, Y. Liu, J. Wang, H. Shen, W. Xu, H. Xu, HMGB1 facilitated macrophage reprogramming towards a proinflammatory M1-like phenotype in experimental autoimmune myocarditis development, Sci. Rep. 6 (2016) 21884.
[25] E. Penton-Arias, D.D. Haines, Natural products: immuno-rebalancing therapeutic approaches, in: Immune Rebalancing, Elsevier, 2016, pp. 229–249.
[26] K. Rahimi, A. Ahmadi, K. Hassanzadeh, Z. Soleimani, T. Sathyapalan, A. Mohammadi, A. Sahebkar, Targeting the balance of T helper cell responses by curcumin in inflammatory and autoimmune states, Autoimmun. Rev. 18 (2019) 738–748.
[27] L. Navarini, D.P.E. Margiotta, G. Gallo Afflitto, A. Afeltra, Cannabinoids in autoimmune and rheumatic diseases, in: Mosaic of Autoimmunity in the Novel Factors of Autoimmune Diseases, Elsevier, 2019, pp. 417–429.
[28] C. González-García, I.M. Torres, R. García-Hernández, L. Campos-Ruíz, L.R. Esparragoza, M.J. Coronado, A.G. Grande, A. García-Merino, A.J.S. López, Mechanisms of action of cannabidiol in adoptively transferred experimental autoimmune encephalomyelitis, Exp. Neurol. 298 (2017) 57–67.
[29] Z.Z. Al-Ghezi, P.B. Busbee, H. Alghetaa, P.S. Nagarkatti, M. Nagarkatti, Combination of cannabinoids, delta-9-tetrahydrocannabinol (THC) and cannabidiol (CBD), mitigates experimental autoimmune encephalomyelitis (EAE) by altering the gut microbiome, Brain Behav. Immun. 82 (2019) 25–35.
[30] W.-S. Lee, K. Erdelyi, C. Matyas, P. Mukhopadhyay, Z.V. Varga, L. Liaudet, G. Haskó, D. Čiháková, R. Mechoulam, P. Pacher, Cannabidiol limits T cell–mediated chronic autoimmune myocarditis: implications to autoimmune disorders and organ transplantation, Mol. Med. 22 (2016) 136–146.
[31] M.A. Neag, A. Mocan, J. Echeverría, R.M. Pop, C.I. Bocsan, G. Cris, A.D. Buzoianu, Berberine: botanical occurrence, traditional uses, extraction methods, and relevance in cardiovascular, metabolic, hepatic, and renal disorders, Front. Pharmacol. 9 (2018) 557.
[32] W. Chueh, J. Lin, Protective effect of isoquinoline alkaloid berberine on spontaneous inflammation in the spleen, liver and kidney of non-obese diabetic mice through downregulating gene expression ratios of pro-/anti-inflammatory and Th1/Th2 cytokines, Food Chem. 131 (2012) 1263–1271.

[33] X. Liu, X. Zhang, L. Ye, H. Yuan, Protective mechanisms of berberine against experimental autoimmune myocarditis in a rat model, Biomed. Pharmacother. 79 (2016) 222–230.
[34] H.M. Shaheen, A.A. Alsenosy, Nuclear factor kappa B inhibition as a therapeutic target of nutraceuticals in arthritis, osteoarthritis, and related inflammation, in: Bioactive Food as Dietary Interventions for Arthritis and Related Inflammatory Diseases, second ed., Academic Press, Elsevier, 2019, pp. 437–453.
[35] D. Wang, S.P. Li, J.-S. Fu, L. Bai, L. Guo, Resveratrol augments therapeutic efficiency of mouse bone marrow mesenchymal stem cell-based therapy in experimental autoimmune encephalomyelitis, Int. J. Dev. Neurosci. 49 (2016) 60–66.
[36] D.M.L. Pinheiro, A.H.S. de Oliveira, L.G. Coutinho, F.L. Fontes, R.K.d.M. Oliveira, T.T. Oliveira, A.L.F. Faustino, V. Lira, J.T.A.d.M. Campos, T.B.P. Lajus, S.J.d. Souza, L.F. Agnez-Lima, Resveratrol decreases the expression of genes involved in inflammation through transcriptional regulation, Free Radic. Biol. Med. 130 (2019) 8–22.
[37] Y. Yoshida, T. Shioi, T. Izumi, Resveratrol ameliorates experimental autoimmune myocarditis, Circ. J. 71 (2007) 397–404.
[38] Z.C. Song, Z.S. Wang, J.H. Bai, Z. Li, J. Hu, Emodin a naturally occurring anthraquinone, ameliorates experimental autoimmune myocarditis in rats, Tohuko J. Exp. Med. 227 (2012) 225–230.
[39] P. Zhao, G. Sub, X. Xiao, E. Hao, X. Zhu, J. Ren, Chinese medicinal herb Radix Astragali suppresses cardiac contractile dysfunction and inflammation in a rat model of autoimmune myocarditis, Toxicol. Lett. 182 (2008) 29–35.
[40] J. Suzuki, M. Ogawa, H. Futamatsu, H. Kosuge, Y.M. Sagesaka, M. Isobe, Tea catechins improve left ventricular dysfunction, suppress myocardial inflammation and fibrosis, and alter cytokine expression in rat autoimmune myocarditis, Eur. J. Heart Fail. 9 (2007) 152–159.
[41] H. Zempo, J. Suzuki, M. Ogawa, R. Watanabe, Y. Tada, C. Takamura, M. Isobe, Chlorogenic acid suppresses a cell adhesion molecule in experimental autoimmune myocarditis in mice, Immunol. Endocr. Metab. Agents Med. Chem. 13 (2013) 232–236.
[42] R. Martín, C. Cordova, J.A. San Román, B. Gutierrez, V. Cachofeiro, M.L. Nieto, Oleanolic acid modulates the immune-inflammatory response in mice with experimental autoimmune myocarditis and protects from cardiac injury. Therapeutic implications for the human disease, J. Mol. Cell. Cardiol. 72 (2014) 250–262.
[43] S. Arumugam, S. Mito, R.A. Thandavarayan, V.V. Giridharan, V. Pitchaimani, V. Karuppagounder, M. Harima, M. Nomoto, K. Suzuki, K. Watanabe, Mulberry leaf diet protects against progression of experimental autoimmune myocarditis to dilated cardiomyopathy via modulation of oxidative stress and MAPK-mediated apoptosis, Cardiovasc. Ther. 31 (2013) 352–362.
[44] A. Nair, D. Chattopadhyay, B. Saha, Plant-Derived Immunomodulators in New Look to Phytomedicine, Elsevier, 2019, pp. 435–499.

4

Multitarget approach for natural products in inflammation

Shintu Jude and Sreeraj Gopi
R&D Centre, Aurea Biolabs (P) Ltd, Kolenchery, Cochin, Kerala, India

4.1 Introduction

Inflammation is a basic defense mechanism of higher organisms, which occurs when the living tissues are infected/injured. Simply, inflammation is a partner of immunity system—a trick of body to isolate the damaged/infected part from the rest and to fasten the healing process. It can be triggered by a number of factors such as pathogens, toxins, injuries, and infarction [1]. Whenever a damage happens, a chemical signaling pathway will be initiated, and in response, there will be changes in the permeability of blood vessel walls, microcirculation, and migration of leukocytes and phagocytes [2]. The white blood cells release chemicals that make the blood vessels to bring fluid toward the affected areas. This causes redness, warmth, and swelling in the location and thereby isolation of the damaged tissues to prevent further contact of the foreign matter with other tissues and to destroy the pathogen at the site of action. Thus inflammation appears with its classic signs—swelling, redness, temperature, immobility, and pain [3]. To a particular extent, inflammation is essential for health and well-being. In normal case, acute inflammation is an immediate, short-term process, which helps for damage repair and tissue homeostasis. But prolonged/uninhibited inflammation leads to disorders, causing various diseases to occur. Prolonged inflammation plays the role of origin or mediator of many diseases such as asthma, cancer, otitis, myopathies, allergy, rheumatoid arthritis, rheumatic fever, colitis, atherosclerosis, diabetes, transplant rejection, inflammatory bowel disease, etc. So the role of inflammation is considerable in disease management. The nature and extent of inflammation mostly depend on the kind of stimuli and location of incident.

There are many therapeutic solutions available for the ailments related to inflammation. Many of them are taken model from natural compounds. Natural remedies are more than a folklore custom, nowadays. Concerns on safety, adverse effects, and efficacy draw attention

Inflammation and Natural Products. https://doi.org/10.1016/B978-0-12-819218-4.00004-3
© 2021 Elsevier Inc. All rights reserved.

toward natural products. So, recent studies put emphasis on herbal solutions for every possible aspect. Plant-derived products are more complex in nature, and their points of actions are numerous. Hence the multitarget approach is well suitable for them in antiinflammatory angle.

4.2 Mechanisms: Mediators and pathways behind inflammation

Inflammation is a complicated process in molecular level and can be outlined as a series of actions take place after the incident of stimuli as follows: pattern receptors on the cell surface recognizes the stimuli, activation of inflammatory pathways, release of inflammatory markers and mediators, and finally the inflammatory cells set in action [4]. The pathophysiological events related to inflammation can result in different types—organ specific, chronic or acute, and reversible or irreversible. But the pathways and processes are following the similar way, to some extent. Only the triggers and control system work differently [3].

Inflammatory disorders can happen in any part of the body, and the cardinal signs of inflammation are the results of changes in cellular and molecular levels. Besides, these changes are associated with many substances that interact with the inflammatory responses. These substances, termed as "inflammatory mediators," alter the functionality of cells and tissues and control inflammatory pathways. Also the overproduction of many molecular inflammatory mediators like cytokines such as tumor necrosis factor (TNF-α), interleukins (ILs), prostaglandins (PGs), nitric oxide (NO), and reactive oxygen species (ROS) are associated with degenerative diseases and disorders [4]. The activation of immune and inflammatory cells can affect the inflammatory signaling pathways such as nuclear factor kappa-B (NF-κB), arachidonic acid (AA), Janus kinase signal transducer and activator of transcription (JAK-STAT), and mitogen-activated protein kinase (MAPK) pathways. Inflammatory pathways have specific targets and play vital roles in many biological activities such as immune response, apoptosis, cell proliferation, cell survival, and growth factors and are associated with pathogenesis of a number of diseases. Medications applied to control/relieve the inflammation are implemented by considering all these facts, and each of them follows the different target for the action. Nonetheless, during medications, the targets are selected on the basis of a number of factors such as diseases, affected organs, and mode of administration, along with the involved mediators, pathways, and the necessary factors of their existence and activity.

4.2.1 Activation of pattern recognition receptors (PRRs) on cells

Different types of PRR receptors include the toll-like receptors (TLRs), retinoic acid-inducible gene (RIG)-I-like receptors (RLRs), C-type lectin receptors (CLRs), and NOD-like receptors (NLR). The germline-encoded PRRs are activated by either the microbial structures known as pathogen-associated molecular patterns (PAMPs) or endogenous signals known as the danger-associated molecular patterns (DAMPS), which are corresponding to a noninfectious damage of tissue or cell [5].

The transmission of PAMPs and DAMPs that occurs through TLRs is mediated by myeloid differentiation factor-88 (MyD88). This receptor activation triggers a signaling cascade including transcription factor translocations (NF-κB, AP-1, etc.), which in turn leads to the production of mediators, initiation of intracellular pathways, etc. An alternate pathway mediated through toll/interleukin-1 receptor (TIR) domain-containing adapter-inducing interferon-β (TRIF) results in the translocation of interferon regulatory factor 3 (IRF3) (Fig. 4.1) [3].

4.2.2 NF-κB pathway

There are many pathways established as involved in inflammation. Most of them contain NF-κB as a common factor (Fig. 4.2). NF-κB is a heterodimer, consisting of two protein subunits, namely, p65 and p50. Usually, it is present in the cytoplasm, in an inactive state by binding to the inhibitor of Kappa B (IκB), which prevents NF-κB from entering into the nuclei. The enzymes IκB kinases (IKK) are activated on receiving stimuli, which in turn phosphorylate IκB. This results in the ubiquitination and proteasomal degradation of IκB and subsequent activation of NF-κB, which is translocated into the nucleus and binds to specific sequences of DNA to activate the transcription of target genes [3]. NF-κB

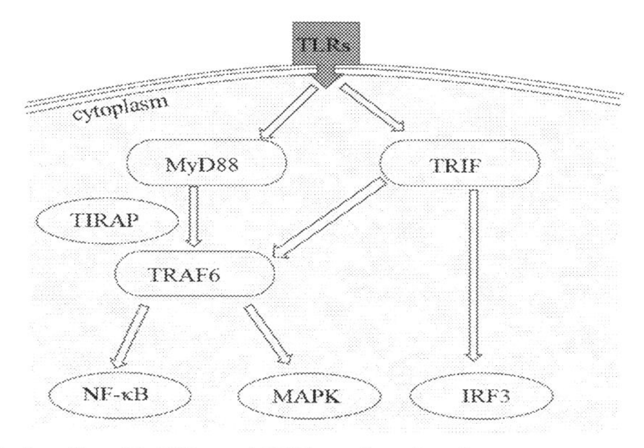

Fig. 4.1 TLR signaling: MyD88- and TRIF-mediated pathways.

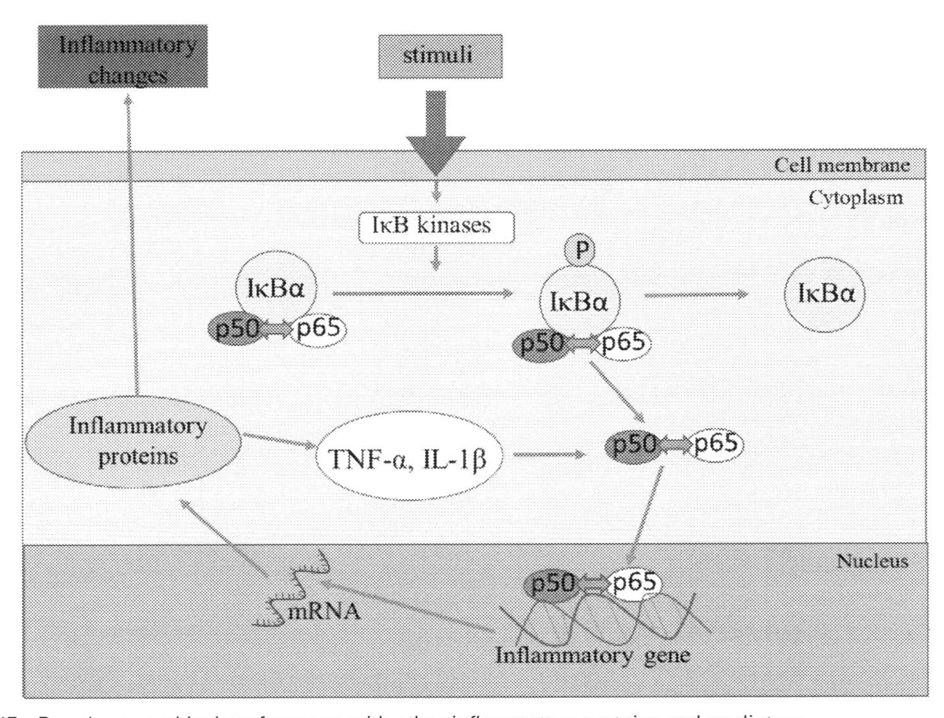

Fig. 4.2 NF-κB pathway and its interferences with other inflammatory proteins and mediators.

pathway regulates the inflammatory responses by controlling proinflammatory cytokine production. Once the inflammatory stimuli have been activated, NF-κB starts to control the gene transcriptions related to inflammation. In chronic inflammatory diseases the NF-κB is found to increase the expressions of the genes for many cytokines, chemokines, adhesion molecules, enzymes, cyclooxygenase − 2 (COX-2), NO, etc. It includes the genes for inducible nitric oxide synthase [6] and cyclooxygenase-2 [7]. NF-κB can be considered as the pathway behind amplification of the disease-specific inflammatory process, by coordinating the activation of inflammatory genes.

In regard to antiinflammatory action, NF-κB pathway can be altered by targeting IKK inhibition and reducing IκB phosphorylation. NF-κB inhibitory properties of the antiinflammatory drugs, glucocorticoids, etc. are found to be effective in terms of these activities. In the cytoplasm, glucocorticoid activates its receptors, which may then bind to the activated NF-κB, preventing it from binding with κB sites [8]. Besides, glucocorticoids can increase the transcription of the gene for the protein IκBα, increasing the binding of activated NF-κB. Also the IκBα protein activates the dissociation of NF-κB from κB sites [9, 10].

4.2.3 MAPK (ERK) pathway

Mitogen-activated protein kinases (MAPKs) are a family of protein kinases (kinase = phosphorylating enzyme), which regulate a number of stimuli to produce corresponding responses. Extracellular signal-regulated protein kinase (ERK) 1/2, p38 MAP kinase, and c-Jun N-terminal kinases (JNK) are different MAPKs and are specifically activated by different stimuli. MAPK pathway contains at least three components, including one MAPK, one MAPK kinase, and one MAPK kinase kinase, as well as three stages of action named Ras activation, enzyme (kinase) cascades, and gene transcription [11]. When external stimuli activate the receptor-linked tyrosine kinase, its cytoplasmic domain gets activated and phosphorylated. The phosphotyrosine residues of an activated receptor bind with docking proteins, which in turn bind to guanine nucleotide exchange factor SOS. So, once the docking occurs, SOS gets activated and promotes the Ras protein to swap its GDP for a GTP and becomes active. This activated Ras promotes the activation of Raf kinase, which in turn activates the mitogen-activated kinase kinase (MEK), which then activates MAPK. MAPK regulates the translation of mRNA and levels as well as activities of many transcription factors to produce inflammatory responses. Forwarding medically, Raf kinase inhibitors and MEK inhibitors are the targets in action to alter the MAPK pathway [12].

4.2.4 JAK-STAT pathway

Janus kinase (JAK) is a group of intracellular, nonreceptor tyrosine kinases, which, being part of the JAK-STAT pathway, transduces the cytokine (especially type I and type II) signals. There are four members in the JAK family, which subsist as a pair of polypeptides. When a cytokine binds with its receptor, this ligation causes conformational changes in the receptor to bring the associated JAK molecules close enough to produce an autophosphorylation and thereby activate JAKs, which in turn phosphorylate and activate the transcription factors named signal transducer and activator of transcription (STATs). After activation, STATs dissociate from the receptor and form dimers and then translocate to the cell nucleus to alter the respective gene transcription [4].

4.2.5 Nitric oxide (NO)

Nitric oxide (NO) is a signaling molecule that comes under the category of reactive oxygen species (ROS). Under normal physiological conditions, NO plays a positive role in many functions including neurotransmission, immune responses, apoptosis regulation, and inflammation. But its overproduction changes the tone to proinflammatory mediator and results in the involvement in inflammatory disorders regardless the sites, such as the cardiovascular system, nervous system, respiratory system, gut, and joints. Besides, due to the small size and solubility in lipids, NO possess high membrane permeability.

As evidenced, inside the body, NO is synthesized from the guanidine nitrogen of arginine by the enzyme nitric oxide synthase (NOS), in the presence of oxygen and NADPH. In mammalian cells, NOS can be of endothelial NOS (eNOS or NOS-3) or neuronal NOS (nNOS or NOS-1), whose activity to produce NO is controlled by calcium levels (which interact with calmodulin-binding domain of NOS) or other stimuli such as shear stress. In contrast, inducible nitric oxide synthase (iNOS or NOS-2) is capable of binding with calmodulin independent of the cellular calcium levels, and so the resultant NO production lasts much longer and produces more NO. Under inflammatory stimuli, moieties such as proinflammatory cytokines and endotoxin induce the expressions of iNOS and COX-2, resulting in the increased production of NO, which in turn upregulate the COX- 2 formation and subsequent production of prostaglandin E2 (PGE2). The resulting cross talks between PGE2/NO and COX-2/iNOS contribute toward a proven inflammatory pathway (Fig. 4.3). So, inflammatory disease management by regulating NO can be achieved by aiming at inhibition of NO biosynthesis by means of inhibition of iNOS, COX-2, arginine availability, etc. [3, 12].

4.2.6 Eicosanoids and arachidonic acid (AA) pathway

Eicosanoids are a group of bioactive lipids derived from phospholipids and act as strong mediators of inflammation. Without any discrimination, every cell in the body can produce eicosanoids from AA, which is a polyunsaturated fatty acid and an important mediator of inflammatory functions. In the AA inflammatory pathway, many significant eicosanoids are appeared, including prostaglandins, prostacyclins, thromboxanes, leukotrienes (LTs), and lipoxins. AA is usually found in a bound form as cell membrane phospholipid, than

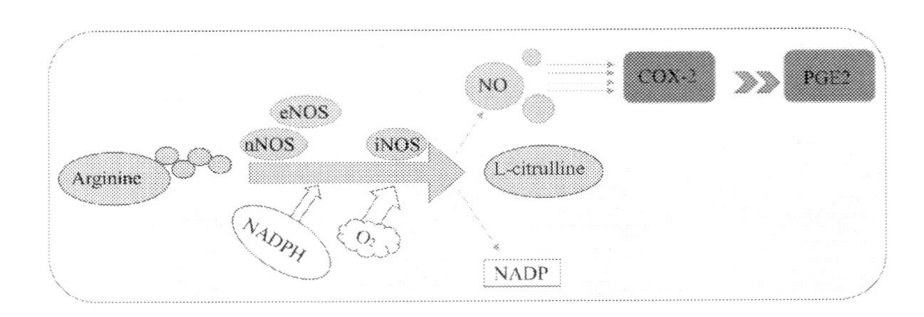

Fig. 4.3 NO pathway.

in the free form. After disruption of cell membrane, these phospholipids release AA and lysophospholipids, by the action of enzymes like phospholipase A2 (PLA_2). Lysophospholipids are converted into lyso-platelet activating factor (Lyso-PAF), followed by the release of platelet-activating factor (PAF), which is an eicosanoid and strong inducer of many inflammatory reactions [13]. AA can undergo metabolism with two different enzymes—COXs and lipoxygenase (LOs). In the presence of COX, AA is converted into prostaglandin endoperoxide H_2, which in turn produces (i) prostaglandin I_2 (prostacyclin) by the action of prostacyclin synthase, (ii) thromboxane A_2 and B_2 in the presence of thromboxane synthase, and (iii) different significant prostaglandins such as PGD_2, PGE_2, and PGF_2 [3]. In LOs-mediated pathway, AA is oxidized into different hydroperoyeicosatetraenoic acids (HETE), depending on the specific LO. Especially with 5-LO, an array of important inflammatory mediators are produced, including LTs. 5-LO has been activated in the presence of Ca^{2+} and ATP, and then, it catalyzes the formation of two intermediates named 5-(*S*)-hydroperoxy-6,8,11,14-eicosatetraenoic acid (5-HPETE) and LTA_4. 5-HPETE is converted into 5-(*S*)-hydroxy-6,8,11,14-eicosatetraenoic acid (5-HETE). LTA_4 is metabolized by specific enzymes into different bioactive compounds such as LTB_4 and LTC_4 [3, 14], which are shown in Fig. 4.4.

Eicosanoids produced during AA pathway are the important chemical mediators of inflammatory reactions. PAF plays an important role in asthma, anaphylaxis, renal inflammation, contraction of

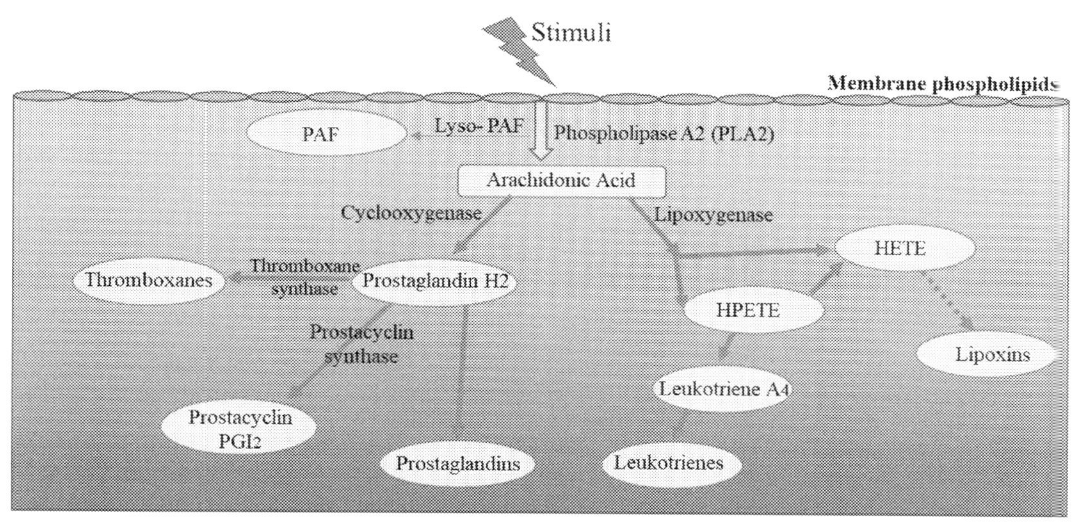

Fig. 4.4 Arachidonic acid pathway. The arrows indicate the cascade progression. Dotted arrow indicates the probability of action.

smooth muscle, increase in vascular permeability and platelet aggregation, exocytosis, etc. [15]. Prostaglandins are proved to be exhibiting inflammatory and anaphylactic reactions. Thromboxanes second the vasoconstriction activities. LTs are a significant part of asthmatic and allergic reactions. Specifically, LTB_4 act as chemotactic substance, contributing toward lymphocyte activity. These mediators are proved to be part of many physiological activities and conditions such as increased blood vessel permeability, chemotaxis of granulocytes, blood vessel dilation, and cartilage degradation. Besides, LTs and prostaglandins play behind the activation of inflammatory cells and production of mediators [3]. But prostacyclin—PGI_2 exhibits distinct antiinflammatory properties among the group, being part of relaxation of blood vessels, inhibition of platelet aggregation, etc. [16]. Similarly, lipoxins also contribute toward antiinflammatory actions by promoting the resolution of inflammation [3]. In addition, they inhibit the chemotaxis of certain leukocytes [17].

Different enzymes such as PLA_2, thromboxane synthase, LTA_4 hydrolase, LTC synthase, and different lipoxygenase influence the AA pathway and biosynthesis of inflammatory eicosanoids. Hence the production and activities of eicosanoids can be altered by targeting these molecules. Also, 5-LO exhibits its maximum activity in the presence of Ca^{2+} and ATP. So, modulation of their levels or availability can regulate the 5-LO activity.

4.2.7 Other inflammatory markers

The predictive substances that correlate the pathways and reasons of inflammatory diseases are termed as inflammatory markers. Like other biomarkers, these are also significantly utilized as indicators during the therapeutic inventions.

4.2.7.1 Cyclooxygenase-2 (COX-2)

COX, also termed as prostaglandin-endoperoxide synthase (PTGS) that is an important rate-limiting enzyme, takes part in the AA inflammatory pathway. It exists in two distinct isozymes, COX-1 and COX-2, of which COX-2 is primarily expressed at sites of extracellular stimuli and inflammation and produces proinflammatory eicosanoids whereas COX-1 is constitutively expressed. The prostaglandins are produced by COX-1 functions for normal cellular processes, while that of COX-2 are involved in inflammation and mitogenic processes (Fig. 4.4). COX pathway is considered to be the major target for nonsteroidal antiinflammatory drugs (NSAIDs), to counteract inflammation, pain, and fever. But their long-term use is associated with renal irritations and gastrointestinal (GI) complications such as ulceration.

For this reason, COX-2-selective inhibitors (COXIBs) have been developed recently as antiinflammatory agents to minimize the risk of GI toxicity. It is proved that COX-2 gene induction can be selectively inhibited by the treatment with glucocorticoids. In such cases selective inhibition of COX-2 without reducing COX-1-mediated thromboxane production could alter the balance between prostacyclin and thromboxane and promote a prothrombotic state, thereby creating a possibility of cardiovascular risk, which has been observed with some COX-2-selective inhibitors [17]. Anyway, molecular investigations on COX-2 are considered to be important in the new research era of inflammation [7].

4.2.7.2 Cytokines

Cytokines are small proteins, secreted from cells, involving in cellular communications and inflammation. They are named based on their "cells of birth" (such as lymphokines, monokines, etc.) or functioning (such as interleukines). Cytokines exhibit pleiotropy and are capable of involving in autocrine, paracrine, and endocrine signaling. Cytokines are related to the regulation of inflammation, pain, immune response, etc. and are grouped into proinflammatory or antiinflammatory cytokines [3]. In another way, they are categorized into type l, which enhances the cellular immune responses, and type 2, which has an upregulating action on antibody responses. Each cytokine has matching cell surface receptor, binding with which allows the intercellular signaling cascades and significant deviations in gene transcription factors resulting in altered cell functions. These can be ended up in triggering the release of other cytokines or development of cytokine receptors. IL-1β released after receiving stimuli by a number of cells are found to be responsible for the overproduction of substance P and PGE2 [18]. IL-6 has neuronal response functions during nerve injury and has been involved in the pain-related behavior. There are proofs for the IL-6-mediated tactile allodynia. TNF-α binds to two of the surface receptors to take part in many inflammatory pathways through the activation of NF-κB and stress-activated protein kinases (SAPKs) and sometimes induces Wallerian degeneration. All of these cytokines are found to allocate hyperalgesia. Chemokines deal with chemotaxis and are upregulated in many situations of neural incidents [19]. High-mobility group box-1 (HMGB1) is a 30-kDa cytosolic protein, which has been considered as a key cytokine mediator of local inflammation, which is released by the innate immune cells. They play roles in the secretion enhancement of endothelial adhesion molecules, epithelial cell barrier failure, and activation of macrophages/monocytes, which in turn release proinflammatory cytokines [20].

As stated previously, antiinflammatory cytokines act to control the responses of proinflammatory cytokines. Cytokines such as IL-1 receptor antagonist, IL-4, IL-10, IL-11, and IL-13 and the receptors specific for IL-1, TNF-α, and IL-18 were identified to inhibit proinflammatory cytokine activities. Besides, there are cytokines such as IL-6, transforming growth factor-beta (TGF-β), and interferon-α (IFN-α), either enhancing or downregulating the inflammatory responses, depending on the circumstances [21].

There are many cases in which single cytokine shows multiple bioactivities and various cytokines exhibit the same activity as well. This specific characteristic of cytokines formulates a scene of requirement for anticytokine agents to act on more than one targets. Specific IL-1β receptor antagonists can downregulate the cytokine binding and respective IL-1β-mediated cellular functional changes. Similarly, TNF-binding proteins and other anti-TNF compounds result in inhibition of further inflammatory responses [22].

4.2.8 Other inflammatory proteins and enzymes

Activator protein-1 (AP-1) is the transcription factor behind many cellular processes, and it regulates the inflammatory gene expression for a number of stimuli. It triggers the activation of JNK, which in turn leads to the phosphorylation of Jun and Fos proteins. Besides, AP-1 plays a major role in the regulation of many of the cytokine and chemokine genes [23]. Heme oxygenase-1 (HO-1) is an Nrf2- regulated gene and plays a critical antiinflammatory functions, in its lower doses [24].

Matrix metalloproteinases (MMPs) are proteolytic enzymes, which take part in the pathogenesis of several inflammatory diseases. Their expression and activity are controlled by the physiological conditions. They involve in the pathways of both anti- and proinflammatory pathways. There are a number of proofs that MMPs can regulate the leukocyte influx, either through the modulation of barrier function and regulation of cytokine/chemokine activity or by the gradient formation [25].

Intercellular adhesion molecule-1 (ICAM-1) alias human rhinovirus receptor is an adhesion molecule induced by several cytokines in a cell-specific way, and its expression is regulated transcriptionally. ICAM-1 on endothelium plays a key role in the migration of leukocytes to sites of inflammation. It acts as a costimulatory molecule for the activation of T cell-mediated host defense system [26].

Sphingosine kinase 1 (SphK1) is considered as a key enzyme to regulate the rate of many biological processes such as cell proliferation, differentiation, migration, and angiogenesis. Activation of SphK1 leads to the phosphorylation of sphingosine, to form sphingosine-1-phosphate (S1P), which triggers the intracellular pathways for these aforementioned biological processes and expressions of nuclear transcription factors such as AP-1 and NF-κB and increases TGF-β expressions, which together lead to inflammation [27].

Adenosine monophosphate (AMP)-activated protein kinase (AMPK) is a serine/threonine protein kinase present in all eukaryotic cells, which has been highly conserved during evolution and plays key roles in the regulation of lipid biosynthesis and cellular energy homeostasis. In mammals, it appears as multiple isoforms and encoded as 12 different heterodimers. It can be activated by means of both allosteric and phosphorylation modes and is implicated in many diseases. Once activated, AMPK enhances the activation of Nrf2 and peroxisome proliferator-activated receptor gamma (PPARγ) and inhibits the NF-κB signaling and MMP-9 and Ccr2 expressions. Hence, many AMPK activators were found to be antiinflammatory and are considered as targets in therapeutic aids [28, 29].

4.2.9 Antiinflammatory Kelch-like ECH-associated protein 1 (Keap1)-nuclear factor erythoid-2-related factor 2 (Nrf2)-antioxidant response elements (ARE) pathway

Nuclear factor erythoid-2-related factor 2 (Nrf2) is a basic leucine zipper protein, which is a transcription factor, kept in the cytoplasm. In normal situation, it is occurring together with its inhibitor protein named Kelch-like ECH-associated protein-1 (Keap-1), which keeps Nrf2 inactive. Under the incidence of stimuli, Nrf2-Keap-1 interactions are resolved, and the free Nrf2 translocates toward the nucleus, and there, it makes complexes (heterodimers) with one small Maf protein. These complexes may activate the antioxidant response element (ARE)-dependent gene expressions of proteins, involving antiinflammatory actions. In addition, Nrf2 supports the synthesis of NADPH [30]. Also, there are studies suggesting the counteraction of Nrf2 toward the NF-κB-related inflammatory responses [31]. Thus triggering the Keap-1-Nrf2-ARE pathway enhances antiinflammatory protocols.

4.3 Distinct pathways behind neuroinflammation; Alzheimer's disease as a case

The inflammatory responses proceed within the brain and spinal cord follow a slightly different mode of action and are termed as neuroinflammation. In the nervous system the inflammation is advanced through inflammatory mediators produced by microglia, astrocytes, endothelial cells, etc. The pathways and progression of neuroinflammation can be better understood with Alzheimer's disease (AD) as an example, since it comes across with most of the pathways.

AD is a neurodegenerative disease, which is experienced by the destruction of memory and thinking skills. It is characterized by the accumulation of amyloid-β (Aβ) plaques in the brain, and the formation of neurofibrillary tangles inside the neurons. There are many hypotheses that explain the signs and pathways of Alzheimer's disease, namely, cholinergic hypothesis (i.e., AD is caused by the reduction in the neurotransmitter, acetyl choline), amyloid hypothesis (Aβ is responsible for AD), Tau hypothesis (tau protein abnormalities initiate the disease), neurovascular hypothesis (poor functioning of blood-brain barrier results in AD), etc. Most of them are holding inflammation as the key factor [32–35].

Studies have proved the relation between the possibility of AD and the genes that regulate the innate immune system. Also, there are results suggesting the key role of systemic inflammation in AD-specific pathology, proved by the presence of increased levels of inflammatory markers in patient's blood [36]. Proinflammatory cytokines released during systemic inflammation may trigger the central nervous system to induce signaling within the brain. Also, these cytokines can bind to endothelial cell receptors and produce signaling for the opening of tight junctions between endothelial cells, and the cytokines release into the bloodstream and in the brain. Proinflammatory cytokines and endokines in the periphery can also stimulate the vagus nerve, and this in turn regulates the stimulation of neural immune cells, which in turn induce the release of cytokines and chemokines into the central nervous system, which then activate the microglia and astrocytes to release proinflammatory cytokines. Thus the inflammatory signaling inside CNS is increased by many pathways, and the prolonged activation of microglia and astrocytes can promote neurodegeneration and cognitive decline [37]. These possibilities are seconded by the findings in another study that the possibility of cognitive decline is more in subjects having higher levels of inflammatory proteins in blood, even decades before the expected age of dementia [36].

Inflammation follows another important pathway toward neurodegeneration by altering the levels of kynurenines (metabolite of amino acid, tryptophan), which can alter the brain physiology by targeting neurotransmitter receptors. Also, in the presence of proinflammatory cytokines, the indoleamine 2,3-dioxygenases (IDO) activity is increased, which elevates the production of a number of neurotoxic and cytotoxic metabolites [38]. In addition, there are supporting evidences for the role of oxidative imbalance and oxidative stress in the pathophysiology of AD [39]. These molecular actions related to inflammation are believed to initiate or aggravate the neurodegenerative processes. Hence the treatment for neuroinflammation targets for the microglial inflammatory activation, proinflammatory cytokine production, etc.

4.4 Multitarget approach for antiinflammatory action

Drug development procedures are focused toward specific drugs with high therapeutic potential on the targets and minimal side effects. But in some cases more than one medication has to be used for a single disease, where the reason behind the condition holds more than one factor. Almost all the inflammation-related diseases could be triggered or enhanced by more than one factor. For example, though inflammation alone can play the major villain role in pathogenesis of many diseases, for example, the combination of "oxidation-inflammation" is more dreadful, especially in cases such as cancer and AD. While coming to pathogenesis, every single disease associated with inflammation follows its own pathways and actions apart from the common, classic, signature pathways of inflammation. At the same time, most of the inflammatory pathogenic pathways are closely interrelated, and the members in the pathways are overlapped and can sometimes be replaced by other participants. In almost all the cases discussed earlier, it can be observed that the targets and pathways follow a multiple channel of action. Hence the therapeutic aids showing multitarget action can be more effective. Besides, many of the antiinflammatory drugs have ended up in unmet results with unacceptable safety characteristics while targeting only one of the pathways [29]. Considering all these facts, it is important to take enough care toward all the overlapping mechanisms that are carried out for antiinflammation purpose, such as the fabrication/modification of pathways, regulation of inflammatory markers, and control over the mediators. So the play here turns toward multitarget drugs.

Multitarget therapeutic aid can be described as the one, which can act on multiple points (members, supporting factors, mediators, etc.) involved in a pathway or biological process. The complex mechanisms involved in chronic inflammation need to be addressed with multiple targets for better medical recovery. Certainly, there are compounds and their combinations available with the preferred functions. Yet the changed concerns in the field have given more value to the natural products. Also the combination of drugs may result in the interactions between themselves and the synergism of side effects as well [40]. Natural multitarget compounds provide interactions with more than one target, with lesser side effects than a combination drug. Even so, they need optimization of a few things such as extent of action toward each target, unwanted interactions of natural matrix, and bioavailability of the compound. Also the multitarget antiinflammatory compounds can be of two types—those handling multiple channels/targets of action of inflammation and those that deal inflammation as one of its targets along with other therapeutic aspects.

There are many phytochemicals that have been profiled, having multiple targets of action and proven therapeutic effects in concern with antiinflammation. Databases and study results handling antiinflammatory natural compounds, their targets of action and structural, and physicochemical properties are now available. In every database the nature of providing information is different in terms of the described peculiarities and classification of compounds. But many of them have the same voice, regarding multitarget antiinflammatory compounds. In a dataset, dealing with the antiinflammatory natural compounds, their biological targets, all atomic structures, and scaffolds, more than 70% of the involved compounds were identified to have more than one target of action [41]. Temml and Schuster have reviewed the computational techniques in the natural products, in a view to elucidate multitarget drugs [42]. A system level procedure for the multitarget compound discovery from different herbs for cardiovascular diseases has been narrated in a study where the antiinflammatory compounds hold a major part [43]. Now, we can move on to some natural compound that leads to antiinflammatory therapeutics.

4.5 Natural compounds with multitarget in antiinflammation

Almost all the plants possessed with their own defense mechanisms and resulting active phytochemicals. The traditional medical systems made use of these secondary metabolites extensively, and many of the compounds were studied for their properties. According to their mode of action, structural peculiarities, or pharmacokinetics, these active compounds are classified under different norms. A feasible, significant classification is based on the chemical structures, which is followed by several studies. Accordingly the natural compounds can be classified as phenolic compounds, alkaloids, terpenes, steroidal compounds, proteins, enzymes, etc.

1. **Phenolics** is a general term, used to mention a number of groups of compounds, including flavonoids, tannins, and lignans. They can interfere with inflammatory mediators such as NO, cytokines, and ILs; pathways including NF-κB; and mediating enzymes such as LO and LT synthases [3].

2. **Alkaloids**, the secondary metabolites found in many plants, are considered as medicinal compounds to treat the ailments ranging from pain to neurodegenerative diseases. These compounds act by reducing the IL-6 and TNF-α levels in plasma and IKKβ and IKKα activities, along with the inhibition of NF-κB activation [3].
3. **Terpenes** constitute the major part of essential oils and impart the flavor of the same. Rather, they were identified as the key active therapeutics of the oils. Studies have proved that different terpenes can prevent/treat inflammation effectively. For example, the terpenes from *Tripterygium wilfordii* were effective in inhibiting the production of IL-1, COX-1, COX-2, PLA2, 5-LO, and 12-LO. Likewise the terpenes, especially 3-*O*-acetyl-11-keto-β-boswellic acid (AKBA), isolated from *Boswellia serrata* inhibited the action of inflammatory enzymes [44, 45].
4. **Flavonoids** are a group of plant-derived active compounds, including flavones, flavonones, flavonols, and isoflavonoids. The antiinflammatory capacity of flavonoids was fulfilled by the regulation of transcription factors such as NF-κB, AP-1, and Nrf2. The inhibition of NO, COX-2, and PGE2 levels and IL-1β, IL-6, and TNF-α production were observed in many studies [3].

As discussed earlier the number of compounds reported to have antiinflammatory properties is huge, and discussion of which is beyond the scope of a single chapter. The chapter presents an overview on some significant multitarget compounds, which are either studied enormously or novel to appear in the field.

4.5.1 Curcuminoids

Curcuminoids, generally termed as curcumin, are one of the most studied natural moiety, obtained from the rhizomes of *Curcuma longa*, which is proved to be a remedy for many health problems. Multiple channels of studies have been conducted on the grounds of the therapeutic effects of curcumin. The antiinflammatory activities of curcumin are attributed to the alteration of different pathways and modulation of the inflammatory mediators. As reviewed by Fadus and crew, it inhibits the production of TNF-α and ILs; downregulates the activity of COX-2, LOs, NF-κB, and iNOS; and alters the JAK-STAT and NF-κB pathways. Curcuminoids' therapeutic action on the inflammatory diseases such as cancer, arthritis, diabetes, neurodegenerative disorders, and infections is proved doubtlessly [46]. There are many clinical trials conducted successfully on the therapeutic effects of curcumin. Let us discuss some interesting studies, among them.

Considering the medication of any major diseases, a crucial snag in the way is the side effects, mediated through inflammation and oxidative stress. Cisplatin-induced renal injury, mediated through NF-κB and

TNF-α, was proved to relieve with curcumin with a significant decrease in serum TNF-α level [47]. Curcuminoids' protective effect on skin inflammation due to ultraviolet (UV) radiations, pollutants from the atmosphere, some drugs/food, etc. is proved doubtlessly in many works. The pretreatment of curcumin could significantly downregulates the expressions of IL-6, IL-8, and TNF-α, along with altering the MAPK and NF-κB pathways [48]. The COX-2 expression, even at the transcriptional level, was decreased by curcuminoid treatment through the inhibition of p38 MAPK and JNK in the irradiated cell line. Along the curcuminoid treatment could suppress DNA binding of AP-1 transcription factor in the HaCaT cells [49]. Curcumin can even bind to Aβ plaques and thus downregulate the progression of AD [46]. Curcumin regulates the gene expressions related to carcinogenesis, cell proliferation, invasion, cell survival, and angiogenesis by inhibiting NF-κB, AP-1, STAT proteins, PPARγ, etc. [50]. Thus curcuminoids can act on different inflammatory pathways and mediators, so as to deliver antiinflammatory results.

4.5.2 Fucoxanthin

Fucoxanthin, a major carotenoid found in brown algae and diatoms, is found to possess antioxidant and antiinflammatory activities and proved to exhibit protective effects toward neuroblastoma and cortical neurons. This carotenoid could effectively inhibit the $A\beta_{42}$-induced microglial activation, indicated by the downregulated production and gene expression of proinflammatory cytokine in BV2 microglial cells. Its antiinflammatory activity could deal with the inhibition of NF-κB and MAPK signaling pathways and are found to be effective in the treatment of AD [51]. The immunosuppressive activity of gliotoxin—a metabolite from *Aspergillus fumigatus*—was reported in an in vivo study. Gliotoxin is a toxic epipolythiodioxopiperazine metabolite and exerts a negative effect in the body. But the pathway includes the inhibition of NF-κB activation. In concentrations of nanomolar level, the compound exerts a specific inhibition toward the activation of NF-κB. In high concentrations the inhibition occurs as a result of preventing the binding of NF-κB with DNA. Along, in intact cells, gliotoxin prevents the IκB-α degradation, thereby reducing NF-κB activation [52].

4.5.3 Quercetin

Quercetin is a plant flavonol, available in many fruits, vegetables, seeds, leaves, and grains, especially in red onions and kale. They are noted for their antioxidant and free radical scavenging effects. The reactions observed in heat stroke mice indicate the multifocal antiinflammatory effects of quercetin, fulfilled by the reversal of poststroke increase in proinflammatory cytokines such as TNF-α and IL-6 and decrease in antiinflammatory cytokine IL-10. Besides, there was an attenuation in the amount of inflammatory cells [53]. By following the same mode of action, quercetin downregulated the traumatic brain injury in rats [54]. Quercetin has proved to influence the inflammatory pathways in the liver by affecting the TLR expression and thus inhibiting the signaling cascade progressing toward MAPKs and NF-κB, mediated via Myd88 [55]. There are further evidences for the liver protection capacity of quercetin in terms of TNF-α level reduction by hindering MAPK, NF-κB and STAT1 pathways or PI3K/AKT pathway [56, 57]. Inhibition of NF-κB by quercetin is attributed to the downregulation of p65 subunit along with the suppression of expressions for different TLR and MyD88 and the inhibition of phosphorylation of p38, JNK1/2, ERK1/2, and IkB kinase (the last one is mediated by the AKT phosphorylation) [57–60].

4.5.4 Resveratrol

Resveratrol is an important phytoalexin, rich in grape seed and red wine. The compound is well known for its therapeutic capabilities, and its efficacy has been verified for many ailments. Resveratrol found to act as a potent activator of AMP-activated protein kinase (AMPK), which exhibit neuroprotective mechanisms against inflammation. The transcript expression levels of IKK, mRNA and protein levels of NF-κB, and the iNOS and COX-2 levels in the Aβ-treated human neural stem cells (hNSCs) were also restored by resveratrol treatment [61]. A dose-dependent inhibition of 12-O-tetradecanoylphorbol-13-acetate (TPA)-induced COX-2 expression has resulted with the resveratrol treatment in ICR mice dorsal skin. It was supported by the decrease in the phosphorylation of ERK, the reduced catalytic activity of both ERK and p38 MAPK, and the inhibition of DNA binding of AP-1 [62]. Alteration of the phosphorylated ERK levels in the brain region with resveratrol treatment was verified in studies [63]. Age-related inflammatory changes in the brain were effectively counteracted with resveratrol, by regulating the levels of IL-1β and TNF-α in the mRNA, serum, and adipose tissues [64].

The pathogenesis of nonalcoholic steatohepatitis (NASH)—a chronic liver disease—involves inflammation as one of a major cause. Administration of resveratrol clearly downregulated the increased inflammatory cell infiltration and injury in high fat diet-induced NASH, mediated through the modulation of endocannabinoids especially the expressions of CB1 and CB2. Here the lipopolysaccharide (LPS) level was downregulated through the TLR4 signaling pathway, which in turn modulates the activation of focal adhesion kinase (FAK)/myeloid differentiation primary response 88 (MyD88)/IL-1R-associated kinase 4 (IRAK4) signaling pathway. Resveratrol influenced the TNF-α, IL-1β, IL-6, and antiinflammatory cytokine IL-10 expressions also [65]. Moreover, diabetes-induced cytokine production was restored by the treatment of resveratrol [66]. Furthermore, resveratrol alleviates the increased activity of SphK1; expressions of S1P, iNOS, and ICAM-1; and activation of NF-κB and AP-1, induced by high glucose condition [67].

4.5.5 Ginsenosides

Ginsenosides are major compounds in most of the organs of ginseng and are known to be therapeutically potential, irrespective to the organ of availability. Among the known ginsenosides, Rb1, Rd., and Rg1 have been studied for their effects on neurodegenerative diseases, in many modes. Rd. ginsenoside therapeutically works in the treatment of lipopolysaccharide (LPS)-induced Parkinson's disease (PD) involving the downregulation of NO and PGE2 levels, by altering the iNOS and COX-2 expressions [68]. Rg1 was proved to suppress the microglial activation by decreasing the production of proinflammatory cytokines, NO, and TNF-α, following a LPS injection. The explained mechanisms touched MAPK and NF-κB pathways through the reduction in expression of iNOS; phosphorylation levels of IκB, p38, ERK1/2, and JNK; and the nuclear translocation of p65 of NF-κB [69]. Ginsenoside Rf exhibits antiinflammatory effects against hypoxia-induced COX-2 expression by binding to PPARγ and thereby inhibiting the COX-2 expression [70]. Administration of the *Panax* (*P.*) *ginseng* (Meyer Korean red ginseng) extract rich in ginsenosides has shown potent protective activities toward the neurotoxicity and neurological dysfunction related to PD in a mouse model induced with 1-methyl-4-phenyl-1,2,3,6-tetrahydropyridine (MPTP). The extract contained a considerable quantity of almost all the ginsenosides such as Rb1, Rb2, Rc, Rd, Re, Rf, Rg1, Rg2s, Rg3s, and Rh1 as well as minor ginsenosides and have shown these positive effects by downregulating the MAPK and NF-κB pathways, where the expressions of proinflammatory cytokines, COX-2, and iNOS were altered [71]. Even the minor ginsenosides such as ginsenoside LS_1 and 5,6-didehydroginsenoside Rg_3 from the *P. ginseng* leaves were capable of considerably inhibiting NO production, and the latter one has exhibited a significant decrease in secretion of PGE2 and TNF-α, along with a suppressed protein expression of iNOS and COX-2 [72]. Likewise the minor triterpenoid saponin named pseudoginsenoside RT_8, isolated from the seeds, also effectively controls the IL-1, IL-6, iNOS, and COX-2 in an antiinflammatory way [73].

4.5.6 Phlorotannins

Phlorotannins, the key health components of brown seaweeds, are finding importance nowadays in the food, animal feed, fertilizer, and drug industries. The term "phlorotannins" denotes a group of compounds that majorly include dieckol, eckol, dioxinodehydroeckol (also known as eckstolonol), phlorofucofuroeckol A, 7-phloroeckol, 8,8′-bieckol, fucofuroeckol A, and 6,6′-bieckol, 1-(3′,5′-dihydroxyphenoxy)-7-(2″,4″,6″-trihydroxyphenoxy)-2,4,9,-trihydroxydibenzo-1, 4-dioxin [74]. Phlorotannins produced by *Ecklonia cava* Kjell. were found to inhibit the production of NO and PGE2, in a dose-dependent manner. Besides, these compounds could suppress the expression of iNOS and COX-2 and the release of IL-1β and IL-6. These effects were comparable for a fermented, bioavailable version of phlorotannin extract with an unfermented fraction [75, 76]. Along with the reduction in proinflammatory cytokines, there is a reduction in the proinflammatory cytokine mediators such as TNF-α and HMGB1 and inhibition of the NF-κB pathway. Also the Nrf2/HO-1 pathway was also altered by *E. cava* phlorotannin extract [77]. In arthritis joints the extract was potent enough to inhibit the level, activity, and m-RNA gene expression of MMP and the iNOS and COX-2 responses, mediated by altering the downstream of the MAPK signaling pathway [78, 79]. Likewise, extracts of many of the seaweeds provide antiinflammatory actions more or less in a similar way. For example, extracts of *Ecklonia stolonifera, Eisenia bicyclis, Eisenia arborea, Fucus vesiculosus, Ascophyllum nodosum* L., *Fucus distichus* L., *Sargassum fusiforme* (Harv.) Setch., *Ecklonia kurome* Okam, *Ecklonia arborea*, etc. containing phlorotannins provided a dose-dependent inhibition of NO and PGE2 production. The LPS-stimulated inflammatory response was defended by many of these extracts, following a cascade of suppression of iNOS and COX-2 expression, inhibition of NF-κB transcription factor activation, and reduction in corresponding proinflammatory cytokine production [80–89].

The individual phlorotannin compounds also offer a better solution in antiinflammatory manner. For example, the microglial activation was effectively regulated by the administration of dieckol, by the inhibition of proinflammatory cytokine levels, and iNOS and COX-2 expressions in BV-2 microglia cells. The antiinflammatory action included the downregulation of NF-κB and MAPK signaling pathways [90]. 6,6′-Bieckol and 8,8′-bieckol acted on the transcriptional level iNOS expression and effectively inhibited the production of NO [91]. Alongside, they reduced the production of IL-6 through the reduction of encoding gene expression [92]. They contribute toward the reduction in transactivation of NF-κB and the nuclear translocation of the same [91]. But these phloroglucinol hexamers differ in their mechanism toward the inhibitory activity of PGE_2: 6,6′-bieckol suppresses the COX-2 expression in the mRNA and proteins, while 8,8′-bieckol reduces the activity of COX-2 [91, 92]. Likewise, phlorofucofuroeckol A also followed a similar pattern of reduction in NO and PGE2 production through the reduction of iNOS and COX-2 levels and downregulation of IL-1β, IL-6, and TNF-α. Besides the treatment also downregulated the iNOS and COX-2 expressions, supported with the control of NF-κB and MAPK actions [93]. Phlorofucofuroeckol B, on the other hand, attenuated the release of TNF-α, IL-1β, and IL-6 and the COX-2 and iNOS expressions, mediated through the associated inhibition of NF-κB activation and Akt/ERK/JNK phosphorylation [94]. Interestingly, *Propionibacterium acnes*-induced expressions of iNOS, COX-2, and TNF-α were reversed by eckol. Besides the translational level inhibition of NF-κB and AKT was also achieved [95]. Collectively, phlorotannins offer a multimode platform for the antiinflammatory actions, both as single compound and in combination.

4.5.7 Catechin and derivatives

Catechin and its derivatives form a major team of active polyphenolic components from plants. There are many compounds come under the category, such as catechin, epicatechin (EC), epigallocatechin (EGC), and epigallocatechin gallate (EGCG). Catechin exerts its pharmacological effects by reducing IL-5 and IL-13 levels and suppressing the NF-κB signal pathway [96]. Besides a pretreatment of catechin significantly ameliorated the activation of NF-κB and the expressions of COX-2, TNF-α, and IL-6 in benzo(*a*)pyrene-induced cells [97]. The mRNA expression levels of TLR4/NFκB-dependent inflammatory genes were lowered with the treatment with catechin [98]. EGCG was proved to target the COX, LO, IKK, AP-1, and NF-κB, during the antiinflammatory action. In addition, EGCG regulates MMP-2, COX-2, IL-6, TNF-α, and chemokines [99, 100]. Interestingly, EGCG could inhibit the action of TGF-β-activated MAP kinase (TAK1) and thus abrogate the production of IL-6 and IL-8 [100]. In a comparison, EGCG was the more effective inhibitor of IL-6 and IL-8, than EGC and EC. EGC and EGCG were effective in the downregulation of IL-1β-induced MMP-2 activity. Specific inhibition of COX-2 was observed with both EGC and EGCG. Together, these polyphenolic compounds exert synergic therapeutic benefits. Product rich in these polyphenols, such as green tea, is identified to have better beneficial effects [100].

4.5.8 Withaferin A

Withaferin A (WA) is a steroidal lactone, mostly found in the plant *Withania somnifera,* commonly known as Ashwagandha or Indian ginseng. The recent studies have proved the antiinflammatory properties of Ashwagandha extract, giving special emphasis on WA. There are evidences of its activity toward inhibition of the cytokine secretion, IL-6 and TNF-α levels, and NF-κB nuclear translocation, along with the modification of NF-κB cysteine residue and modulation of Nrf2/

NFκB signaling [101–103]. Evidences on the activity of WA as Nrf2 inducer explain its protective role in liver injuries [104].

We have passed through only eight compounds, not because they are the only moieties, with relevant antiinflammatory activities. We chose them as they represent different classes of compounds working with different modes of mechanisms and bring light on the research pathways of multitarget antiinflammatory diseases. There are many other compounds that exhibit antiinflammatory potential and use successfully in the therapeutic field.

4.6 Combination of phytochemicals

Studies have proved that the combination of different potent phytochemicals produces a synergic effect on the chronic inflammation and can be a miraculous solution for the initiation and progression of a number of diseases. For example, curcumin, resveratrol, and [6]-gingerol were studied for their antiinflammatory activities related to prostate cancer (PCa) treatment. These phytochemicals were found to decrease the cytokine-induced p-38-dependent proinflammatory changes, mediated by the upregulation of mitogen-activated protein kinase phosphatase-5 (MKP-5). MKP-5 overexpression resulted in decrease of cytokine-induced NF-κB activation and expression of proinflammatory genes COX-2, IL-6, and IL-8. Apart from increasing the MKP-5 expression, curcumin, resveratrol, and [6]-gingerol guided the studied PCa cell lines (DU 145, PC-3, LNCaP, and LAPC-4) to retain the ability to upregulate MKP-5 [105].

A faulty immune response such as hypersensitivity, false recognition of self-antigen, and attack against alloantigen is another trigger for inflammatory diseases. Many of the immunosuppressive drugs influence the diseases by regulating cytokine secretion and regulatory T-cell frequency and modulating the expression of CD28, CD80, and CTLA-4. The effect of curcumin and resveratrol on the activity of lymphocytes and their mechanism of action were investigated in a study. They were found to act as immunosuppressant as they could downregulate the proliferation of T cells and secretion of interferon-gamma (IFN-γ) and IL-4; suppress the proliferation of B cells and the release of immunoglobulin G1 (IgG1) and IgG2a; inhibit the proinflammatory cytokines such as IL-1, IL-6, and TNF-α; enhance the antiinflammatory cytokine IL-10; and attenuate the expression of CD80 and CD28. The first two of these findings were assumed to result from the inhibition in the activity of Th1 and Th2 cells from both T and B cells. Besides, curcumin increased the expression of CTLA-4. Altogether, curcumin and resveratrol contribute toward the specific immune suppression and for immunotherapy [106].

The synergic effects of different polyphenols from grape powder could effectively reduce the inflammation by altering the mediators such as ERK and JNK and the transcription factors like NF-κB and cJun [107]. At the same time the red grape pomace suppressed the activation of NF-κB, iNOS, and COX-2 [108]. A Mexican traditional remedy for inflammation is the infusion of *Buddleia scordioides*, *Chamaemelum nobile*, and *Litsea glaucescens*, which is proved to regulate COX-2, TNF-α, NF-κB, and IL-8 cytokines in studies [30]. Likewise, green tea polyphenols exhibit a synergic effect on the reduction of MMP-9/2 expression induced by TNF-α or LPS [109]. Also the levels of inflammation markers such as COX-2, PGE2, proliferating cell nuclear antigen, and cyclin D1 [99] and the proinflammatory cytokines such as TNF-α, IL-6, and IL-1β were significantly reduced by the consumption of green tea polyphenols. In addition, there are evidences for the regulation of the PI3K/AKT pathway by the extract [110]. As per a molecular docking study, green tea polyphenols might inhibit the enzymes, namely, β-secretase, γ- secretase, GSK-3β, BuChE, and AChE, hence providing a multiverse solution for AD [111]. A combination of curcuminoids, with essential oil and water extract of turmeric, has proved to improve the bioavailability of curcumin and manage the rheumatoid arthritis, without any side effects [112]. These bioavailable curcuminoids in combination with β-caryophyllene, AKBA, and *Kaempferia galangal* flavonoids could remarkably relieve the symptoms and pain of osteoarthritis [113].

There are much more compounds and combinations to explore, on the way. The manner of activities and mechanisms may differ for each of them. Also the associated side effects and synergism will vary, depending on the administration, combinations, amount, etc. Anyhow the natural product researches and clinical studies verify the multitarget approach of natural products on antiinflammation. Inflammation itself is not a disease, but in extended course, it delivers a chronic nature and disease cause. So the important part in the attempts for therapeutic solutions is the wise use of remedies in perfect amount.

4.7 Conclusion

Inflammation is considered as the defense mechanism of body. But in chronic condition, it is identified as the factor behind many diseases and researches on its physiological roles made an impact in the therapeutic solutions of many pathophysiological ailments. As the mechanisms and pathways of inflammation are revealed to be different, antiinflammatory medications are developed to target them accordingly. Hence a single compound or formulation with potential to manage more targets is drawing attention. Rather, natural compounds with these multitarget actions are more interesting. We have discussed different inflammatory targets of action, concerns on multitarget approach, and different natural compounds showing this multitarget antiinflammatory potential.

References

[1] K. Lingappan, NF-κB in oxidative stress, Curr. Opin. Toxicol. 7 (2018) 81–86.
[2] P.W. Baumert, Acute inflammation after injury, Postgrad. Med. 97 (1995) 35–49.
[3] G.D. Lin, R.W. Li, Natural products targeting inflammation processes and multiple mediators, in: Natural Products and Drug Discovery, Elsevier, 2018, pp. 277–308.
[4] L. Chen, H. Deng, H. Cui, J. Fang, Z. Zuo, J. Deng, Y. Li, X. Wang, L. Zhao, Inflammatory responses and inflammation-associated diseases in organs, Oncotarget 9 (2018) 7204.
[5] A.V. Gudkov, E.A. Komarova, p53 and the carcinogenicity of chronic inflammation. Cold Spring Harb. Perspect. Med. 6 (11) a026161.
[6] Q.W. Xie, Y. Kashiwabara, C. Nathan, Role of transcription factor NF-kappa B/Rel in induction of nitric oxide synthase, J. Biol. Chem. 269 (1994) 4705–4708.
[7] K. Yamamoto, T. Arakawa, N. Ueda, S. Yamamoto, Transcriptional roles of nuclear factor B and nuclear factor-interleukin-6 in the tumor necrosis factor-dependent induction of cyclooxygenase-2 in MC3T3-E1 cells, J. Biol. Chem. 270 (1995) 31315–31320.
[8] I.M. Adcock, C.R. Brown, P.J. Barnes, Tumour necrosis factor α causes retention of activated glucocorticoid receptor within the cytoplasm of A549 cells, Biochem. Biophys. Res. Commun. 225 (1996) 545–550.
[9] I.M. Adcock, C.R. Brown, O. Kwon, P.J. Barnes, Oxidative stress induces NFkB DNA binding and inducible NOS mRNA in the human epithelial cell line A549, Biochem. Soc. Trans. 22 (1994) 186S.
[10] R.I. Scheinman, P.C. Cogswell, A.K. Lofquist, A.S. Baldwin, Role of transcriptional activation of IκBα in mediation of immunosuppression by glucocorticoids, Science 270 (1995) 283–286.
[11] E.K. Kim, E.J. Choi, Pathological roles of MAPK signaling pathways in human diseases, Biochim. Biophys. Acta, Mol. Basis Dis. 1802 (2010) 396–405.
[12] J.N. Sharma, A. Al-Omran, S.S. Parvathy, Role of nitric oxide in inflammatory diseases, Inflammopharmacology 15 (2007) 252–259.
[13] J.R. Burke, K.R. Gregor, R.A. Padmanabha, A b-lactam inhibitor of cytosolic phospholipase A2 which acts in a competitive, reversible manner at the lipid/water interface, J. Enzyme Inhib. 13 (1998) 195–206.
[14] B. Samuelsson, S.E. Dahle'n, J.A. Lindgren, C.A. Rouzer, C.N. Serhan, Leukotrienes and lipoxins: structures, biosynthesis, and biological effects, Science 237 (1987) 1171–1176.
[15] W.L. Smith, Prostanoid biosynthesis and mechanisms of action, Am. J. Physiol. 263 (1992) 181–191.
[16] S. Moncada, R. Gryglewski, S. Bunting, J.R. Vane, An enzyme isolated from arteries transforms prostaglandin endopreoxides to an unstable substance that inhibit platelet aggregation, Nature 263 (1976) 663–665.
[17] S.P. Khanapure, D.S. Garvey, D.R. Janero, L. Gordon Letts, Eicosanoids in inflammation: biosynthesis, pharmacology, and therapeutic frontiers, Curr. Top. Med. Chem. 7 (2007) 311–340.
[18] L. Neeb, P. Hellen, C. Boehnke, J. Hoffmann, S. Schuh-Hofer, U. Dirnagl, U. Reuter, IL-1β stimulates COX-2 dependent PGE 2 synthesis and CGRP release in rat trigeminal ganglia cells, PLoS One 6 (2011), e17360.

[19] J.M. Zhang, J. An, Cytokines, inflammation and pain, Int. Anesthesiol. Clin. 45 (2007) 27.
[20] H. Wang, H. Yang, K.J. Tracey, Extracellular role of HMGB1 in inflammation and sepsis, J. Intern. Med. 255 (2004) 320–331.
[21] S.M. Opal, V.A. DePalo, Anti-inflammatory cytokines, Chest 117 (2000) 1162–1172.
[22] J.R. Foster, The functions of cytokines and their uses in toxicology, Int. J. Exp. Pathol. 82 (2001) 171–192.
[23] R. Zenz, R. Eferl, C. Scheinecker, K. Redlich, J. Smolen, H.B. Schonthaler, L. Kenner, E. Tschachler, E.F. Wagner, Activator protein 1 (Fos/Jun) functions in inflammatory bone and skin disease, Arthritis Res. Ther. 10 (2008) 201.
[24] A. Paine, B. Eiz-Vesper, R. Blasczyk, S. Immenschuh, Signaling to heme oxygenase-1 and its anti-inflammatory therapeutic potential, Biochem. Pharmacol. 80 (2010) 1895–1903.
[25] A.M. Manicone, J.K. McGuire, Matrix metalloproteinases as modulators of inflammation, in: Seminars in Cell & Developmental Biology, 19, Academic Press, 2008, pp. 34–41.
[26] A. Van de Stolpe, P.T. Van der Saag, Intercellular adhesion molecule-1, Int. J. Mol. Med. 74 (1996) 13–33.
[27] M.L. Ng, N.S. Yarla, M. Menschikowski, O.A. Sukocheva, Regulatory role of sphingosine kinase and sphingosine-1-phosphate receptor signaling in progenitor/stem cells, World J. Stem Cells 10 (2018) 119.
[28] A. Salminen, M.T. Hyttinen, K. Kaarniranta, AMP-activated protein kinase inhibits NF-κB signaling and inflammation: impact on healthspan and lifespan, J. Mol. Med. 89 (2011) 667–676.
[29] C.G. Honnappa, U.M. Kesavan, A concise review on advances in development of small molecule anti-inflammatory therapeutics emphasising AMPK: An emerging target, Int. J. Immunopathol. Pharmacol. 29 (2016) 562–571.
[30] E. Herrera-Carrera, M.R. Moreno-Jimenez, N.E. Rocha-Guzman, J.A. Gallegos-Infante, J.O. Diaz-Rivas, C.I. Gamboa-Gomez, R.F. Gonzalez-Laredo, Phenolic composition of selected herbal infusions and their anti-inflammatory effect on a colonic model in vitro in HT-29 cells, Cogent. Food Agric. 1 (2015) 1059033.
[31] I. Bellezza, A.L. Mierla, A. Minelli, Nrf2 and NF-κB and their concerted modulation in Cancer pathogenesis and progression, Cancers (Basel). 2 (2010) 483–497.
[32] P.T. Francis, A.M. Palmer, M. Snape, G.K. Wilcock, The cholinergic hypothesis of Alzheimer's disease: a review of progress, J. Neurol. Neurosurg. Psychiatry 66 (1999) 137–147.
[33] J. Hardy, D. Allsop, Amyloid deposition as the central event in the aetiology of Alzheimer's disease, Trends Pharmacol. Sci. 12 (1991) 383–388.
[34] A. Mudher, S. Lovestone, Alzheimer's disease—do tauists and baptists finally shake hands? Trends Neurosci. 25 (2002) 22–26.
[35] R. Deane, B.V. Zlokovic, Role of the blood-brain barrier in the pathogenesis of Alzheimer's disease, Curr. Alzheimer Res. 4 (2007) 191–197.
[36] K.A. Walker, R.F. Gottesman, A. Wu, D.S. Knopman, A.L. Gross, T.H. Mosley, E. Selvin, B.G. Windham, Systemic inflammation during midlife and cognitive change over 20 years: the ARIC study, Neurology 92 (2019) e1256–e1267.
[37] K.A. Walker, N.F. Bronte, W. Reyhan, Understanding the role of systemic inflammation in Alzheimer's disease, ACS Chem. Nerosci. 10 (2019) 3340–3342.
[38] R. Schwarcz, J.P. Bruno, P.J. Muchowski, H.Q. Wu, Kynurenines in the mammalian brain: when physiology meets pathology, Nat. Rev. Neurosci. 13 (2012) 465–477.
[39] D.A. Butterfield, M. Perluigi, R. Sultana, Oxidative stress in Alzheimer's disease brain: new insights from redox proteomics, Eur. J. Pharmacol. 545 (2006) 39–50.

[40] J.D.O. Viana, M.B. Félix, M.D.S. Maia, V.D.L. Serafim, L. Scotti, M.T. Scotti, Drug discovery and computational strategies in the multi-target drugs era, Braz. J. Pharm. Sci. 54 (2018).
[41] R. Zhang, J. Lin, Y. Zou, X.J. Zhang, W.L. Xiao, Chemical space and biological target network of anti-inflammatory natural products, J. Chem. Inf. Model. 59 (2018) 66–73.
[42] V. Temml, D. Schuster, Computational studies on natural products for the development of multi-target drugs, in: Multi-Target Drug Design Using Chem-Bioinformatic Approaches, Humana Press, New York, NY, 2018, pp. 187–201.
[43] C. Zheng, J. Wang, J. Liu, M. Pei, C. Huang, Y. Wang, System-level multi-target drug discovery from natural products with applications to cardiovascular diseases, Mol. Divers. 18 (2014) 621–635.
[44] F.C. Huang, W.K. Chan, K.J. Moriarty, D.C. Zhang, M. Chang, W. He, K.T. Yu, A. Zilberstein, Novel cytokine release inhibitors. Part I: triterpenes, Bioorg. Med. Chem. Lett. 8 (1998) 1883–1886.
[45] R.W. Li, G.D. Lin, S.P. Myers, D.N. Leach, Anti-inflammatory activity of Chinese medicinal vine plants, J. Ethnopharmacol. 85 (2003) 61–67.
[46] M.C. Fadus, C. Lau, J. Bikhchandani, H.T. Lynch, Curcumin: An age-old anti-inflammatory and anti-neoplastic agent, J. Tradit. Complement. Med. 7 (2017) 339–346.
[47] G. Ramesh, W.B. Reeves, TNF-α mediates chemokine and cytokine expression and renal injury in cisplatin nephrotoxicity, J. Clin. Investig. 110 (2002) 835–842.
[48] J.W. Cho, K.S. Lee, Anti-inflammatory effect of curcumin on UVB-induced inflammatory cytokines in HaCaT cells, Korean J. Dermatol. 47 (2009) 121–126.
[49] J.W. Cho, K. Park, G.R. Kweon, B.C. Jang, W.K. Baek, M.H. Suh, C.W. Kim, K.S. Lee, S.I. Suh, Curcumin inhibits the expression of COX-2 in UVB-irradiated human keratinocytes (HaCaT) by inhibiting activation of AP-1: p38 MAP kinase and JNK as potential upstream targets, Exp. Mol. Med. 37 (2005) 186–192.
[50] S. Shishodia, T. Singh, M.M. Chaturvedi, Modulation of transcription factors by curcumin, Adv. Exp. Med. Biol. 595 (2007) 127–148.
[51] R. Pangestuti, T.S. Vo, D.H. Ngo, S.K. Kim, Fucoxanthin ameliorates inflammation and oxidative reponses in microglia, J. Agric. Food Chem. 61 (2013) 3876–3883.
[52] H.L. Pahl, B. Krauss, K. Schulze-Osthoff, T. Decker, E.B.M. Traenckner, M. Vogt, C. Myers, T. Parks, P. Warring, A. Mühlbacher, A.P. Czernilofsky, The immunosuppressive fungal metabolite gliotoxin specifically inhibits transcription factor NF-kappa B, J. Exp. Med. 183 (1996) 1829–1840.
[53] X. Lin, C.H. Lin, T. Zhao, D. Zuo, Z. Ye, L. Liu, M.T. Lin, Quercetin protects against heat stroke-induced myocardial injury in male rats: Antioxidative and antiinflammatory mechanisms, Chem. Biol. Interact. 265 (2017) 47–54.
[54] T. Yang, B. Kong, J.W. Gu, Y.Q. Kuang, L. Cheng, W.T. Yang, X. Xia, H.F. Shu, Anti-apoptotic and anti-oxidative roles of quercetin after traumatic brain injury, Cell. Mol. Neurobiol. 34 (2014) 797–804.
[55] L.F. Vázquez-Flores, S. Casas-Grajales, E. Hernández-Aquino, E.E. Vargas-Pozada, P. Muriel, Antioxidant, antiinflammatory, and antifibrotic properties of quercetin in the liver, in: Liver Pathophysiology, Academic Press, 2017, pp. 653–674.
[56] C.M. Liu, J.Q. Ma, W.R. Xie, S.S. Liu, Z.J. Feng, G.H. Zheng, A.M. Wang, Quercetin protects mouse liver against nickel-induced DNA methylation and inflammation associated with the Nrf2/HO-1 and p38/STAT1/NF-κB pathway, Food Chem. Toxicol. 82 (2015) 19–26.
[57] S. Pisonero-Vaquero, Á. Martínez-Ferreras, M.V. García-Mediavilla, S. Martínez-Flórez, A. Fernández, M. Benet, J.L. Olcoz, R. Jover, J. González-Gallego, S. Sánchez-Campos, Quercetin ameliorates dysregulation of lipid metabolism

genes via the PI3K/AKT pathway in a diet-induced mouse model of nonalcoholic fatty liver disease, Mol. Nutr. Food Res. 59 (2015) 879–893.
[58] E. Marcolin, B. San-Miguel, D. Vallejo, J. Tieppo, N. Marroni, J. González-Gallego, M.J. Tuñón, Quercetin treatment ameliorates inflammation and fibrosis in mice with nonalcoholic steatohepatitis, J. Nutr. 142 (2012) 1821–1828.
[59] W.Y. Chen, S.Y. Lin, H.C. Pan, S.L. Liao, Y.H. Chuang, Y.J. Yen, S.Y. Lin, C.J. Chen, Beneficial effect of docosahexaenoic acid on cholestatic liver injury in rats, J. Nutr. Biochem. 23 (2012) 252–264.
[60] J.Q. Ma, Z. Li, W.R. Xie, C.M. Liu, S.S. Liu, Quercetin protects mouse liver against CCl4-induced inflammation by the TLR2/4 and MAPK/NF-κB pathway, Int. Immunopharmacol. 28 (2015) 531–539.
[61] M.C. Chiang, C.J. Nicol, Y.C. Cheng, Resveratrol activation of AMPK-dependent pathways is neuroprotective in human neural stem cells against amyloid-beta-induced inflammation and oxidative stress, Neurochem. Int. 115 (2018) 1–10.
[62] J.K. Kundu, K.S. Chun, S.O. Kim, Y.J. Surh, Resveratrol inhibits phorbol ester-induced cyclooxygenase-2 expression in mouse skin: MAPKs and AP-1 as potential molecular targets, Biofactors 21 (2004) 33–39.
[63] W.J. Chen, J.K. Du, X. Hu, Q. Yu, D.X. Li, C.N. Wang, X.Y. Zhu, Y.J. Liu, Protective effects of resveratrol on mitochondrial function in the hippocampus improves inflammation-induced depressive-like behaviour, Physiol. Behav. 182 (2017) 54–61.
[64] S. Im Jeong, J.A. Shin, S. Cho, H.W. Kim, J.Y. Lee, J.L. Kang, E.M. Park, Resveratrol attenuates peripheral and brain inflammation and reduces ischemic brain injury in aged female mice, Neurobiol. Aging 44 (2016) 74–84.
[65] M. Chen, P. Hou, M. Zhou, Q. Ren, X. Wang, L. Huang, S. Hui, L. Yi, M. Mi, Resveratrol attenuates high-fat diet-induced non-alcoholic steatohepatitis by maintaining gut barrier integrity and inhibiting gut inflammation through regulation of the endocannabinoid system, Clin. Nutr. 39 (2020) 1264–1275.
[66] S.S. Gocmez, T.D. Şahin, Y. Yazir, G. Duruksu, F.C. Eraldemir, S. Polat, T. Utkan, Resveratrol prevents cognitive deficits by attenuating oxidative damage and inflammation in rat model of streptozotocin diabetes induced vascular dementia, Physiol. Behav. 201 (2019) 198–207.
[67] Y. Deng, W. Gong, Q. Li, X. Wu, L. Wu, X. Zheng, W. Chen, H. Huang, Resveratrol inhibits high glucose-induced activation of AP-1 and NF-κB via SphK1/S1P2 pathway to attenuate mesangial cells proliferation and inflammation, J. Funct. Foods 55 (2019) 86–94.
[68] W.M. Lin, Y.M. Zhang, R. Moldzio, W.D. Rausch, Ginsenoside Rd attenuates neuroinflammation of dopaminergic cells in culture, in: M. Gerlach, J. Deckert, K. Double, E. Koutsilieri (Eds.), Neuropsychiatric Disorders An Integrative Approach. Journal of Neural Transmission. Supplemental, Vol. 72, Springer, Vienna, 2007.
[69] J.F. Hu, X.Y. Song, S.F. Chu, J. Chen, H.J. Ji, X.Y. Chen, Y.H. Yuan, N. Han, J.T. Zhang, N.H. Chen, Inhibitory effect of ginsenoside Rg1 on lipopolysaccharide-induced microglial activation in mice, Brain Res. 1374 (2011) 8–14.
[70] H. Song, J. Park, K. Choi, J. Lee, J. Chen, H.J. Park, B.I. Yu, M. Iida, M.R. Rhyu, Y. Lee, Ginsenoside Rf inhibits cyclooxygenase-2 induction via peroxisome proliferator-activated receptor gamma in A549 cells, J. Ginseng Res. 43 (2019) 319–325.
[71] J.H. Choi, M. Jang, S.Y. Nah, S. Oh, I.H. Cho, Multi-target effects of Korean red ginseng in animal model of Parkinson's disease: antiapoptosis, antioxidant, anti-inflammation, and maintenance of blood–brain barrier integrity, J. Ginseng Res. 42 (2018) 379–388.
[72] F. Li, Y. Cao, Y. Luo, T. Liu, G. Yan, L. Chen, L. Ji, L. Wang, B. Chen, A. Yaseen, A.A. Khan, Two new triterpenoid saponins derived from the leaves of Panax ginseng and their antiinflammatory activity, J. Ginseng Res. 43 (2019) 600–605.

[73] T. Rho, H.W. Jeong, Y.D. Hong, K. Yoon, J.Y. Cho, K.D. Yoon, Identification of a novel triterpene saponin from Panax ginseng seeds, pseudoginsenoside RT8, and its antiinflammatory activity, J. Ginseng Res. 44 (2020) 145–153.
[74] M. Barbosa, G. Lopes, P.B. Andrade, P. Valentão, Bioprospecting of brown seaweeds for biotechnological applications: Phlorotannin actions in inflammation and allergy network, Trends Food Sci. Technol. 86 (2019) 153–171.
[75] W.A.J.P. Wijesinghe, G. Ahn, W.W. Lee, M.C. Kang, E.A. Kim, Y.J. Jeon, Anti-inflammatory activity of phlorotannin-rich fermented Ecklonia cava processing by-product extract in lipopolysaccharide-stimulated RAW 264.7 macrophages, J. Appl. Phycol. 25 (2013) 1207–1213.
[76] S.Y. Kim, E.A. Kim, M.C. Kang, J.H. Lee, H.W. Yang, J.S. Lee, T.I. Lim, Y.J. Jeon, Polyphenol-rich fraction from Ecklonia cava (a brown alga) processing by-product reduces LPS-induced inflammation in vitro and in vivo in a zebrafish model, Algae 29 (2014) 165–174.
[77] Y.I. Yang, J.H. Woo, Y.J. Seo, K.T. Lee, Y. Lim, J.H. Choi, Protective effect of brown alga phlorotannins against hyper-inflammatory responses in lipopolysaccharide-induced sepsis models, J. Agric. Food Chem. 64 (2016) 570–578.
[78] B. Ryu, Y. Li, Z.J. Qian, M.M. Kim, S.K. Kim, Differentiation of human osteosarcoma cells by isolated phlorotannins is subtly linked to COX-2, iNOS, MMPs, and MAPK signaling: implication for chronic articular disease, Chem. Biol. Interact. 179 (2009) 192–201.
[79] H.C. Shin, H.J. Hwang, K.J. Kang, B.H. Lee, An antioxidative and antiinflammatory agent for potential treatment of osteoarthritis from Ecklonia cava, Arch. Pharm. Res. 29 (2006) 165–171.
[80] R. Wei, M.S. Lee, B. Lee, C.W. Oh, C.G. Choi, H.R. Kim, Isolation and identification of anti-inflammatory compounds from ethyl acetate fraction of Ecklonia stolonifera and their anti-inflammatory action, J. Appl. Phycol. 28 (2016) 3535–3545.
[81] T.H. Kim, S.K. Ku, T. Lee, J.S. Bae, Vascular barrier protective effects of phlorotannins on HMGB1-mediated proinflammatory responses in vitro and in vivo, Food Chem. Toxicol. 50 (2012) 2188–2195.
[82] Y. Sugiura, Y. Kinoshita, M. Abe, N. Murase, R. Tanaka, T. Matsushita, M. Usiu, K. Hanaoka, M. Miyata, Suppressive effects of the diethyl ether fraction from a brown alga Sargassum fusiforme on allergic and inflammatory reactions, Fish. Res. 82 (2016) 369–377.
[83] Y. Sugiura, K. Nagayama, Y. Kinoshita, R. Tanaka, T. Matsushita, The anti-allergic effect of the ethyl acetate fraction from an *Ecklonia kurome* extract, Food Agr. Immunol. 26 (2015) 181–193.
[84] Y. Sugiura, R. Tanaka, H. Katsuzaki, K. Imai, T. Matsushita, The anti-inflammatory effects of phlorotannins from Eisenia arborea on mouse ear edema by inflammatory inducers, J. Funct. Foods 5 (2013) 2019–2023.
[85] Y. Sugiura, M. Usui, H. Katsuzaki, K. Imai, M. Kakinuma, H. Amano, M. Miyata, Orally administered phlorotannins from Eisenia arborea suppress chemical mediator release and cyclooxygenase-2 signaling to alleviate mouse ear swelling, Mar. Drugs 16 (2018) 267.
[86] Y. Sugiura, M. Usui, H. Katsuzaki, K. Imai, M. Miyata, Anti-inflammatory effects of 6,6′-bieckol and 6,8′-bieckol from Eisenia arborea on mouse ear swelling, J. Food Sci. Technol. 23 (2017) 475–480.
[87] M.C. Zaragozá, D. López, M.P. Sáiz, M. Poquet, J. Pérez, P. Puig-Parellada, F. Màrmol, P. Simonetti, C. Gardana, Y. Lerat, P. Burtin, C. Inisan, I. Rousseau, M. Besnard, M.T. Mitjavila, Toxicity and antioxidant activity in vitro and in vivo of two Fucus vesiculosus extractsJ, J. Agric. Food Chem. 56 (2008) 7773–7780.
[88] M. Dutot, R. Fagon, M. Hemon, P. Rat, Antioxidant, anti-inflammatory, and anti-senescence activities of a phlorotannin-rich natural extract from brown seaweed *Ascophyllum nodosum*, Appl. Biochem. Biotechnol. 167 (2012) 2234–2240.

[89] J. Kellogg, D. Esposito, M.H. Grace, S. Komarnytsky, M.A. Lila, Alaskan seaweeds lower inflammation in RAW 264.7 macrophages and decrease lipid accumulation in 3T3-L1 adipocytes, J. Funct. Foods 15 (2015) 396–407.
[90] W.K. Jung, S.J. Heo, Y.J. Jeon, C.M. Lee, Y.M. Park, H.G. Byun, Y.H. Choi, S.G. Park, I.W. Choi, Inhibitory effects and molecular mechanism of dieckol isolated from marine brown alga on COX-2 and iNOS in microglial cells, J. Agric. Food Chem. 57 (2009) 4439–4446.
[91] Y.I. Yang, H.C. Shin, S.H. Kim, W.Y. Park, K.T. Lee, J.H. Choi, 6,6′-Bieckol, isolated from marine alga Ecklonia cava, suppressed LPS-induced nitric oxide and PGE-production and inflammatory cytokine expression in macrophages: the inhibition of NF-κB, Int. Immunopharmacol. 12 (2012) 510–517.
[92] Y.I. Yang, S.H. Jung, K.T. Lee, J.H. Choi, 8,8′-Bieckol, isolated from edible brown algae, exerts its anti-inflammatory effects through inhibition of NF-κB signaling and ROS production in LPS-stimulated macrophages, Int. Immunopharmacol. 23 (2014) 460–468.
[93] A.R. Kim, M.S. Lee, T.S. Shin, H. Hua, B.C.J. Jang, S. Choi, D.S. Byun, T. Utsuki, D. Ingram, H.R. Kim, Phlorofucofuroeckol A inhibits the LPS-stimulated iNOS and COX-2 expressions in macrophages *via* inhibition of NF-κB, Akt, and p38 MAPK, Toxicol. In Vitro 25 (2011) 1789–1795.
[94] D.K. Yu, B. Lee, M. Kwon, N. Yoon, T. Shin, N.G. Kim, J.S. Choi, H.R. Kim, Phlorofucofuroeckol B suppresses inflammatory responses by down-regulating nuclear factor κB activation via Akt, ERK, and JNK in LPS-stimulated microglial cells, Int. Immunopharmacol. 28 (2015) 1068–1075.
[95] S.H. Eom, E.H. Lee, K. Park, J.Y. Kwon, P.H. Kim, W.K. Jung, Y.M. Kim, Eckol from Eisenia bicyclis inhibits inflammation through the Akt/NF-κB signaling in Propionibacterium acnes-induced human keratinocyte Hacat cells, J. Food Biochem. 41 (2017), e12312.
[96] Z. Pan, Y. Zhou, X. Luo, Y. Ruan, L. Zhou, Q. Wang, Y.J. Yan, Q. Liu, J. Chen, Against NF-κB/thymic stromal lymphopoietin signaling pathway, catechin alleviates the inflammation in allergic rhinitis, Int. Immunopharmacol. 61 (2018) 241.
[97] A. Shahid, R. Ali, N. Ali, S.K. Hasan, P. Bernwal, S.M. Afzal, A. Vafa, S. Sultana, Modulatory effects of catechin hydrate against genotoxicity, oxidative stress, inflammation and apoptosis induced by benzo(a)pyrene in mice, Food Chem. Toxicol. 92 (2016) 64–74.
[98] J.K. Hodges, J. Zhu, Z. Yu, Y. Vodovotz, G. Brock, G.Y. Sasaki, P. Dey, R.S. Bruno, Intestinal-level anti-inflammatory bioactivities of catechin-rich green tea: rationale, design, and methods of a double-blind, randomized, placebo-controlled crossover trial in metabolic syndrome and healthy adults, Contemp. Clin. Trials Commun. 17 (2020) 100495.
[99] S. Riegsecker, D. Wiczynski, M.J. Kaplan, S. Ahmed, Potential benefits of green tea polyphenol EGCG in the prevention and treatment of vascular inflammation in rheumatoid arthritis, Life Sci. 93 (2013) 307–312.
[100] S. Fechtner, A. Singh, M. Chourasia, S. Ahmed, Molecular insights into the differences in anti-inflammatory activities of green tea catechins on IL-1β signaling in rheumatoid arthritis synovial fibroblasts, Toxicol. Appl. Pharmacol. 329 (2017) 112–120.
[101] L. Gambhir, R. Checker, D. Sharma, M. Thoh, A. Patil, M. Degani, V. Gota, S.K. Sandur, Thiol dependent NF-κB suppression and inhibition of T-cell mediated adaptive immune responses by a naturally occurring steroidal lactone Withaferin A, Toxicol. Appl. Pharmacol. 289 (2015) 297–312.
[102] S. Tekula, A. Khurana, P. Anchi, C. Godugu, Withaferin-A attenuates multiple low doses of Streptozotocin (MLD-STZ) induced type 1 diabetes, Biomed. Pharmacother. 106 (2018) 1428–1440.

[103] R. Maitra, M.A. Porter, S. Huang, B.P. Gilmour, Inhibition of NFκB by the natural product Withaferin A in cellular models of cystic fibrosis inflammation, J. Inflamm. 6 (2009) 15.

[104] D.L. Palliyaguru, D.V. Chartoumpekis, N. Wakabayashi, J.J. Skoko, Y. Yagishita, S.V. Singh, T.W. Kensler, Withaferin A induces Nrf2-dependent protection against liver injury: role of Keap1-independent mechanisms, Free Radic. Biol. Med. 101 (2016) 116–128.

[105] L. Nonn, D. Duong, D.M. Peehl, Chemopreventive anti-inflammatory activities of curcumin and other phytochemicals mediated by MAP kinase phosphatase-5 in prostate cells, Carcinogenesis 28 (2007) 1188–1196.

[106] S. Sharma, K. Chopra, S.K. Kulkarni, J.N. Agrewala, Resveratrol and curcumin suppress immune response through CD28/CTLA-4 and CD80 co-stimulatory pathway, Clin. Exp. Immunol. 147 (2007) 155–163.

[107] C.C. Chuang, W. Shen, H. Chen, G. Xie, W. Jia, S. Chung, M.K. McIntosh, Differential effects of grape powder and its extract on glucose tolerance and chronic inflammation in high-fat-fed obese mice, J. Agric. Food Chem. 60 (2012) 12458–12468.

[108] S. Nishiumi, R. Mukai, T. Ichiyanagi, H. Ashida, Suppression of lipopolysaccharide and galactosamine-induced hepatic inflammation by red grape pomace, J. Agric. Food Chem. 60 (2012) 9315–9320.

[109] R. Arcone, M. Palma, V. Pagliara, G. Graziani, M. Masullo, G. Nardone, Green tea polyphenols affect invasiveness of human gastric MKN-28 cells by inhibition of LPS or TNF-α induced matrix Metalloproteinase-9/2, Biochim. Open 3 (2016) 56–63.

[110] G. Chen, L. He, P. Zhang, J. Zhang, X. Mei, D. Wang, Y. Zhang, X. Ren, Z. Chen, Encapsulation of green tea polyphenol nanospheres in PVA/alginate hydrogel for promoting wound healing of diabetic rats by regulating PI3K/AKT pathway, Mater. Sci. Eng. C 110 (2020) 110686.

[111] M.K. Mazumder, S. Choudhury, Tea polyphenols as multi-target therapeutics for Alzheimer's disease: An in silico study, Med. Hypotheses 125 (2019) 94–99.

[112] A. Amalraj, K. Varma, J. Jacob, C. Divya, A.B. Kunnumakkara, S.J. Stohs, S. Gopi, A novel highly bioavailable curcumin formulation improves symptoms and diagnostic indicators in rheumatoid arthritis patients: a randomized, double-blind, placebo-controlled, two-dose, three-arm, and parallel-group study, J. Med. Food 20 (2017) 1022–1030.

[113] A. Amalraj, J. Jacob, K. Varma, A.B. Kunnumakkara, C. Divya, S. Gopi, Acujoint™, a highly efficient formulation with natural bioactive compounds, exerts potent anti-arthritis effects in human osteoarthritis—a pilot randomized double blind clinical study compared to combination of glucosamine and chondroitin, J. Herb. Med. 17-18 (2019) 100276.

Antiinflammatory activity of natural dietary flavonoids

Matheus Chiaradia de Souza, Ana Laura Tironi de Castilho, Carolina Mendes Tarran, Mateus Souza Zabeu, Leonardo de Liori Teixeira, and Ariane Leite Rozza

Department of Structural and Functional Biology, São Paulo State University (UNESP), Botucatu, SP, Brazil

5.1 Introduction

Inflammation is a response of the organism to the breakdown of tissue homeostasis [1]. It can be caused by pathogen infection (bacteria, fungi, or viruses), injury or tissue malfunction (cuts, burns, or autoimmune diseases), or even irritation caused by chemicals. The immediate and short-lived inflammatory response, also called acute, begins with increased vascular permeability and leukocyte migration [2], especially neutrophils that will try to eliminate pathogenic organisms by the release of reactive oxygen species (ROS), reactive nitrogen species (RNS), and some proteases [3]. In addition, neutrophils release inflammatory mediators (especially chemokines, such as interleukin-8—IL-8) that will attract macrophages to the site of inflammation, thereby increasing the inflammatory response.

These macrophages, when activated by recognition of pathogenic endotoxins such as lipopolysaccharides (LPS) through their toll-like receptors (TLR), increase their phagocytic activity [4] and produce a whole cascade of inflammatory mediators (prostaglandins and leukotrienes) and cytokines, such as interleukins IL-1, IL-6, and tumor necrosis factor-α (TNF-α) [5]. If the inflammatory agent is not eliminated, the chronic inflammatory response begins, which may last from months to years. The high concentration of macrophages and lymphocytes at the site of inflammation releases mediators that in the long term become harmful not only to the inflammatory agent but also to the person's tissues. It may even lead to the formation of granulomas, which are a nodule of inflammatory tissue composed of agglomerated macrophages and activated

Inflammation and Natural Products. https://doi.org/10.1016/B978-0-12-819218-4.00008-0
© 2021 Elsevier Inc. All rights reserved.

T lymphocytes [6]. Many diseases are related to the process of chronic inflammation, such as Alzheimer's diseases [7], rheumatoid arthritis [8], and atherosclerosis [9].

Among the conventional treatments, there are nonsteroidal anti-inflammatory drugs (NSAIDs), which basically act to inhibit cyclooxygenases (COX), enzymes that act in the synthesis of prostaglandins, important inflammatory mediators, and there are also drugs that inhibit the activity of cytokines [10]. Although they demonstrate effectiveness in treating inflammation, these treatments may have adverse effects, such as increased risk of gastrointestinal bleeding when using nonselective NSAIDs [11] and increase in cholesterol levels when using Tocilizumab (IL-6 inhibitor) together with other antirheumatic drugs [12].

The pharmaceutical industry is watching the advent of a people profile more concerned with conscious consumption in food and health, increasingly questioning the uncontrolled use of synthetic drugs and searching more natural alternatives. Currently the use of medicinal plants and herbal medicines is spread worldwide and encouraged by the World Health Organization (WHO), which published a series of monographs containing scientific information on medicinal plants with acknowledged medicinal benefits and the best means of using them [13–16]. Thus the importance of research searching for natural products as a medicinal alternative to conventional treatments is reinforced, aiming to minimize adverse effects and to avoid its recurrence, besides the cure of the disease. The antiinflammatory activities of medicinal plants are generally exerted through their secondary metabolites (organic compounds that do not directly contribute in the vital processes of the plant, such as growth, development, and proliferation). They play an important role in relation to the environment and can provide protection against low temperatures [17], ultraviolet radiation [18], air pollution [19], and predation by herbivores [20] among others.

There are three major classes of secondary metabolites: terpenes, phenolic compounds, and alkaloids. Flavonoids are phenolic compounds, and among their natural sources are fruits, vegetables, seeds, and some medicinal plants [21]. As the main natural sources of these classes, we have flavones (parsley and celery), flavanones (citrus fruits), flavonols (onions, broccoli, and blueberries), isoflavones (soybeans), flavanols (apricots and tea), and anthocyanidins (cherry and blackberries) [22, 23]. There is no standardized method of extracting flavonoids from their natural sources, so each sample should be analyzed in a particular way [24]. Structurally, define the flavonoids as having a common structure of C6-C3-C6, where both extremities are aromatic rings interconnected by a linear chain of three carbons [25]. According to the difference in C3 structure, flavonoids can be divided into six main classes: flavones, flavanones, flavonols, isoflavones, flavanols, or catechins and anthocyanidins. Several studies have demonstrated a wide diversity of biological activities of flavonoids, such as antioxidant, antimicrobial, antiviral, anticarcinogenic, and antiinflammatory (which is the focus of this chapter). It is also known that antiinflammatory activity of flavonoids can occur by mechanisms similar to that of NSAIDs or by inhibition of proinflammatory cytokines [26].

Thus this chapter approached the antiinflammatory potential of flavonoids from the natural diet, considering WHO data about the top causes of death in 2016 [27], focusing on the major diseases that caused death (cardiovascular diseases, diabetes, and gastrointestinal diseases). Scientific publications covering the period from 2010 to 2019 were considered.

5.2 Flavonoids and cardiovascular diseases

According to the World Health Organization, cardiovascular diseases make up for the most common causes of death all around the world, with ischemic heart disease alone killing over 9 million people in 2016. This statement is true to all countries, except for those with very low-income rates, where infectious diseases, nutritional or neonatal deficiencies, and vertical transmitted pathogens still lead the statistics [27]. Natural products, mostly extracted or derived from compounds found in plants, have been used for the purpose of treating or aiding in treatment of numerous diseases for centuries, either in the form of teas, functional foods, herbal supplements, or semisynthetic drugs. For cardiovascular diseases, scientists are searching for natural molecules with effects on oxidative stress, inflammation, and intermediary cytokines for the treatment or protection of the cardiovascular system.

A flavonoid that was found to regulate key factors in cardiovascular diseases is baicalin, an active compound found in the radix of *Scutellaria baicalensis*, which has been used in traditional Chinese medicine for centuries. Chen et al. [28] have tested the protective effect of baicalin on isoproterenol-induced acute myocardial infarction in male Wistar rats. Baicalin treatment induced a decrease in the infarct size in cardiac muscle and diminished inflammatory mediators, such as iNOS protein expression and the levels of TNF-α and IL-6. Baicalin also showed an antioxidant effect, by reducing the levels of SOD and MDA. These results evidenced that baicalin is a promising molecule to be used in the prevention of myocardial infarction.

In a similar way, Li et al. [29] investigated the cardioprotective effect of puerarin-V, a newly developed crystal form of puerarin. Puerarin is found in the roots of the Chinese herb kudzu. As a result, puerarin-V avoided mice mortality and the upregulation of inflammatory cytokines TNF-α, IL-1β, and IL-6, which was associated with the normalization in gene expression of PPAR-γ and PPAR-γ/NF-κB phosphorylation. The decrease of inflammatory cytokine production was also observed in vitro, using human coronary artery endothelial cells. Puerarin-V is an option for the treatment of myocardial infarction.

Puerarin has also been tested in clinical studies. Zhang et al. [30] investigated the effects of puerarin in coronary artery disease patients with stable angina pectoris. There was a control group, receiving the conventional treatment (nitrate, β-receptor blockers, calcium antagonists, statins, and aspirin), while the treatment group received additional daily doses of puerarin. The treatment with puerarin decreased the duration of angina pectoris and improved parameters of the electrocardiogram. The levels of TNF-α, IL-6, and hypersensitive C-reactive protein in the serum were decreased by puerarin. The combination of decreasing inflammatory factors and improving clinical symptoms evidenced the cardioprotective effect of puerarin.

The effects of quercetin in cardiovascular disease have also been studied. Quercetin can be found in many daily foods and beverages, such as wine, tea, onions, and berries. Quercetin has been tested in patients with coronary artery disease [31]. These patients presented high levels of IL-1β, TNF-α, and IL-10, but these levels were decreased after oral treatment with quercetin. The expression of IkBα gene, which is involved in the inflammation cascade, was also decreased by the treatment. There was a clinical improvement in these patients, which can be explained by the antiinflammatory effects of quercetin.

5.3 Flavonoids and diabetes mellitus

Diabetes mellitus is a disease that may occur in any age and affects more than 300 million people worldwide. It is characterized by high glucose levels in the blood, which results from the nonsecretion of insulin or insulin resistance [32]. In recent studies, oxidative stress and inflammation were recognized as one of the most important causes of diabetes complications [33], both being caused by high glucose levels and free fatty acids in the blood. This hyperglycemic state leads to quickly formation of advanced glycation end products (AGEs) that are responsible for severe complications associated with it, such as nephropathy, arteriosclerosis, retinopathy, and neuropathy [34]. Moreover the current treatments lead to several side effects; thus alternative treatments have been studied, like using the extract of plants to reduce complications associated with hyperglycemia or to reduce the hyperglycemia itself. In this context, flavonoids have been highlighted in many studies in vivo and in vitro that prove their efficiency in the treatment of inflammatory state caused by diabetes.

Diabetic liver injury is a diabetes consequence, and it is characterized by inflammatory responses, liver fibrosis, and lipid accumulation. Thus antiinflammatory agents stood out like possible treatment of it. In this context, Yin et al. [35] tested the antiinflammatory effects of baicalein, a flavonoid isolated from *Scutellariae radix*, in diabetic mice and in cell culture. For in vivo tests, C57BL/KsJ-db/db mice were used and proinflammatory cytokynes IL-1β and IL-6, and TNF-α levels were measured using ELISA kits. In vitro tests were made with HepG2 cells culture, which had its viability carried out by MTT experiments, and it was analyzed using Western blot to quantify effects of baicalein on HMGB1/TLR4/NF-κB signaling pathway. Firstly, oral glucose tests showed that glucose levels decreased in the group treated with baicalein. Besides that, histological analyses suggest that these treatments effectively inhibit lipid accumulation in liver. The antiinflammatory effect was evaluated in both tests, in vitro and in vivo, since mice treated with baicalein had decreased IL-1β, IL-6, and TNF-α levels compared with those in model mice. In the HepG2 cells, these levels were decreased after treatment with different concentrations of baicalein. Another relevant point is that the pathway HMGB1/TLR4/NF-Kb that participates in inflammatory signaling cascade by inducing the production of proinflammatory cytokines [36] had HMGB1, TLR4, Myd88, NF-κB, and IκBi expression blocked by the treatment with baicalein. In the vitro assay the treatment reversed the HMGB1, TLR4, Myd88, NF-κB, and IκB increase that happened in control group. Based on this result, authors compared baicalein with metformin, a drug currently used to treat inflammatory effects of diabetes, which affect HMGB1/TLR4/NF-κB pathway in diabetic db/db mice.

Gaur et al. [37] tested the rhizome of *Glycyrrhiza glabra* L. (Fabaceae) in diabetic mice. The rhizome contains isoliquiritigenin (ISL) and liquiritigenin (LTG) being the first a precursor of flavonoids and the second a flavonoid itself. These compounds already had antidiabetic activity reported, but this study was the first done with tests in vivo. Both compounds were isolated from chloroform/ethyl acetate fraction of the rhizomes. The test used Swiss albino male mice with diabetes mellitus induced by streptozotocin (type I diabetes), and the results showed that isoliquiritigenin and their precursors were able to decrease the glucose levels in the blood, acting as antihyperglycemic agents. Moreover the lipid levels in diabetic mice are commonly increased, but, in this test, the diabetic mice presented normal levels. Other symptom from diabetes mellitus is a severe loss in body weight caused by the increased muscle destruction or degradation of structural proteins [38], but the animals have body weight gain. Hence, this extract was shown as a candidate to develop new drugs to treat diabetes.

Wang et al. [39] investigated the antiinflammatory effect of formononetin, an isoflavone present majorly in the bioactive extracted of

Trifolium pratense L. (red clover). The tests were made in streptozotocin-induced diabetic male mice on cognitive impairment to prove extract antiinflammatory effect and in vitro using glycyrrhizin in high glucose-stimulated SH-SY5Y cells, to prove high-mobility group box 1 (HMGB1) importance in inflammatory cascades. Not only the causes of cognitive disorder in diabetics are not totally understood, but also several studies support that overexpression of cytokines related with inflammation, such as interleukin (IL)-1β, tumor necrosis factor (TNF)-α, and interleukin (IL)-6 is associated with it. High-mobility group box 1 (HMGB1) is responsible to initiate the cascade that phosphorylates nuclear factor kappa B (NF-κB), translocating it to nucleus, promoting the expression of proinflammatory cytokines [40], thus inducing inflammation. Moreover, previous studies demonstrate that NLRP3 is related to accelerate cognitive impairment [30]. In this study, levels of IL-6, IL-1β, and TNF-α in serum and hippocampus were measured using ELISA kits, and results showed that the diabetic mice had its levels increased, as expected, and the group treated with formononetin had its levels decreased, suggesting its antiinflammatory effect. Western blot was used to elucidate HMGB1/TLR4/NF-κB pathway, and the results proved that HMGB1 have an important role in this pathway. This research provides evidences that formononetin can slow down cognitive impairments by decreasing proinflammatory cytokines.

Studies using nicotinamide/streptozotocin-induced male athymic nude mice were responsible to prove the antidiabetic effect of chrysin [41]. Enzymatic-photocolorimetric methods for glucose and lipid quantification showed that 50 mg/kg of chrysin is able to reduce hyperglycemia as 120 mg/kg of metformin, a drug currently used to treat diabetes. Comparing these results with previous studies, authors concluded that chrysin acts as an insulin sensitizer [42, 43]. As diabetics has increased inflammatory state, tests were made to prove antiinflammatory chrysin effects, so this study used cytokine quantification assay to quantify the proinflammatory cytokines IL-1β, IL-6, and TNF-α serum concentration. The results showed that the production of IL-1β and TNF-α were reduced, while the IL-6 level was not diminished. So, this compound has ability as antiinflammatory by inhibiting the production of proinflammatory molecules. However, more tests have to be done to test the toxicity of this compound to classify it as a possible medication.

Puerarin is a natural isoflavone from *Pueraria lobata* and has already been cataloged with cardioprotective, neuroprotective, antioxidative, antiinflammatory, and many other effects. Eastern Asia countries already consume this plant as food and herb; however, its safety on human is not clarified yet. In animals, this herb shows nontoxic activity, and despite its low efficacy caused partially by pharmacokinetics profile, it is highlighted as a possible adjuvant to diabetes treatment and diabetes complication treatment, due its antiinflammatory effects. Puerarin also has registers of hypoglycemic, antioxidant, and inhibition of AGE formation [44]. Tests in vitro with LPS-induced Raw 264.7 macrophages proved puerarin as responsible for inhibiting, at mRNA and protein level expression, iNOS, COX-2, and C-reactive protein and responsible for inhibiting NF-κB pathway, a mediator of the inflammatory cascade [45]. Other inflammatory processes as ERK phosphorylation and mRNA expression of TNF-α and iNOS were inhibited using puerarin, this time testing it in free fatty acid (FFA)-induced Raw 264.7 macrophages [46]. In LPS-primed ARPE-19 cells, puerarin activated Nrf2/HO-1 and inhibit IRE1 and PERK phosphorylation and ATF6α nuclear expression, which led to relevant reverse amyloid β (AβÞ1-40-induced NLRP3) inflammasome activation [47]. Tests using STZ-induced diabetic rat models also proved antiinflammatory effects, which happened probably because of the inhibition NADPH oxidase-derived ROS generation and NF-κB activation. In this context, further studies are necessary to validate puerarin antidiabetic activity in humans but with the ancient studies give then a good direction.

Further studies showed that microalgae interfere in carbohydrate metabolism [48]. *Enteromorpha prolifera*, green algae, has already been used as healthcare food in East Asia, which led to studies on physicochemical and functional effects. This study validated *E. prolifera*'s ability to regulate glucose metabolism and its antiinflammatory, antiviral, anticoagulant, and antioxidant activities; thereby the authors used water-ethanol extract from green macroalgae (EPW) and its flavonoid-rich fraction less than 3 kDa (EPW3) to test its antidiabetic effect in streptozotocin-induced diabetes Kunming male ICR mice on a high-sucrose/high-fat diet through investigation of EPW effect on glucose metabolism-related mechanism. Six animals were randomly chosen as normal group, and diabetes were induced in other 18 mice, which were divided in model group, EPW-treated group and EPW3-treated group. Metabolites present in the EPW3 were detected with liquid chromatographic and mass spectrographic analysis, and it showed a high content of flavonoids. The animals had its body weight measured, and, in the beginning, treated group had lowered body weight compared with normal group, but after 60 days, EPW/EPW3 had no alteration compared with model group. Regarding glucose levels on mice's blood, model group had it increased, which means that diabetes was successfully established. EPW and EPW3 group had increased blood glucose levels, and, after 2 weeks, the hyperglycemic state was attenuated. The glucose tolerance in model group was similar to that on EPW-treated group, and both groups showed apparent low glucose tolerance. In contrast, EPW3 were able to alleviate the abnormal glucose tolerance of diabetic mice. The study used histopathology to analyze injuries of liver and kidneys, and the results showed that model group presents abnormality on liver cells, while EPW and EPW3 groups didn't present histopathologic chances. Diabetic mouse kidney also presents histopathological alterations, as elevation of inflammatory cells foci, but it was lessen in EPW and EPW3 group, especially in EPW3. In contrast, other changes, such as thicken of glomerular basilar, were not ameliorated with the treatment. Moreover, qPCR was used to quantify insulin signal pathways to better understanding EPW antihyperglycemic effect, and the results proved that the extract was able to regulate insulin signaling pathway by activation of IRS1/PI3K/AKT and inhibition of JNK1/2. *E. prolifera* extracts also activated PI3K/Akt and inhibited JNK signaling pathways, which could contribute to the improvement of glucose uptake in liver. The results supported the hypoglycemic activity of green seaweed *E. prolifera* flavonoids, suggesting that EPW and EPW3 are good candidates to diabetes therapy. As exposed, many kinds of flavonoids present antidiabetic effect by different pathways or mode of action, and the majority of studies affirmed that the antidiabetic effect is caused or influenced by the antiinflammatory action of flavonoids.

5.4 Flavonoids and gastrointestinal diseases

Licoflavone is the major compost present in *Glycyrrhiza* spp., a dry root commonly used in traditional Chinese medicine, which was identified as having antiinflammatory properties. Yang et al. [49] studied this natural medicine's potential; the aim of his study was to clarify the possible mechanisms of licoflavone in gastric ulcer treatment. Sprague-Dawley rats were submitted to acetic acid-induced gastric ulcer. After 3 days the rats were randomly divided into seven groups (control, model, ranitidine, omeprazole, high dose, middle dose, and low dose of licoflavone) and were orally treated during 7 days. The authors concluded that licoflavone could cure gastric ulcer, especially the middle dose group, and that these results could be related to the metabolites found in the study and their respective pathways; the vast majority of them were associated with inflammatory mediators and amino acids. Since numerous studies have already shown that ulcer's development and treatment could be associated with inflammation, the reduction of inflammatory mediators releases configures as an effective measure for the treatment.

Camellia japonica is an Asian plant whose fruits are used in phytomedicine to treat several inflammatory diseases. Akanda et al. [50] aimed to elucidate the immunopharmacological activities of *C. japonica* fruit extract. Using phytochemical analysis the authors observed that flavonoids were the main compounds present in the fruits, more precisely quercetin, quercetin-3-*O*-glucoside, quercitrin, and kaempferol, besides flavonoid glycosides. The researchers treated one group of mice with *C. japonica* and the other one with ranitidine; after 1 h, they induced gastric ulcer in both groups using acidified ethanol; the third group (control) was treated with normal saline. It is known that inflammatory cytokines as TNF-α, IL-6, and IL-1β and the enzymes iNOS and COX-2 are frequently associated with gastric damage. Applying a qPCR analysis, they concluded that the gene expression level of TNF-α, IL-6, IL-1β, and iNOS was significantly downregulated in *C. japonica* and ranitidine group when compared with saline-treated group. Lastly, COX-2 expression was blocked in the group treated with *C. japonica*. Taken together, these results evidence an antiinflammatory activity of the fruit extract, which is due to the presence of flavonoids.

Arab et al. [51] developed a study analyzing the potential protective effects of diosmin, a natural citrus flavone, against ethanol-induced gastric ulcer in rats, using ethanol-induced gastric ulcer model. The results obtained by the researchers imply in a significant decrease in levels of mieloperoxidase (MPO), a marker of neutrophil infiltration, and TNF-α, a proinflammatory cytokine in the gastric tissue. There was also an increase in IL-10 activity, an antiinflammatory cytokine, besides the preservation of histological architecture of the stomach. Therefore the authors concluded that the antiinflammatory effect could make diosmin a successful natural medicine against gastric ulcer.

Inflammatory bowel diseases, such as Chron's disease and ulcerative colitis, are characterized as a prominent infiltration of inflammatory cells in the intestine, and there is no cure known yet. Abron et al. [52] developed a study using soy isoflavones 4′,5,7-trihydroxyisoflavone (genistein) in the treatment of dextran sodium sulfate (DSS)-induced colitis. One of the cells affected by this type of disease is macrophages, which can present two phenotypes: M1 and M2. M1 is a proinflammatory phenotype and therefore produces high levels of IL-12 and IL-23, and M2 is antiinflammatory profile, so it produces IL-10 and IL-13 cytokines. Mice orally received genistein during 14 days. Using an ELISA assay the researchers discovered that the levels of TNF-α, IL-6, MCP-1, and IL-1β presented a significant reduction as compared with the vehicle-treated group. The authors observed an increase in the number of M2 macrophages when comparing the genistein-treated group with the vehicle-treated group. In conclusion, genistein may be useful for the treatment of IBD.

He et al. [53] studied the effects of dried fruits of *Citrus aurantium* L. in spite of elucidating its regulatory effects on inflammatory bowel disease and explain the mechanisms of its active compounds naringenin, nobiletin, and hesperetin, which are flavonoids. The animals treated with *C. aurantium* exhibited amelioration in regard of weight loss, diarrhea, and bloody stool, suggesting that the treatment may be responsible for relieving the effects of TNBS-induced acute colitis. Naringenin, nobiletin, and hesperetin showed antiinflammatory effects in vitro, modulating the activities of markers of inflammation, such as TNF-α, COX-2, iNOS, and NF-κB. These effects evidenced that these three flavonoids are promising molecules to be used in the treatment of IBD.

5.5 Conclusion and perspectives

The naturally occurring flavonoids discussed in this review exert antiinflammatory activity through distinct pathways, helping to maintain an integral health and fighting different diseases. These findings reinforce the importance of ethnopharmacological research, in preclinical and clinical studies, which could led to the development of a cheap, efficient and safe alternative to the treatment of inflammatory diseases.

References

[1] R. Medzhitov, Origin and physiological roles of inflammation, Nature 454 (2008) 428–435.

[2] A. Di Lorenzo, C. Fernández-Hernando, G. Cirino, W.C. Sessa, Akt1 is critical for acute inflammation and histamine-mediated vascular leakage, Proc. Natl. Acad. Sci. 106 (2009) 14552–14557.

[3] C. Nathan, Neutrophils and immunity: challenges and opportunities, Nat. Rev. Immunol. 6 (2006) 173–182.

[4] P.D. Smith, T.T. MacDonald, R.S. Blumberg, S. Richard, Society for Mucosal Immunology, Principles of Mucosal Immunology, Garland Science, 2013.

[5] C.A. Feghali, T.M. Wright, Cytokines in acute and chronic inflammation, Front. Biosci. 2 (1997) d12–d26.

[6] C.A. Bonham, M.E. Strek, K.C. Patterson, From granuloma to fibrosis: sarcoidosis associated pulmonary fibrosis, Curr. Opin. Pulm. Med. 22 (2016) 484–491.

[7] S.L. Gardener, S.R. Rainey-Smith, R.N. Martins, Diet and inflammation in Alzheimer's disease and related chronic diseases: a review, J. Alzheimers Dis. 50 (2015) 301–334.

[8] E. Siouti, E. Andreakos, The many facets of macrophages in rheumatoid arthritis, Biochem. Pharmacol. 165 (2019) 152–169.

[9] L. Groh, S.T. Keating, L.A.B. Joosten, M.G. Netea, N.P. Riksen, Monocyte and macrophage immunometabolism in atherosclerosis, Semin. Immunopathol. 40 (2018) 203–214.

[10] G. Schreiber, M.R. Walter, Cytokine-receptor interactions as drug targets, Curr. Opin. Chem. Biol. 14 (2010) 511–519.

[11] A. Lanas, L.A. García-Rodríguez, M.T. Arroyo, F. Gomollón, F. Feu, A. González-Pérez, E. Zapata, G. Bástida, L. Rodrigo, S. Santolaria, M. Güell, C.M. de Argila, E. Quintero, F. Borda, J.M. Piqué, Asociación Española de Gastroenterología, Risk of upper gastrointestinal ulcer bleeding associated with selective cyclo-oxygenase-2 inhibitors, traditional non-aspirin non-steroidal anti-inflammatory drugs, aspirin and combinations, Gut 55 (2006) 1731–1738.

[12] M.C. Genovese, J.D. McKay, E.L. Nasonov, E.F. Mysler, N.A. da Silva, E. Alecock, T. Woodworth, J.J. Gomez-Reino, Interleukin-6 receptor inhibition with tocilizumab reduces disease activity in rheumatoid arthritis with inadequate response to disease-modifying antirheumatic drugs: the tocilizumab in combination with traditional disease-modifying antirheumatic drug therapy study, Arthritis Rheum. 58 (2008) 2968–2980.
[13] WHO, Monographs on Selected Medicinal Plants, vol. 1, 1999.
[14] WHO, Monographs on Selected Medicinal Plants, vol. 2, 2003.
[15] WHO, Monographs on Selected Medicinal Plants, vol. 3, 2007.
[16] WHO, Monographs on Selected Medicinal Plants, vol. 4, 2009.
[17] X. Zhu, J. Liao, X. Xia, F. Xiong, Y. Li, J. Shen, B. Wen, Y. Ma, Y. Wang, W. Fang, Physiological and iTRAQ-based proteomic analyses reveal the function of exogenous γ-aminobutyric acid (GABA) in improving tea plant (*Camellia sinensis* L.) tolerance at cold temperature, BMC Plant Biol. 19 (2019) 43–54.
[18] S. Takshak, S.B. Agrawal, Defense potential of secondary metabolites in medicinal plants under UV-B stress, J. Photochem. Photobiol. B Biol. 193 (2019) 51–88.
[19] A. Singh, M. Agrawal, Effects of ambient and elevated CO2 on growth, chlorophyll fluorescence, photosynthetic pigments, antioxidants, and secondary metabolites of *Catharanthus roseus* (L.) G Don. grown under three different soil N levels, Environ. Sci. Pollut. Res. 22 (2015) 3936–3946.
[20] A.A. Agrawal, M.G. Weber, On the study of plant defence and herbivory using comparative approaches: how important are secondary plant compounds, Ecol. Lett. 18 (2015) 985–991.
[21] A.N. Panche, A.D. Diwan, S.R. Chandra, Flavonoids: an overview, J. Nutr. Sci. 5 (2016) e47–e58.
[22] C. Manach, A. Scalbert, C. Morand, C. Rémésy, L. Jiménez, Polyphenols: food sources and bioavailability, Am. J. Clin. Nutr. 79 (2004) 727–747.
[23] M.H. Pan, C.S. Lai, S. Dushenkov, C.T. Ho, Modulation of inflammatory genes by natural dietary bioactive compounds, J. Agric. Food Chem. 57 (2009) 4467–4477.
[24] C.D. Stalikas, Extraction, separation, and detection methods for phenolic acids and flavonoids, J. Sep. Sci. 30 (2007) 3268–3295.
[25] C. Santos-Buelga, A.S. Feliciano, Molecules flavonoids: from structure to health issues, Molecules 22 (2017) 477–481.
[26] N. Leyva-López, E. Gutierrez-Grijalva, D. Ambriz-Perez, J. Heredia, Flavonoids as cytokine modulators: a possible therapy for inflammation-related diseases, Int. J. Mol. Sci. 17 (2016) 921.
[27] World Health Organization, The Top 10 Causes of Death, 2018, Retrieved from http://www.who.int/news-room/fact-sheets/detail/the-top-10-causes-of-death.
[28] H. Chen, Y. Xu, J. Wang, W. Zhao, H. Ruan, Baicalin ameliorates isoproterenol-induced acute myocardial infarction through iNOS, inflammation and oxidative stress in rat, Int. J. Clin. Exp. Pathol. 8 (2015) 10139–10147.
[29] X. Li, T. Yuan, D. Chen, Y. Chen, S. Sun, D. Wang, L. Fang, Y. Lu, D. Du, Cardioprotective effects of puerarin-V on isoproterenol-induced myocardial infarction mice is associated with regulation of PPAR-gamma/NF-kB pathway, Molecules 23 (2018) 3322–3336.
[30] S. Zhang, L. Chen, Z. Zhou, W. Fan, S. Liu, Effects of Puerarin on clinical parameters, vascular endothelial function, and inflammatory factors in patients with coronary artery disease, Med. Sci. Monit. 25 (2019) 402–408.
[31] N. Chekalina, Y. Burmak, Y. Petrov, Z. Borisova, Y. Manusha, Y. Kazakov, I. Kaidashev, Quercetin reduces the transcriptional activity of NF-kB in stable coronary artery disease, Indian Heart J. 70 (2018) 593–597.
[32] American Diabetes Association (ADA), Classification and diagnosis of diabetes, Diab. Care 40 (2017) S11–S24.

[33] S.P. Palem, P. Abraham, A study on the level of oxidative stress and inflammatory markers in type 2 diabetes mellitus patients with different treatment modalities, J. Clin. Diagn. Res. 9 (2015) 4–7.
[34] A.J. Vieira, F.P. Beserra, M.C. Souza, B.M. Totti, A.L. Rozza, Limonene: aroma of innovation in health and disease, Chem. Biol. Interact. 283 (2018) 97–106.
[35] H. Yin, L. Huang, T. Ouyang, L. Chen, Baicalein improves liver inflammation in diabetic db/db mice by regulating HMGB1/TLR4/NF-κB signaling pathway, Int. Immunopharmacol. 55 (2018) 55–62.
[36] J.S. Park, F. Gamboni-Robertson, Q. He, D. Svetkauskaite, J.Y. Kim, D. Strassheim, J.W. Sohn, S. Yamada, I. Maruyama, A. Banerjee, A. Ishizaka, E. Abraham, High mobility group box 1 protein interacts with multiple Toll-like receptors, Am. J. Physiol. Cell Physiol. 290 (2006) C917–C924.
[37] R. Gaur, K.S. Yadav, R.K. Verma, N.P. Yadav, R.S. Bhakuni, In vivo anti-diabetic activity of derivatives of isoliquiritigenin and liquiritigenin, Phytomedicine 21 (2014) 415–422.
[38] S. Stephen Irudayaraj, C. Sunil, V. Duraipandiyan, S. Ignacimuthu, Antidiabetic and antioxidant activities of *Toddalia asiatica* (L.) Lam. leaves in Streptozotocin induced diabetic rats, J. Ethnopharmacol. 143 (2012) 515–523.
[39] J. Wang, L. Wang, J. Zhou, A. Qin, Z. Chen, The protective effect of formononetin on cognitive impairment in streptozotocin (STZ)-induced diabetic mice, Biomed. Pharmacother. 106 (2018) 1250–1257.
[40] D. Xueyang, M. Zhanqiang, M. Chunhua, H. Kun, Fasudil, an inhibitor of Rho-associated coiled-coil kinase, improves cognitive impairments induced by smoke exposure, Oncotarget 7 (2016) 78764–78772.
[41] J.J. Ramírez-Espinosa, J. Saldaña-Ríos, S. García-Jiménez, R. Villalobos-Molina, G. Ávila-Villarreal, A.N. Rodríguez-Ocampo, G. Bernal-Fernández, S. Estrada-Soto, Chrysin induces antidiabetic, antidyslipidemic and anti-inflammatory effects in athymic nude diabetic mice, Molecules 23 (2018) 2–9.
[42] M. Torres-Piedra, R. Ortiz-Andrade, R. Villalobos-Molina, N. Singh, J.L. Medina-Franco, S.P. Webster, M. Binnie, G. Navarrete-Vázquez, S. Estrada-Soto, A comparative study of flavonoid analogues on streptozotocin-nicotinamide induced diabetic rats: quercetin as a potential antidiabetic agent acting via 11β-hydroxysteroid dehydrogenase type 1 inhibition, Eur. J. Med. Chem. 45 (2010) 2606–2612.
[43] H.M. El-Bassossy, S.M. Abo-Warda, A. Fahmy, Chrysin and luteolin attenuate diabetes-induced impairment in endothelial-dependent relaxation: effect on lipid profile, AGEs and NO generation, Phytother. Res. 27 (2013) 1678–1684.
[44] X. Chen, J. Yu, J. Shi, Management of diabetes mellitus with Puerarin, a natural isoflavone from *Pueraria lobata*, Am. J. Chin. Med. 46 (2018) 1771–1789.
[45] W. Hu, X. Yang, C. Zhe, Q. Zhang, L. Sun, K. Cao, Puerarin inhibits iNOS, COX-2 and CRP expression via suppression of NF-κB activation in LPS-induced RAW264.7 macrophage cells, Pharmacol. Rep. 63 (2011) 781–789.
[46] Y. Tu, C.X. Gong, L. Ding, X.Z. Liu, T. Li, F.F. Hu, S. Wang, C.P. Xiong, S.D. Liang, H. Xu, A high concentration of fatty acids induces TNF-α as well as NO release mediated by the P2X4 receptor, and the protective effects of puerarin in RAW264.7 cells, Food Funct. 8 (2017) 4336–4346.
[47] K. Wang, X. Zhu, K. Zhang, Y. Yao, M. Zhuang, C. Tan, F. Zhou, L. Zhu, Puerarin inhibits amyloid β-induced NLRP3 inflammasome activation in retinal pigment epithelial cells via suppressing ROS-dependent oxidative and endoplasmic reticulum stresses, Exp. Cell Res. 357 (2017) 335–340.
[48] X. Yan, C. Yang, G. Lin, Y. Chen, S. Miao, B. Liu, C. Zhao, Antidiabetic potential of green seaweed *Enteromorpha prolifera* flavonoids regulating insulin signaling pathway and gut microbiota in type 2 diabetic mice, J. Food Sci. 84 (2019) 165–173.

[49] Y. Yang, S. Wang, Y.R. Bao, T.J. Li, G.L. Yang, X. Chang, X.S. Meng, Anti-ulcer effect and potential mechanism of licoflavone by regulating inflammation mediators and amino acid metabolism, J. Ethnopharmacol. 199 (2017) 175–182.
[50] M.R. Akanda, B.Y. Park, Involvement of MAPK/NF-κB signal transduction pathways: *Camellia japonica* mitigates inflammation and gastric ulcer, Biomed. Pharmacother. 95 (2017) 1139–1146.
[51] H.H. Arab, S.A. Salama, H.A. Omar, E.S.A. Arafa, I.A. Maghrabi, Diosmin protects against ethanol-induced gastric injury in rats: novel anti-ulcer actions, PLoS One 3 (2015) e0122417.
[52] J.D. Abron, N.P. Singh, R.L. Price, M. Nagarkatti, P.S. Nagarkatti, U.P. Singh, Genistein induces macrophage polarization and systemic cytokine to ameliorate experimental colitis, PLoS One 7 (2018) 0199631.
[53] W. He, Y. Li, M. Liu, H. Yu, Q. Chen, Y. Chen, J. Ruan, Z. Ding, Y. Zhang, T. Wang, *Citrus aurantium* L. and its flavonoids regulate TNBS-induced inflammatory bowel disease through anti-inflammation and suppressing isolated jejunum contraction, Int. J. Mol. Sci. 19 (2018) 3057–3071.

6

Antiinflammatory effects of turmeric (*Curcuma longa*) and ginger (*Zingiber officinale*)

Thahira Banu Azeez and Janeline Lunghar
School of Sciences, Department of Home Science, The Gandhigram Rural Institute-Deemed to be University, Gandhigram, Dindigul, Tamil Nadu, India

6.1 Introduction

The inflammatory process of the body gets encapsulated as a response to injury, infection, or irritation. As a result of damage or injury and exposure to bacteria and virus, it can also be a result of the ischemia, inadequate intake of nutritional factors, toxins, and exposure to extreme temperatures or radiation. The typical inflammation is identified by the signs of redness, swelling, pain, and heat in the body. However, other types of inflammatory processes occur internally in our body, and the kind of inflammation is referred to as the "silent inflammation," and this kind of inflammatory process occurs in the body with or without signs or symptoms perhaps for many years before they appear. Classically, inflammation is characterized by an increase in the blood flow, reddening of the affected part due to increased erythrocyte accumulation and edema [1].

Today, inflammation is far more complicated than might first appear from the simple description and is a significant responsibility of the immune system to tissue damage and infection. However, not all disease gives rise to inflammation. Inflammation ranges from the acute inflammation associated with *Staphylococcus aureus* infection of the skin (the boil), through to chronic inflammatory processes resulting in remodeling of the artery wall in atherosclerosis, the bronchial wall in asthma and chronic bronchitis, and the debilitating destruction of the joints associated with rheumatoid arthritis. These processes involve the significant cells of the immune system, including neutrophils, basophils, mast cells, T cells, and B-cells. Inflammation is a well-orchestrated response to harmful stimuli, including tissue injury and infection [2]. Seventy to eighty percent of the world's population depend on herbal sources in their primary health care [3, 4]. Especially

Inflammation and Natural Products. https://doi.org/10.1016/B978-0-12-819218-4.00011-0
© 2021 Elsevier Inc. All rights reserved.

in developing nations where the cost of consulting a physician and fee of medicine is beyond the edge of most people, thereby the demand is increasing day by day. These drugs are antiinflammatory and used to ease pain in various conditions, including arthritis, muscle, and ligament pains [5, 6]; among the full range of spices and herbs available, turmeric and ginger are tapped for its therapeutic property.

6.2 Turmeric

Turmeric (*Curcuma longa*) is a plant distributed throughout tropical and subtropical regions of the world and cultivated in Asian countries, mainly in China and India. Turmeric belongs to the family *Zingiberaceae*, and it has been traditionally used for centuries in Asia for medicinal purposes and in cuisine. India is the largest producer, consumer, and exporter of turmeric in terms of dried and ground tuber as a spice product. Turmeric is a plant that has a very long history of medicinal use, dating back nearly 4000 years. In Southeast Asia, turmeric is used not only as a principal spice but also as a component in religious ceremonies. Because of its brilliant yellow color (Fig. 6.1), turmeric is known as "Indian saffron."

6.2.1 Morphology and taxonomy of genus *Curcuma*

The genus *Curcuma* is morphologically highly variable for the different conventional taxonomic traits. *Curcuma* is perennial rhizomatous herbs, 50–200 cm tall, and the leaf shoot dying back during the dry period of tropical areas. Rhizomes are ovoid without branches or branched, fleshy, and aromatic. Rhizomes are usually light brown externally, but they can be of different shades of yellow, white, light to deep orange, bluish to deep blue, yellow with greenish borders internally [7]. India is the largest producer, followed by Thailand and

Fig. 6.1 Turmeric (*Curcumin longa*)

other significant producers like Southeast Asian countries, Central and Latin America, and Taiwan. The global production of turmeric is around 11 lakh tonnes per annum. India dominates the production in global scenario contributing 78%, followed by China (8%), Myanmar (4%) and Nigeria, and Bangladesh together, and contributing to 6% of the global production. India produces nearly the world's entire turmeric crop and consumes 80% of it. With its inherent qualities and high content of the essential bioactive compound curcumin, Indian turmeric is to be the best in the world. Erode, a city in the South Indian state of Tamil Nadu, is the world's largest producer of and the most important trading center for turmeric. It is also known as "Yellow City," "Turmeric City," or "Textile City." Sangli, a city of Maharashtra, is second only to Erode in size and importance as a production and trading site for turmeric.

Turmeric is one of the most useful herbal medicinal plants. Copious researches had proven that most of the turmeric activities are due to curcumin. The rhizome of turmeric is used in indigenous medicine since the early 1970s, and there is no doubt to the efficacy of turmeric, in its medicinal properties, especially of its antiinflammatory use.

The turmeric rhizomes have to undergo processing before being used. Rhizomes are boiled or steamed to remove the raw odor, gelatinize the starch, and produce a more uniformly colored product. In the traditional Indian process, rhizomes are placed in pans or earthenware filled with water and then covered with leaves and a layer of cow dung. The ammonia in the cow dung reacted with the turmeric to give the final product. For hygienic reasons, this method has been discouraged. In present-day processing, rhizomes are placed in shallow pans in large iron vats containing 0.05%–0.1% alkaline water (e.g., solution of sodium bicarbonate). The rhizomes then boiled for between 40 and 45 minutes or more depending on the variety. The rhizomes are removed from the water and dried in the sun immediately to prevent overcooking. The final moisture content should be between 8% and 10% (wet basis). The powder maintains its coloring properties indefinitely, although the flavor may diminish over time. Protecting the turmeric powder from sunlight retards the rate of deterioration.

6.2.2 Composition of turmeric

More than 100 components are isolated from turmeric. Volatile oil, turmerone, and the coloring pigment curcuminoid are the major components of the root. Curcuminoids consist of curcumin demethoxycurcumin, 5′-methoxycurcumin, and dihydrocurcumin, found to be natural antioxidants [8, 9]. In a standard form, turmeric contains moisture (> 9%), curcumin (5%–6.6%), extraneous matter (< 0.5% by weight), mold (< 3%), and volatile oils (< 3.5%). Volatile oils include D-α-phellandrene, D-sabinene, cineol, borneol, zingiberene, and sesquiterpenes [10]. The components responsible for the aroma of turmeric are turmerone, turmerone, and zingiberene. The rhizomes contain four new polysaccharides—ukonans, stigmasterol, β-sitosterol, cholesterol, and 2-hydroxymethyl anthraquinone [11, 12]. Turmeric is also a good source of the ω-3 fatty acid and α-linolenic acid. The nutrient composition of turmeric is given in Table 6.1

Indian Ayurvedic system of treating diseases used turmeric for antiinflammation before any western research study had ever done. Number of research studies on curcumin, a natural compound present in the rhizomes of plant *C. longa* Linn., demonstrated that it has antiinflammatory action [13]. A *Curcuma* paste made from powdered turmeric mixed with slaked lime or water and applied locally is an ancient household that gave a quick remedy for muscular pain, inflamed joints, and sprains.

Curcumin (1,7-bis(4-hydroxy-3-methoxyphenyl)-1,6-heptadiene-3,5-dione), also called diferuloylmethane, is the leading natural polyphenol found in the rhizome of *C. longa* (turmeric) and known as an antiinflammatory agent [14]. Curcumin, a polyphenol, has been shown to target multiple signaling molecules while also demonstrating activity at the cellular level, which has helped to support its numerous health benefits [15]. Curcumin has shown to benefit inflammatory conditions [16] metabolic syndrome [17] pain and to assist in the management of inflammatory and degenerative eye conditions [18, 19].

Table 6.1 Nutrient composition of turmeric/100 g.

Constituents	Quantity
Energy	390 kcal
Carbohydrates	69.9 g
Dietary fiber	21 g
Protein	8 g
Fat	10 g
Calcium	0.2 g
Phosphorus	0.26 g
Sodium	10 mg
Potassium	2.5 g
Iron	47.5 mg
Thiamine	0.9 mg
Riboflavin	0.19 mg
Niacin	4.8 mg
Ascorbic acid	50 mg

6.2.3 Mechanisms of action

Tumor necrosis factor-α (TNF-α) is a significant mediator of inflammation in most diseases and regulated by the activation of a transcription factor and nuclear factor (NF)-κB. Whereas TNF-α is said to be the most potent NF-κB activator, the expression of TNF-α is also regulated by NF-κB. In addition to TNF-α, NF-κB is stimulated by most inflammatory cytokines; gram-negative bacteria; various disease-causing viruses; environmental pollutants; chemical, physical, mechanical, and psychological stress; high glucose; fatty acids; ultraviolet radiation; cigarette smoke; and other disease-causing factors. Agents that downregulate NF-κB and NF-κB-regulated gene products have potential efficacy against several of these diseases. Curcumin has shown to block NF-κB activation that increased by several inflammatory stimuli [17]. Curcumin plays a vital role in pathological conditions. The antiinflammatory effect of curcumin has interfered through its ability to inhibit cyclooxygenase-2 (COX-2), lipoxygenase (LOX), and inducible nitric oxide synthase (iNOS) the enzymes involved in mediating the inflammatory processes. The pathology of certain types of human cancer and inflammatory disorders is associated with improper upregulation of COX-2 and iNOS. Curcumin, with its potent antiinflammatory property, exerts chemopreventive effects on carcinogenesis. Probable other mechanisms by which curcumin induces its antiinflammatory effects are through peroxisome proliferator–activated receptor gamma (PPAR-γ). PPARs belong to the superfamily of nuclear receptors consisting of three genes that give rise to three different subtypes, PPAR-α, PPAR-δ, and PPAR-γ. Among them, PPAR-γ is the most widely studied form. Upon ligand binding, PPAR-γ forms heterodimers with the retinoid X receptor and binds to a peroxisome proliferation response element (PPRE) in a gene promoter leading to regulation of gene transcription [20].

A recent study reported that gene and protein levels of PPAR-γ in the liver decreased by approximately 50% at 20 hours after the onset of sepsis. Pretreatment with curcumin for 3 days at 0.24 μmol/kg body weight in these septic rats produced 45% and 65% increase in PPAR-γ mRNA and protein levels, respectively. The mRNA and protein levels of PPAR-γ in the treatment group were similar to sham controls [21]. To confirm that the beneficial effect of curcumin in sepsis, mediated through the PPAR-γ pathway, a separate group of animals is treated for 3 days with PPAR-γ antagonist, GW9662, at 1.5 mg/kg along with curcumin at 0.24 μmol/kg body weight. Rats were subjected to sepsis by CLP and 20 hours after surgery, collected blood and tissue samples. Concurrent administration of curcumin and GW9662 in the septic rats completely abolished the effects of curcumin on serum levels of the liver enzymes, ALT and AST, lactate, and TNF-α [20]. Furthermore, in vitro using RAW 264.7 cells, pretreatment with 50- and 100-μM curcumin increased

PPAR-γ mRNA levels by 86% and 125%, respectively, compared with LPS treatment alone. Thus the beneficial effect of curcumin appears to be mediated by the upregulation of PPAR-γ [21].

Numerous studies have shown the importance of curcumin as a potent immunomodulatory agent in T cells, B cells, neutrophils, natural killer cells, dendritic cells, and macrophages [22]. Curcumin induces apoptosis in human neutrophils [23]. Neutrophils are the first line of host immune defense against foreign substances and regulate the biological activities by apoptosis. Delayed neutrophil apoptosis is associated with acute lung injury and sepsis [24–26]. The effect of curcumin on spontaneous neutrophil apoptosis by performing myeloperoxidase activity and migration assays was studied. The outcomes showed that curcumin increased constitutive neutrophil apoptosis and abrogated the transbilayer migration-induced delay in neutrophil apoptosis. Curcumin treatment decreased neutrophil migration and myeloperoxidase release indicating a reduction in neutrophil activation. A study on the effect of curcumin on p38 mitogen-activated protein kinase and caspase-3 activity showed a marked increase in p38 phosphorylation and caspase-3 activity in the presence of curcumin. Treatment of p38-specific inhibitor, SB203580, suppressed both curcumin-induced apoptosis and caspase-3 activation. From this study, that curcumin induces apoptosis in human neutrophil, and its effect mediated by the activation of p38 and caspase-3 activity is understood. Curcumin appears to block the synthesis of certain prostaglandins through inhibition of the COX enzyme [27].

6.2.4 Molecular mechanism and biochemical changes

Zhang et al. investigated whether curcumin inhibited chenodeoxycholate (CD) or phorbol ester (phorbol 12-myristate 13-acetate, PMA)-mediated induction of COX-2 in several gastrointestinal cell lines (SK-GT-4, SCC450, IEC-18, and HCA- 7) [28]. Treatment with curcumin suppressed CD- and PMA-mediated induction of COX-2 protein and synthesis of prostaglandin E2. Curcumin also suppressed the induction of COX-2 mRNA by CD and PMA [28]. Curcumin inhibited the cell growth of HT-29 cells in a level and time-dependent manner. There was a marked inhibition of mRNA and protein expression of COX-2, but not COX-1. Kim et al. demonstrated that the inhibitory action of curcumin on Janus kinase (JAK)-STAT signaling could contribute to its antiinflammatory activity in the brain [29]. In both rat primary microglia and murine BV2 microglial cells, curcumin effectively suppressed the ganglioside, lipopolysaccharide (LPS), or interferon (IFN-γ)-stimulated induction of COX-2 and inducible NO synthase, essential enzymes that mediate inflammatory methods. Curcumin markedly inhibited the phosphorylation of STAT1 and STAT3, as well as JAK1 and JAK2 in microglia, activated with gangliosides, LPS, or IFN-gamma, thus attenuating inflammatory response of microglial brain cells [29].

Several NSAIDs like ketoprofen and suprofen inhibit the release of lysosomal enzymes from the neutrophils [30]. The role of lysosomal enzymes, that is, acid phosphatase and cathepsin D as the mediator of inflammation, is well documented [31, 32]. A comparison of the effect of curcumin on stabilization of lysosomal enzymes and ibuprofen showed that curcumin (200 mg/kg) prevented the increase by 50% while ibuprofen (20 mg/kg) prevented it by 61%. In an in vitro study, curcumin had a more significant lysosomal membrane stabilization effect than ibuprofen [33]. Joe et al. demonstrated that curcumin and capsaicin lower the release of lysosomal enzymes and eicosanoids in rat peritoneal macrophages [34]. Oxidative stress implicated in many chronic diseases, and its pathological processes are closely related to those of inflammation, in that one can be easily induced by another. The inflammatory cells liberate several reactive species at the site of inflammation leading to oxidative stress [35].

Inflammation is identified in the development of many chronic diseases and conditions [17, 36, 37]. These diseases include Alzheimer's disease (AD), Parkinson's disease, multiple sclerosis, epilepsy, cerebral injury, cardiovascular disease, metabolic syndrome, cancer, allergy, asthma, bronchitis, colitis, arthritis, renal ischemia, psoriasis, diabetes, obesity, depression, fatigue, and acquired immune deficiency syndrome [17]. Turmeric, a yellow-colored phenolic pigment, has 70%–76% curcumin present along with 16% demethoxycurcumin and 8% bisdemethoxycurcumin. Extensive scientific researches on curcumin have demonstrated a broad spectrum of therapeutic effects as an antiinflammatory agent. Also, copious reports found that it also has potential healing in autoimmune deficiency syndrome (AIDS) [38–40]. Curcuminoids, namely, curcumin demethoxycurcumin and bisdemethoxycurcumin [41], are extensively studied, and the chemical properties have yielded jaw-dropping results on inflammatory effects and even inhibition of specific cancerous cells. Curcumin is a potent antiinflammatory agent with specific lipoxygenase- and COX-2-inhibiting properties. In vitro and in vivo studies have demonstrated its effects at decreasing both acute and chronic inflammation.

Curcumin has inhibited edema at doses between 50 and 200 mg/kg, in mice. A 50% reduction in edema achieved with a treatment of 48 mg/kg body weight, with curcumin nearly as effective as cortisone and phenylbutazone at similar doses. In rats a lower dose of 20–80 mg/kg decreased paw inflammation and edema. Curcumin also inhibited formaldehyde-induced arthritis in rats at a dose of 40 mg/kg and demonstrated no acute toxicity at doses up to 2 g/kg/day [42]. Animal study on the effect of curcumin on rheumatoid arthritis showed that intraperitoneal injection of turmeric extract containing 4-mg total curcuminoids/kg/day for 4 days before induction of arthritis inhibited joint inflammation in both acute (75%) and chronic (68%) phases. To test the efficacy of oral preparation, a 30-fold higher dose of the curcuminoid preparation, given to rats 4 days before arthritis induction, reduced joint inflammation by 48% [43].

The effect of curcumin on reducing inflammation of the lungs and the airways of asthmatic mice was studied. It decreased the response of specific proinflammatory cytokines with the activation of the Nrf2/HO-1 (nuclear factor erythroid 2–related factor/heme oxygenase-1) signaling pathway that occurs in the processes of inflammation [44]. Nuclear factor erythroid 2–related factor (Nrf2) is a cytoprotective factor that regulates the expression of the gene coding for antioxidant, antiinflammatory, and detoxifying proteins. Pancreatitis is another inflammatory process associated with the secretion of NF-κB cytokines. Curcumin significantly reduces the activation of this cytokine and AP-1 (activator protein 1) and reduces the mRNA induction of iNOS (nitric oxide synthase), TNF-α, and IL-6 cytokines in the pancreas [45]. Studies on the effects of curcumin on allergies have shown that curcumin inhibits the NF-κB in the airway, together with the transcription factor GATA3; reduces IgE in serum; and inhibits the Notch1-GTA3 signaling pathway [46]. Curcumin reduces the inflammation associated with colitis by significantly reducing the activity of myeloperoxidase and TNF-α [47]. Several studies also have shown the antiarthritic effects of curcumin in humans with osteoarthritis (OA) and rheumatoid arthritis (RA) [48, 49].

Despite its reported benefits via inflammatory and antioxidant mechanisms, one of the significant problems with ingesting curcumin by itself is its poor bioavailability [50], which appears to be primarily due to poor absorption, rapid metabolism, and quick elimination. Several agents tested to improve curcumin's bioavailability by addressing these various mechanisms. Most of them block the metabolic pathway of curcumin; however, when combined with black pepper [51], it is reported to increase the bioavailability of curcumin by 2000% [52]. Curcumin is available in several forms including capsules, tablets, ointments, energy drinks, soaps, and cosmetics [15]. Curcuminoids are approved by the US Food and Drug Administration (FDA) as "generally recognized as safe" (GRAS) due to its safety profiles on clinical trials. Doses between 4000 and 8000 mg/day [47] and of treatments up to 12,000 mg/day of 95% concentration of the curcuminoids are safe [36].

Turmeric is useful in healing peptic ulcers. In phase II clinical trial, 45 patients with peptic ulcer received capsule-filled turmeric orally in the dose of 2 capsules (300 mg each) 5 times daily. After 4 weeks of treatment, ulcers are absent in 48% of cases. After 12 weeks of treatment, ulcer-free cases increased to 76% [53]. A study of 62 patients with external cancerous lesions supplemented with, ethanol extract of turmeric, indicated that it produced remarkable symptomatic relief in smell, lesion size, and itching [54]. A study on eight healthy subjects showed that the presence of turmeric in curry increases bowel motility and activates hydrogen-producing bacterial flora in the colon, thereby increasing the concentration of breath hydrogen [55]. Turmeric paste is used to heal wounds or to protect against infection. In certain parts of Bangladesh, turmeric is the most common application on the cut umbilical cord after delivery [56]. A total of 100 hemodialysis patients who also experienced treatment-resistant uremic pruritus for at least 6 weeks were randomized in a double-blind trial to receive 1500 mg of turmeric daily or placebo for 8 weeks. Both groups experienced significantly reduced total pruritus scores after treatment ($P = .0001$ for each). The authors reported that the mean reduction in the turmeric group (13.6) was significantly better than the mean reduction in the placebo group (7.2) ($P = .001$). Additionally, the mean decrease in C-reactive protein was significantly greater with turmeric than placebo (− 0.8 mg/L vs 0.4 mg/L, respectively; $P = .012$) [57]. In a randomized controlled study ($n = 90$) in the management of acute inflammation pain after removal of impacted molars the group supplemented with curcumin (400 mg three times daily × 24 hours) were effective in reducing postoperative inflammatory pain than in the control (mefenamic acid) supplemented patients . Additionally, more patients in the curcumin group reported higher pain score reductions of either six or seven than the control group (57% vs 24%, respectively, for treatment vs control) [58].

6.2.5 Safety of curcumin

A large number of studies on curcumin identified included studies on the antioxidant, antiinflammatory, antiviral, and antifungal properties of curcuminoids. Studies on the toxicity and antiinflammatory properties of curcumin have included in vitro, animal, and human studies. A phase 1 human trial with 25 subjects using up to 8000 mg of curcumin per day for 3 months found no toxicity from curcumin. Five other human trials using 1125–2500 mg of curcumin per day have also found it to be safe. These social studies have found some evidence of the antiinflammatory activity of curcumin. Curcumin has been demonstrated to be safe in six human trials and has shown antiinflammatory activity. It may exert its antiinflammatory activity by inhibition of several different molecules that play a role in inflammation. However, according to the Recommended Daily Intake and Dietary Reference Intake (DRI), curcumin can be utilized 400–600 mg three times a day [59], recommended to be taken with food to avoid gastrointestinal distress. Curcumin, the active ingredient of the rhizome, has been proved to play a vital role in the treatment of various proinflammatory chronic diseases. The enormous therapeutic properties of turmeric rhizome and the absence of any significant toxicity have added to its increasing demand for drug manufacturing. Thus researchers suggest that curcumin can help in the management of oxidative and inflammatory conditions, metabolic syndrome, arthritis, anxiety, and hyperlipidemia. It may also help in the management of exercise-induced inflammation and muscle soreness, henceforth enhancing recovery and subsequent performance of inactive people.

6.3 Ginger (*Zingiber officinale*)

Ginger (Fig. 6.2) has been used by traditional Chinese and Indian medicine for centuries. In recent times, ginger is introduced into various tropical countries. Also, ginger (*Zingiber officinale*) is widely consumed as a spice and used in food preservation. The beneficial health effects of ginger are well documented. Ginger (*Z. officinale*) is among the leading herbs considered for an array of applications in traditional medicines like Chinese and Ayurveda. It possesses certain pharmacological activities like cardiovascular protection, antioxidant, antiinflammatory, glucose-lowering, and anticancer activities [60]. Various scientific investigations have documented composition and the biological activities of ginger extracts. The ginger rhizome is utilized for many centuries in different food products [61].

6.3.1 Morphology

Ginger rhizome is collected from the underground stems that are surrounded by the sheathing bases of the two-ranked leaves. It is usually an erect perennial growing plant that is grown in the ground from 1 to 3 ft in height. Rhizomes are 7–15 cm long and 1–1.5 cm broad and

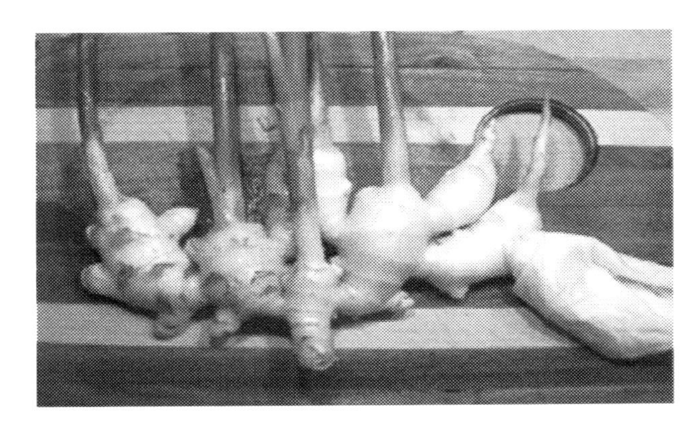

Fig. 6.2 Rhizome *Z. officinale*.

laterally compressed. The branches that arise obliquely from the rhizome are about 1–3 cm long and terminate in depressing scars or undeveloped buds. The outer surface is buff-colored and longitudinally striated or fibrous [62]. The fractured surface shows a narrow cortex, a well-marked endodermis, and a wide stele [63].

6.3.2 Taxonomic position

The family Zingiberaceae is the largest family of Zingiberales and is one of the 10 largest monocotyledonous families in India. It occurs chiefly in the tropics with about 52 genera and 1400 species with the highest concentration in the Indo-Malayan region of Asia and represented by 22 genera and 178 species in India.

6.3.3 Cultivation

Z. officinale, also known as ginger, belongs to the family Zingiberaceae and is a slender perennial plant firstly cultivated in China and then spread to India, Southeast Asia, West Africa, and the Caribbean. The ginger plant reaches to the height of 2 ft and has greenish-yellow flowers resembling orchids with an aromatic pungent taste. It is a tropical plant, and its underground stem is used for culinary spice and medicinal purposes. The rhizome of *Z. officinale* is one of the most widely used species of the Zingiberaceae family and is a common condiment for various foods and beverages. Table 6.2 shows the world's top 10 ginger-producing countries according to recent data by FAO.

6.3.4 Stability

Ginger is widely used in pharmaceuticals, nutraceuticals, and cosmetics. The principal constituent 6-gingerol in the ginger extract acts as an antioxidant. However, 6-gingerol is unstable due to the acidic environment or as a result of the increase in temperature; presence of light, air, and heat; and long-term storage. The stability of 6-gingerol is enhanced by nanoemulsion formulation [64].

6.3.5 Nutrient profile and active components

Ginger possesses a high nutritional value. The medicinal plants and their constituents cure various diseases such as antioxidant, anti-inflammatory, anticancer, antidiabetic, and antitumor effects. Ginger is widely used in a variety of foods because of its nutritional composition and flavoring compounds. Ginger rhizomes are a rich source of carbohydrates, vitamins, minerals, and iron. The different vitamins, minerals, and phytochemicals content that are present in ginger rhizomes are shown in Tables 6.3–6.5.

Table 6.2 World's top 10 ginger producers in 2008.

Country	Production (tonnes) (FAO estimate)
India	420,000
China	285,000
Indonesia	177,000
Nepal	158,905
Nigeria	138,000
Bangladesh	57,000
Japan	42,000
Thailand	34,000
Philippines	28,000
Sri Lanka	8270
World	**1,387,445**

Courtesy: Statistical Division in Economic and Social Development Unit, FAO (2008).

Table 6.3 Nutrient composition of ginger (per 100 g 3.5 oz).

Constituents	Ginger root (ground)	Ginger root (raw)
Energy	1404 kJ (336 kcal)	333 kJ (80 kcal)
Carbohydrates	71.6 g	17.7 g
Sugars	3.39 g	1.7 g
Dietary fiber	14.1 g	2.0 g
Fats	4.24 g	0.75 g
Protein	8.98 g	1.82 g

Courtesy: USDA, Nutrient data for 2013, Spices, Ginger. 2013.

According to the literature the ginger rhizome has a spicy tart flavor of *Z. officinale* that is due to the presence of a large number of essential oils and phenolic compounds such as gingerol, shogaol, zingerone, paradol, and capsaicin [65–67].

6.3.6 Chemical composition

Phytochemical studies show that ginger rhizome contains several biologically active compounds with medicinal property, namely, essential oils, phenolic compounds, flavonoids, alkaloids, glycosides, saponins, steroids, terpenoids, and tannins

Table 6.4 Vitamin content of ginger (per 100 g).

Constituents	Ginger root (ground)	Ginger root (raw)
Thiamine (B1)	0.046 mg	0.025 mg
Riboflavin (B2)	0.17 mg	0.034 mg
Niacin (B3)	9.62 mg	0.75 mg
Pantothenic acid (B5)	0.477 mg	0.203 mg
Vitamin B6	0.626 mg	0.16 mg
Folate (B9)	13 μg	11 μg
Vitamin C	0.7 mg	5 mg
Vitamin E	0.0	0.26 mg

Courtesy: USDA, Nutrient data for 2013, Spices, Ginger. 2013.

Table 6.5 Minerals content of ginger (per 100 g).

Constituents	Ginger root (ground)	Ginger root (raw)
Calcium	114 mg	16 mg
Iron	19.8 mg	0.6 mg
Magnesium	214 mg	43 mg
Manganese	33.3 mg	0.229 mg
Phosphorus	168 mg	34 mg
Potassium	1320 mg	415 mg
Sodium	27 mg	13 mg
Zinc	3.64 mg	0.34 mg

Courtesy: USDA, Nutrient data for 2013, Spices, Ginger. 2013.

6.3.7 Dosage

Ginger can be taken safely in the form of powder, extract, tincture, capsules, and oils, up to 2 g in three divided doses per day or up to four cups of tea daily.

6.3.8 Mechanism of action of ginger and its constituents

Ginger contains various active phytoconstituents such as [6]-gingerol and [6]-paradol, shogaols, zingerone, and galanals A and B41–43. According to the availability of % phytocontent, they play a considerable role in preventing and curing various diseases through different mechanisms. Ginger has antioxidant activity through inhibiting free radicals and oxidative stress, and it has antiinflammatory activity through inhibiting nuclear factor κB and COX1. Ginger modulates the genetic pathways such as apoptosis, activates the tumor suppressor gene, and inhibits VEFG that shows antitumor activity. Ginger induces apoptosis and activates p53 that are responsible for the cancer prevention [68] Ginger inhibits the different types of a pathogen, such as bacteria and microbes, as it possesses the antibacterial and antimicrobial activity [69].

6.3.9 Antiinflammatory effect of ginger

Recent studies have shown that ginger has different pharmacological effects, due to its various components such as gingerols and shogaols. So far, more than 40 antioxidant compounds are detected in ginger. The pharmacological actions of ginger and its isolated compounds include immunomodulatory, antitumorogenesis, antiinflammatory, antiapoptosis, glucose, and lipid-lowering effect and antiemetic [70]. Researchers speculate that the antiinflammatory activity of ginger is due to the inhibition of cyclooxygenase and lipooxygenase pathways by its biologically active components [71]. Specifically, they are 6-gingerol, 10-dihydrogingerdione, and 10-gingerdione [72]. The major pungent constituents of ginger are 6-gingerol, and 6-shogaol has many interesting pharmacological effects, such as antioxidant, antitumor-promoting, and antiinflammatory effects [73]. The valuable pharmacological substances in *Z. officinale* rhizome give a reason to consider these raw materials as a promising cure to treat the inflammatory and painful conditions due to their ability for specific blocking of the inflammation of the pain-sensitive receptors and ion channels, TRP receptors in particular [74, 75].

A case study reported that ginger (*Z. officinale*) is given to a rheumatic arthritis disorder with a dose of 50-g fresh ginger for 3 months. A result is taken from the patients who consumed the ginger daily cooked with vegetables and various meats. After 3 months of the ginger regimen, the patient was completely free of pain, swelling, and inflammation. Additionally, Ayurvedic medicine reports ginger to be useful for rheumatic disorders. The theoretical mechanism of ginger in the context of treating rheumatic disorders is it acts as a dual inhibitor of cyclooxygenase (prostaglandins) and lipoxygenase (leukotrienes) pathways. Ginger is speculated to inhibit the formation of free radicals (superoxide), which may alleviate chronic inflammation in arthritis [76].

Thus far, several studies have speculated on the effect of ginger on blood glucose and lipids levels. Mahluji et al. studied the effect of daily intake of 2 g of powdered ginger in type 2 diabetic patients. After

2 months, insulin, HOMA (insulin resistance score), TG, and LDL decreased significantly in the ginger group compared with the placebo group, with no significant changes observed in FPG, HbA1C, total cholesterol, and HDL levels [77].

Also, in another study, 4 g of ginger powder were given to healthy subjects, patients with coronary artery disease (CAD), and type 2 diabetic patients with or without CAD. After 3 months, no significant changes are observed in blood sugar and serum lipids [78]. Diabetes is an inflammatory disease that is associated with metabolic disorders [79].

The antiinflammatory effect of gingerols, shogaols, and diarylheptanoids in ginger is possibly through inhibition of cyclooxygenase, inducible nitric oxide synthase, and lipoxygenase activity and suppression of prostaglandin synthesis and interference in cytokine signaling. It is shown that the consumption of ginger powder for 12 weeks can cause a significant reduction in CRP in patients with type 2 diabetes. The use of 1 g of powdered ginger daily for 10 weeks made a 27.6% reduction in mean CRP levels in obese men [80].

Chronic, low-grade inflammation and activation of the innate immune system are intimately involved in the pathogenesis of diabetes [79]. Ginger, antiinflammatory properties known for centuries, can supplement to reduce the inflammation [70].

Gingerol is homologs of 1-(3-methoxy-4-hydroxyphenyl)-3-keto-5-hydroxyhexane and includes the subgroup methyl gingerols; gingerol or groups of gingerols (4, 6, 8, and 10 gingerols) are the most active constituent of fresh ginger that are structurally related polyphenolic compounds, which are responsible for providing different pharmacological actions. 6-Shogaol inhibited cell proliferation by inducing cells to autophagy cell death through AKT/mTOR inhibition in human non–small cell lung cancer A549 cells and hence can be a promising chemopreventive agent [81].

Another study also reported that 6-shogaol has a strong antitissue or expectorant effect and could help reduce blood pressure. It has also shown antiallergic effects, inhibits the release of histamine from mast cells, and inhibits cancer growth in ovarian cancer. Ginger not only shows a significant antiinflammatory effect but also indicates adverse and side effects such as a gastric ulcer. Ginger inhibits COX and 5-lipoxygenase, essential for arachidonate metabolism and inturn the synthesis of pro-inflammatory cytokines such as IL1, TNFα, and IL81 that are essential for resolving inflammation [82] and blocked the elevated expression of TNFα in liver cancer in animal [83]. It regulates the induction of inflammatory genes [70, 84] and the production of NF-κB gene involved in cellular proliferation and angiogenesis [85].

A recent study documented the ability of a hexane fraction of dried ginger methanolic extract to suppress proinflammatory gene expression in LPS-activated BV2 microglial cells, thus indicating antineuroinflammatory activity [86]. Gingerol and structurally related pungent principles of ginger, including shogaol, exert inhibitory effects on the biosynthesis of prostaglandins and leukotrienes through suppression of prostaglandin synthase or 5-lipoxygenase [18]. Several reports have addressed the antiinflammatory effects of whole ginger extract on the production of NO/iNOS, PGE2/COX-2, TNF-a, IL-1b, and macrophage chemoattractant protein-1 (MCP-1) in murine macrophages, such as RAW264.7 cells and J774.1 cells, as well as human monocytes, U937 cells [87].

6.4 Conclusion

Medicinal plants have provided a reliable source for the preparation of new drugs as well as combating diseases, from the dawn of civilization. Turmeric and ginger are the different sources of various types of chemical compounds, which are responsible for a variety of activities. Although a lot of experiments are done, more investigations are needed to exploit other therapeutic utility to combat diseases. The effect of curcumin on different models of tissue injury should be investigated. Thus more in vitro and preclinical researches are needed to render curcumin or its metabolites and ginger as therapeutic supplements.

References

[1] C.N. Lumeng, A.R. Saltiel, Inflammatory links between obesity and metabolic disease, J. Clin. Invest. 121 (2011) 2111–2117.

[2] N.A. Punchard, C.J. Whelan, I. Adcock, The Journal of Inflammation, J. Inflamm. 1 (2004) 1, https://doi.org/10.1186/1476-9255-1-1.

[3] S. Sosa, M.J. Balick, R. Arvigo, R.G. Esposito, C. Pizza, G. Altinier, A. Tubaro, Screening of the topical anti-inflammatory activity of some Central American plants, J. Ethnopharmacol. 81 (2000) 211–215.

[4] K. Rohini, P.S. Srikumar, J. Saxena, K.A. Mahesh, Alteration in the levels of micronutrients in tuberculosis patients, Int. J. Biol. Med. Res. 4 (2014) 2958–2961.

[5] P. Roy, S. Amdekar, A. Kumar, V. Singh, A preliminary study of the antioxidant properties of flowers and roots of *Pyrostegia venusta* (Ker Gawl) Miers, BMC Complement. Altern. Med. 11 (2011) 69–75.

[6] K. Rohini, P.S. Srikumar, Therapeutic role of coumarins and coumarin-related compounds, J. Thermodyn. Catal. 5 (2014) 130–140.[7]https://shodhganga.inflibnet.ac.in.https://shodhganga.inflibnet.ac.in.

[8] A.J. Ruby, G. Kuttan, K.D. Babu, K.N. Rajasekharan, R. Kuttan, Anti-tumour and antioxidant activity of natural curcuminoids, Cancer Lett. 94 (1995) 79–83.

[9] R. Selvam, L. Subramanian, R. Gayathri, N. Angayarkanni, The antioxidant activity of turmeric (*Curcuma longa*), J. Ethnopharmacol. 47 (1995) 59–67.

[10] M. Ohshiro, M. Kuroyanag, A. Keno, Structures of sesquiterpenes from *Curcuma longa*, Phytochemistry 29 (1990) 2201–2205.

[11] L.D. Kapoor, Handbook of Ayurvedic Medicinal Plants, CRC Press, Boca Raton, FL, 1990.

[12] K.R. Kirtikar, B.D. Basu, E. Blatter, J.F. Caius, K.S. Mhaskar, Indian Medicinal Plants, second ed., vol. II, Lalit Mohan Basu, 1993, p. 1182.

[13] K.V. Balakrishnan, P.N. Ravindran, K. Nirmal Babu, Postharvest technology and processing of turmeric, 2007, pp. 193–200.

[14] B.B. Aggarwal, A. Kumar, A.C. Bharti, Anticancer potential of curcumin: preclinical and clinical studies, Anticancer Res. 23 (2003) 363–398.

[15] S.C. Gupta, S. Patchva, B.B. Aggarwal, Therapeutic roles of curcumin: lessons learned from clinical trials, AAPS J. 15 (2013) 195–218.

[16] B.B. Aggarwal, K.B. Harikumar, Potential therapeutic effects of curcumin, the anti-inflammatory agent, against neurodegenerative, cardiovascular, pulmonary, metabolic, autoimmune and neoplastic diseases, Int. J. Biochem. Cell Biol. 41 (2009) 40–59.

[17] Y. Panahi, M.S. Hosseini, N. Khalili, E. Naimi, L.E. Simental-Mendia, M. Majeed, A. Sahebkar, Effects of curcumin on serum cytokine concentrations in subjects with metabolic syndrome; a post-hoc analysis of a randomized controlled trial, Biomed. Pharmacother. 82 (2016) 578–582.

[18] V. Kuptniratsaikul, P. Dajpratham, W. Taechaarpornkul, M. Buntragulpoontawee, P. Lukkanapichonchut, C. Chootip, J. Saengsuwan, K. Tantayakom, S. Laongpech, Efficacy and safety of *Curcuma domestica* extracts compared with ibuprofen in patients with knee osteoarthritis, a multicenter study, Clin. Interv. Aging 9 (2014) 451–458.

[19] F. Mazzolani, S. Togni, Oral administration of a curcumin-phospholipid delivery system for the treatment of central serous chorioretinopathy: a 12-month follow-up study, Clin. Ophthalmol. 7 (2013) 939–945.

[20] B.M. Forman, J. Chen, R.M. Evans, The peroxisome proliferator-activated receptors: ligands and activators, Ann. N. Y. Acad. Sci. 804 (1996) 266–275.

[21] A.M. Siddiqui, X. Cui, R. Wu, W. Dong, M. Zhou, M. Hu, H.H. Simms, P. Wang, The anti-inflammatory effect of curcumin in an experimental model of sepsis is mediated by the up-regulation of peroxisome proliferator-activated receptor-γ, Crit. Care Med. 34 (2006) 1874–1882.

[22] G.C. Jagetia, B.B. Aggarwal, "Spicing up" of the immune system by curcumin, J. Clin. Immunol. 27 (2007) 19–35.

[23] M. Hu, Q. Du, I. Vancurova, X. Lin, E.J. Miller, H.H. Simms, P. Wang, Proapoptotic effect of curcumin on human neutrophils: activation of the p38 mitogen-activated protein kinase pathway, Crit. Care Med. 33 (2005) 2571–2578.

[24] D.C. Angus, W.T. Linde-Zwirble, J. Lidicker, G. Clermont, J. Carcillo, M.R. Pinsky, Epidemiology of severe sepsis in the United States: analysis of incidence, outcome, and associated costs of care, Crit. Care Med. 29 (2001) 1303–1310.

[25] A. Ayala, C.S. Chung, J.L. Lomas, G.Y. Song, L.A. Doughty, S.H. Gregory, W.G. Cioffi, B.W. LeBlanc, J. Reichner, H.H. Simms, P.S. Grutkoski, Shock-induced neutrophil mediated priming for acute lung injury in mice: divergent effects of TLR-4 and TLR-4/FasL deficiency, Am. J. Pathol. 161 (2002) 2283–2294.

[26] R. Taneja, J. Parodo, S.H. Jia, A. Kapus, O.D. Rotstein, J.C. Marshall, Delayed neutrophil apoptosis in sepsis is associated with the maintenance of mitochondrial transmembrane potential and reduced caspase-9 activity, Crit. Care Med. 32 (2004) 1460–1469.

[27] K.C. Srivastava, A. Bordia, S.K. Verma, Curcumin, a major component of food spice turmeric (*Curcuma longa*) inhibits aggregation and alters eicosanoid metabolism in human blood platelets, Prostaglandins Leukot. Essent. Fat. Acids 52 (1995) 223–227.

[28] F. Zhang, N.K. Altorki, J.R. Mestre, K. Subbaramaiah, A.J. Dannenberg, Inhibition of cyclo-oxygenase 2 expression in colon cells by the chemopreventive agent curcumin involves inhibition of NF-kappaB activation via the NIK/IKK signaling complex, Carcinogenesis 20 (1999) 445–451.

[29] H.Y. Kim, E.J. Park, E.H. Joe, I. Jou, Curcumin suppresses Janus kinase-STAT inflammatory signaling through activation of Src homology 2 domain-containing tyrosine phosphatase 2 in brain microglia, J. Immunol. 171 (2003) 6072–6079.

[30] R.J. Smith, Nonsteroidal anti-inflammatory agents: regulators of the phagocytic secretion of lysosomal enzymes from guinea pig neutrophils, J. Pharmacol. Exp. Ther. 207 (1978) 618–629.

[31] G. Weismann, The role of lysozymes in inflammation and disease, Annu. Rev. Med. 18 (1967) 97–112.

[32] E.L. Becker, P.M. Henson, In vitro studies of immunologically induced secretion of mediators from the cell and related phenomena, Adv. Immunol. 17 (1973) 93–193.

[33] R. Srivastava, R.C. Srimal, Modification of certain inflammation-induced biochemical changes by curcumin, Indian J. Med. Res. 81 (1985) 215–223.

[34] B. Joe, B.R. Lokesh, Effect of curcumin and capsaicin on arachidonic acid metabolism and lysosomal enzyme secretion by rat peritoneal macrophages, Lipids 32 (1997) 1173–1180.

[35] S.K. Biswas, Does the interdependence between oxidative stress and inflammation explain the antioxidant paradox? Oxidative Med. Cell. Longev. 2016 (2016), 5698931.

[36] C.D. Lao, M.T. Ruffin, D. Normolle, D.D. Heath, S.I. Murray, J.M. Bailey, M.E. Boggs, J. Crowell, C.L. Rock, D.E. Brenner, Dose escalation of a curcuminoid formulation, BMC Complement. Altern. Med. 6 (2006) 10.

[37] J.S. Jurenka, Anti-inflammatory properties of curcumin, a major constituent of *Curcuma longa*, A review of preclinical and clinical research, Altern. Med. Rev. J. Clin. Ther. 14 (2009) 141–153.

[38] A. Mazumder, S. Wang, N. Neamati, M. Nicklaus, S. Sunder, J. Chen, G.W. Milne, W.G. Rice, T.R. Burke Jr., Y. Pommier, Antiretroviral agents as inhibitors of both human immunodeficiency virus type 1 integrase and protease, J. Med. Chem. 39 (1996) 2472–2481.

[39] Z. Sui, R. Salto, J. Li, C. Craik, P.R. Ortiz de Montellano, Inhibition of the HIV-1 and HIV-2 proteases by curcumin and curcumin boron complexes, Bioorg. Med. Chem. 1 (1993) 415–422.

[40] J.S. James, Curcumin update: could food spice be low-cost antiviral? AIDS Treat. News 176 (1993) 1–3.

[41] Cayman Chem, Demethoxycurcumin and Bisdemethoxycurcumin, 2014, Retrieved from: https://www.caymanchem.com/app/template/Product.vm/catalog/1096.

[42] C.V. Rao, A. Rivenson, B. Simi, B.S. Reddy, Chemoprevention of colon carcinogenesis by dietary curcumin, a naturally occurring plant phenolic compound, Cancer Res. 55 (1995) 259–266.

[43] B. Shpitz, N. Giladi, E. Sagiv, S. Lev-Ari, E. Liberman, D. Kazanov, N. Arber, Celecoxib and curcumin additively inhibit the growth of colorectal cancer in a rat model, Digestion 74 (2006) 140–144.

[44] L. Liu, Y. Shang, M. Li, X. Han, J. Wang, Curcumin ameliorates asthmatic airway inflammation by activating nuclear factor-E2-related factor 2/haem oxygenase (HO)-1 signaling pathway, Clin. Exp. Pharmacol. Physiol. 42 (2015) 520–529.

[45] A. Gulcubuk, D. Haktanir, A. Cakiris, D. Ustek, O. Guzel, M. Erturk, M. Karabagli, I. Akyazi, H. Cicekci, K. Altunatmaz, Effects of curcumin on proinflammatory cytokines and tissue injury in the early and late phases of experimental acute pancreatitis, Pancreatology 13 (2013) 347–354.

[46] L. Chong, W. Zhang, Y. Nie, G. Yu, L. Liu, S. Wen, L. Zhu, C. Li, Protective effect of curcumin on acute airway inflammation of allergic asthma in mice through Notch1-GATA3 signaling pathway, Inflammation 37 (2014) 1476–1485.

[47] P. Basnet, N. Skalko-Basnet, Curcumin, an anti-inflammatory molecule from a curry spice on the path to cancer treatment, Molecules 16 (2011) 4567–4598.

[48] Y. Henrotin, F. Priem, A. Mobasheri, Curcumin, a new paradigm and therapeutic opportunity for the treatment of osteoarthritis: curcumin for osteoarthritis management, Springerplus 2 (2013) 56.
[49] G. Belcaro, M.R. Cesarone, M. Dugall, L. Pellegrini, A. Ledda, M.G. Grossi, S. Togni, G. Appendino, Product-evaluation registry of Meriva®, a curcumin-phosphatidylcholine complex, for the complementary management of osteoarthritis, Panminerva Med. 52 (2010) 55–62.
[50] P. Anand, A.B. Kunnumakkara, R.A. Newman, B.B. Aggarwal, Bioavailability of curcumin: problems and promises, Mol. Pharm. 4 (2007) 807–818.
[51] H.K. Han, The effects of black pepper on the intestinal absorption and hepatic metabolism of drugs, Expert Opin. Drug Metab. Toxicol. 7 (2011) 721–729.
[52] G. Shoba, D. Joy, T. Joseph, M. Majeed, R. Rajendran, P.S. Srinivas, Influence of piperine on the pharmacokinetics of curcumin in animals and human volunteers, Planta Med. 64 (1998) 353–356.
[53] C. Prucksunand, B. Indrasukhsri, M. Leethochawalit, K. Hungspreugs, Phase II clinical trial on the effect of the long turmeric (*Curcuma longa* Linn) on the healing of the peptic ulcer, Southeast Asian J. Trop. Med. Public Health 32 (2001) 208–215.
[54] R. Kuttan, P.C. Sudheeran, C.D. Josph, Turmeric and curcumin as topical agents in cancer therapy, Tumori 73 (1987) 29–31.
[55] A. Shimouchi, K. Nose, M. Takaoka, H. Hayashi, T. Kondo, Effect of dietary turmeric on breath hydrogen, Dig. Dis. Sci. 54 (2009) 1725–1729.
[56] M.A. Alam, N.A. Ali, N. Sultana, L.C. Mullany, K.C. Teela, N.U. Khan, P.J. Winch, Newborn umbilical cord and skincare in Sylhet District, Bangladesh: implications for the promotion of umbilical cord cleansing with topical chlorhexidine, J. Perinatol. 28 (Suppl. 2) (2008) S61–S68.
[57] M. Pakfetrat, F. Basiri, L. Malekmakan, J. Roozbeh, Effects of turmeric on uremic pruritus in end-stage renal disease patients: a double-blind randomized clinical trial, J. Nephrol. 27 (2014) 203–207.
[58] T. Mauilina, H. Diana, A. Cahyanto, A. Amaliya, The efficacy of curcumin in managing acute inflammation pain on the post-surgical removal of impacted third molars patients: a randomized controlled trial, J. Oral Rehabil. 45 (2018) 677–683.
[59] NIH, The Dietary Supplement Label Database, 2014, Retrieved from: http://www.dsld.nlm.nih.gov/dsld/index.jsp.
[60] Y. Shukla, M Singh, Cancer preventive properties of ginger: a brief review, Food Chem. 45 (2007) 683–690.
[61] H. McGee, in: H. McGee (Ed.), On Food and Cooking, the Science and Lore of the Kitchen, second ed., Scribner, New York, 2004, pp. 425–426.
[62] W.C. Evans, Trease and Evans Pharmacognosy, sixteenth ed., Saunders Elsevier, 2002, pp. 289–292.
[63] M. Ali, Textbook of Pharmacognosy, second ed., CBS Publishers and Distributors, 1998, pp. 258–262.
[64] U. Chutamas, I. Ratana, A. Rathapon, Development of nanoemulsion formulations of ginger extract, Adv. Mater. Res. 684 (2013) 12–15.
[65] S.B. Dugasani, M.R. Pichika, V.D. Nadarajah, Comparative antioxidant and anti-inflammatory effects of [6]-gingerol, [8]-gingerol, [10]-gingerol and [6]-shogaol, J. Ethnopharmacol. 127 (2009) 515–520.
[66] R.B. Van Breemen, Y. Tao, L. Wenkui, Cyclooxygenase-2 inhibitors in ginger (*Zingiber officinale*), Fitoterapia 82 (2011) 38–43.
[67] A. Hazan, R. Kumar, H. Matzner, A. Priel, The pain receptor TRPV1 displays agonist-dependent activation stoichiometry, Sci. Rep. 5 (2015) 1–13.
[68] Y. Liu, R.J. Whelan, B.R. Pattnaik, K. Ludwig, E. Subudhi, H. Rowland, N. Claussen, N. Zucker, S. Uppal, D.M. Kushner, M. Felder, M.S. Patankar, A. Kapur, Terpenoids from *Zingiber officinale* (ginger) induce apoptosis in endometrial cancer cells through the activation of p53, PLoS One 7 (2012) e53178.

[69] A. Giriraju, G.Y. Yunus, Assessment of the antimicrobial potential of 10% ginger extract against Streptococcus mutants, *Candida albicans*, and *Enterococcus faecalis* an in vitro study, Indian J. Dent. Res. 24 (2013) 397–400.
[70] B.H. Ali, G. Blunden, M.O. Tanira, A. Nemmar, Some phytochemical, pharmacological and toxicological properties of ginger (*Zingiber officinale* Roscoe): a review of recent research, Food Chem. Toxicol. 46 (2008) 409–420.
[71] J.M. Jellin, P.J. Gregor, F. Batz, K. Hitchens, Pharmacist's Letter/Prescriber's Letter Natural Medicines Comprehensive Database, fourth ed., Therapeutic Research Faculty, Stockton, CA, 2002, pp. 584–586.
[72] C.A. Newall, L.A. Anderson, J.D. Philipson, Herbal Medicine, a Guide for Health Care Professionals, The Pharmaceutical Press, 1996, pp. 135–137.
[73] H.Y. Young, Y.L. Luo, H.Y. Cheng, W.C. Hsieh, J.C. Liao, W.H. Peng, Analgesic and anti-inflammatory activities of [6]-gingerol, J. Ethnopharmacol. 96 (2005) 207–210.
[74] R. Brito, S. Sheth, D. Mukherjea, TRPV1 a potential drug target for treating various diseases, Cells 3 (2014) 517–545.
[75] B. Veronesi, M. Oortgiesen, The TRPV1 receptor: target of toxicants and therapeutics, Toxicol. Sci. 89 (2006) 1–3.
[76] K.C. Srivastava, T. Mustafa, Ginger (*Zingiber officinale*) and rheumatic disorders, Med. Hypothesis 29 (1989) 25–28.
[77] S. Mahluji, V.E. Attari, M. Mobasseri, L. Payahoo, A. Ostadrahimi, S.E. Golzari, Effects of ginger (*Zingiber officinale*) on plasma glucose level, HbA1c and insulin sensitivity in type 2 diabetic patients, Int. J. Food Sci. Nutr. 64 (2013) 682–686.
[78] A. Bordia, S.K. Verma, K.C. Srivastava, Effect of ginger (*Zingiber officinale* Rosc.) and fenugreek (*Trigonella foenumgraecum* L.) on blood lipids, blood sugar and platelet aggregation in patients with coronary artery disease, Prostaglandins Leukot. Essent. Fatty Acids 56 (1997) 379–384.
[79] J.F. Navarro, C. Mora, Diabetes, inflammation, proinflammatory cytokines, and diabetic nephropathy, Sci. World J. 6 (2006) 908–917.
[80] S. Atashak, M. Peeri, A. Jafari, M.A. Azarbayijani, Effects of 10-week resistance training and ginger consumption on C-reactive protein and some cardiovascular risk factors in obese men, Physiol. Pharmacol. 14 (2010) 318–328.
[81] J.Y. Hung, Y.L. Hsu, C.T. Li, Y.C. Ko, W.C. Ni, M.S. Huang, 6-Shogaol, an active constituent of dietary ginger, induces autophagy by the AKT/mTOR pathway in human non-small cell lung cancer A549 cells, J. Agric. Food Chem. 57 (2009) 9809–9816.
[82] S.K. Verma, M. Singh, P. Jain, A. Bordia, Protective effect of ginger, *Zingiber officinale* Rosc on experimental atherosclerosis in rabbits, Indian J. Exp. Biol. 42 (2004) 736–738.
[83] M.H. Pan, M.C. Hsieh, J.M. Kuo, C.S. Lai, H. Wu, S. Sang, C.T. Ho, [6]-Shogaol induces apoptosis in human colorectal carcinoma cells via ROS production, caspase activation, and GADD 153 expression, Mol. Nutr. Food Res. 52 (2008) 527–537.
[84] S. Tripathi, D. Bruch, D.S. Kittur, Ginger extract inhibits LPS induced macrophage activation and function, BMC Complement. Altern. Med. 8 (2008) 1–7.
[85] L. Nonn, D. Duong, D.M. Peehl, Chemopreventive anti-inflammatory activities of curcumin and other phytochemicals mediated by MAP kinase phosphatase 5 in prostate cells, Carcinogenesis 28 (2007) 1188–1196.
[86] H.W. Jung, C.H. Yoon, K.M. Park, H.S. Han, Y.K. Park, Hexane fraction of Zingiberis Rhizoma Crudus extract inhibits the production of nitric oxide and proinflammatory cytokines in LPS-stimulated BV2 microglial cells via the NFkappaB pathway, Food Chem. Toxicol. 47 (2009) 1190–1197.
[87] R.C. Lantz, G. Chen, M. Sarihan, A.M. Solyom, S.D. Jolad, B.N. Timmermann, The effect of extracts from ginger rhizome on inflammatory mediator production, Phytomedicine 14 (2007) 123–128.

7

Antiinflammatory activity of *Boswellia*

Karthik Varma[a], Józef T. Haponiuk[b], and Sreeraj Gopi[a]
[a]R&D Centre, Aurea Biolabs (P) Ltd, Kolenchery, Cochin, Kerala, India,
[b]Chemical Faculty, Gdansk University of Technology, Gdańsk, Poland

7.1 Introduction

From ancient days onward, plants and plant-derived phytochemicals have been used in the treatment of various health disorders. Natural products have many scientific superiorities like low side effects and assured biological activities. *Boswellia serrata* belonging to the Burseraceae family is also referred as Olibanum, Frankincense, and Salai guggul and is found in India, Middle East, and African nations [1]. Till date, there are almost 21 species of *Boswellia* spread across the world. The name *Boswellia* was given in honor of Johan Boswell, who discovered some species of *Boswellia*. Egyptians were the first to refer the use of Frankincense as an herbal medicine where it was used for the treatment of tumors, inflammatory disorders, and asthma [2]. Traditional Indian Ayurvedic system of medicine referred *Boswellia* as "Salai guggul" in Ayurvedic literatures like Charaka Samhita in 1st and 2nd century AD for the treatment of various disorders [3]. Scholars over Middle East have also mentioned the use of frankincense in the treatment of various disorders like wounds, arthritis, and gastrointestinal inflammation [4].

The majority of the studies of *Boswellia* species are concentrated toward the antiinflammatory activities. In addition, *Boswellia* is used for controlling asthma [5], age-related disorders [6], neurorecovery [7], skin disorder [8], cancer [9], antidepressant [10], etc. The phytochemical constituents that are responsible for the pharmacological activities include essential oil terpenes and the so-called boswellic acids. Several clinical and preclinical studies have evaluated the various antiinflammatory activities.

The chapter presents an overview about the phytochemical constituents, pharmacokinetics, and different clinical and preclinical study data to provide an overview on the antiinflammatory activities of various species of *Boswellia*.

Inflammation and Natural Products. https://doi.org/10.1016/B978-0-12-819218-4.00010-9
© 2021 Elsevier Inc. All rights reserved.

7.2 Taxonomy and phytochemistry

There are nearly 20 *Boswellia* species, which are distributed throughout globally, but most spread in India, Africa, and Middle East countries. The 4–5-m-long tree upon peeling yields droplets of creamy white gummy oleoresin that poses a strong balsamic odor [11]. The nonvolatile fraction of the *Boswellia* species includes boswellic acids as the important components. The history of isolation and characterization of boswellic acid dates back to the 1890 when Alexander Tschirch and Oscar Halbey isolated the neutral and basic fractions of the Frankincense [12], which was further confirmed by Winterstein suggested that the compounds as boswellic acids (BA) [13]. The major components of the extract include volatile oils (5%–10%), resin (50%–60%), and mucus (10%–15%). Out of the resinous matter, boswellic acids (BA) constitute 30%–40% [14]. The structural elucidation elaborates that boswellic acids are pentacyclic triterpenes that exist in α-configuration (geminal methyl groups at C-20) or in a β-configuration (vicinal methyl groups at C19, C20). The different boswellic acids that are normally present in the resin include α-boswellic acid, 3-*o*-acetyl-11-keto-beta-boswellic acid (AKBA), 11-keto-beta boswellic acid (KBA), acetyl-α-boswellic acid, β-boswellic acid, acetyl β-boswellic acid, 9,11-dehydro-α-boswellic acid, acetyl-9,11-dehydro-α-boswellic acid, 9,11-dehydro-β-boswellic acid, acetyl-9,11-dehydro-β-boswellic acid, lupeolic acid, acetyl lupeolic acid, 11-keto-β-boswellic acid, and acetyl-11-keto-β-boswellic acid [15]. The volatile fraction of the *Boswellia* species mainly includes different sesquiterpenes like α-pinene, α-thujene, β-pinene, sabinene, myrcene, limonene, Þ-cymene, α-copaene, β-caryophyllene, α-humulene, δ-cadinene, and β-caryophyllene oxide [16]. The phytochemical structures of some of the major boswellic acids and bioactive are represented in Fig. 7.1.

7.3 Pharmacological activities of Boswellia

7.3.1 Molecular targets of boswellic acids and mechanism of action of boswellic acids

There are several mechanisms of action that contributes to the antiinflammatory and immunomodulatory activities. BA has the ability to inhibit the formation of leukotrienes (LT), which act as potent mediators of inflammatory disorders, arthritis management, and immunomodulatory disorders [17]. The mechanism by which boswellic acids inhibit LT synthesis can be attributed to the ability of boswellic acid to interact with 5-lipoxygenase (5-LO) and suppress the activity [18]. Studies also reported that elevation of intracellular Ca^{2+} is one of the key factors involved in the signaling event for activation of cells

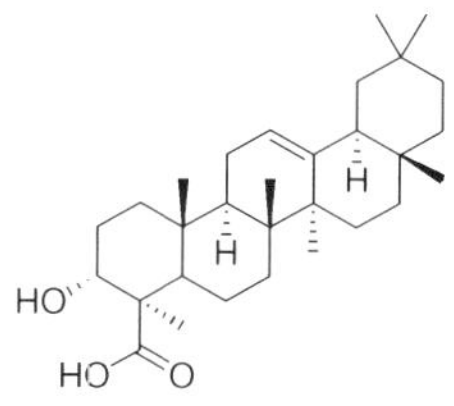

Alpha Boswellic acid

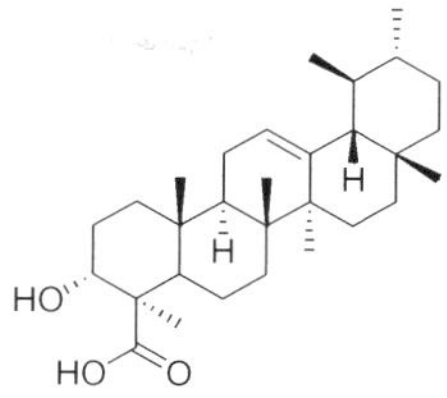

Beta Boswellic acid

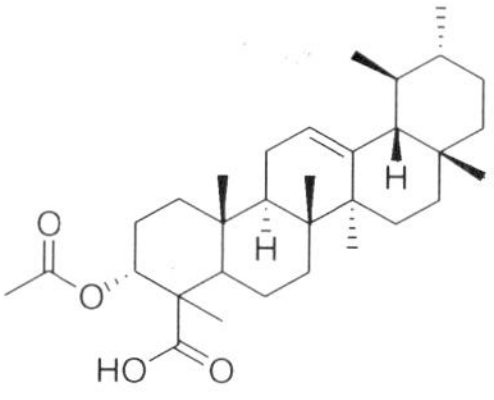

3-Acetyl-beta-boswellic acid

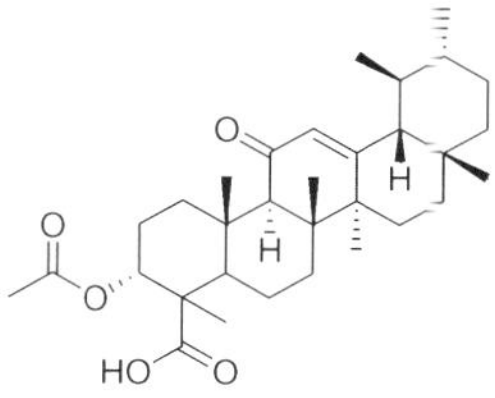

AKBA

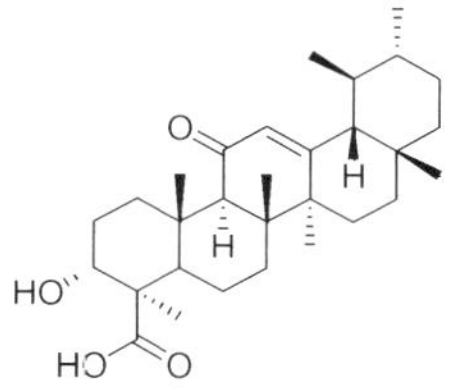

11-Keto-beta-boswellic acid

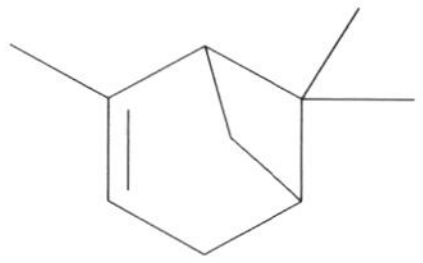

Alpha pinene

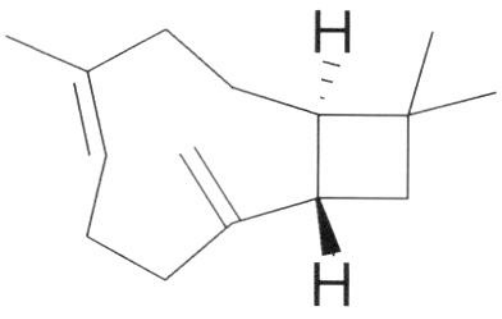

Beta Caryophyllene

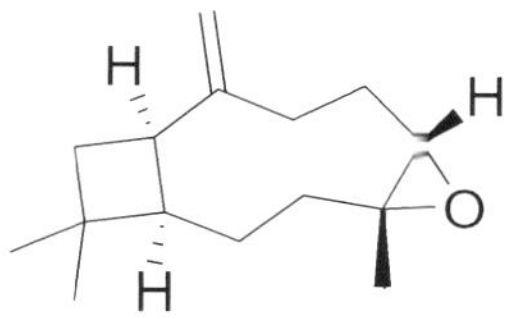

Beta Caryophyllene Oxide

Fig. 7.1 Structures of major bioactives in *Boswellia* species.

and initialization of functional parameters in inflammatory leukocytes and found that boswellic acids induced Ca^{2+} mobilization [19]. AKBA and KBA have the ability to mobilize the activation of mitogen-activated protein kinases (MAPK), one of the contributory factors in the inflammatory responses [20]. Primary studies evaluated that BA inhibit the generation of LTB_4 particularly AKBA that was identified the most potent one [21]. BA play a predominant role in the attenuation of connective tissue metabolites like hydroxyproline, hexosamine and uronic acid in urine of rats. Human leukocyte elastase (HLE) is a serine protease secreted by the polymorphonuclear leukocytes (PMNL) and is highly destructive. HLE is supposed to play a predominant role in inflammatory disorders like arthritis, and Safayhi et al. showed that AKBA reduced the HLE activity in vitro with an IC_{50} value of approximately 15 μM [22]. Another predominant factor that causes disruption to the cell tissues is the propagation of oxygen radicals. Heil et al. studied that AKBA and the extract were found capable of inhibition of nicotinamide adenine dinucleotide phosphate (NADPH) oxidase activity in rats and found that actives can inhibit the generation of reactive oxygen radicals [23].

7.3.2 Bioavailability

To prove the efficacy of any drug, bioavailability is a major concern. All of the drugs even though they have attributed pharmacological activities, they may not be available in bioavailable form. Many studies have presented the data on the bioavailability of KBA and AKBA, the most predominant of the bioactives. The pharmacokinetics can be evealed by the measurement of the levels of actives in the serum, saliva, etc. In a rat study, KBA and AKBA were proved to be bioavailable in female albino rats following oral administration [24]. An open uncontrolled trial involving 12 male healthy volunteers involving administration of a branded formulation containing 6.44% KBA, 2% AKBA, 18.51% βBA, 8. 58% 3-*o*-acetyl-β-BA, 6.93% α-BA and 1.85% 3-*o*-acetyl-α-BA were used. The study found that the maximum concentration of KBA was found after a time duration of 4.5 h of 2.72 μM, and a study has also shown that the elimination half-life was 5.97 h [25]. Several other studies have also reported some corresponding values for the absorption of KBA and AKBA. Food intake can also play a dominant role in evaluating the bioavailability values. Sterk et al. studied the role of food intake in the bioavailability of boswellic acids. The subjects were divided into two groups, and one group was fasted 10 h before and 4 h after drug administration. The second group received high fat meal together with the drug. It has been found that C_{max} of KBA is about 2.7 times and AKBA 4.8 times higher than in the fasted group [26].

7.4 Preclinical studies

Several preclinical studies have confirmed most of the biological activities of the extract and related active components of boswellic acids and other volatile oil constituents. Despite the exact mechanism of action of the extract, there are many attributed mechanisms of action of the boswellic acids, which are well cited in many literatures. Several preclinical studies, including in vitro and in vivo, have confirmed the antiinflammatory activities using different models (Table 7.1).

The first experimental studies regarding the antiinflammatory and analgesic activities of the *B. serrata* extract was proven by a couple of studies, which was done four decades ago [43, 44]. Conventional nonsteroidal antiinflammatory drugs generally referred as NSAID may have the relief effects, but longtime consumption of these may lead to adverse side effects like hepatotoxicity and damage of body parts. *B. serrata* is not only the species that has the antiinflammatory and analgesic activities, other varieties like *Boswellia carterii* and *Boswellia elongata* have reported antiinflammatory activities proved by both preclinical and clinical studies [27, 28]. A study conducted by Mothana et al. discussed the antiinflammatory activity of an extract of *B. elongata* in different models such as carrageenan-induced rat paw edema, cotton pellet granuloma in rats, acetic acid–induced abdominal writing, and hot plate test model in mice in a dose-dependent manner. The extract at a dosage of 400 mg/kg was found to deliver significant

Table 7.1 Preclinical studies on the antiinflammatory activities of various *Boswellia* extracts.

Compound	Disease condition	Reference
Liquisolid AKBA	Gastrointestinal ulcer	[27]
Boswellic acids: β-boswellic acids, α-boswellic acids, lupeolic acids	Inflammation	[28]
Extract of *B. elongata* and *Jatropha unicostata*	Inflammation	[29]
B. frereana extract	Cartilage degradation	[30]
Methanol extract of *B. dalzielii*	Arthritis management	[31]
Extract of *C. myrrha* and *B. carterii*	Analgesic	[32]
Boswellic acids	Nonalcoholic fatty liver disease	[33]
Alpha-boswellic acid	Ethanol induced gastric ulcer	[34]
Boswellic acid	Colonic cancer	[35]
AKBA	Renal fibrosis	[36]
B. serrata extract	Immune homeostasis	[37]
Boswellic acid	Inflammation	[38]
8-Cembranoids	Ulcerative colitis	[39]
Tetra- and pentacyclic terpenic acids	Inflammation	[40]
12-Uursene-2-diketone	Inflammation	[41]
AKBA	Ileitis	[42]

decrease in the rat paw and cotton pellet granuloma, lower acetic acid-induced abdominal constrictions, etc. [29]. In addition, the extract also showed significant antioxidant activity. Similarly, *Boswellia frereana* was also evaluated for their significant antiinflammatory activity in an in vitro model of cartilage degradation. The mechanism of action was suggested to be blockage of propagation and initiation of proinflammatory mediators and matrix metalloproteinase, thus evaluating the role of the extract in the management of arthritis [30]. Studies have also evaluated the antiinflammatory activities of *Boswellia dalzielii*. The methanolic bark extracts of the herb showed significant antiinflammatory activity in rats [31]. *Boswellia* extracts are also used in combination with multiple herbs in traditional practices of medicine. An in vivo study has used a combination of *B. carterii* and *Commiphora myrrha* for their antiinflammatory and analgesic effects. The study included the individual extracts and the combined extracts; activities of both the extracts were evaluated and found that the combined extracts exhibited stronger activity than the individual extracts [32]. Several preclinical studies have also confirmed the various antiinflammatory actions of different extracts of *B. serrata* and *B. carterii* in nonalcoholic fatty liver disease, ethanol-induced gastric ulcer, colonic cancer, ileitis, multiple sclerosis, renal fibrosis, etc. [33–38].

Moreover the antiinflammatory activity of the genus *Boswellia* cannot be attributed to boswellic acids alone. The antiulcerative activity of 8-cembranoids and incensole were evaluated and has proved the beneficiality of the actives [39]. A couple of other bioactive like incensole acetate, 12-ursene-2-diketone, 3-oxo-tirucallic acid, and synthetic derivatives of AKBA were also evaluated for various antiinflammatory activities [39–42] (Table 7.1).

7.5 Clinical studies of the antiinflammatory action

The antiinflammatory and antiarthritic activities of the boswellic acids and other valuable phytochemicals are not only proved by preclinical studies; many human clinical trials also contribute to the substantiating pharmacological activities. Most of the studies constitute either a mixture of boswellic acids or single compounds like AKBA and KBA and some studies have concentrated on essential oil sesquiterpenes (Table 7.2). The studies evaluated the antiinflammatory activities in different dosages. The antiinflammatory activities of the bioactives are mostly evaluated arthritis management studies. In a three arm, parallel group, randomized, double-blind clinical study involving 201 patients supplemented with a combination of curcumin (350 mg) and boswellic acid (150 mg) for a study period of 12 weeks showed significant improvement in the results. The study outcome was evaluated using osteoarthritis (OA) physical function, WOMAC scores, and assessing disease severity. A significant improvement in primary study outcomes signifies the antiinflammatory pathway of boswellic acids [45]. Some of the clinical studies have concentrated the antiinflammatory activity AKBA, one of the most bioactive of the boswellic acid isomers. A commercial product with 30% AKBA content had been clinically studied in 75 subjects. Subjects were randomly divided into three different groups and where supplemented with either placebo, 100-mg test product or 250-mg test product, for a study duration of 90 days. The study product showed significant improvement in the overall study outcomes as shown by the significant improvement in the WOMAC scores and other clinical parameters [46]. The most effective supplement in the management of knee osteoarthritis is glucosamine sulfate. In a randomized clinical trial, methylsulfonylmethane had been used in combination with boswellic acids in comparison with glucosamine sulfate. One hundred twenty subjects who were affected by knee OA were randomly assigned to two different groups and were subjected with the either test product or reference material for 60 days. The subjects were supplemented with either 5 g of MSM and 7.2 mg of boswellic acid or 1500 mg of glucosamine sulfate. Significant improvements were shown in the visual analog scale

Table 7.2 Clinical studies involving the antiinflammatory potential of various *Boswellia* species.

Trial protocol	Extract/active	Reference
Randomized, double-blind, placebo-controlled 201 subjects	Combination of curcumin + boswellic acid	[45]
Randomized, double-blind, placebo-controlled 50 subjects	AKBA	[46]
Randomized trial, 120 subjects	Methylsulfonylmethane (MSM) plus boswellic acids (BA)	[47]
Randomized, double-blind, cross over, 12 volunteers	*B. serrata* extract	[48]
Randomized, double-blind, 24 subjects	Polyherbal formulation comprising *K. galanga, C. longa, B. serrata*, and *P. nigrum*	[49]
Randomized, double-blind,78 subjects	Conventional drugs plus *B. serrata* extract/placebo	[50]
Investigator initiated, bicentric, phase IIa, baseline to treatment, 28 subjects	Frankincense extract	[51]
Randomized, 80 subjects	*B. papyrifera*	[43]
Randomized, double-blind, 44 subjects	Radiotherapy + *B. serrata* extract	[52]
One subject with urinary bladder cancer	*B. sacra* oleo gum resin	[53]
One subject with breast cancer	*B. serrata* extract	[54]
12 subjects with brain tumor	*B. serrata* extract	[55]
29 subjects with glioma	Boswellic acid	[56]
One subject with malignant glioma	Oleo gum resin extract	[57]

(VAS), Lequesne index (LI), and the use of antiinflammatory scores [47]. *B. serrata* commonly referred as Shallaki in Ayurvedic practices was analyzed for their analgesic activity in healthy volunteers using mechanical pain model. Twelve subjects have obtained written consent, and they were randomly divided into two different groups, which were administered either 125 mg of the test product or either placebo. Mean baseline changed in the pain threshold force, and pain tolerance force showed significant changes ($p < 0.01$) in comparison with the placebo [48]. *B. serrata* was also used in combination with polyherbs like *Kaempferia galanga, Curcuma longa*, and *Piper nigrum* and was clinically evaluated for their effects in relieving the symptoms of OA. In a 3-month study involving subjects with knee OA, the subjects supplemented with the test product showed significant improvement in the WOMAC scores [49]. Multiple clinical studies have also evaluated the role of boswellic acids either alone or in combination for the treatment of rheumatoid arthritis (RA). In a study conducted by Sander et al. where they used a combination of conventional drugs and boswellic acids versus placebo, there was considerable difference in the study outcomes as indicated by the response scores [50]. Boswellic acids are also found to be useful in many other inflammatory and immunological diseases also. In an investigation initiated of bicentric phase, open-label, baseline-to-treatment pilot study, frankincense extract was evaluated for the treatment of multiple sclerosis. The study evaluated that the extract was beneficial in reducing the contrast-enhancing lesions, brain atrophy, and upregulation of T-cell markers [51]. *Boswellia papyrifera* was also evaluated their beneficiary in multiple sclerosis in a randomized clinical trial involving 80 patients for 8weeks and observed in increasing visuospatial memory [43]. Another important area of antiinflammatory activity is the role of the extract in various cancer disorders. In a randomized clinical trial, 4200mg/day of *B. serrata* extract was supplemented in combination with radiotherapy in 44 patients with primary or secondary brain tumors and observed that a subsequent reduction in the cerebral edema of greater than 75% was observed in 60% of the patients supplemented with the extract when compared with placebo [52]. The hydrodistillates of the gum resin extract of *B. sacra* were used to study its effect on urinary bladder. Even though the correct mechanism is not signified, the subject was tumor free after the surgery of the second occurrence [53]. Similarly, many studies also concentrated on multiple cancer studies, like breast cancer [54], brain tumors [55], glioma [56], and malignant glioma [57]. Various extracts of different species of *Boswellia* had been used in a number of human clinical trials involving subjects with bronchial asthma [5], ulcerative colitis [58], morbus Crohn [59], collagenous colitis [60], plague-induced gingivitis [61], and analgesic studies [48] and have obtained very promising results through multiple mode of actions.

7.6 Toxicity and side effects in clinical evaluations

Even though the *Boswellia* extracts have been found to be promising for multiple inflammatory conditions, studies have also reported some experimental related toxicology also. Before the development of any formulation, toxicology is a major concern. Many studies have concentrated the experimental and clinical toxicology of the frankincense extracts. *B. serrata* and *Boswellia ovalifoliolata* have been considered safe in multiple in vivo and not showing any symptoms of weight loss, skin irritation, and damage of any tissues. The LD50 values were observed to be above 5000mg/kg body weight for oral toxicity and 2000mg/kg body weight for dermal toxicity. Studies also indicated low genotoxicity for the extracts [62–64].

Some of the studies have also reported some experimental related occasional or rare side effects like nausea, increased joint pain, epigastric pain, allergic contact dermatitis, abdominal pain, and gastrointestinal symptoms. In contrast, some of the studies have reported no side effects and no sever adverse events [65, 66].

7.7 Conclusion

Traditional systems of medicine had always backed the beneficial use of *Boswellia* species in different forms because of their innumerable pharmacological activities. The majority of the antiinflammatory formulations include *Boswellia* as one of the ingredients. The antiinflammatory activities of the species had been well established by the scientific communities based on various studies including preclinical and clinical. Most of the studies have concentrated on the antiinflammatory potential of boswellic acids and many mechanisms of action and bioavailability of KBA and AKBA. More studies need to be generated to demonstrate the exact molecular targets and the antiinflammatory pathways.

References

[1] S.P. Ambasta, The Wealth of India: Raw Material, Council of Scientific and Industrial Research, New Delhi, 1988, pp. 203–205.

[2] H.P.T. Ammon, Boswellic acids in chronic inflammatory diseases, Planta Med. 72 (2006) 1100–1116.

[3] L. Cseke, A. Kirakosyan, P. Kaufman, S. Warber, J. Duke, H. Brielmann, Natural Products From Plants, Taylor & Francis, Boca Raton, FL, 2006.

[4] N.K. Roy, D. Parama, K. Banik, D. Bordoloi, A.K. Devi, K. Thakur, G. Padmavathi, M. Shakibaei, L. Fan, G. Sethi, A.B. Kunnumakkara, An update on pharmacological potential of boswellic acids against chronic diseases, Int. J. Mol. Sci. 20 (2019) 4101.

[5] I. Gupta, V. Gupta, A. Parihar, S. Gupta, R. Ludtke, H. Safayhi, H.P. Ammon, Effects of *Boswellia serrata* gum resin in patients with bronchial asthma: results of a double blind, placebo-controlled, 6-week clinical study, Eur. J. Med. Res. 3 (1998) 511–514.

[6] A. Pedretti, R. Capezzera, C. Zane, E. Facchinetti, P. Calzavara-Pinton, Effects of topical boswellic acid on photo and age-damaged skin: clinical, biophysical, and echographic evaluations in a double blind, randomized, split-face study, Planta Med. 76 (2010) 555–560.

[7] A. Rajabian, H. Sadeghnia, S. Fanoudi, A. Hosseini, Genus Boswellia as a new candidate for neurodegenerative disorders, Iran. J. Basic Med. Sci. 23 (2020) 277–286.

[8] S. Togni, G. Maramaldi, F. Di Pierro, M. Biondi, A cosmeceutical formulation based on boswellic acids for the treatment of erythematous eczema and psoriasis, Clin. Cosmet. Investig. Dermatol. 7 (2014) 321–327.

[9] Y. Yuan, S.X. Cui, Y. Wang, H.N. Ke, R.Q. Wang, H.X. Lou, Z.H. Gao, X.J. Qu, Acetyl-11-keto-beta-boswellic acid (AKBA) prevents human colonic adenocarcinoma growth through modulation of multiple signalling pathways, Biochim. Biophys. Acta 1830 (2013) 4907–4916.

[10] A. Moussaieff, N. Rimmerman, T. Bregman, A. Straiker, C.C. Felder, S. Shoham, Incensole acetate, an incense component, elicits psychoactivity by activating TRPV3 channels in the brain, FASEB J. 22 (2008) 3024–3034.

[11] A. Sultana, K. Raheman, A.R. Padmaja, S. Rahman, *Boswellia serrata* roxb. A traditional herb with versatile pharmacological activity: a review, Int. J. Pharm. Sci. Res. 4 (2013) 2106–2117.
[12] A. Tschirch, O. Halbey, Untersuchungen über die Sekrete. 28. U ber das Olibanum, Arch. Pharm. 236 (1898) 487–503.
[13] A. Winterstein, G. Stein, Untersuchungen in der Saponinreihe X. Zur Kenntnis der Mono-oxy-triterpenesäuren, Z. Physiol. Chem. 208 (1932) 9–25.
[14] K. Gerbeth, J. Meins, S. Kirste, F. Momm, M. Schubert-Zsilavecz, M. Abdel-Tawab, Determination of major boswellic acid in plasma by high-pressure liquid chromatography/mass spectrometry, J. Pharm. Biomed. Anal. 56 (2011) 998–1005.
[15] A. Sharma, S. Chhikara, S. Ghodekar, S. Bhatia, M. Kharya, V. Gajbhiye, A.S. Mann, A.G. Namdeo, K.R. Mahadik, Phytochemical and pharmacological investigations on *Boswellia serrata*, Pharmacogn. Rev. 3 (2009) 206–215.
[16] A. Al-Harrasi, S. Al-Saidi, Phytochemical analysis of the essential oil from botanically certified oleogum resin of *Boswellia sacra* (Omani Luban), Molecules 13 (2008) 2181–2189.
[17] M.P. Wymann, R. Schneiter, Lipid signalling in disease, Nat. Rev. Mol. Cell Biol. 9 (2008) 162–176.
[18] E.R. Sailer, L.R. Subramanian, B. Rall, R.F. Hoernlein, H.P. Ammon, H. Safayhi, Acetyl-11-keto-beta-boswellic acid (AKBA): structure requirements for binding and 5-lipoxygenase inhibitory activity, Br. J. Pharmacol. 117 (1996) 615–618.
[19] A. Altmann, D. Poeckel, L. Fischer, M. Schubert-Zsilavecz, D. Steinhilber, O. Werz, Coupling of boswellic acid-induced Ca2+ mobilisation and MAPK activation to lipid metabolism and peroxide formation in human leucocytes, Br. J. Pharmacol. 141 (2004) 223–232.
[20] A. Altmann, L. Fischer, M. Schubert-Zsilavecz, D. Steinhilber, O. Werz, Boswellic acids activate p42(MAPK) and p38 MAPK and stimulate Ca (2+) mobilization, Biochem. Biophys. Res. Commun. 290 (2002) 185–190.
[21] H. Safayhi, T. Mack, J. Sabieraj, M.I. Anazodo, L.R. Subramanian, H.P. Ammon, Boswellic acids: novel, specific, nonredox inhibitors of 5-lipoxygenase, J. Pharmacol. Exp. Ther. 261 (1992) 1143–1146.
[22] H. Safayhi, B. Rall, E.R. Sailer, H.P. Ammon, Inhibition by boswellic acids of human leukocyte elastase, J. Pharmacol. Exp. Ther. 281 (1997) 460–463.
[23] K. Heil, H.P. Ammon, H. Safayhi, Inhibition of NADPH-oxidase by AKBA in intact PMNs, Naunyn Schmiedebergs Arch. Pharmacol. 363S (2001) R14.
[24] K. Reising, J. Meins, B. Bastian, G. Eckert, W.E. Mueller, M.S. Zsilavecz, M. Abdel Tawab, Determination of boswellic acids in brain and plasma by high-performance liquid chromatography/tandem mass spectrometry, Anal. Chem. 77 (2005) 6640–6645.
[25] S. Sharma, V. Thawani, L. Hingorani, M. Shrivastava, V.R. Bhate, R. Khiyani, Pharmacokinetic study of 11-keto-beta-boswellic acid, Phytomedicine 11 (2004) 255–260.
[26] V. Sterk, B. Büchele, T. Simmet, Effect of food intake on the bioavailability of boswellic acids from an herbal preparation in healthy volunteers, Planta Med. 70 (2004) 1155–1160.
[27] D.M. Mostafa, N.M. Ammar, S.H. Abd El-Alim, A. Kassem, R. Hussein, G. Awad, S. Abdul-Wanees El-Awdan, *Boswellia carterii* liquisolid systems with promoted anti-inflammatory activity, Curr. Drug Deliv. 12 (2015) 454–463.
[28] N. Banno, T. Akihisa, K. Yasukawa, Anti-inflammatory activities of the triterpene acids from the resin of *Boswellia carteri*, J. Ethnopharmacol. 107 (2006) 249–253.
[29] R.A. Mothana, Anti-inflammatory, antinociceptive and antioxidant activities of the endemic Soqotraen *Boswellia elongata* Balf. f. and *Jatropha unicostata* Balf. f. in different experimental models, Food Chem. Toxicol. 49 (2011) 2594–2599.

[30] E.J. Blain, A.Y. Ali, V.C. Duance, *Boswellia frereana* (frankincense) suppresses cytokine-induced matrix metalloproteinase expression and production of pro-inflammatory molecules in articular cartilage, Phytother. Res. 24 (2010) 905–912.
[31] M. Mbiantcha, J. Almas, A.D. Atsamo, G. Ateufack, S.U. Shabana, D.F. Bomba, W.Y. Nana, D. Nida, Anti-inflammatory and anti-arthritic effects of methanol extract of the stem bark of *Boswellia dalzielii* Hutch (Burseraceae) in rats, Inflammopharmacology 26 (2018) 1383–1398.
[32] S. Su, Y. Hua, Y. Wang, W. Gu, W. Zhou, J. Duan, H. Jiang, T. Chen, Y. Tang, Evaluation of the anti-inflammatory and analgesic properties of individual and combined extracts from *Commiphora myrrha*, and *Boswellia carterii*, J. Ethnopharmacol. 139 (2012) 649–656.
[33] S.A. Zaitone, B.M. Barakat, S.E. Bilasy, M.S. Fawzy, E.Z. Abdelaziz, N.E. Farag, Protective effect of boswellic acids versus pioglitazone in a rat model of diet-induced non-alcoholic fatty liver disease: influence on insulin resistance and energy expenditure, Naunyn Schmiedebergs Arch. Pharmacol. 388 (2015) 587–600.
[34] Y. Zhang, J. Jia, Y. Ding, Y. Ma, P. Shang, T. Liu, G. Hui, L. Wang, M. Wang, Z. Zhu, Y. Li, A. Wen, Alpha-boswellic acid protects against ethanol-induced gastric injury in rats: involvement of nuclear factor erythroid-2-related factor 2/heme oxygenase-1 pathway, J. Pharm. Pharmacol. 68 (2016) 514–522.
[35] S. Singh, A. Khajuria, S.C. Taneja, R.K. Khajuria, J. Singh, R.K. Johri, G.N. Qazi, The gastric ulcer protective effect of boswellic acids, a leukotriene inhibitor from *Boswellia serrata*, in rats, Phytomedicine 15 (2008) 408–415.
[36] J.J. Liu, R.D. Duan, LY294002 enhances boswellic acid-induced apoptosis in colon cancer cells, Anticancer Res. 29 (2009) 2987–2991.
[37] D. Beghelli, G. Isani, P. Roncada, G. Andreani, O. Bistoni, M. Bertocchi, G. Lupidi, A. Alunno, Antioxidant and ex vivo immune system regulatory properties of *Boswellia serrata* extracts, Oxidative Med. Cell. Longev. 2017 (2017), 7468064.
[38] A. Henkel, N. Kather, B. Mönch, H. Northoff, J. Jauch, O. Werz, Boswellic acids from frankincense inhibit lipopolysaccharide functionality through direct molecular interference, Biochem. Pharmacol. 83 (2012) 115–121.
[39] J. Ren, Y.G. Wang, A.G. Wang, L.Q. Wu, H.J. Zhang, W.J. Wang, Y.L. Su, H. Qin, Cembranoids from the gum resin of *Boswellia carterii* as potential antiulcerative colitis agents, J. Nat. Prod. 78 (2015) 2322–2331.
[40] M. Verhoff, S. Seitz, M. Paul, S.M. Noha, J. Jauch, D. Schuster, O. Werz, Tetra-and pentacyclic triterpene acids from the ancient anti-inflammatory remedy frankincense as inhibitors of microsomal prostaglandin E(2) synthase-1, J. Nat. Prod. 7 (2014) 1445–1451.
[41] B. Gayathri, N. Manjula, K.S. Vinaykumar, B.S. Lakshmi, A. Balakrishnan, Pure compound from *Boswellia serrata* extract exhibits anti-inflammatory property in human PBMCs and mouse macrophages through inhibition of TNF alpha, IL-1beta, NO and MAP kinases, Int. Immunopharmacol. 7 (2007) 473–482.
[42] C.F. Krieglstein, C. Anthoni, E.J. Rijcken, M. Laukötter, H.U. Spiegel, S.E. Boden, S. Schweizer, H. Safayhi, N. Senniger, G. Shcurmann, Acetyl-11-keto-beta-boswellic acid, a constituent of a herbal medicine from *Boswellia serrata* resin, attenuates experimental ileitis, Int. J. Color. Dis. 16 (2001) 88–95.
[43] B. Sedighi, A. Pardakhty, H. Kamali, K. Shafiee, B.N. Hasani, Effect of *Boswellia papyrifera* on cognitive impairment in multiple sclerosis, Iran. J. Neurol. 13 (2014) 149–153.
[44] A. Kar, M.K. Menon, Analgesic effect of the gum resin of *Boswellia serrata* Roxb, Life Sci. 8 (1969) 1023–1028.
[45] A. Haroyan, V. Mukuchyan, N. Mkrtchyan, N. Minasyan, S. Gasparyan, A. Sargsyan, Efficacy and safety of curcumin and its combination with boswellic acid in osteoarthritis: a comparative, randomized, double-blind, placebo-controlled study, BMC Complement. Altern. Med. 18 (2018) 7.

[46] K. Sengupta, K.V. Alluri, A.R. Satish, S. Mishra, T. Golakoti, K.V. Sarma, D. Dey, S.P. Raychaudhuri, A double blind, randomized, placebo controlled study of the efficacy and safety of 5-Loxin for treatment of osteoarthritis of the knee, Arthritis Res. Ther. 10 (2008) R85.

[47] A. Notarnicola, G. Maccagnano, L. Moretti, V. Pesce, S. Tafuri, A. Fiore, B. Moretti, Methylsulfonylmethane and boswellic acids versus glucosamine sulfate in the treatment of knee arthritis: randomized trial, Int. J. Immunopathol. Pharmacol. 29 (2016) 140–146.

[48] K. Prabhavathi, U.S. Chandra, R. Soanker, P.U. Rani, A randomized, double blind, placebo controlled, cross over study to evaluate the analgesic activity of *Boswellia serrata* in healthy volunteers using mechanical pain model, Indian J. Pharmacol. 46 (2014) 475–479.

[49] A. Amalraj, J. Jacob, K. Varma, A.B. Kunnumakkara, C. Divya, S. Gopi, Acujoint™, a highly efficient formulation with natural bioactive compounds, exerts potent anti-arthritis effects in human osteoarthritis—a pilot randomized double blind clinical study compared to combination of glucosamine and chondroitin, J. Herb. Med. 17 (2019) 100276.

[50] O. Sander, G. Herborn, R. Rau, Is H15 (resin extract of *Boswellia serrata*, "incense") a useful supplement to established drug therapy of chronic polyarthritis? Results of a double-blind pilot study, Z. Rheumatol. 57 (1998) 11–16.

[51] K.H. Stürner, N. Verse, S. Yousef, R. Martin, M. Sospedra, Boswellic acids reduce Th17 differentiation via blockade of IL-1β-mediated IRAK1 signaling, Eur. J. Immunol. 44 (2014) 1200–1212.

[52] S. Kirste, M. Treier, S.J. Wehrle, G. Becker, M. Abdel-Tawab, K. Gerbeth, M. Abdel-Tawab, K. Gerbeth, M.J. Hug, B. Lubrich, A. Grosu, F. Momm, *Boswellia serrata* acts on cerebral edema in patients irradiated for brain tumors: a prospective, randomized, placebo-controlled, double-blind pilot trial, Cancer 117 (2011) 3788–3795.

[53] L. Xia, D. Chen, R. Han, Q. Fang, S. Waxman, Y. Jing, Boswellic acid acetate induces apoptosis through caspase-mediated pathways in myeloid leukemia cells, Mol. Cancer Ther. 4 (2005) 381–388.

[54] D.F. Flavin, A lipoxygenase inhibitor in breast cancer brain metastases, J. Neuro-Oncol. 82 (2007) 91–93.

[55] J.R. Streffer, M. Bitzer, M. Schabet, J. Dichgans, M. Weller, Response of radio chemotherapy-associated cerebral edema to a phytotherapeutic agent, H15, Neurology 56 (2001) 1219–1221.

[56] D.W. Böker, M. Winking, The role of boswellic acids in the therapy of malignant glioma, Dtsch. Ärztebl. 94 (1997). A-1197/B-965/C-873 (German).

[57] M. Winking, D.K. Böker, T. Simmet, Boswellic acid as an inhibitor of the perifocal edema in malignant glioma in man, J. Neurooncol. 30 (1996) 39.

[58] I. Gupta, A. Parihar, P. Malhotra, S. Gupta, R. Lüdtke, H. Safayhi, H.P. Ammon, Effects of gum resin of *Boswellia serrata* in patients with chronic colitis, Planta Med. 67 (2001) 391–395.

[59] W. Holtmeier, S. Zeuzem, J. Preiss, W. Kruis, S. Böhm, C. Maaser, A. Raedler, C. Schmidt, J. Schnitker, J. Schwarz, M. Zeitz, W. Caspary, Randomized, placebo controlled, double-blind trial of *Boswellia serrata* in maintaining remission of Crohn's disease: good safety profile but lack of efficacy, Inflamm. Bowel Dis. 17 (2011) 573–582.

[60] A. Madisch, S. Miehlke, O. Eichele, J. Mrwa, B. Bethke, E. Kuhlisch, E. Bastlein, G. Wilhems, A. Morgner, B. Wigginghaus, M. Stolte, *Boswellia serrata* extract for the treatment of collagenous colitis. A double blind, randomized, placebo-controlled, multicenter trial, Int. J. Colorectal. Dis. 22 (2007) 1445–1451.

[61] M. Khosravi Samani, H. Mahmoodian, A. Moghadamnia, A.P.B. Mir, M. Chitsazan, The effect of frankincense in the treatment of moderate plaque induced gingivitis: a double-blinded randomized clinical trial, Daru 19 (2011) 288–294.

[62] R. Sharma, S. Singh, G.D. Singh, A. Khajuria, T. Sidiq, S.K. Singh, G. Chashoo, S.S. Pagoch, A. Kaul, A.K. Saxena, R.K. Johri, S.C. Taneja, In vivo genotoxicity evaluation of a plant based antiarthritic and anticancer therapeutic agent boswellic acids in rodents, Phytomedicine 16 (2009) 1112–1118.

[63] K. Hientz, A. Mohr, D. Bhakta-Guha, T. Efferth, The role of p53 in cancer drug resistance and targeted chemotherapy, Oncotarget 8 (2017) 8921–8946.

[64] M. Asad, M. Alhomoud, Proulcerogenic effect of water extract of *Boswellia sacra* oleo gum resin in rats, Pharm. Biol. 54 (2016) 225–230.

[65] V.K. Alluri, S. Dodda, E.K. Kilari, T. Golakoti, K. Sengupta, Toxicological assessment of a standardized *Boswellia serrata* gum resin extract, Int. J. Toxicol. 38 (2019) 423–435.

[66] P. Singh, K.M. Chacko, M.L. Aggarwal, B. Bhat, R.K. Khandwal, S. Sultana, B.T. Kuruvilla, A-90 day gavage safety assessment of *Boswellia serrata* in rats, Toxicol. Int. 19 (2012) 273–278.

8

Antiinflammatory activity of galangal

Karthik Varma[a], Józef T. Haponiuk[b], and Sreeraj Gopi[a]

[a]*R&D Centre, Aurea Biolabs (P) Ltd, Kolenchery, Cochin, Kerala, India,*
[b]*Chemical Faculty, Gdansk University of Technology, Gdańsk, Poland*

8.1 Introduction

Inflammation is the body's first line of immune response to fight external pathogens and respond to harmful damaged cells, toxic materials, and toxic radiation [1]. The human body responds to inflammation by removing the injurious responses and initiates the healing process. There are different sorts of inflammatory conditions, which include mild to moderate and then lead to acute inflammatory conditions [2]. The onset of inflammatory conditions includes damage to cells and eventually cell death and the worst mortality. Inflammation is often characterized by redness in the body part, pain, heat sensation, swelling, loss of body functions, etc. [3]. There are several treatment methods including physiotherapy, supplementation of drugs, and surgical treatment methods. Mainly nonsteroidal antiinflammatory drugs (NSAIDs) are the most commonly used in inflammatory disorders. But long-term consumption of NSAIDs can often lead to several adverse side effects, including an increased risk of cardiovascular diseases.

Natural products and derived extracts have gained many accolades from ancient days onward because of their numerable pharmacological activities [4]. One of the most important and versatile species with multiple pharmacological activities is the Zingiberaceae family [5]. *Alpinia galanga* wild (*A. galanga*) and *Kaempferia galanga* (*K. galanga*) belong to the Zingiberaceae family and are potent herbs with attributed pharmacological activities. Commonly known as greater galangal and lesser galangal, both species have miraculous pharmacological activities [6, 7]. *Alpinia officinarum* Hance and *Alpinia calcarata* are also commonly referred to as lesser galangal and have a similar phytochemistry equivalent to *A. galanga*. The species are cultivated and grown in Asian countries such as India, Indo-China, Malaysia, Bangladesh, and China [8].

Inflammation and Natural Products. https://doi.org/10.1016/B978-0-12-819218-4.00001-8
© 2021 Elsevier Inc. All rights reserved.

The extracts of the rhizome and essential oil components are well explored on account of the pharmacological activities and the presence of phytochemicals. Traditional systems of medicines such as Ayurveda and Unani as well as folk medicines describe the use of the herb in the treatment of disorders such as diabetes [9], neurodegenerative diseases [10], amnesia [11], abdominal discomfort [12], etc.

The chapter presents an overview of the phytochemistry, pharmacological activities, antiinflammatory activity, and preclinical and clinical studies.

8.2 Phytochemistry

Phytochemical evaluations of the species have laid the foundation for the explanation of the numerous pharmacological activities of *K. galanga* and *A. galangal*. Phytochemicals can be classified into alkaloids, terpenes, and phenolic acids based on their metabolic biosynthetic pathway [13, 14]. Several studies have evaluated the presence of phytochemicals such as flavonoids, terpenoids, saponins, phenolic compounds, and volatile oils [15, 16]. The major bioactives found in *A. galanga* are galangin, kaempferol, galangal acetate, and 1,8-cineole [17–20]. Phenolic compounds and their derivatives such as *p*-hydroxy benzoic acid, vanillic acid, ferulic acid, kaempferol-3-*O*-methyl ether, kaempferol, apigenin, luteolin, and 1′-acetoxyeugenol acetate were also qualified. Multiple bioactives such as cineol, borneol, 3-carene-5-one, camphene, kaempferol, kaempferide, and cinnamaldehyde have been isolated from the rhizomes of *K. galanga* [21]. The seeds of *A. galanga* were extracted with 95% ethanol and subjected to purification with petroleum ether, ethyl acetate, and water fractions. Their phytochemical profiling evaluated the presence of 1′S-1′-acetoxyeugenol acetate, 1′S-1′-acetoxychavicol acetate, isocoranarin D, and caryolane-1,9β-diol; several bioactive components were quantified in the ethanolic extract [22]. The methanolic extract of *A. galangal* yielded several bioactives when evaluated using preparative high-performance liquid chromatography (HPLC) and nuclear magnetic resonance (NMR) spectroscopic analysis [23]. Several extracts such as 80% aqueous acetone, hexane, etc., yielded different fractions with various phytochemicals, of which some are novel [24–26]. Similarly, the hexane, chloroform, methanol, 50% ethanol, and supercritical fluid extraction (SCFE) extracts also gave information on the different amounts of phytochemicals present in *A. officinarum* [26–30]. The essential oil components of the rhizomes of *A. galanga* showed the presence of multiple essential oil sesquiterpenes [31]. The structural details of some of the bioactive components are given in Fig. 8.1 and described in Table 8.1.

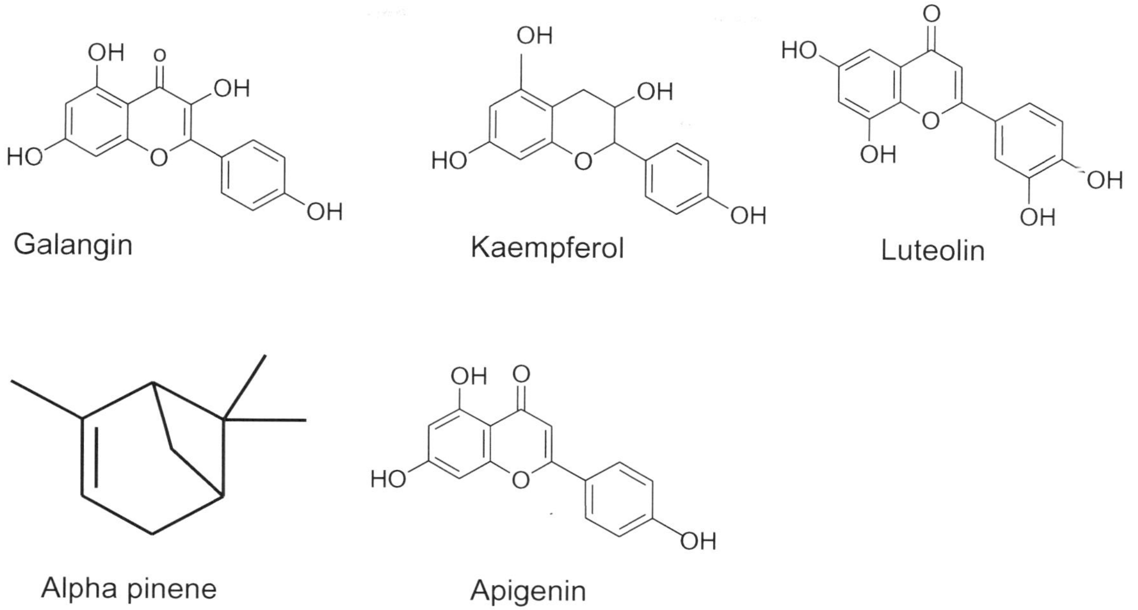

Galangin

Kaempferol

Luteolin

Alpha pinene

Apigenin

Fig. 8.1 Structures of bioactives.

Table 8.1 Bioactives in various extracts.

Herbs	Extract	Bioactives	Reference
A. galanga		*p*-Hydroxy benzoic acid, vanillic acid, ferulic acid, kaempferol-3-*O*-methyl ether, kaempferol, apigenin, luteolin, 1′-acetoxyeugenol acetate	[21]
A. galanga seeds	95% ethanol	1′S-1′-acetoxyeugenol acetate, 1′S-1′-acetoxychavicol acetate, isocoranarin D, and caryolane-1,9β-diol	[22]
A. galanga	Methanol	Galangol A, galangol B, galangol C	[23]
A. galanga	80% aqueous acetone	Galangalditerpene A, galangalditerpene B, galangalditerpene C, clovane-2β,9α-diol	[24]
A. galanga	80% aqueous acetone	Galangol D diacetate, 1′S-1′-acetoxy-chavicol acetate, 1′S-1′-acetoxteugenol acetate, 1′S-1′-hydroxychavicol acetate, 1′S-1′-hydroxyeugenol acetate, trans-*p*-coumaryl acetate, trans-*p*-acetoxycinnamoyl alcohol, trans-*p*-coumaryl alcohol, trans-*p*-coumaryl aldehyde	[25]
A. officinarum	Hexane	1′S-1′-Acetoxychavicol	[26]

Continued

Table 8.1 Bioactives in various extracts—cont'd

Herbs	Extract	Bioactives	Reference
A. officinarum	Sequential extraction with hexane, chloroform, and methanol	Galangin, kaempferide, kaempferide-3-*O*-β-D-glucoside	[27]
A. officinarum	Methanol	Nootkatone, yakuchinone A, diarylheptanoid, hannokinol, hexahydrocurcumin, galangin, pinocembrin, isorhamnetin, luteolin, rutin, apigenin, quercetin, acacetin, chrysin, tectochrysin, izalpinin, kaempferol, kaempferide	[28]
A. galanga	Hydrodistillation	α-Pinene, camphene, sabinene, β-pinene, β-myrcene, α-terpinene, 1,8-cineole 1,3,6-octatriene, γ-terpinene, α-terpinolene, borneol, α-terpineol, terpinen-4-ol, 2-butenal, chavicol, 4-allylphenyl acetate, eugenol, geranyl acetate, methyl eugenol, β-selinene, β-farnesene, germacrene-D, β-bisaboloene, β-sesquiphellandrene	[29]

8.3 Mechanism of antiinflammatory pathway

There are many mechanisms of action that contribute to the antiinflammatory activities. Pharmacological studies have reported that the combined extracts of *Cinnamomum cassia* Blume, *Anemarrhena asphodeloides* Bunge, and *A. officinarum* were shown to exhibit antiinflammatory activity by inhibiting nitric oxide (NO) production in a dose-dependent manner, probably by the inhibition of the NF-κB pathway [32]. Similar studies have also evaluated various extracts such as the aqueous acetone extract of the dried rhizome, which exhibited a dose-dependent inhibition of lipopolysaccharide (LPS)-induced nitric oxide synthase (IC_{50} 57 μg/mL) [33]. Similar studies showed significant suppression of nitrites and proinflammatory cytokines IL-1β and TNF-α via the modulation of mitogen-activated protein kinase (MAPK) and NF-κB. Molecular docking studies revealed that multiple compounds, galangin, kaempferide, isorhamnetin, and two diarylheptanoids isolated from the rhizomes of *A. officinarum* exhibited antiinflammatory activity as investigated in HepG2 cells stimulated by lipopolysaccharide. The molecular docking studies revealed that phytochemicals have the ability to downregulate the induced gene expression, thereby opening a pathway for the antiinflammatory activity of the herb [34]. Studies also showed that the phytochemical compound galangin downregulated mRNA levels of cytokines, including IL-1β and IL-6, and proinflammatory genes such as iNOS in LPS-activated macrophages in a dose-dependent manner. In addition, galangin supplementation decreased the protein expression levels of iNOS in activated macrophages and was found to exhibit antiinflammatory activity by inhibiting the extracellular signal-regulated kinase (ERK) and NF-κB-p65 phosphorylation. The study also showed that galangin downregulated the activity of IL-1β production in LPS-activated macrophages [35, 36].

8.4 Pharmacological activities

Many of the traditional systems of medicine and modern formulations have scientifically proven the antiinflammatory activity of various species of galangal to be used for different antiinflammatory formulations and the phytochemicals contributing to this. The antiinflammatory activity of various extracts was proved by both preclinical and clinical models.

8.4.1 Preclinical studies

In the traditional Indian system of medicine, the dried rhizome and leaves of *K. galanga* are used to treat headaches, swelling, stomach aches, toothaches, and rheumatism [37]. The antiinflammatory and analgesic activities of different doses of 600 and 1200 mg/kg of the *K. galanga* extract were elucidated in animal models such as the carrageenan and cotton pellet models that showed significant antiinflammatory activity and tail flick while the hot plate model showed significant analgesic activity [38]. When supplemented subcutaneously in doses of 30, 100, and 300 mg/kg, the aqueous extracts of *K. galanga* leaves show a significant antiinflammatory effect in rats in a dose-dependent manner [39]. *K. galanga* extracts also exhibited time- and dose-dependent antiinflammatory activities. The methanolic extract of the extract when supplemented at a dosage of 100 and 200 mg/kg exhibited significant antiinflammatory activity [40]. Ethyl *p*-methoxy cinnamate, an important phytochemical that is obtained from the extracts of *K. galanga*, showed significant antiinflammatory activity, thereby giving an idea of the pharmacological activity of some of the phytochemicals present in the extract [41].

In another study, the acute and chronic antiinflammatory activities of the root extract of *A. galanga* were studied in albino rats of either sex (150–200 g). The acute antiinflammatory activity was evaluated using rat paw edema carrageenan-, bradykinin-, and 5-HT-induced rat paw edema and the chronic antiinflammatory activity was evaluated using formaldehyde-induced rat paw edema. The efficacy of both the acute and chronic activities showed significant improvement in comparison with the standard drugs [42]. *K. galanga* has been used as a traditional medicine for antirheumatic activities because of its antiinflammatory activity. Various extracts such as alcoholic and petroleum ether extracts were tested against adjuvant-induced chronic inflammation in rats. The study suggested that the extract effectively reduced the progression of acute and chronic inflammation in rats [43]. The antinociceptive and antiinflammatory activities of the aqueous extracts of *K. galanga* were evaluated using female Balb/c mice and Sprague-Dawley rats. The extract showed significant antiinflammatory activity when assessed using the carrageenan-induced paw edema test [44]. The phytochemical analysis also revealed the antiinflammatory activities of *A. galanga*. The antiinflammatory activity of the ethanolic extract of *A. galanga* showed that the extract possessed significant antiinflammatory activity in carrageenan-induced pleurisy rats. The results showed that extracts at different dosages of 100, 200, and 400 mg exhibited significant ($P < .005$) antiinflammatory activity [42]. Various extracts of *A. galanga* such as alcohol, water, etc., were found to be effective in chronic arthritis in albino rats and the antiinflammatory activity was similar to β-methazone [45]. A concentrated extract of galangal rhizome has been reported to be effective at controlling the symptoms related to rheumatic knee pain [16]. A study conducted by Yu e al. concentrated on the isolation of multiple components from the extract, which has the ability to suppress the activity of the T-helper cell; this is reported to be the possible factor responsible for the inflammatory conditions [46]. The isolated chavicol analogues showed a suppressive effect on the inflammatory immune disorders, which resulted from the overactivation of cytokine production [47]. Similar studies using the total aqueous and alcoholic extract from alpinia rhizomes when evaluated using different models such as acute and subacute in rat models showed significant antiinflammatory activities [48]. The extract of *A. galanga* was also found to be effective in topical applications. The methanolic extract showed significant antiinflammatory activity in carrageenan induced rat paw edema when evaluated using positive controls of Piroxicam and methyl salicylate. The degree of inhibition of edema showed significant improvement when compared to the positive control [49]. The antiinflammatory activity of the methanolic and aqueous methanolic (1:1) of the rhizomes of *A. galangal* was investigated in carrageenan-induced paw edema in Wistar rats and compared with the positive drug control ibuprofen. The methanolic extract showed a maximum inhibition of 79.51% on carrageenan-induced rat paw edema [50]. Cancer is also a sort of inflammatory condition affecting body parts. Studies investigated the potential of the *A. galanga* rhizome to induce cytotoxic and apoptotic cells in the cultured human breast carcinoma cell line (MCF-7), when compared to nonmalignant (MRC-5) cells cultured.

The extent of apoptosis was measured using flow cytometry [51]. Similarly, the active component 1S′-1′-acetoxychavicol acetate exhibited inhibitory activity in oral squamous cell carcinoma by the regulation of NF-κB and IKK α/β activation [52]. Inflammatory bowel disease (IBD) is a sort of chronic inflammatory disease that affects the digestive tract. Multiple studies have found that galangin showed significant improvement in the clinical parameters, thereby contributing to the antiinflammatory activity [36, 53]. Similarly, various extracts are found to be beneficial in different sorts of inflammatory conditions.

8.4.2 Clinical studies

The proven preclinical antiinflammatory activity of the galanga species has become eye opening for human studies, either as a single extract or as a combination of herbs. The antiinflammatory studies include various sort of arthritis conditions such as rheumatoid arthritis, osteoarthritis, etc. The *K. galanga* rhizome extract was used to compare the efficacy of the extract against inflammatory markers such as tumor necrosis factor (TNF-α) and prostaglandin (PGE-2) with meloxicam in patients with knee osteoarthritis. The study was a two-phase study involving the preparation of the extracts, followed by double-blind clinical trials involving pre- and postdesign involving 18 subjects. The clinical efficacy was evaluated using WOMAC scores. The test results didn't reveal any significant changes in the pain scores, stiffness, and physical function between the intervention and control groups. The study didn't reveal significant differences in the parameters of inflammatory markers (TNF-α) and PGE-2 among the groups. The study thus gives initial information on the possibility of exploring the extract as an antiinflammatory drug [54].

Similarly, the pharmacological and clinical effectiveness of *Zingiber officinale* and *A. galanga* extract were studied. During the 12-week study, the subjects were supplemented with capsules containing ginger extract (150 mg) and powdered galangal rhizome dry extract (125 mg). The study population involving 40 subjects was randomly divided into two groups with different age populations and the primary outcomes were assessed. The primary outcomes included the response of the subjects in experiencing at least a 15-mm reduction in pain between the baseline and the final examination for knee pain upon standing and walking. The secondary outcomes included the measurement of the mean ± SE of measurements for knee pain upon standing and walking. All parameters showed positive results in the all the primary and secondary outcomes [55]. A similar combination was evaluated for its efficacy in the treatment of knee OA. A total of 261 subjects with knee OA and moderate-to-severe pain were enrolled in a randomized, double-blind, placebo-controlled, multi-center, parallel group, 6-week study. The primary outcome was the proportion of responders experiencing a reduction in knee pain upon standing using an intent-to-treat analysis. A responder was defined by a reduction in pain of greater than or = 15 mm on a visual analog scale (VAS). In the group of 247 subjectable patients, the percentage of responders experiencing a reduction in knee pain was significant in the test group compared with the control group (63% versus 50%; $P = .048$). An analysis of the secondary efficacy variables also indicated a greater response in the test group compared with the control group when analyzing mean values: reduction in knee pain upon standing (24.5 mm versus 16.4 mm; $P = .005$), reduction in knee pain after walking 50 feet (15.1 mm versus 8.7 mm; $P = .016$), and reduction in the Western Ontario and McMaster Universities osteoarthritis composite index (12.9 mm versus 9.0 mm; $P = .087$). But as an outcome, the patients receiving ginger extract experienced more mild gastrointestinal (GI) adverse events than did the placebo group (59 patients versus 21 patients) [56]. From ancient days onward, herbal preparations including multiple ingredients have been used for the treatment of various inflammatory conditions, including osteoarthritis (OA). A polyherbal formulation containing *K. galanga*, *Boswellia serrata*, *Curcuma longa*, and *Piper nigrum* has been clinically evaluated for its effect in relieving the symptoms of OA. The primary outcomes of the study included evaluating the Western Ontario and McMaster Universities osteoarthritis index (WOMAC) score, the pain score, the functional ability score, the visual analog scale (VAS), and the Lequesene functional index. The randomized, 90-day, double-blind, active-controlled, single-center, clinical study carried out using glucosamine (1500 mg) and chondroitin (1200 mg) showed significant improvement in the primary and secondary outcomes [57].

8.5 Usage in traditional systems

Both *A. galanga* and *K. galanga* are used in various forms of traditional systems of medicine such as Ayurvedic preparations. Commonly referred to as "Kulanjan," *A. galanga* has been used in various Ayurvedic preparations such as Maha Narayana oil, Rasna sapthakaya, Dasamul Iguradi Kwathaya, Kumara Guliya, etc., of which most are used for inflammatory disorders [58]. *K. galanga* is used in almost 59 Ayurvedic preparations [59]. The herb is one of the prior ingredients in some Ayurvedic preparations such as Gandha Thailam and Rasnairandadi Kashayam, which is most commonly used in rheumatic diseases, fractures, and sprains.

8.6 Toxicity studies

The development cycle of any new drug will be completed only if the toxicity studies are conducted and the results need to be on the safer side. Most of the studies have concentrated on acute and chronic studies. In acute (24 h) and chronic (90 days) situations, oral toxicity studies on the ethanolic extract of the rhizomes of *A. galanga* and *Curcuma longa* were evaluated in mice. The acute dosages included 0.5, 1.0, and 3 g/kg body weight whereas the chronic dosage was 100 mg/kg/day of the extract. The primary outcomes of the study included measurement of all the external morphological, hematological, and spermatogenic changes. Changes in the body weight and vital organs were also reviewed. The rats treated with *A. galanga* extract showed significant difference in weight gain. Hematological studies also revealed a significant rise in the RBC level, a gain in the weights of the sexual organs, and increased sperm motility and sperm count. The extract didn't showed any spermatotoxic effects [60]. Various extracts of the *K. galanga* rhizome such as the ethanolic extract were screened for the Hippocratic screening test as well as acute and subacute toxicity studies in rats. The hexane extract was tested for dermal irritation. In the acute toxicity test, rats supplemented with 5 g/kg of the extract didn't produce either mortality or significant changes in the body and organ weights. Moreover, neither gross abnormalities nor histopathological changes were detected. Rats supplemented with 25, 50, or 100 mg/kg of the ethanolic extract didn't show any mortality during the 28 days. The differential leukocyte content showed only a slight decrease in rats supplemented with 50 and 100 mg/kg supplementation. Blood parameters also showed no changes, thereby substantiating the safety of the herb. The dermal studies of the extract also showed no irritating effects [8].

8.7 Conclusion

Natural products and derived phytochemicals possess research interest these days. The herbs in the Zingiberaceae family are among the most widely explored. *Alpinia galanga* and *Kaempferia galanga* are the most common used herbs for antiinflammatory activities. Both the essential oil and extracts are widely used for antiinflammatory activities. The exact mechanism of the antiinflammatory pathway is unknown. Many preclinical and clinical studies have confirmed the effects, but more studies are needed to predict the exact molecular pathway behind the antiinflammatory activities.

References

[1] R. Medzhitov, Inflammation 2010: new adventures of an old flame, Cell 140 (2010) 771–776.

[2] O. Takeuchi, S. Akira, Pattern recognition receptors and inflammation, Cell 140 (2010) 805–820.

[3] A. Murakami, H. Ohigashi, Targeting NOX, INOS, COX-2 in inflammatory cells: chemoprevention using food phytochemicals, Int. J. Cancer 121 (2007) 2357–2363.

[4] B.V. Patel, A Report of the Seminar on, Herbal Drugs: Present Status and Future Prospects, Perd Centre, Ahmedabad, 2001.

[5] T. Wu, K. Larsen, Zingibearcea, Flora China 24 (2000) 322–377.

[6] A. Bhatt, O.B. Kean, C.L. Keng, Sucrose, benzylaminopurine and photoperiod effects on in vitro culture of *Kaempferia galanga* Linn, Plant Biosyst. Int. J. Deal. Asp. Plant Biol. 146 (4) (2012) 900–905.

[7] C.R. Achuthan, J. Padikkala, Hypolipidemic effect of *Alpinia galanga* (Rasna) and *Kaempferia galanga* (Kachoori), Indian J. Clin. Biochem. 12 (1) (1997) 55–58.

[8] D. Kanjanapothi, A. Panthong, N. Lertprasertsuke, C. Rujjanawate, D. Kaewpinit, R. Sudthayakorn, W. Choochote, U. Chaithong, A. Jitpakdi, B. Pitasawat, Toxicity of crude rhizome extract of *Kaempferia galanga* L. (Proh Hom), J. Ethnopharmacol. 90 (2–3) (2004) 359–365.

[9] R.K. Verma, G. Mishra, P. Singh, K.K. Jha, R.L. Khosa, Anti-diabetic activity of methanolic extract of *Alpinia galanga* Linn. aerial parts in streptozotocin induced diabetic rats, Ayu 36 (2015) 91–95.

[10] H.J.C. Singh, V. Alagarsamy, P.V. Diwan, S. Sathesh Kumar, J.C. Nisha, Y. Narsimha Reddy, Neuroprotective effect of *Alpinia galanga* (L.) fractions on Aβ(25-35) induced amnesia in mice, J. Ethnopharmacol. 138 (1) (2011) 85–91.

[11] H.J.C. Singh, V. Alagarsamy, S. Sathesh Kumar, Y. Narsimha Reddy, Neurotransmitter metabolic enzymes and antioxidant status on Alzheimer's disease induced mice treated with *Alpinia galanga* (L.) Willd, Phytother. Res. 25 (2011) 1061–1067.

[12] S. Athamaprasangsa, U. Buntrarongroj, P. Dampawan, N. Ongkavoranan, V. Rukachaisirikul, S. Sethijinda, M. Sornnarintra, P. Sriwub, W.C. Taylor, A 1,7 diarylheptanoid from *Alpinia conchigera*, Phytochemistry 37 (1994) 871–873.

[13] R. Croteau, T.M. Kutchan, N.G. Lewis, Natural products (secondary metabolites), in: B. Buchanan, W. Gruissem, R. Jones (Eds.), Biochemistry & Molecular Biology of Plants, American Society of Plants, Rockville, MD, 2015, pp. 1250–1318.

[14] E.M. Yahia, Fruit and Vegetable Phytochemicals: Chemistry and Human Health, second ed., vol. 2, John Wiley & Sons, Hoboken, NJ, 2017.

[15] N. Aziman, N. Abdullah, Z.M. Noor, W.S.S.W. Kamarudin, K.S. Zulkifli, Phytochemical profiles and antimicrobial activity of aromatic Malaysian herb extracts against food-borne pathogenic and food spoilage microorganisms, J. Food Sci. 79 (2014) M583–M592.

[16] A. Chudiwal, D. Jain, R. Somani, *Alpinia galanga* Wild.–an overview on phytopharmacological properties, Indian J. Nat. Prod. Resour. 1 (2010) 143–149.

[17] S. Ghosh, L. Rangan, Alpinia: the gold mine of future therapeutics, 3 Biotech 3 (2013) 173–185.

[18] A. Hamad, A. Alifah, A. Permadi, D. Hartanti, Chemical constituents and antibacterial activities of crude extract and essential oils of *Alpinia galanga* and *Zingiber officinale*, Int. Food Res. 23 (2016) 837–841.

[19] S. Jaju, N. Indurwade, D. Sakarkar, N. Fuloria, M. Ali, S. Das, S.P. Base, Galango flavonoid isolated from rhizome of *Alpinia galanga* (L) sw (Zingiberaceae), Trop. J. Pharm. Res. 8 (2009) 545–550.

[20] P. Ravindran, G. Pillai, I. Balachandran, M. Divakaran, Galangal, in: Handbook of Herbs and Spices, Elsevier, 2012, pp. 303–318.

[21] G.C. Huang, C.L. Kao, W.J. Li, S.T. Huang, H.T. Li, C.Y. Chen, A new phenylalkanoid from the rhizomes of Alpinia galanga, Chem. Nat. Compd. 54 (2018) 1072–1075.

[22] Q.H. Zeng, C.L. Lu, X.W. Zhang, J.G. Jiang, Isolation and identification of ingredients inducing cancer cell death from the seeds of *Alpinia galanga*, a Chinese spice, Food Funct. 6 (2015) 431–443.

[23] M.Q. Bian, J. Kang, H.Q. Wang, Q.J. Zhang, C. Liu, R.Y. Chen, Three new norsesquiterpenoids from the seeds of Alpinia galanga, J. Asian Nat. Prod. Res. 16 (2014) 459–464.

[24] Y. Manse, K. Ninomiya, R. Nishi, Y. Hashimoto, S. Chaipech, O. Muraoka, T. Morikawa, Labdane-type diterpenes, galangalditerpenes A–C, with melanogenesis inhibitory activity from the fruit of *Alpinia galanga*, Molecules 22 (2017) 2279.

[25] Y. Manse, K. Ninomiya, R. Nishi, I. Kamei, Y. Katsuyama, T. Imagawa, Melanogenesis inhibitory activity of a 7-O-9′-linked neolignan from *Alpinia galanga* fruit, Bioorg. Med. Chem. 24 (2016) 6215–6224.

[26] R.G. Baradwaj, M.V. Rao, T. Senthil Kumar, Novel purification of 1'S-1'-acetoxychavicol acetate from *Alpinia galanga* and its cytotoxic plus anti-proliferative activity in colorectal adenocarcinoma cell line SW480, Biomed. Pharmacother. 91 (2017) 485–493.

[27] G. Eumkeb, S. Sakdarat, S. Siriwong, Reversing β-lactam antibiotic resistance of Staphylococcus aureus with galangin from *Alpinia officinarum* Hance and synergism with ceftazidime, Phytomedicine 18 (2010) 40–45.

[28] B.B. Zhang, Y. Dai, Z.X. Liao, L.S. Ding, Three new antibacterial active diarylheptanoids from *Alpinia officinarum*, Fitoterapia 81 (2010) 948–952.

[29] L.P. Köse, I. Gülçin, A.C. Gören, J. Namiesnik, A.L. Martinez-Ayala, S. Gorinstein, LC–MS/MS analysis, antioxidant and anticholinergic properties of galanga (*Alpinia officinarum Hance*) rhizomes, Ind. Crop. Prod. 74 (2015) 712–721.

[30] J.C. Luo, W. Rui, M.M. Jiang, Q.L. Tian, X. Ji, Y.F. Feng, Separation and identification of diarylheptanoids in supercritical fluid extract of *Alpinia officinarum* by UPLC-MS-MS, J. Chromatogr. Sci. 48 (2010) 795–801.

[31] N. Khumpirapang, S. Pikulkaew, S. Anuchapreeda, S. Okonogi, *Alpinia galanga* oil—a new natural source of fish anaesthetic, Aquac. Res. 49 (2018) 1546–1556.

[32] M.Y. Jeong, J.S. Lee, J.D. Lee, N.J. Kim, J.W. Kim, S. Lim, A combined extract of *Cinnamomi Ramulus*, *Anemarrhenae Rhizoma* and *Alpiniae Officinari* Rhizoma suppresses production of nitric oxide by inhibiting NF-κB activation in RAW 264.7 cells, Phytother. Res. 22 (2008) 772–777.

[33] H. Matsuda, S. Ando, T. Kato, T. Morikawa, M. Yoshikawa, Inhibitors from the rhizomes of *Alpinia officinarum* on production of nitric oxide in lipopolysaccharide-activated macrophages and the structural requirements of diarylheptanoids for the activity, Bioorg. Med. Chem. 14 (2006) 138–142.

[34] A.A. Elgazar, N.M. Selim, N.M. Abdel-Hamid, M.A. El-Magd, H.M. El Hefnawy, Isolates from *Alpinia officinarum* Hance attenuate LPS-induced inflammation in HepG2: evidence from in silico and in vitro studies, Phytother. Res. 32 (7) (2018) 1273–1288.

[35] Y.C. Jung, M.E. Kim, J.H. Yoon, P.R. Park, H. Youn, H. Lee, J.S. Lee, Anti-inflammatory effects of galangin on lipopolysaccharide-activated macrophages via ERK and NF-κB pathway regulation, Immunopharmacol. Immunotoxicol. 36 (6) (2014) 426–432.

[36] R. Sangaraju, N. Nalban, S. Alavala, V. Rajendran, M.K. Jerald, R. Sistla, Protective effect of galangin against dextran sulfate sodium (DSS)-induced ulcerative colitis in Balb/c mice, Inflamm. Res. 68 (8) (2019) 691–704.

[37] R. Mitra, J. Orbell, M.S. Muralitharan, Agriculture — medicinal plants of Malaysia, Asia Pac. Biotech. News 11 (2007) 105–110.

[38] A.M. Vittalrao, T. Shanbhag, M. Kumari, K.L. Bairy, S. Shenoy, Evaluation of anti-inflammatory and analgesic activities of alcoholic extract of *Kaempferia galanga* in rats, Indian J. Physiol. Pharmacol. 55 (2011) 13–24.

[39] M.R. Sulaiman, Z.A. Zakaria, I.A. Daud, F.N. Ng, Y.C. Ng, M.T. Hidayat, Antinociceptive and anti-inflammatory activities of the aqueous extract of *K. Galanga* leaves in animal models, J. Nat. Med. 62 (2008) 221–227.
[40] W. Ridtitid, C. Sae-wong, W. Reanmongkol, M. Wongnawa, Anti-inflammatory activity of the methanol extract of Kaempferia galanga Linn. in experimental animals, Planta Med. (2009) 75.
[41] M.I. Umar, M.Z. Asmawi, A. Sadikun, I.J. Atangwho, M.F. Yam, R. Altaf, A. Ahmed, Bioactivity-guided isolation of ethyl-p-methoxycinnamate, an anti-inflammatory constituent, from *Kaempferia galanga* L. extracts, Molecules 17 (2012) 8720–8734.
[42] K.R. Subash, G. Bhanu Prakash, K. Vijayachandra Reddy, K. Manjunath, K. Umamaheswara Rao, Anti-inflammatory activity of ethanolic extract of *Alpinia galanga* in carrageenan induced pleurisy rats, Natl. J. Physiol. Pharm. Pharmacol. 6 (2016) 468–470.
[43] P.C. Jagadish, K.P. Latha, J. Mudgal, G.K. Nampurath, Extraction, characterization and evaluation of *Kaempferia galanga* L. (Zingiberaceae) rhizome extracts against acute and chronic inflammation in rats, J. Ethnopharmacol. 24 (2016) 434–439.
[44] M.R. Sulaiman, Z.A. Zakaria, I.A. Daud, F.N. Ng, Y.C. Ng, M.T. Hidayat, Antinociceptive and anti-inflammatory activities of the aqueous extract of Kaempferia galanga leaves in animal models, J. Nat. Med. 62 (2008) 221–227.
[45] G.P. Sharma, P.V. Sharma, Experimental study of anti-inflammatory activity of some raasna drugs, J. Res. Indian Med. Yoga Homoeopath. 12 (1978) 18–21.
[46] E.S. Yu, H.J. Min, K. Lee, J.W. Nam, E.K. Seo, J.H. Hong, E.S. Hwang, Anti-inflammatory activity of p-coumaryl alcohol-γ-Omethylether is mediated through modulation of interferon-γ-production in Th cells, Brit. J. Pharmacol. 156 (2009) 1107–1114.
[47] H.J. Min, J.W. Nam, E.S. Yu, J.H. Hong, E.K. Seo, E.S. Hwang, Effect of naturally occurring hydroxychavicol acetate on the cytokine production in T helper cells, Int. Immunopharmacol. 9 (2009) 448–454.
[48] R. Satish, R. Dhananjayan, Evaluation of the anti-inflammatory potential of rhizome of *Alpinia galanga*, Biomedicine 23 (2003) 91–96.
[49] M. Nagashekhar, H. Shivprasad, Anti-inflammatory and analgesic activity of the topical preparation of *Alpinia galanga* wild, Biomed. Contents 1 (2006) 63–69.
[50] A. Unnisa, T.D. Parveen, Anti-inflammatory and acute toxicity studies of the extracts from the rhizomes of *Alpinia galanga* Wild, Der Pharm. Sin. 2 (2) (2011) 361–367.
[51] S. Samarghandian, M.A. Hadjzadeh, J.T. Afshari, M. Hosseini, Antiproliferative activity and induction of apoptotic by ethanolic extract of *Alpinia galanga* rhizome in human breast carcinoma cell line, BMC Complement. Altern. Med. 14 (2014) 192.
[52] L.L.A. In, N.M. Arshad, H. Ibrahim, M.N. Azmi, K. Awang, N.H. Nagoor, 1'-Acetoxychavicol acetate inhibits growth of human oral carcinoma xenograft in mice and potentiates cisplatin effect via proinflammatory microenvironment alterations, BMC Complement. Altern. Med. 12 (1) (2012) 179.
[53] X. Xuan, A. Ou, S. Hao, J. Shi, X. Jin, Galangin protects against symptoms of dextran sodium sulfate-induced acute colitis by activating autophagy and modulating the gut microbiota, Nutrients 12 (2020) 347.
[54] N.A. Taslim, M.N. Djide, Y. Rifai, A.N. Syahruddin, Y.R. Rampo, M. Mustamin, S. Angriawan, Double blind randomized clinical trial of *kaempferia galanga* l extract as an anti-inflammation (prostaglandin e2 and tumor necrosis factor alpha) on osteoarthritis, Asian J. Pharm. Clin. Res. 12 (2019) 63–66.
[55] G. Selga, M. Sauka, L. Aboltina, A. Davidova, P. Kaipainen, D. Kheder, T. Westermarck, F. Atroshi, Pharmacological and clinical effectiveness of *Zingiber officinale* and *Alpinia galanga* in patients with osteoarthritis, in: Pharmacology and Nutritional Intervention in the Treatment of Disease, IntechOpen, 2014. 7.

[56] R.D. Altman, K.C. Marcussen, Effects of a ginger extract on knee pain in patients with osteoarthritis, Arthritis Rheum. 44 (2001) 2531–2538.
[57] A. Amalraj, J. Jacob, K. Varma, A.B. Kunnumakkara, C. Divya, S. Gopi, Acujoint™, a highly efficient formulation with natural bioactive compounds, exerts potent anti-arthritis effects in human osteoarthritis – a pilot randomized double blind clinical study compared to combination of glucosamine and chondroitin, J. Herb. Med. 17 (2019) 100276.
[58] M.S.M. Shiffa, N. Fahamiya, M.U.Z.N. Farzana, P. Miriyalini, Kulanjan (Alpinia Galanga) from the perspective of Unani medicine, Int. J. Univers. Pharm. Bio Sci. 5 (2016) 1–9.
[59] V.V. Sivarajan, I. Balachandran, Ayurvedic Drugs and Their Plant Sources, Oxford and IBH Publishing Co. Pvt. Ltd., New Delhi, 1994.
[60] S. Qureshi, A.H. Shah, A.M. Ageel, Toxicity studies on *Alpinia galanga* and *Curcuma longa*, Planta Med. 58 (2) (1992) 124–127.

9

Antiinflammatory natural products from marine algae

Ayman M. Mahmoud[a,b], May Bin-Jumah[c], and Mohammad H. Abukhalil[d]

[a]Physiology Division, Department of Zoology, Faculty of Science, Beni-Suef University, Beni-Suef, Egypt, [b]Biotechnology Department, Research Institute of Medicinal and Aromatic Plants, Beni-Suef University, Beni-Suef, Egypt, [c]Department of Biology, College of Science, Princess Nourah Bint Abdulrahman University, Riyadh, Saudi Arabia, [d]Department of Biology, Faculty of Science, Al-Hussein Bin Talal University, Ma'an, Jordan

9.1 Introduction

The oceans represent a vast area of the earth that play a central role in its dynamic and in supporting the human well-being. The crucial role of the oceans ranges from providing livelihoods, food, and recreational opportunities to regulating the global climate [1, 2]. The marine environment has been well acknowledged as an interesting source of natural compounds with unique and uncommon chemical properties. Based on the amazing features of these compounds, a great progress in the molecular modeling and chemical synthesis of new therapeutics with high specificity and efficacy might be achieved [2–4]. Marine microorganisms represent a valuable source for bioactive secondary metabolites that might represent promising leads in the development of effective therapeutic agents [5].

In the last few years, multiple studies have been conducted to explore the chemical structure, biochemical properties, and physical features, as well as the biotechnological applications of bioactive compounds derived from marine organisms [6]. Among marine organisms, algae that can be defined as aquatic eukaryotic photosynthetic organisms diversified in size from microalgae to macroalgae represent a reservoir of structurally diverse bioactive compounds with beneficial biological effects [7]. The marine algae vary greatly in size and range from few microns in the microalgae to several meters in the giant kelps [8]. Macroalgae, also called seaweeds, include the green, red, and brown algae and occupy the littoral zone,

Inflammation and Natural Products. https://doi.org/10.1016/B978-0-12-819218-4.00012-2
© 2021 Elsevier Inc. All rights reserved.

whereas microalgae occur in both littoral and benthic habitats and also throughout the ocean [2, 5]. Owing to their richness in bioactive compounds, marine algae have attracted great attention in the fields of biomedicine and natural products. Sterols, sulfated polysaccharides, brominated phenols, fucoxanthin, amino acids and amines, vitamins, guanidine derivatives, kainic acid, nitrogen heterocyclics, and brominated oxygen heterocyclics are among the active ingredients responsible for various beneficial effects of algae [8–11]. Marine algae–derived compounds have been reported to exhibit antiinflammatory, antidiabetic, antioxidant, anticoagulant, antiviral, antiallergic, antimicrobial, and anticarcinogenic effects [12–24].

Macroalgae are a group of multicellular plant-like protists that can be classified into green (*Chlorophyta*), red (*Rhodophyta*), and brown (*Phaeophyta*). Chlorophyll *a* and *b*, carotenes, and xanthophylls are the pigments responsible for the green color of *Chlorophyta*, whereas phycobilin and fucoxanthin are the pigments behind the red and brown colors of *Rhodophyta* and *Phaeophyta*, respectively [25]. Marine macroalgae could be utilized as functional foods providing multiple health benefits. Seaweeds are rich sources of various bioactive compounds and many minerals, including potassium, iodine, iron, and sulfur [26]. The content of biochemical elements, such as carbohydrates, proteins, lipids, minerals, and vitamins, in different seaweeds varies depending on seasonal conditions and the geographic area [27]. Seaweeds have been utilized as foods and crude drugs for the treatment of iodine deficiency, hypercholesterolemia, diabetes, and vermifuges and as a source of vitamins. Additionally, some seaweeds have been employed as ointments and dressings [8].

Microalgae are photosynthetic microscopic organisms, which constitute a major part of marine water phytoplankton. Microalgae are able to adapt to adverse environmental conditions such as extreme temperature and hydrothermal vents. Hence, these organisms represent good candidates for drug discovery because of their unique compounds developed for survival and defense. Given their potential health benefits, microalgae and their bioactive constituents have gained interest as shown in different studies. With the great diversity of their species, strains, and phytochemical constituents, microalgae could give a clue to the huge interest in the development of effective therapeutic agents for the treatment of chronic diseases. Microalgae are rich sources of several biologically active compounds, including polysaccharides, phenolics, fatty acids, carotenoids, and sterols [28]. Previous studies have demonstrated the antiinflammatory, antioxidant, hypocholesterolemic, antiviral, and anticancer activities of microalgal-derived extracts and active constituents [28, 29].

9.2 Inflammation

Inflammation is a response of the immune system to stimulation by invading pathogens or endogenous signals such as damaged cells. The developed inflammatory response can result in tissue repair or sometimes pathological changes when the response goes unchecked [30, 31]. Inflammation is an essential mechanism in human health and disease. Various pathogenic factors, such as infection, tissue injury, or cardiac infarction, can induce inflammation by causing tissue damage [31–33]. The process of inflammation delivers leukocytes and proteins to foreign invaders, such as microbes, and to damaged or necrotic tissues, and it activates the recruited cells and molecules, which then function to eliminate the harmful or unwanted substances [31, 34]. A close link between inflammation and multiple chronic diseases, including cancer, asthma, metabolic syndrome, diabetes, cardiovascular disease, inflammatory bowel disease (IBD), rheumatoid arthritis, and chronic obstructive lung disease, has been demonstrated [33]. The inflammatory response developed through a series of sequential steps is as follows: (1) cell surface pattern receptors recognize detrimental stimuli, (2) circulating leukocytes and proteins are recruited to the site where the offending agent is located, (3) the leukocytes and proteins are activated and work in concert to eliminate the offending substance, (4) the reaction is controlled and terminated, and (5) the injured tissue is repaired [32]. Hence, inflammation is a dynamic process with many mediators, including interleukins (IL), tumor necrosis factor (TNF)-α, and vascular endothelial growth factor (VEGF) playing central roles [31].

Inflammation is generally classified into acute and chronic. Acute inflammation is of short duration and typically develops within minutes or hours and lasts for several hours or a few days [35]. It is characterized by the exudation of plasma proteins and fluids (edema) and the emigration of leukocytes, predominantly neutrophils. This process causes the cardinal signs of acute inflammation, such as redness, heat, swelling, and pain [36]. When acute inflammation achieves its desired goal of eliminating the offenders, the inflammatory response can then decrease and resolve. However, if the initial response failed to eliminate the stimulus, the reaction progresses to a protracted type of inflammation that is called chronic inflammation [37]

Chronic inflammation can be defined as slow, long-term inflammation lasting for prolonged periods and is associated with the presence of macrophages, lymphocytes, and plasma cells, and the proliferation of blood vessels, fibrosis, and more tissue destruction [32, 38]. It may follow acute inflammation or may begin insidiously as a low-grade smoldering response without any manifestations of an acute reaction [36, 37]. Chronic inflammation has been demonstrated as the main culprit behind the tissue damage in many common and disabling human diseases, such as tuberculosis, atherosclerosis, rheumatoid arthritis, and pulmonary fibrosis [38–40]. Additionally, it has been implicated in the progression of cancer and degenerative diseases, such as Alzheimer's disease [33, 35].

9.2.1 Antiinflammatory agents

The molecular understanding of inflammatory response has introduced many potential therapeutic targets for controlling inflammation and preventing its deleterious effects. In this context, nonsteroidal antiinflammatory drugs (NSAIDs), such as aspirin and ibuprofen, are among the most effective and widely used antiinflammatory agents. NSAIDs decrease fever, prevent blood clots, reduce pain, and in higher doses reduce inflammation. This class of drugs works by inhibiting the activity of cyclooxygenases (COX-1 and COX-2), enzymes involved in the synthesis of prostaglandins (PGs) and thromboxanes (TXs). PGs and TXs are active mediators involved in inflammation and blood clotting, respectively [41]. However, NSAIDs have many side effects depending on the specific drug, in particular, increased risk of gastrointestinal bleeds and ulcers, nephrotoxicity, and heart attack [42, 43]

Glucocorticoids are powerful antiinflammatory drugs modeled upon cortisol, which is the principal human glucocorticoid [44]. These synthetic drugs have been employed in the treatment of septic shock, multiple sclerosis, systemic lupus erythematosus, IBD, asthma, allergy, rheumatoid arthritis, and many other inflammatory and autoimmune diseases [45]. However, the therapeutic application of antiinflammatory steroids has been associated with serious adverse effects particularly with high dosage and prolonged use [44].

Moreover, several biologicals have been developed for treating inflammation and inflammatory diseases. These newly developed antiinflammatory biologicals include anticytokine therapies, which reduce specific cytokines or block their receptors. These anticytokine agents have found a place in the treatment of psoriasis, rheumatoid arthritis, multiple sclerosis, IBD, and other autoimmune diseases. However, anticytokine therapies can decrease the host immune defense against infection and cancer [45]. Other antiinflammatory agents preventing the binding of monocyte-lymphocyte costimulatory molecules or depleting B lymphocytes have also been developed [31]. Furthermore, peroxisome proliferator-activated receptor (PPAR) agonists, histone deacetylase inhibitors, and small RNAs are antiinflammatory agents that are currently in use or under development [45]. Although different classes of effective antiinflammatory agents are available, the development of more effective and less toxic antiinflammatory therapeutics is a challenge. Therefore natural products have attracted considerable attention due to their safety and studies being conducted to evaluate their antiinflammatory activities.

9.3 Algal natural products with antiinflammatory activity

9.3.1 Polysaccharides

Polysaccharides are the main components in marine both micro- and macroalgae. Given their health-promoting effects, a large number of studies aimed to isolate and characterize marine algae–derived polysaccharides [46]. Most of the marine polysaccharides are sulfated and have been proven to exert antiinflammatory, immunomodulatory, antitumor, antidiabetic, antihypercholesterolemia, and anticoagulant effects [3]. These beneficial effects making marine algae polysaccharides promising bioactive products and biomaterials with a wide range of applications particularly in drug development [3]. Molecular size, type, and ratio of the constituent monosaccharides and the nature of glycosidic linkages are the main chemical properties that are closely related to the biological activities of polysaccharides [47].

Numerous marine algae–derived polysaccharides have been isolated, and their antiinflammatory activity has been well documented. Sulfated polysaccharide isolated from *Lobophora variegata* showed antiinflammatory effects in acute inflammatory conditions via inhibition of inducible nitric oxide synthase (iNOS) and COX activities [48]. Sulfated polysaccharides from *Porphyra haitanensis* regulated the imbalance of the Th1/Th2 immune response and modified serum immunoglobulin (Ig) E and IgG in a mouse model of tropomyosin-induced allergy [49]. In tropomyosin-elicited splenic lymphocytes in vitro, *P. haitanensis*-sulfated polysaccharides suppressed IL-4, IL-5, and IL-13 mRNA expression and augmented interferon (IFN)-γ and IL-10 [49]. Sulfated polysaccharides from *Caulerpa mexicana* mitigated carrageenan-induced paw edema and myeloperoxidase (MPO) and alleviated histamine and dextran, but not serotonin in paw edema, pointing to histamine as the main target [50]. Polysaccharides from the marine red alga *Digenea simplex* attenuated inflammation provoked by dextran and histamine and decreased neutrophil in paws and TNF-α and IL-1β levels in the peritoneal cavity of a murine model of carrageenan-triggered edema [51].

Fucoidan, ulvan, and laminarin sulfate are marine algae polysaccharides with promising antiinflammatory properties [52]. In addition to fucoidan, other marine polysaccharides, such as chitin and its derivatives, alginate, and porphyrin have been demonstrated as downregulators of allergic responses [53]. Alginic acid is an anionic polysaccharide isolated from *Sargassum wightii* and can reduce inflammatory process in arthritic rats through reduction of activities of COX, MPO, lipoxygenase (LPO), and the levels of C-reactive protein (CRP), ceruloplasmin, and rheumatoid factor (RF) [54]. The in vitro antiinflammatory effect of alginic acid from *Sargassum horneri* against urban aerosol-induced inflammation has been recently demonstrated [55]. In aerosol-induced keratinocytes, alginic acid suppressed COX-2, IL-6, and TNF-α and inhibited the key mediators of the nuclear factor-kappaB (NF-κB) and mitogen-activated protein kinase (MAPK) pathways. In addition, alginic acid reduced proinflammatory cytokines, iNOS, NO, COX-2, and PGE2 in aerosol-induced macrophages [55].

Fucan, isolated from brown algae *Sargassum vulgare*, is a sulfated polysaccharide that exhibited a strong antiinflammatory activity in a model of carrageenan-induced acute inflammation and paw edema [56]. The antiinflammatory activity of fucan has been demonstrated by the reduced edema and cellular infiltration [56]. Sulfated fucans from *Laminaria hyperborea* showed antiproliferative activity and decreased VEGF secretion from uveal melanoma and retinal pigment epithelium cell lines as recently reported by Dörschmann et al. [57].

Fucoidan is a complex sulfated polysaccharide derived from the cell walls of brown seaweeds and some marine invertebrate tissues. Fucoidan can significantly inhibit the release NO induced by bacterial lipopolysaccharide (LPS) and reduce inflammation [58]. Fucoidan is also a ligand for macrophage scavenger receptor A, which can be taken up by macrophages and inhibits NO production. Inhibition of leukocytes migration to the inflamed tissues is another manifestation of the antiinflammatory activity of fucoidan [59]. It was reported that fucoidan from wakame reduces the expression of COX-2 in rabbit articular chondrocytes in a dose-dependent manner, thereby exerting antiarthritis effects [60]. In addition, fucoidan can synergistically enhance the efficacy of antiinflammatory drugs. Fucoidan-coated ciprofloxacin-loaded chitosan nanoparticles can effectively treat intracellular and biofilm infections of *Salmonella* [61]. In a rat model of peptone-induced inflammation, intravenous administration of fucoidan prevented inflammation particularly when administered 15 min after peptone injection where P-selectin is maximally expressed [62]. In support of the in vivo findings, fucoidan showed a strong binding affinity toward P-selectin when compared with the P-selectin ligand SiaLea/x-PAA-biot [62]. Fucoidan has also been applied as a P-selectin targeting agent in the imaging of aorta in $ApoE^{-/-}$ [63] and arterial thrombi in rat models of infective endocarditis or abdominal aortic aneurysms [64].

Fucoidan reduced inflammatory cytokine expression in the lung of a mouse model of radiation-induced pneumonitis and lung fibrosis [65]. In a *Porphyromonas gingivalis*-infected mouse model of periodontitis, fucoidan inhibited gingival inflammation, neutrophil recruitment, and expression of proinflammatory cytokines [66]. Recently, fucoidan derived from *Padina commersonii* has been reported to inhibit LPS-induced inflammation in macrophages in vitro [67]. The antiinflammatory effect of fucoidan has been mediated via blocking toll-like receptor (TLR)/NF-κB signaling pathway [67]. Fucoidan from the seaweed *Fucus vesiculosus* decreased inflammatory cell infiltration, serum IL-4 and IgE, and infiltration of $CD4^+$ T cells in skin lesions and ameliorated ear swelling in 2,4-dinitrofluorobenzene-treated mice [68].

Laminarin is a polysaccharide composed of (1,3)-β-D-glucan with β(1,6) branching and occurs abundantly in *Laminaria* species. Laminarin demonstrated antibacterial, chemopreventive, and prebiotic activities. It showed a potential to modulate gut microbiota, which in turn can regulate neuroinflammation [69]. In a rat model of LPS-induced hepatitis, laminarin modulated the immune response, reduced recruitment of inflammatory cells, and decreased secretion of the proinflammatory mediator TNF-α [70]. In a dextran sodium sulfate (DSS)-induced porcine model, laminarin and/or fucoidan reduced colonic IL-6 mRNA abundance and ameliorated inflammation [71].

9.3.2 Fatty acids and lipid derivatives

Marine algae contain several types of lipids, including glycolipids, phospholipids, betaine lipids, and glycerolipids [72]. The lipid content of marine macroalgae is generally low, and most of it is represented by long-chained polyunsaturated fatty acids (PUFA), mainly omega-3 and omega-6 fatty acids. The content of PUFA in macroalgae differs according to the temperature where it is higher in those living in cold habitats. PUFAs possess multiple health benefits, including regulation of blood clotting and blood pressure [73]; improving the development of nervous system [74]; lowering the risk of obesity, diabetes, and arthritis [74–76]; and regulation of microglial signaling and suppressing neuroinflammation [74]. In addition, *n*-3 PUFAs have been shown to inhibit NF-κB, a key transcription factor regulating the expression of proinflammatory mediators [77]

Marine microalgae are responsible for the biosynthesis of *n*-3 PUFAs and have recently become a good alternative source of the PUFAs eicosapentaenoic acid (EPA) and docosahexaenoic acid (DHA) (Fig. 9.1). EPA and DHA are fatty acids that showed in vivo antiinflammatory activity, and clinical evidence demonstrates their beneficial effects as a dietary supplement in IBD, asthma, and rheumatoid arthritis [78–81]. EPA and DHA were reported to exert antiinflammatory effects by decreasing PGs, leukotrienes, and proinflammatory cytokines [82, 83], increasing

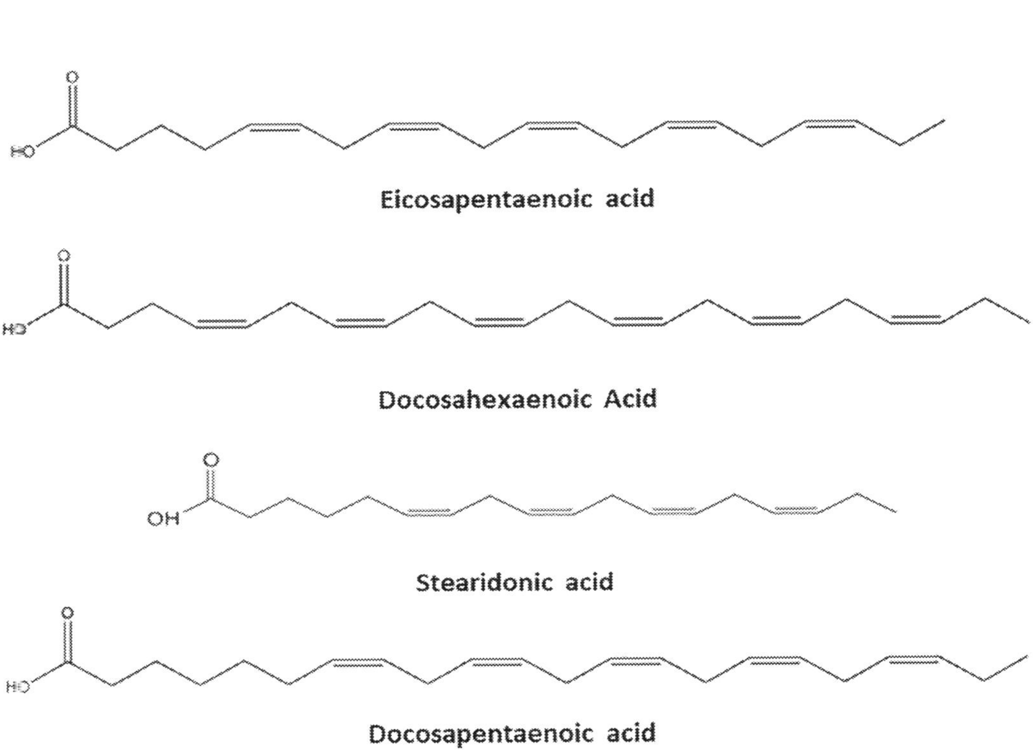

Eicosapentaenoic acid

Docosahexaenoic Acid

Stearidonic acid

Docosapentaenoic acid

Fig. 9.1 Chemical structures of some antiinflammatory fatty acids isolated from marine algae.

antiinflammatory mediators [84], inhibiting the expression of adhesion molecules [85], and suppressing leukocyte chemotaxis [86].

Microalgae-derived fatty acids decreased the percentage of $CD4^+$ T cells producing IFNγ and TNF-α and increased those producing IL-17A and IL-12 from two different mouse strains [87]. Antiinflammatory effects have also been observed in studies examining macroalgal PUFA. Khan et al. isolated EPA and stearidonic acid (SDA) (Fig. 9.1) from the seaweed *Undaria pinnatifida* and reported that these PUFAs reduced phorbol myristate acetate (PMA)-induced inflammation in mice when applied topically [88]. The brown seaweed *Ishige okamurae* was shown to alleviate allergic inflammation by reducing histamine release and modulating inflammatory cytokine production in human basophilic cells, and the efficacious component was confirmed to be the PUFAs [89]. Another study showed that algal oils containing DHA and omega-6 docosapentaenoic acid (DPA*n*-6) (Fig. 9.1) exhibited antiinflammatory activity and prevented LPS-stimulated IL-1β and TNF-α secretion by human peripheral blood mononuclear cells (PBMCs) [90]. In an in vitro study, EPA and DHA inhibited NF-κB activity and IL-1β secretion and activated ERK1/2 MAPK in the colon adenocarcinoma cell line Caco-2 [91]. In the same context, DHA and DPA*n*-6 reduced the production of TNF-α, IL-1β, and PGE2 and downregulate COX-2 expression in human PBMCs [92]. The antiinflammatory effect of algal DHA and DPA*n*-6 was supported by their ability to reduce paw edema in rats to an extent similar to indomethacin [92]. The α-, and β-unsaturated carbonyl moiety have been shown to enhance the antiinflammatory activity of algal-oxygenated fatty acids as revealed by the structure–activity relationship and in vitro studies [93].

7-Methoxy-9-methylhexadeca-4,8-dienoic acid is a methoxylated fatty acid isolated from *Ishige okamurae*. This fatty acid derivative inhibited the activity of bacterial phospholipase A2 in vitro and edema and erythema induced by PMA in BALB/c mice [94]. In the same context, acetylene-containing fatty acid derivative from *Liagora farinosa* has been demonstrated to exert a phospholipase A2 inhibitory activity [95]. Lee et al. have reported the antiinflammatory efficacy of (*E*)-10-oxooctadec-8-enoic acid and (*E*)-9-oxooctadec-10-enoic acid isolated from the red alga *Gracilaria verrucosa* [96]. These enoic acid derivatives suppressed NF-κB in LPS-induced macrophages, resulting in inhibition of NO, TNF-α, and IL-6 production [96]. The same alga has been studied by Dang et al. who showed that the isolated (9*E*,12*E*)-11-oxo-9,12-octadecadienoic acid and (8*E*,11*E*)-10-oxooctadeca-8,11-dienoic acid inhibited NO, IL-6, and TNF-α production from macrophages challenged with LPS [93]. In a study on *Scytonema julianum*, a glyco-analog of phosphatidylglycerol, acyl-acetylated sphingosine, and a phosphoglyco-analog of acyl-sphingosine were isolated, and their antiinflammatory activity has been investigated [97]. These isolated lipid derivatives inhibited platelet-activating factor (PAF)-induced rabbit platelet aggregation [97].

9.3.3 Proteins and peptides

Algae have been consumed in the human diet due to their high protein content, nutritional value, and health promoting properties [98]. Some marine algae are known to contain a high protein content similar to those of traditional protein sources [99, 100]. For instance, *Porphyra* spp. were reported to contain relatively higher amounts of proteins. Proteins of seaweeds are rich in glutamic and aspartic amino acids, whereas their cysteine content is relatively low [98]. Glutamic acid contributes to the typical taste of marine seaweeds. Carnosine, taurine, mycosporine-like and other bioactive amino acids, and peptides are constituents of the macroalgae [69, 90]. Mycosporine-like amino acids isolated from *Chlamydomonas hedleyi* have been demonstrated to possess antiinflammatory activity evidenced by reduced COX-2 expression in HaCat cells [101]. The amino acid carnosadine isolated from the red alga *Grateloupia carnosa* has shown an antiinflammatory activity [102].

The peptides produced through enzymatic hydrolysis of seaweeds protein exhibited antioxidant, antiinflammatory, immunomodulatory, and antimicrobial activities [103]. The bioactive peptide isolated from *Pyropia yezoensis* reduced the release of the proinflammatory mediators IL-1β, TNF-α, iNOS, and COX-2 in LPS-induced macrophages in a dose-dependent manner [104]. In addition, the antiinflammatory activity of *Pyropia yezoensis* peptide was associated with downregulation of the MAPKs p38, ERK, and JNK [104]. The antiinflammatory activity of protein hydrolysates from *Ulva* spp. has also been investigated in splenic macrophages and lymphocytes. These protein hydrolysates exerted a modulatory effect on TLR4/NFκB/MAPK pathways and the production of cytokines [105]. Another study showed a novel antiinflammatory action of purified peptides from enzymatic hydrolysate of the edible microalgae *Spirulina maxima* mediated via preventing the release of histamine from antigen-stimulated mast cells [106].

Lectins are glycoproteins isolated from marine algae and showed antiinflammatory and antiviral and many other biological activities. The functionalities of lectins depend on the composition, sequence, and number of the amino acids [107]. Lectin isolated from the marine green alga *Caulerpa cupressoides* exerted antiinflammatory effect manifested by the reduced neutrophil number and inhibition of paw edema in mice [108]. Lectin from *Pterocladiella capillacea* inhibited leukocyte migration induced by carrageenan in male Wistar rats, and its antinociceptive effect has been suggested to occur via peripheral rather than a central-acting mechanism [109]. In another study, lectin agglutinin isolated from *Hypnea cervicornis* inhibited carrageenan- and antigen-induced hypernociception and prevented neutrophil recruitment to the plantar tissue when administered intravenously [110]. A recent study demonstrated that lectin obtained from the red seaweed *Bryothamnion triquetrum* showed antiinflammatory effects through decreasing neutrophil migration, MPO activity, and TNF-α and IL-1β production in carrageenan-induced peritonitis in mice [111]. In the same line a lectin from *Caulerpa cupressoides* reduced mechanical hypernociception and inflammation in the rat temporomandibular joint during zymosan-induced arthritis through the reduction of leukocyte influx and expression of IL-1β and TNF-α [112].

The antiinflammatory potential of the pigment-protein complex phycocyanin has been investigated in many studies. Phycocyanin isolated from *Arthrospira maxima* suppressed carrageenan-induced paw edema in rats, arachidonic acid–induced ear edema in mice, cotton pellet granuloma, and acetic acid–induced colitis in rats [113]. C-phycocyanin isolated from *Spirulina platensis* reduced the expression of COX-2 and IL-6 and prevented chemically induced tumor promotion in mouse skin [114]. C-phycocyanin from blue-green algae has also been reported to prevent glucose oxidase–induced inflammation in mouse paw [115].

9.3.4 Phenolic compounds

Phenolic compounds are a diverse group of secondary metabolites characterized by the presence of hydroxyl groups attached to aromatic hydrocarbon rings. Phenolics are very well-known phytochemicals produced in plants through the acetate-malonate and shikimate acid pathways [116]. These compounds could be simple phenols or complex molecules depending on their structural properties. This class of compounds includes phenolic acids, coumarins, flavonoids, stilbenes, lignans, lignins, phlorotannins, and their derivatives [117]. These metabolites have attracted great attention due to their extensive biological activities, including antioxidant, antiinflammatory, anticarcinogenic, antidiabetic, and hepatoprotective effects [118–126].

Phenols are largely represented in the plant kingdom; however, the phenolic compounds present in marine algae are different when compared with those produced by terrestrial plants [70, 127]. Phloroglucinols and phlorotannins (eckols, phlorethols, fuhalols, fucols fucophlorethols, and ishofuhalols) are the common phenolics in marine algae (Fig. 9.2). Phlorotannins are found in *Phaeophyta*, whereas bromophenols, phenolic acids, and flavonoids occur in *Chlorophyta* and *Rhodophyta* [69, 128, 129]. Given their rich content of phenolic compounds, brown seaweeds have extensively been investigated in terms of their bioactive constituents and health-promoting effects [130]. In this context, phlorotannins have shown potent inhibitory effects toward TNF-α, IL-6, IL-1β, COX-2, and iNOS in LPS-induced microglia [130].

The antiinflammatory activity of edible brown alga *Eisenia bicyclis* and its constituents fucosterol and phlorotannins has been investigated by Jung et al. [131]. Different fractions of *Eisenia bicyclis* inhibited LPS-induced NO, tert-butyl hydroperoxide (t-BHP)–induced oxidative stress, and iNOS and COX-2 expression in murine macrophages [131]. The dichloromethane fraction showed a strong antiinflammatory activity, and phytochemical analysis revealed the presence of fucosterol. In addition, eckol, dieckol, phloroglucinol, dioxinodehydroeckol, phlorofucofuroeckol A, and 7-phloroeckol are six known phlorotannins that were isolated from the ethyl acetate fraction of *Eisenia bicyclis*. These isolated compounds inhibited LPS-induced NO generation in a dose-dependent manner [131].

Eckol, dieckol, phloroglucinol, phlorofucofuroeckol-A, and 8,8′-bieckol are phlorotannins purified from *Eisenia bicyclis*, and their antiinflammatory potential has been investigated against porcine pancreas phospholipase A2, soybean LPO, COX-1, and COX-2 [132]. The antiinflammatory effects of phlorotannins have also been investigated using mast cells. Treatment of the histamine-releasing RBL-2H3 mast cells with different phlorotannins suppressed inflammation as

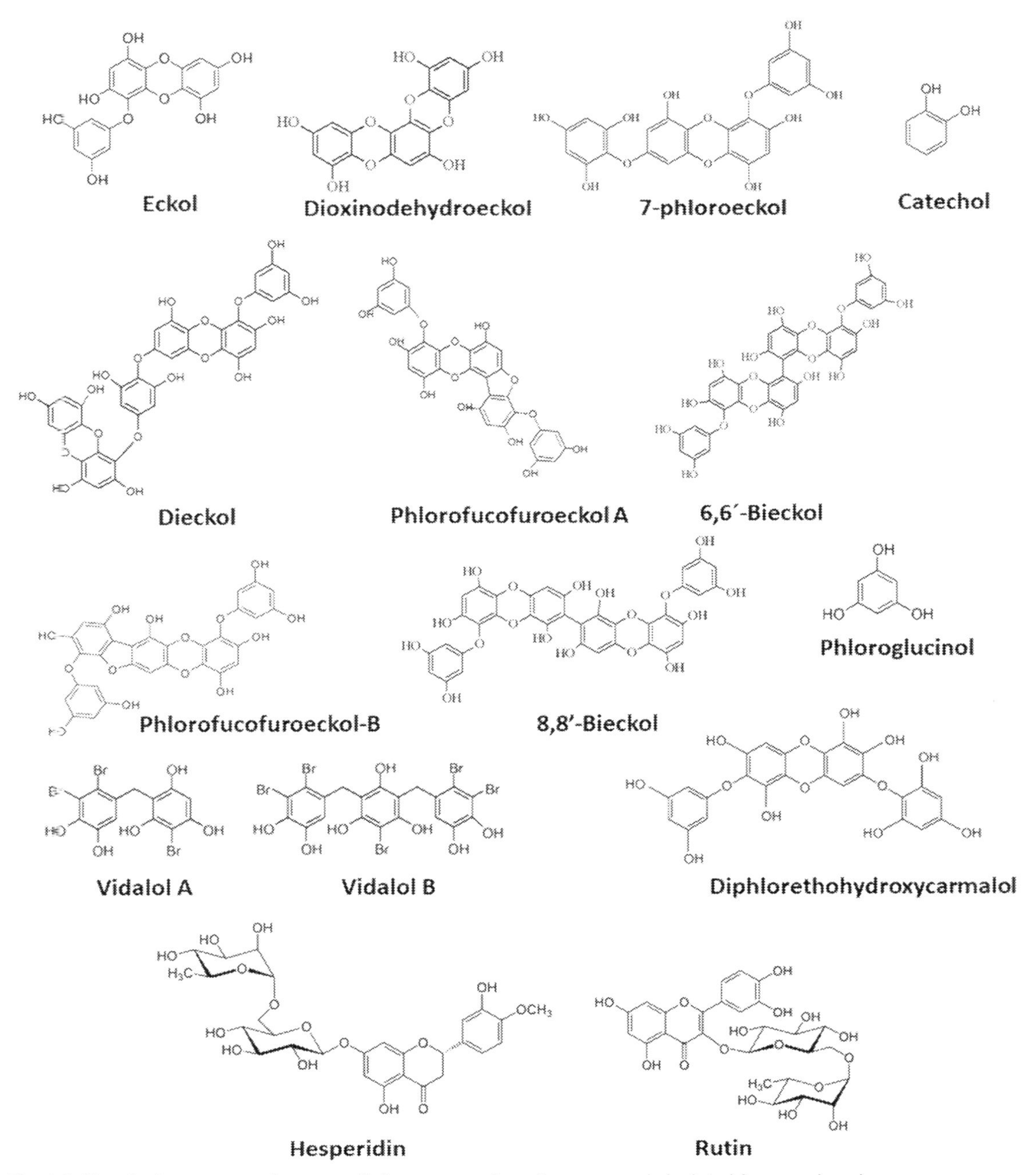

Fig. 9.2 Chemical structures of some antiinflammatory phenolic compounds isolated from marine algae.

evidenced by the decreased COX-2 both mRNA abundance and activity, where 6,6′-bieckol, 8,8′-bieckol, and phlorofucofuroeckol-A exerted the highest activity [133].

Phlorofucofuroeckol-B is a bioactive phenolic isolated from the edible brown seaweed *Eisenia arborea* and showed an ability to inhibit the release of histamine from rat basophile leukemia [134]. The crude extract of *Porphyra dentata* and catechol, hesperidin, and rutin isolated from this red alga showed antiinflammatory effects and prevented NO production from LPS-induced macrophages. Catechol and rutin showed an inhibitory effect on the expression of NF-κB and iNOS [135].

Vidalols A and B isolated from *Vidalia obtusiloba*, a Caribbean marine red alga, inhibited phospholipase A2 and ear inflammation induced by phorbol ester in mice [136]. Dieckol isolated from *Ecklonia cava* has been demonstrated to suppress NO and PGE2 production and iNOS and COX-2 expression dose-dependently in murine microglia. In addition, dieckol downregulated NF-κB, p38 MAPK, TNF-α, and IL-1β induced by LPS in microglia in vitro [137]. Phlorofucofuroeckol A and B from the brown seaweed *Ecklonia stolonifera* exerted an antiinflammatory activity manifested by the suppressed iNOS expression and PGE2 production in LPS-stimulated macrophages [138].

The ethanolic extract of *Ishige okamurae* has been reported to inhibit the production of TNF-α, IL-1β, and PGE2 from LPS-induced macrophages via inactivating NF-κB [22]. Further studies on *Ishige okamurae* resulted in isolation of the biologically active phlorotannin diphlorethohydroxycarmalol (DPHC) that possesses potent antiinflammatory activities [139–141]. DPHC downregulated iNOS, COX-2, and VEGF receptor 2 (VEGFR-2) expression and suppressed NF-κB in macrophages and human umbilical vein endothelial cells [139, 140]. In LPS-induced macrophages, DPHC inhibited IL-6 release and suppressed the phosphorylation or the nuclear translocation of NF-κB in a dose-dependent manner [141]. In a mouse model of atopic dermatitis, DPHC decreased IgE, inflammatory cell infiltration, and ear thickness [141]. Octaphlorethol A isolated from another species, *Ishige foliacea*, inhibited the inflammatory response induced by CpG in murine bone marrow–derived macrophages and dendritic cells by inhibiting MAPK and NF-κB signaling [142].

9.3.5 Terpenoids, sterols, and carotenoids

Terpenoids are a large group of organic secondary metabolites constructed by isoprene units. Terpenoids are classified into mono-, sesqui-, di-, sester-, tri-, and tetraterpenoids based on the number of isoprene units [143]. Sterols and carotenoids could be categorized under terpenoids. Steroids consist of six isoprene units and are derivatives of triterpenoids, whereas carotenoids are tetraterpenoid derivatives with eight isoprene units [143]. Fucosterol is a sterol with multiple biological activities and predominates in *Phaeophyta*, whereas cholesterol is abundant in *Rhodophyta*, and *Chlorophyta* contains various steroids [144].

The antiinflammatory activity of marine algae–derived terpenoids (Fig. 9.3) has been investigated in numerous studies. Epitaondiol isolated from the brown alga *Stypopodium flabelliforme* has shown antiinflammatory activities mediated through its ability to suppress phospholipase A2, COX, and NF-κB [145]. Rhiphocephalin is a linear sesquiterpene from *Rhipocephalus phoenix* with promising antiinflammatory efficacy. It has shown an ability to inhibit the production of bee venom–derived phospholipase A2 [146]. Neorogioltriol is diterpene extracted from *Laurencia glandulifera* that exerted antiinflammatory effects in LPS-stimulated macrophages mediated via NF-κB inhibition [147]. Fucosterol isolated from the brown seaweed *Eisenia bicyclis* inhibited LPS-induced NO generation in a dose-dependent manner, t-BHP-induced reactive oxygen species (ROS) generation, and iNOS and COX-2 expression [131]. Epitaondiol, pacifenol, and stypotriol triacetate are three terpenoids isolated from marine algae. These compounds inhibited leukocyte accumulation and ear edema in mice challenged with 12-*O*-tetradecanoylphorbol 13-acetate (TPA) [148]. In a study conducted by Caroprese et al., a mixture of ergosterol and 7-dehydroporiferasterol isolated from *Dunaliella tertiolecta* inhibited the release of proinflammatory cytokines TNF-α, IL-6, and IL-1β from sheep PBMCs challenged with LPS [149]. The antiinflammatory activity of sterols isolated from the green alga *Chlorella vulgaris* has been demonstrated by Yasukawa et al. [150]. Ergosterol, ergosterol peroxide, 7-dehydroporiferasterol peroxide, 7-dehydroporiferasterol, and 7-oxocholesterol significantly inhibited TPA-induced inflammation in mice [150].

Fucoxanthin is a major carotenoid present in the chloroplasts of brown algae. It possesses many beneficial biological effects, and its antiinflammatory potential has been reported in many studies. Accordingly, Heo et al. [151] and Kim et al. [152] have demonstrated the inhibitory effects of fucoxanthin on inflammatory mediators in LPS-stimulated macrophages. In these studies, fucoxanthin suppressed the expression of iNOS and COX-2 and reduced NO, PGE2, TNF-α, IL-1β, and IL-6 production, an effect mediated via inhibiting NF-κB and MAPK signaling [151, 152]. Fucoxanthin isolated from *Myagropsis myagroides* and *Ishige okamurae* has also inhibited NO, iNOS, PGE2, and COX-2 and suppressed the gene expression levels of TNF-α, IL-1β, and IL-6 in LPS-induced macrophages [153]. Sakai et al. have reported the potential of fucoxanthin to suppress the degranulation of mast cells through inhibition of antigen-induced aggregation of high affinity IgE receptor, demonstrating its antiinflammatory and

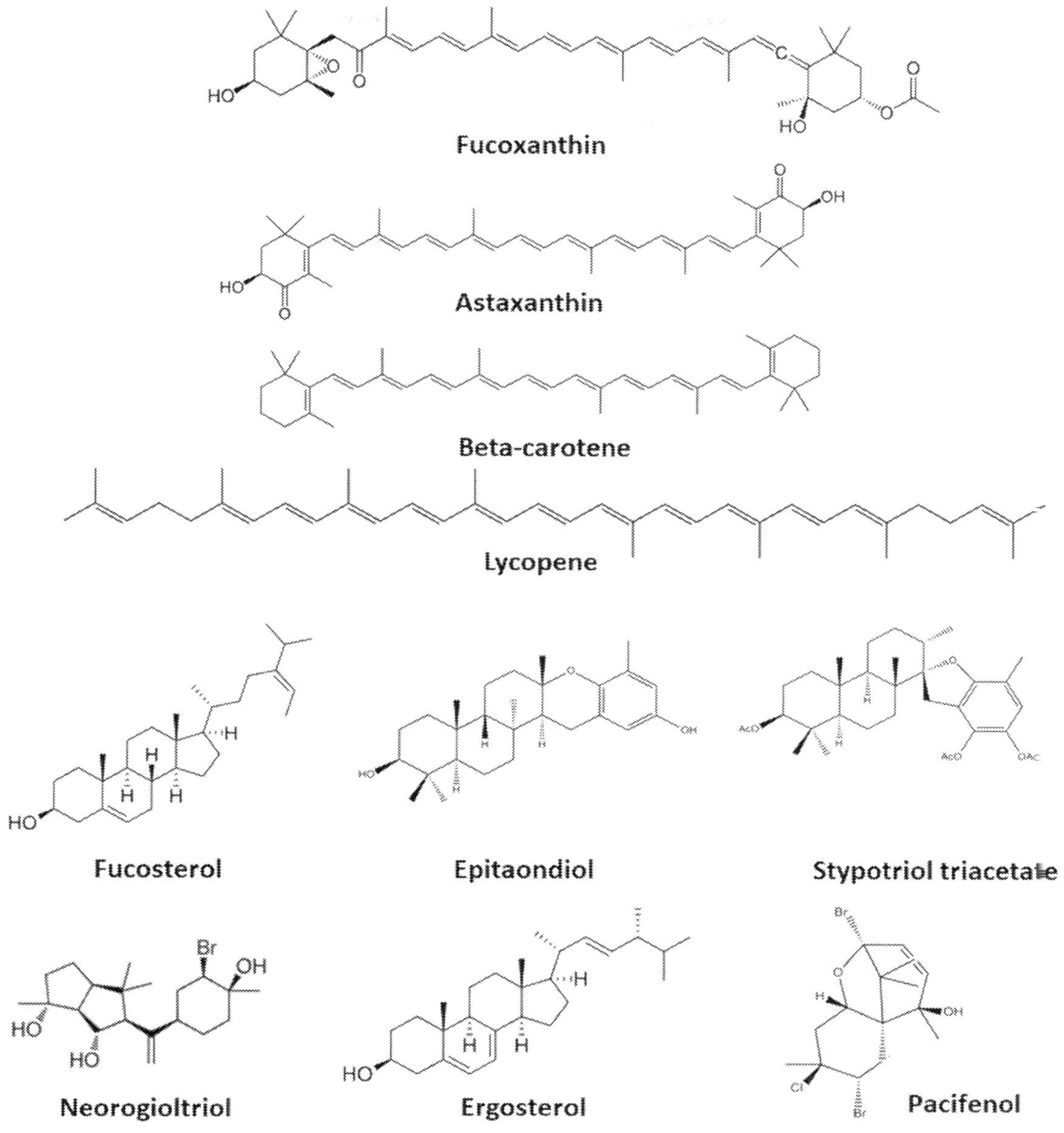

Fig. 9.3 Chemical structures of some antiinflammatory terpenoids isolated from marine algae.

antiallergy efficacies [154]. The antiinflammatory effect of fucoxanthin on DSS-induced colitis in mice has been recently reported by Yuan-Ping et al. [155]. In this study, fucoxanthin ameliorated the colonic histological alterations and suppressed PGE2, COX-2, and NF-κB [155]. In another study, fucoxanthin decreased the expression of IL-6, IL-1β,

TNF-α, and NF-κB phosphorylation in a mouse model of LPS-induced sepsis [156]. Very recently, fucoxanthin combined with rosmarinic acid exerted a potent antiinflammatory effect in UVB-exposed keratinocytes. This combination downregulated components of the NOD-, LRR- and pyrin domain-containing protein 3 (NLRP3) inflammasome system and reduced IL-1β production [157]. Furthermore, fucoxanthin derivatives isolated from marine algae exerted antiinflammatory efficacy. For instance, 9′-cis-(6′R) fucoxanthin and 13-cis and 13′-cis-(6′R) fucoxanthins isolated from *Sargassum siliquastrum* inhibited of NO and PGE2 production and iNOS and COX-2 expression in LPS-stimulated RAW 264.7 cells [158].

The antiinflammatory activity of other marine algae–derived carotenoids, including beta-carotene, astaxanthin, and lycopene, has been investigated. Beta-carotene purified from *Dunaliella bardawil* suppressed mucosal MPO activity and inflammatory mediators in a murine model of acetic acid–induced small bowel inflammation [159]. The consumption of astaxanthin has been associated with decreased plasma CRP and suppressed iNOS, COX, PGF_2, and TNF-α in human subjects [160]. Astaxanthin suppressed oxidative stress and inflammation and blocked preeclampsia progression induced by *N*ω-nitro-L-arginine methyl ester (L-NAME) in rats [161]. The consumption of astaxanthin and β-carotene has demonstrated beneficial effects in suppressing *Helicobacter pylori*–induced gastric inflammation [162]. Recent studies have shown that astaxanthin can inhibit LPS-induced production of TNF-α and IL-6 from bovine endometrial epithelial cells [163] and TLR-4, NF-κB, IL-1β TNF-α, and IL-6 in the lung of ochratoxin-induced mice [164]. Moreover the dietary supplementation of astaxanthin at an early stage of esophageal cancer in F344 rats inhibited oxidative stress and inflammation via suppression of NFκB and COX2 as recently reported by Cui et al. [165]. In a mouse model of LPS-induced acute lung injury and sepsis, astaxanthin inhibited activation of MAPK/NF-κB signaling [166]. Furthermore, lycopene isolated from *Chlorella marina* conferred protection against collagen-induced arthritic effects in rats through the inhibition of MPO, COX, and LPO [167].

9.3.6 Alkaloids

Alkaloids are a large group of naturally occurring organic compounds that contain nitrogen atoms. In a series of complex metabolic pathways involving the Mannich reaction, alkaloids are synthesized from amino acid precursors [168]. Owing to their pharmacological activities, alkaloids have gained special interest, and the alkaloid chemistry has been widely studied in terrestrial plants; however, few studies were conducted on algae [169]. Alkaloids derived from marine algae include phenylethylamine (PEA) and indole and halogenated indole alkaloids. Hordenine and caulerpin were the first reported alkaloids isolated from marine algae [169]. The PEA alkaloid group includes important alkaloids, such as tyramine and hordenine, and found in some brown (*Desmarestia aculeata* and *Desmarestia viridis*), red marine algae (*Polysiphonia urceolata*, *Delesseria sanguine*, *Dumontia incrassata*, and *Polyides rotundus*) [170, 171] and microalgae (*Scenedesmus acutus*) [172]. Members of this alkaloid group act as neuromodulators and antidepressants [173]. Tyramine has been detected in *Laminaria saccharina* and *Chondrus crispus*, whereas hordenine occurs in *Phyllophora nervosa* and *Ahnfeltia paradoxa* [169]. Caulerpin, caulersin, and many other alkaloids have been isolated from marine algae and have been demonstrated to exert a variety of pharmacological effects (reviewed in [169, 174]). However, the antiinflammatory effects of marine algae–derived alkaloids need to be explored.

9.3.7 Other secondary metabolites possessing antiinflammatory activity

In addition to polysaccharides, phenolics, proteins and peptides, terpenoids, and fatty acids and lipid derivatives, many other bioactive secondary metabolites with antiinflammatory activities have been isolated; however, it could not fit into the aforementioned categories. Floridoside and D-isofloridoside isolated from the marine red alga *Laurencia undulata* inhibited oxidative stress and suppressed MPO activity and matrix metalloproteinases [175]. In the study of Bruno et al., monogalactosyldiacylglycerol, digalactosyldiacylglycerol and sulphoquinovosyldiacylglycerol isolated from thermophilic blue-green alga ETS-05 colonies inhibited inflammation and ear edema in croton oil–induced mice [176]. Quinone purified from *Perithalia capillaris* has been demonstrated to inhibit the production of superoxide from human neutrophils, demonstrating antioxidant and antiinflammatory activities [177].

Cymopol and cyclocymopol from *Cymopolia barbata* exhibited an inhibitory effect on bee venom–derived PGA2 production [95]. Very recently, Bousquet et al. have isolated cymopol and other related compounds from the marine green alga *Cymopolia barbata* and investigated their antiinflammatory activities [178]. Cymopol and cymopol quinone exerted both in vitro and in vivo antioxidant and antiinflammatory effects mediated via modulating Nrf2 signaling and colon microbiota composition [178].

γ-Aminobutyric acid purified from *Laminaria japonica* inhibited inflammatory response in LPS-induced murine macrophages [179]. (12*Z*)-Cis-maneonene-D and (12*E*)-cis-maneonene-E isolated from *Laurencia obtuse* prevented inflammatory responses in peripheral blood neutrophils as reported by Ayyad et al. [180]. The antiinflammatory activity of sargachromanol G from *Sargassum siliquastrum* has been evaluated in LPS-stimulated macrophages. Treatment of the LPS-induced cells with sargachromanol G reduced NO, PGE2, and COX-2 production and suppressed the gene expression levels of TNF-α, IL-1β, and IL-6 [181]. Pheophorbide A and pheophytin are two secondary metabolites purified from *Saccharina japonica* and prevented inflammation, NO production, and iNOS expression in LPS-induced murine macrophages [182].

Other compounds isolated from marine algae have been suggested to exert antiinflammatory activities based on in silico studies. For instance, diethyl phthalate, methyl salicylate, benzoic acid, hexadecanoic acid, ethyl ester, and (*E*)-9-octadecenoic acid ethyl ester purified from *Sargassum wightii* have been reported to act as COX-2 inhibitors based on molecular docking studies [183].

9.4 Conclusions

Marine algae extracts and bioactive constituents, including sulfated polysaccharides, fatty acids and lipid derivatives, proteins and peptides, phenolic compounds, terpenes, and many other secondary metabolites have been demonstrated to exert preventive/protective effects against inflammatory conditions. Most of the marine-derived bioactive compounds showed antioxidant activity along with their antiinflammatory effects. In particular the antiinflammatory effects of several phytochemicals purified from marine algae were mediated via inhibiting the key pathways of inflammation, mainly NF-κB and MAPK signaling, resulting in suppressed NO, TNF-α, IL-6, IL-1β, iNOS, and COX-2. Given the role of oxidative stress and inflammation in provoking tissue injury, marine algae-derived compounds with dual antioxidant and antiinflammatory activities represent promising candidates to prevent or attenuate the progression of inflammatory disorders. However, further studies corroborated by clinical trials are needed to investigate the safety and exact mechanisms underlying the antiinflammatory activities of marine algae-derived natural products.

Conflict of interest

None.

References

[1] B.S. Halpern, C. Longo, D. Hardy, K.L. McLeod, J.F. Samhouri, S.K. Katona, K. Kleisner, S.E. Lester, J. O'Leary, M. Ranelletti, An index to assess the health and benefits of the global ocean, Nature 488 (2012) 615–620.

[2] C. Alves, J. Silva, S. Pinteus, H. Gaspar, M.C. Alpoim, L.M. Botana, R. Pedrosa, From marine origin to therapeutics: the antitumor potential of marine algae-derived compounds, Front. Pharmacol. 9 (2018) 1–58.

[3] M. de Jesus Raposo, A. de Morais, R. de Morais, Marine polysaccharides from algae with potential biomedical applications, Mar. Drugs 13 (2015) 2967–3028.

[4] J.P. Atkins, D. Burdon, M. Elliott, A.J. Gregory, Management of the marine environment: integrating ecosystem services and societal benefits with the DPSIR framework in a systems approach, Mar. Pollut. Bull. 62 (2011) 215–226.

[5] H. Kang, C. Seo, Y. Park, Marine peptides and their anti-infective activities, Mar. Drugs 13 (2015) 618–654.

[6] S.-K. Kim, Y.D. Ravichandran, S.B. Khan, Y.T. Kim, Prospective of the cosmeceuticals derived from marine organisms, Biotechnol. Bioproces. Eng. 13 (2008) 511–523.

[7] L. Tchokouaha Yamthe, R. Appiah-Opong, P. Tsouh Fokou, N. Tsabang, F. Fekam Boyom, A. Nyarko, M. Wilson, Marine algae as source of novel antileishmanial drugs: a review, Mar. Drugs 15 (2017) 323–340.

[8] A.A. El Gamal, Biological importance of marine algae, Saudi Pharma. J. 18 (2010) 1–25.

[9] J.W. Blunt, B.R. Copp, W.-P. Hu, M.H. Munro, P.T. Northcote, M.R. Prinsep, Marine natural products, Nat. Prod. Rep. 26 (2009) 170–244.

[10] Z. Montero-Lobato, M. Vázquez, F. Navarro, J. Fuentes, E. Bermejo, I. Garbayo, C. Vílchez, M. Cuaresma, Chemically-induced production of anti-inflammatory molecules in microalgae, Mar. Drugs 16 (2018) 478–490.

[11] U. Lindequist, Marine-derived pharmaceuticals–challenges and opportunities, Biomol. Ther. 24 (2016) 561–572.

[12] S. Khalid, M. Abbas, F. Saeed, H. Bader-Ul-Ain, H.A.R. Suleria, Therapeutic potential of seaweed bioactive compounds, in: Seaweed Biomaterials, IntechOpen, 2018, pp. 1–20.

[13] A.M. Mahmoud, O.E. Hussein, S.A. Ramadan, Amelioration of cyclophosphamide-induced hepatotoxicity by the brown seaweed Turbenaria ornata, Int. J. Clin. Toxicol. 1 (2013) 9–17.

[14] A.M. Mahmoud, A.M. El-Derby, K.N.M. Elsayed, E.M. Abdella, Brown seaweeds ameliorate renal alterations in mice treated with the carcinogen azoxymethane, Int. J. Pharm. Pharm. Sci. 6 (2014) 365–369.

[15] A.M. Mahmoud, E.M. Abdella, A.M. El-Derby, E.M. Abdella, Protective effects of Turbinaria ornata and Padina pavonia against azoxymethane-induced colon carcinogenesis through modulation of PPAR gamma, NF-kappaB and oxidative stress, Phytother. Res. 29 (2015) 737–748.

[16] E. Abdella, A. Mahmoud, A. El-Derby, Brown seaweeds protect against azoxymethane-induced hepatic repercussions through up-regulation of peroxisome proliferator activated receptor gamma and attenuation of oxidative stress, Pharm. Biol. 54 (2016) 2496–2504.

[17] Y.V. Yuan, N.A. Walsh, Antioxidant and antiproliferative activities of extracts from a variety of edible seaweeds, Food Chem. Toxicol. 44 (2006) 1144–1150.

[18] S.K. Chandini, P. Ganesan, N. Bhaskar, In vitro antioxidant activities of three selected brown seaweeds of India, Food Chem. 107 (2008) 707–713.

[19] M. Artan, Y. Li, F. Karadeniz, S.-H. Lee, M.-M. Kim, S.-K. Kim, Anti-HIV-1 activity of phloroglucinol derivative, 6, 6′-bieckol, from Ecklonia cava, Bioorg. Med. Chem. 16 (2008) 7921–7926.

[20] Y. Li, S.-H. Lee, Q.-T. Le, M.-M. Kim, S.-K. Kim, Anti-allergic effects of phlorotannins on histamine release via binding inhibition between IgE and FcεRI, J. Agric. Food Chem. 56 (2008) 12073–12080.

[21] J. Kang, M. Khan, N. Park, J. Cho, M. Lee, H. Fujii, Y. Hong, Antipyretic, analgesic, and anti-inflammatory activities of the seaweed Sargassum fulvellum and Sargassum thunbergii in mice, J. Ethnopharmacol. 116 (2008) 187–190.

[22] M.M. Kim, N. Rajapakse, S.K. Kim, Anti-inflammatory effect of Ishige okamurae ethanolic extract via inhibition of NF-κB transcription factor in RAW 264.7 cells, Phytother. Res. 23 (2009) 628–634.

[23] C.-S. Kong, J.-A. Kim, N.-Y. Yoon, S.-K. Kim, Induction of apoptosis by phloroglucinol derivative from Ecklonia cava in MCF-7 human breast cancer cells, Food Chem. Toxicol. 47 (2009) 1653–1658.
[24] W.A. Pushpamali, C. Nikapitiya, M. De Zoysa, I. Whang, S.J. Kim, J. Lee, Isolation and purification of an anticoagulant from fermented red seaweed Lomentaria catenata, Carbohydr. Polym. 73 (2008) 274–279.
[25] M. Øverland, L.T. Mydland, A. Skrede, Marine macroalgae as sources of protein and bioactive compounds in feed for monogastric animals, J. Sci. Food Agric. 99 (2019) 13–24.
[26] M.L. Wells, P. Potin, J.S. Craigie, J.A. Raven, S.S. Merchant, K.E. Helliwell, A.G. Smith, M.E. Camire, S.H. Brawley, Algae as nutritional and functional food sources: revisiting our understanding, J. Appl. Phycol. 29 (2017) 949–982.
[27] A.R. Circuncisão, M.D. Catarino, S.M. Cardoso, A. Silva, Minerals from macroalgae origin: health benefits and risks for consumers, Mar. Drugs 16 (2018) 400.
[28] R. Sathasivam, R. Radhakrishnan, A. Hashem, E.F. Abd_Allah, Microalgae metabolites: a rich source for food and medicine, Saudi J. Biol. Sci. 26 (2017) 709–722.
[29] T.A. Olasehinde, A.O. Olaniran, A.I. Okoh, Therapeutic potentials of microalgae in the treatment of Alzheimer's disease, Molecules 22 (2017) 480–498.
[30] D.D. Chaplin, Overview of the immune response, J. Allergy Clin. Immunol. 125 (2010) S3–S23.
[31] L. Chen, H. Deng, H. Cui, J. Fang, Z. Zuo, J. Deng, Y. Li, X. Wang, L. Zhao, Inflammatory responses and inflammation-associated diseases in organs, Oncotarget 9 (2018) 7204–7218.
[32] P. Libby, Inflammatory mechanisms: the molecular basis of inflammation and disease, Nutr. Rev. 65 (2007) S140–S146.
[33] J. Zhong, G. Shi, Regulation of inflammation in chronic disease, Front. Immunol. 10 (2019) 737–747.
[34] G.-J. Wu, S.-M. Shiu, M.-C. Hsieh, G.-J. Tsai, Anti-inflammatory activity of a sulfated polysaccharide from the brown alga Sargassum cristaefolium, Food Hydrocoll. 53 (2016) 16–23.
[35] E. Ricciotti, G.A. FitzGerald, Prostaglandins and inflammation, Arterioscler. Thromb. Vasc. Biol. 31 (2011) 986–1000.
[36] C. Nathan, Points of control in inflammation, Nature 420 (6917) (2002) 846–852.
[37] G. Schett, M.F. Neurath, Resolution of chronic inflammatory disease: universal and tissue-specific concepts, Nat. Commun. 9 (2018) 1–8.
[38] R. Pahwa, I. Jialal, Chronic Inflammation, StatPearls [Internet], StatPearls Publishing, 2018.
[39] G. Schett, D. Elewaut, I.B. McInnes, J.-M. Dayer, M.F. Neurath, How cytokine networks fuel inflammation: toward a cytokine-based disease taxonomy, Nat. Med. 19 (2013) 822–824.
[40] I.B. McInnes, G. Schett, Pathogenetic insights from the treatment of rheumatoid arthritis, Lancet 389 (2017) 2328–2337.
[41] I.L. Meek, M.A. Van de Laar, H.E. Vonkeman, Non-steroidal anti-inflammatory drugs: an overview of cardiovascular risks, Pharmaceuticals 3 (2010) 2146–2162.
[42] A. Lanas, F.K.L. Chan, Peptic ulcer disease, Lancet 390 (2017) 613–624.
[43] M. Bally, N. Dendukuri, B. Rich, L. Nadeau, A. Helin-Salmivaara, E. Garbe, J.M. Brophy, Risk of acute myocardial infarction with NSAIDs in real world use: bayesian meta-analysis of individual patient data, BMJ 357 (2017) j1909.
[44] M.W. Whitehouse, Anti-inflammatory glucocorticoid drugs: reflections after 60 years, Inflammopharmacology 19 (2011) 1–19.
[45] C.A. Dinarello, Anti-inflammatory agents: present and future, Cell 140 (2010) 935–950.

[46] S.-Y. Xu, X. Huang, K.-L. Cheong, Recent advances in marine algae polysaccharides: Isolation, structure, and activities, Mar. Drugs 15 (2017) 388–404.
[47] D.j. Hu, K.l. Cheong, J. Zhao, S.p. Li, Chromatography in characterization of polysaccharides from medicinal plants and fungi, J. Sep. Sci. 36 (2013) 1–19.
[48] R.C. Siqueira, M.S. da Silva, D.B. de Alencar, A.d.F. Pires, N.M. de Alencar, M.G. Pereira, B.S. Cavada, A.H. Sampaio, W.R. Farias, A.M.S. Assreuy, In vivo anti-inflammatory effect of a sulfated polysaccharide isolated from the marine brown algae Lobophora variegata, Pharm. Biol. 49 (2011) 167–174.
[49] C. Shi, T. Pan, M. Cao, Q. Liu, L. Zhang, G. Liu, Suppression of Th2 immune responses by the sulfated polysaccharide from Porphyra haitanensis in tropomyosin-sensitized mice, Int. Immunopharmacol. 24 (2015) 211–218.
[50] J.G. Carneiro, J.A. Rodrigues, E. de Sousa Oliveira Vanderlei, R.B. Souza, A.L. Quindere, C.O. Coura, I.W. de Araujo, H.V. Chaves, M.M. Bezerra, N.M. Benevides, Peripheral antinociception and anti-inflammatory effects of sulphated polysaccharides from the alga Caulerpa mexicana, Basic Clin. Pharmacol. Toxicol. 115 (2014) 335–342.
[51] J.G. Pereira, J.X. Mesquita, K.S. Aragao, A.X. Franco, M.H. Souza, T.V. Brito, J.M. Dias, R.O. Silva, J.V. Medeiros, J.S. Oliveira, C.M. Abreu, R.C. de Paula, A.L. Barbosa, A.L. Freitas, Polysaccharides isolated from Digenea simplex inhibit inflammatory and nociceptive responses, Carbohydr. Polym. 108 (2014) 17–25.
[52] N. Patil, V. Le, A.D. Sligar, L. Mei, D. Chavarria, E.Y. Yang, A. Baker, Algal polysaccharides as therapeutic agents for atherosclerosis, Front. Cardiovasc. Med. 5 (2018) 153–171.
[53] T.S. Vo, D.H. Ngo, K.H. Kang, W.K. Jung, S.K. Kim, The beneficial properties of marine polysaccharides in alleviation of allergic responses, Mol. Nutr. Food Res. 59 (2015) 129–138.
[54] C. Sarithakumari, G. Renju, G.M. Kurup, Anti-inflammatory and antioxidant potential of alginic acid isolated from the marine algae, Sargassum wightii on adjuvant-induced arthritic rats, Inflammopharmacology 21 (2013) 261–268.
[55] I.P.S. Fernando, T.U. Jayawardena, K.K.A. Sanjeewa, L. Wang, Y.J. Jeon, W.W. Lee, Anti-inflammatory potential of alginic acid from Sargassum horneri against urban aerosol-induced inflammatory responses in keratinocytes and macrophages, Ecotoxicol. Environ. Saf. 160 (2018) 24–31.
[56] C.M.P.G. Dore, M.G.d.C.F. Alves, L.S.E.P. Will, T.G. Costa, D.A. Sabry, L.A.R. de Souza Rêgo, C.M. Accardo, H.A.O. Rocha, L.G.A. Filgueira, E.L. Leite, A sulfated polysaccharide, fucans, isolated from brown algae Sargassum vulgare with anticoagulant, antithrombotic, antioxidant and anti-inflammatory effects, Carbohydr. Polym. 91 (2013) 467–475.
[57] P. Dorschmann, G. Kopplin, J. Roider, A. Klettner, Effects of sulfated fucans from laminaria hyperborea regarding VEGF secretion, cell viability, and oxidative stress and correlation with molecular weight, Mar. Drugs 17 (2019) 258–277.
[58] Y.Q. Cui, L.J. Zhang, T. Zhang, D.Z. Luo, Y.J. Jia, Z.X. Guo, Q.B. Zhang, X. Wang, X.M. Wang, Inhibitory effect of fucoidan on nitric oxide production in lipopolysaccharide-activated primary microglia, Clin. Exp. Pharmacol. Physiol. 37 (2010) 422–428.
[59] Y. Wang, M. Xing, Q. Cao, A. Ji, H. Liang, S. Song, Biological activities of fucoidan and the factors mediating its therapeutic effects: a review of recent studies, Mar. Drugs 17 (2019) 183–201.
[60] A.-R. Phull, M. Majid, I.-u. Haq, M.R. Khan, S.J. Kim, In vitro and in vivo evaluation of anti-arthritic, antioxidant efficacy of fucoidan from Undaria pinnatifida (Harvey) Suringar, Int. J. Biol. Macromol. 97 (2017) 468–480.
[61] S. Elbi, T. Nimal, V. Rajan, G. Baranwal, R. Biswas, R. Jayakumar, S. Sathianarayanan, Fucoidan coated ciprofloxacin loaded chitosan nanoparticles for the treatment of intracellular and biofilm infections of Salmonella, Colloids Surf. B: Biointerfaces 160 (2017) 40–47.

[62] M. Preobrazhenskaya, A. Berman, V. Mikhailov, N. Ushakova, A. Mazurov, A. Semenov, A. Usov, N. Nifant'Ev, N. Bovin, Fucoidan inhibits leukocyte recruitment in a model peritonial inflammation in rat and blocks interaction of P-selectin with its carbohydrate ligand, IUBMB Life 43 (1997) 443–451.
[63] M.J. Jacobin-Valat, K. Deramchia, S. Mornet, C.E. Hagemeyer, S. Bonetto, R. Robert, M. Biran, P. Massot, S. Miraux, S. Sanchez, MRI of inducible P-selectin expression in human activated platelets involved in the early stages of atherosclerosis, NMR Biomed. 24 (2011) 413–424.
[64] B.W. Jo, S.-K. Choi, Degradation of fucoidans from Sargassum fulvellum and their biological activities, Carbohydr. Polym. 111 (2014) 822–829.
[65] H.H. Yu, E. Chengchuan Ko, C.L. Chang, K.S. Yuan, A.T.H. Wu, Y.S. Shan, S.Y. Wu, Fucoidan inhibits radiation-induced pneumonitis and lung fibrosis by reducing inflammatory cytokine expression in lung tissues, Mar. Drugs 16 (2018) 392–412.
[66] J. Park, J.D. Cha, K.M. Choi, K.Y. Lee, K.M. Han, Y.S. Jang, Fucoidan inhibits LPS-induced inflammation in vitro and during the acute response in vivo, Int. Immunopharmacol. 43 (2017) 91–98.
[67] K.K. Asanka Sanjeewa, T.U. Jayawardena, H.S. Kim, S.Y. Kim, I.P. Shanura Fernando, L. Wang, D.T.U. Abetunga, W.S. Kim, D.S. Lee, Y.J. Jeon, Fucoidan isolated from Padina commersonii inhibit LPS-induced inflammation in macrophages blocking TLR/NF-kappaB signal pathway, Carbohydr. Polym. 224 (2019) 115–195.
[68] T. Tian, H. Chang, K. He, Y. Ni, C. Li, M. Hou, L. Chen, Z. Xu, B. Chen, M. Ji, Fucoidan from seaweed Fucus vesiculosus inhibits 2,4-dinitrochlorobenzene-induced atopic dermatitis, Int. Immunopharmacol. 75 (2019) 105–823.
[69] M.C. Barbalace, M. Malaguti, L. Giusti, A. Lucacchini, S. Hrelia, C. Angeloni, Anti-inflammatory activities of marine algae in neurodegenerative diseases, Int. J. Mol. Sci. 20 (2019) 3061–3081.
[70] A.M. Neyrinck, A. Mouson, N.M. Delzenne, Dietary supplementation with laminarin, a fermentable marine β (1–3) glucan, protects against hepatotoxicity induced by LPS in rat by modulating immune response in the hepatic tissue, Int. Immunopharmacol. 7 (2007) 1497–1506.
[71] C.J. O'Shea, J.V. O'Doherty, J.J. Callanan, D. Doyle, K. Thornton, T. Sweeney, The effect of algal polysaccharides laminarin and fucoidan on colonic pathology, cytokine gene expression and Enterobacteriaceae in a dextran sodium sulfate-challenged porcine model, J. Nutr. Sci. 5 (2016) 1–9.
[72] P. Kumari, M. Kumar, C. Reddy, B. Jha, Algal lipids, fatty acids and sterols, in: Functional Ingredients From Algae for Foods and Nutraceuticals, Elsevier, 2013, pp. 87–134.
[73] M. Manuelli, L. Della Guardia, H. Cena, Enriching diet with n-3 PUFAs to help prevent cardiovascular diseases in healthy adults: results from clinical trials, Int. J. Mol. Sci. 18 (2017) 1552–1568.
[74] S. Layé, A. Nadjar, C. Joffre, R.P. Bazinet, Anti-inflammatory effects of omega-3 fatty acids in the brain: physiological mechanisms and relevance to pharmacology, Pharmacol. Rev. 70 (2018) 12–38.
[75] V.J. van Ginneken, J.P. Helsper, W. de Visser, H. van Keulen, W.A. Brandenburg, Polyunsaturated fatty acids in various macroalgal species from north Atlantic and tropical seas, Lipids Health Dis. 10 (2011) 104–112.
[76] M.A. Santos, P. Colepicolo, D. Pupo, M.T. Fujii, C.M. de Pereira, M.F. Mesko, Antarctic red macroalgae: a source of polyunsaturated fatty acids, J. Appl. Phycol. 29 (2017) 759–767.
[77] Y. Adkins, D.S. Kelley, Mechanisms underlying the cardioprotective effects of omega-3 polyunsaturated fatty acids, J. Nutr. Biochem. 21 (9) (2010) 781–792.
[78] P.C. Calder, Marine omega-3 fatty acids and inflammatory processes: effects, mechanisms and clinical relevance, Biochim. Biophys. Acta 1851 (2015) 469–484.

[79] A. Belluzzi, C. Brignola, M. Campieri, A. Pera, S. Boschi, M. Miglioli, Effect of an enteric-coated fish-oil preparation on relapses in Crohn's disease, N. Engl. J. Med. 334 (1996) 1557–1560.
[80] R.J. Goldberg, J. Katz, A meta-analysis of the analgesic effects of omega-3 polyunsaturated fatty acid supplementation for inflammatory joint pain, Pain 129 (2007) 210–223.
[81] T. Nagakura, S. Matsuda, K. Shichijyo, H. Sugimoto, K. Hata, Dietary supplementation with fish oil rich in omega-3 polyunsaturated fatty acids in children with bronchial asthma, Eur. Respir. J. 16 (2000) 861–865.
[82] C. Miller, R.Y. Yamaguchi, V.A. Ziboh, Guinea pig epidermis generates putative anti-inflammatory metabolites from fish oil polyunsaturated fatty acids, Lipids 24 (1989) 998–1003.
[83] D.Y. Oh, S. Talukdar, E.J. Bae, T. Imamura, H. Morinaga, W. Fan, P. Li, W.J. Lu, S.M. Watkins, J.M. Olefsky, GPR120 is an omega-3 fatty acid receptor mediating potent anti-inflammatory and insulin-sensitizing effects, Cell 142 (2010) 687–698.
[84] C.N. Serhan, N. Chiang, T.E. Van Dyke, Resolving inflammation: dual anti-inflammatory and pro-resolution lipid mediators, Nat. Rev. Immunol. 8 (2008) 349–361.
[85] H. Yamada, M. Yoshida, Y. Nakano, T. Suganami, N. Satoh, T. Mita, K. Azuma, M. Itoh, Y. Yamamoto, Y. Kamei, M. Horie, H. Watada, Y. Ogawa, In vivo and in vitro inhibition of monocyte adhesion to endothelial cells and endothelial adhesion molecules by eicosapentaenoic acid, Arterioscler. Thromb. Vasc. Biol. 28 (2008) 2173–2179.
[86] C.M. Kirsch, D.G. Payan, M.Y. Wong, J.G. Dohlman, V.A. Blake, M.A. Petri, J. Offenberger, E.J. Goetzl, W.M. Gold, Effect of eicosapentaenoic acid in asthma, Clin. Allergy 18 (1988) 177–187.
[87] L. Gutiérrez-Pliego, B. Martinez-Carrillo, A. Reséndiz-Albor, I. Arciniega-Martínez, J. Escoto-Herrera, C. Rosales-Gómez, R. Valdés-Ramos, Effect of supplementation with n-3 fatty acids extracted from microalgae on inflammation biomarkers from two different strains of mice, J. Lip. 2018 (2018) 119–127.
[88] M.N.A. Khan, J.-Y. Cho, M.-C. Lee, J.-Y. Kang, N.G. Park, H. Fujii, Y.-K. Hong, Isolation of two anti-inflammatory and one pro-inflammatory polyunsaturated fatty acids from the brown seaweed Undaria pinnatifida, J. Agric. Food Chem. 55 (2007) 6984–6988.
[89] T.-S. Vo, J.-A. Kim, I. Wijesekara, C.-S. Kong, S.-K. Kim, Potent effect of brown algae (Ishige okamurae) on suppression of allergic inflammation in human basophilic KU812F cells, Food Sci. Biotechnol. 20 (2011) 1227–1236.
[90] J.M. Nauroth, M. Van Elswyk, Y. Liu, L. Arterburn, Anti-inflammatory activity of algal oils containing docosahexaenoic acid (DHA) and omega-6 docosapentaenoic acid (DPAn-6)(101.5), Am. Assoc. Immnol. (2007).
[91] K. Mobraten, T.M. Haug, C.R. Kleiveland, T. Lea, Omega-3 and omega-6 PUFAs induce the same GPR120-mediated signalling events, but with different kinetics and intensity in Caco-2 cells, Lipids Health Dis. 12 (2013) 101–120.
[92] L. Spencer, C. Mann, M. Metcalfe, M.B. Webb, C. Pollard, D. Spencer, D. Berry, W. Steward, A. Dennison, The effect of omega-3 FAs on tumour angiogenesis and their therapeutic potential, Eur. J. Cancer 45 (2009) 2077–2086.
[93] H.T. Dang, H.J. Lee, E.S. Yoo, P.B. Shinde, Y.M. Lee, J. Hong, D.K. Kim, J.H. Jung, Anti-inflammatory constituents of the red alga Gracilaria verrucosa and their synthetic analogues, J. Nat. Prod. 71 (2008) 232–240.
[94] J.Y. Cho, Y.P. Gyawali, S.H. Ahn, M.N.A. Khan, I.S. Kong, Y.K. Hong, A methoxylated fatty acid isolated from the brown seaweed Ishige okamurae inhibits bacterial phospholipase A 2, Phytother. Res. 22 (2008) 1070–1074.
[95] A.M.S. Mayer, V.J. Paul, W. Fenical, J.N. Norris, M.S. de Carvalho, R.S. Jacobs, Phospholipase A 2 inhibitors from marine algae, Hydrobiologia 260 (1993) 521–529.

[96] H.J. Lee, H.T. Dang, G.J. Kang, E.J. Yang, S.S. Park, W.J. Yoon, J.H. Jung, H.K. Kang, E.S. Yoo, Two enone fatty acids isolated from Gracilaria verrucosa suppress the production of inflammatory mediators by down-regulating NF-κB and STAT1 activity in lipopolysaccharide-stimulated RAW 264.7 cells, Arch. Pharm. Res. 32 (2009) 453–462.
[97] S. Antonopoulou, T. Nomikos, A. Oikonomou, A. Kyriacou, M. Andriotis, E. Fragopoulou, A. Pantazidou, Characterization of bioactive glycolipids from Scytonema julianum (cyanobacteria), Comp. Biochem. Physiol. B 140 (2005) 219–231.
[98] P. MacArtain, C.I.R. Gill, M. Brooks, R. Campbell, I.R. Rowland, Nutritional value of edible seaweeds, Nutr. Rev. 65 (2007) 535–543.
[99] I. Sousa, L. Gouveia, A.P. Batista, A. Raymundo, N.M. Bandarra, Microalgae in novel food products, Food Chem. Res. Dev. (2008) 75–112.
[100] S. Bleakley, M. Hayes, Algal proteins: extraction, application, and challenges concerning production, Foods 6 (2017) 33–67.
[101] S.-S. Suh, J. Hwang, M. Park, H. Seo, H.-S. Kim, J. Lee, S. Moh, T.-K. Lee, Anti-inflammation activities of mycosporine-like amino acids (MAAs) in response to UV radiation suggest potential anti-skin aging activity, Mar. Drugs 12 (2014) 5174–5187.
[102] T. Wakamiya, H. Nakamoto, T. Shiba, Structural determination of carnosadine, a new cyclopropyl amino acid, from red alga Grateloupia carnosa, Tetrahedron Lett. 25 (1984) 4411–4412.
[103] S.-K. Kim, I. Wijesekara, Development and biological activities of marine-derived bioactive peptides: a review, J. Funct. Foods 2 (2010) 1–9.
[104] H.A. Lee, I.H. Kim, T.J. Nam, Bioactive peptide from Pyropia yezoensis and its anti-inflammatory activities, Int. J. Mol. Med. 36 (2015) 1701–1706.
[105] R. Cian, C. Hernández-Chirlaque, R. Gámez-Belmonte, S. Drago, F. Sánchez de Medina, O. Martínez-Augustin, Green alga Ulva spp. hydrolysates and their peptide fractions regulate cytokine production in splenic macrophages and lymphocytes involving the TLR4-NFκB/MAPK pathways, Mar. Drugs 16 (2018) 235–250.
[106] T.-S. Vo, B. Ryu, S.-K. Kim, Purification of novel anti-inflammatory peptides from enzymatic hydrolysate of the edible microalgal Spirulina maxima, J. Funct. Foods 5 (2013) 1336–1346.
[107] H. Korhonen, A. Pihlanto, Bioactive peptides: production and functionality, Int. Dairy J. 16 (2006) 945–960.
[108] E.S.O. Vanderlei, K.K.N.R. Patoilo, N.A. Lima, A.P.S. Lima, J.A.G. Rodrigues, L.M.C.M. Silva, M.E.P. Lima, V. Lima, N.M.B. Benevides, Antinociceptive and anti-inflammatory activities of lectin from the marine green alga Caulerpa cupressoides, Int. Immunopharmacol. 10 (2010) 1113–1118.
[109] L.M.C.M. Silva, V. Lima, M.L. Holanda, P.G. Pinheiro, J.A.G. Rodrigues, M.E.P. Lima, N.M.B. Benevides, Antinociceptive and anti-inflammatory activities of lectin from marine red alga Pterocladiella capillacea, Biol. Pharm. Bull. 33 (2010) 830–835.
[110] J.G. Figueiredo, F.S. Bitencourt, T.M. Cunha, P.B. Luz, K.S. Nascimento, M.R.L. Mota, A.H. Sampaio, B.S. Cavada, F.Q. Cunha, N.M.N. Alencar, Agglutinin isolated from the red marine alga Hypnea cervicornis J. Agardh reduces inflammatory hypernociception: Involvement of nitric oxide, Pharmacol. Biochem. Behav. 96 (2010) 371–377.
[111] T.P.C. Fontenelle, G.C. Lima, J.X. Mesquita, J.L. de Souza Lopes, T.V. de Brito, F.d.C.V. Júnior, A.B. Sales, K.S. Aragão, M.H.L.P. Souza, A.L. dos Reis Barbosa, Lectin obtained from the red seaweed Bryothamnion triquetrum: secondary structure and anti-inflammatory activity in mice, Int. J. Biol. Macromol. 112 (2018) 1122–1130.

[112] R.L. da Conceição Rivanor, H.V. Chaves, D.R. do Val, A.R. de Freitas, J.C. Lemos, J.A.G. Rodrigues, K.M.A. Pereira, I.W.F. de Araújo, M.M. Bezerra, N.M.B. Benevides, A lectin from the green seaweed Caulerpa cupressoides reduces mechanical hyper-nociception and inflammation in the rat temporomandibular joint during zymosan-induced arthritis, Int. Immunopharmacol. 21 (2014) 34–43.
[113] C. Romay, N. Ledón, R. González, Further studies on anti-inflammatory activity of phycocyanin in some animal models of inflammation, J. Inflamm. Res. 47 (1998) 334–338.
[114] N.K. Gupta, K.P. Gupta, Effects of C-phycocyanin on the representative genes of tumor development in mouse skin exposed to 12-O-tetradecanoyl-phorbol-13-acetate, Environ. Toxicol. Pharmacol. 34 (2012) 941–948.
[115] C. Romay, J. Armesto, D. Remirez, R. González, N. Ledon, I. García, Antioxidant and anti-inflammatory properties of C-phycocyanin from blue-green algae, J. Inflamm. Res. 47 (1998) 36–41.
[116] R. Tsao, Chemistry and biochemistry of dietary polyphenols, Nutrients 2 (2010) 1231–1246.
[117] A. Khoddami, M.A. Wilkes, T.H. Roberts, Techniques for analysis of plant phenolic compounds, Molecules 18 (2013) 2328–2375.
[118] A.M. Mahmoud, M.B. Ashour, A. Abdel-Moneim, O.M. Ahmed, Hesperidin and naringin attenuate hyperglycemia-mediated oxidative stress and proinflammatory cytokine production in high fat fed/streptozotocin-induced type 2 diabetic rats, J. Diabetes Complicat. 26 (2012) 483–490.
[119] A.M. Mahmoud, Influence of rutin on biochemical alterations in hyperammonemia in rats, Exp. Toxicol. Pathol. 64 (2012) 783–789.
[120] A.M. Mahmoud, Hematological alterations in diabetic rats - role of adipocytokines and effect of citrus flavonoids, EXCLI J. 12 (2013) 647–657.
[121] A.M. Mahmoud, Hesperidin protects against cyclophosphamide-induced hepatotoxicity by upregulation of PPARγ and abrogation of oxidative stress and inflammation, Can. J. Physiol. Pharmacol. 92 (2014) 717–724.
[122] R.R. Ahmed, A.M. Mahmoud, M.B. Ashour, A.M. Kamel, Hesperidin protects against diethylnitrosamine-induced nephrotoxicity through modulation of oxidative stress and inflammation, Natl. J. Physiol. Pharm. Pharmacol. 5 (2015) 391–397.
[123] S.H. Aladaileh, M.H. Abukhalil, S.A.M. Saghir, H. Hanieh, M.A. Alfwuaires, A.A. Almaiman, M. Bin-Jumah, A.M. Mahmoud, Galangin activates Nrf2 signaling and attenuates oxidative damage, inflammation, and apoptosis in a rat model of cyclophosphamide-induced hepatotoxicity, Biomolecules 9 (2019) 346–366.
[124] O.Y. Althunibat, A.M. Al Hroob, M.H. Abukhalil, M.O. Germoush, M. Bin-Jumah, A.M. Mahmoud, Fisetin ameliorates oxidative stress, inflammation and apoptosis in diabetic cardiomyopathy, Life Sci. 221 (2019) 83–92.
[125] M.S. Aly, S.R. Galaly, N. Moustafa, H.M. Mohammed, S.M. Khadrawy, A.M. Mahmoud, Hesperidin protects against diethylnitrosamine/carbon tetrachloride-induced renal repercussions via up-regulation of Nrf2/HO-1 signaling and attenuation of oxidative stress, J. Appl. Pharm. Sci. 7 (2017) 7–14.
[126] A.M. Mahmoud, S.M. Abd El-Twab, E.S. Abdel-Reheim, Consumption of polyphenol-rich Morus alba leaves extract attenuates early diabetic retinopathy: the underlying mechanism, Eur. J. Nutr. 56 (2017) 1671–1684.
[127] L.C. Coelho, P.M. Silva, V.L. Lima, E.V. Pontual, P.M. Paiva, T.H. Napoleao, M.T. Correia, Lectins, interconnecting proteins with biotechnological/pharmacological and therapeutic applications, Evid. Based Complement. Altern. Med. 2017 (2017), 1594074.
[128] S.-H. Eom, Y.-M. Kim, S.-K. Kim, Antimicrobial effect of phlorotannins from marine brown algae, Food Chem. Toxicol. 50 (2012) 3251–3255.

[129] G. Corona, M.M. Coman, Y. Guo, S. Hotchkiss, C. Gill, P. Yaqoob, J.P. Spencer, I. Rowland, Effect of simulated gastrointestinal digestion and fermentation on polyphenolic content and bioactivity of brown seaweed phlorotannin-rich extracts, Mol. Nutr. Food Res. 61 (2017) 1–31.

[130] N.V. Thomas, S.K. Kim, Potential pharmacological applications of polyphenolic derivatives from marine brown algae, Environ. Toxicol. Pharmacol. 32 (2011) 325–335.

[131] H.A. Jung, S.E. Jin, B.R. Ahn, C.M. Lee, J.S. Choi, Anti-inflammatory activity of edible brown alga Eisenia bicyclis and its constituents fucosterol and phlorotannins in LPS-stimulated RAW264. 7 macrophages, Food Chem. Toxicol. 59 (2013) 199–206.

[132] T. Shibata, K. Nagayama, R. Tanaka, K. Yamaguchi, T. Nakamura, Inhibitory effects of brown algal phlorotannins on secretory phospholipase A 2 s, lipoxygenases and cyclooxygenases, J. Appl. Phycol. 15 (2003) 61–66.

[133] Y. Sugiura, M. Usui, H. Katsuzaki, K. Imai, M. Kakinuma, H. Amano, M. Miyata, Orally administered phlorotannins from Eisenia arborea suppress chemical mediator release and cyclooxygenase-2 signaling to alleviate mouse ear swelling, Mar. Drugs 16 (2018) 267–281.

[134] Y. Sugiura, K. Matsuda, Y. Yamada, M. Nishikawa, K. Shioya, H. Katsuzaki, K. Imai, H. Amano, Isolation of a new anti-allergic phlorotannin, phlorofucofuroeckol-B, from an edible brown alga, Eisenia arborea, Biosci. Biotechnol. Biochem. 70 (2006) 2807–2811.

[135] K. Kazlowska, T. Hsu, C.C. Hou, W.C. Yang, G.J. Tsai, Anti-inflammatory properties of phenolic compounds and crude extract from Porphyra dentata, J. Ethnopharmacol. 128 (2010) 123–130.

[136] D.F. Wiemer, D.D. Idler, W. Fenical, Vidalols A and B, new anti-inflammatory bromophenols from the Caribbean marine red alga Vidalia obtusaloba, Experientia 47 (1991) 851–853.

[137] W.K. Jung, S.J. Heo, Y.J. Jeon, C.M. Lee, Y.M. Park, H.G. Byun, Y.H. Choi, S.G. Park, I.L.W. Choi, Inhibitory effects and molecular mechanism of dieckol isolated from marine brown alga on COX-2 and iNOS in microglial cells, J. Agric. Food Chem. 57 (2009) 4439–4446.

[138] M.S. Lee, M.S. Kwon, J.W. Choi, T. Shin, H.K. No, J.S. Choi, D.S. Byun, J.I. Kim, H.R. Kim, Anti-inflammatory activities of an ethanol extract of Ecklonia stolonifera in lipopolysaccharide-stimulated RAW 264.7 murine macrophage cells, J. Agric. Food Chem. 60 (2012) 9120–9129.

[139] K.H.N.Fernando,H.W.Yang,Y.Jiang,Y.J.Jeon,B.Ryu,Diphlorethohydroxycarmalol Isolated from Ishige okamurae represses high glucose-induced angiogenesis in vitro and in vivo, Mar. Drugs 16 (2018) 375–390.

[140] S.J. Heo, J.Y. Hwang, J.I. Choi, S.H. Lee, P.J. Park, D.H. Kang, C. Oh, D.W. Kim, J.S. Han, Y.J. Jeon, H.J. Kim, I.W. Choi, Protective effect of diphlorethohydroxycarmalol isolated from Ishige okamurae against high glucose-induced-oxidative stress in human umbilical vein endothelial cells, Food Chem. Toxicol. 48 (2010) 1448–1454.

[141] S.-C. Han, N.-J. Kang, G.-J. Kang, Y.-S. Koh, J.-W. Hyun, N.-H. Lee, H.-K. Kang, E.-S. Yoo, 71: Anti-inflammatory effect of diphlorethohydroxycarmalol (DPHC) isolated from lshige okamuarae in vitro and in vivo, Cytokine 70 (2014) 44–45.

[142] Z. Manzoor, V.B. Mathema, D. Chae, H.K. Kang, E.S. Yoo, Y.J. Jeon, Y.S. Koh, Octaphlorethol a inhibits the CPG-induced inflammatory response by attenuating the mitogen-activated protein kinase and nf-κb pathways, Biosci. Biotechnol. Biochem. 77 (2013) 1970–1972.

[143] E.M. Balboa, E. Conde, A. Moure, E. Falqué, H. Domínguez, In vitro antioxidant properties of crude extracts and compounds from brown algae, Food Chem. 138 (2013) 1764–1785.

[144] D.S. Bhakuni, D.S. Rawat, Bioactive Marine Natural Products, Springer, New York, NY, 2010.
[145] S. Terracciano, M. Aquino, M. Rodriquez, M.C. Monti, A. Casapullo, R. Riccio, L. Gomez-Paloma, Chemistry and biology of anti-inflammatory marine natural products: molecules interfering with cyclooxygenase, NF-κB and other unidentified targets, Curr. Med. Chem. 13 (2006) 1947–1969.
[146] H.H. Sun, W. Fenical, Rhipocephalin and rhipocephenal; toxic feeding deterrents from the tropical marine alga rhipocephalus phoenix, Tetrahedron Lett. 20 (1979) 685–688.
[147] R. Chatter, R.B. Othman, S. Rabhi, M. Kladi, S. Tarhouni, C. Vagias, V. Roussis, L. Guizani-Tabbane, R. Kharrat, In vivo and in vitro anti-inflammatory activity of neorogioltriol, a new diterpene extracted from the red algae Laurencia glandulifera, Mar. Drugs 9 (2011) 1293–1306.
[148] B. Gil, M.L. Ferrándiz, M.J. Sanz, M.C. Terencio, A. Ubeda, J. Rovirosa, A. San-Martin, M.J. Alcaraz, M. Payá, Inhibition of inflammatory responses by epitaondiol and other marine natural products, Life Sci. 57 (2) (1995) PL25–PL30.
[149] M. Caroprese, M. Albenzio, M.G. Ciliberti, M. Francavilla, A. Sevi, A mixture of phytosterols from Dunaliella tertiolecta affects proliferation of peripheral blood mononuclear cells and cytokine production in sheep, Vet. Immunol. Immunopathol. 150 (2012) 27–35.
[150] K. Yasukawa, T. Akihisa, H. Kanno, T. Kaminaga, M. Izumida, T. Sakoh, T. Tamura, M. Takido, Inhibitory effects of sterols isolated from Chlorella vulgaris on 12-O-tetradecanoylphorbol-13-acetate-induced inflammation and tumor promotion in mouse skin, Biol. Pharm. Bull. 19 (1996) 573–576.
[151] S.-J. Heo, S.-C. Ko, S.-M. Kang, H.-S. Kang, J.-P. Kim, S.-H. Kim, K.-W. Lee, M.-G. Cho, Y.-J. Jeon, Cytoprotective effect of fucoxanthin isolated from brown algae Sargassum siliquastrum against H 2 O 2-induced cell damage, Eur. Food Res. Technol. 228 (2008) 145–151.
[152] K.-N. Kim, S.-J. Heo, W.-J. Yoon, S.-M. Kang, G. Ahn, T.-H. Yi, Y.-J. Jeon, Fucoxanthin inhibits the inflammatory response by suppressing the activation of NF-κB and MAPKs in lipopolysaccharide-induced RAW 264.7 macrophages, Eur. J. Pharm. 649 (2010) 369–375.
[153] S.J. Heo, W.J. Yoon, K.N. Kim, G.N. Ahn, S.M. Kang, D.H. Kang, A. Affan, C. Oh, W.K. Jung, Y.J. Jeon, Evaluation of anti-inflammatory effect of fucoxanthin isolated from brown algae in lipopolysaccharide-stimulated RAW 264.7 macrophages, Food Chem. Toxicol. 48 (2010) 2045–2051.
[154] S. Sakai, T. Sugawara, K. Matsubara, T. Hirata, Inhibitory effect of carotenoids on the degranulation of mast cells via suppression of antigen-induced aggregation of high affinity IgE receptors, J. Biol. Chem. 284 (2009) 28172–28179.
[155] Y.-P. Yang, Q.-Y. Tong, S.-H. Zheng, M.-D. Zhou, Y.-M. Zeng, T.-T. Zhou, Anti-inflammatory effect of fucoxanthin on dextran sulfate sodium-induced colitis in mice, Nat. Prod. Res. (2018) 1–5.
[156] J. Su, K. Guo, J. Zhang, M. Huang, L. Sun, D. Li, K.-L. Pang, G. Wang, L. Chen, Z. Liu, Fucoxanthin, a marine xanthophyll isolated from Conticribra weissflogii ND-8: preventive anti-inflammatory effect in a mouse model of sepsis, Front. Pharmacol. 10 (2019) 906–923.
[157] A. Rodríguez-Luna, J. Ávila-Román, H. Oliveira, V. Motilva, E. Talero, Fucoxanthin and rosmarinic acid combination has anti-inflammatory effects through regulation of NLRP3 inflammasome in UVB-exposed HaCaT keratinocytes, Mar. Drugs 17 (2019) 451–465.
[158] S.J. Heo, W.J. Yoon, K.N. Kim, C. Oh, Y.U. Choi, K.T. Yoon, D.H. Kang, Z.J. Qian, I.W. Choi, W.K. Jung, Anti-inflammatory effect of fucoxanthin derivatives isolated from Sargassum siliquastrum in lipopolysaccharide-stimulated RAW 264.7 macrophage, Food Chem. Toxicol. 50 (2012) 3336–3342.

[159] A. Lavy, Y. Naveh, R. Coleman, S. Mokady, M.J. Werman, Dietary Dunaliella bardawil, a β-carotene-rich alga, protects against acetic acid-induced small bowel inflammation in rats, Inflamm. Bowel Dis. 9 (2003) 372–379.

[160] J.S. Park, J.H. Chyun, Y.K. Kim, L.L. Line, B.P. Chew, Astaxanthin decreased oxidative stress and inflammation and enhanced immune response in humans, Nutr. Metab. 7 (2010) 18–28.

[161] R.R. Xuan, T.T. Niu, H.M. Chen, Astaxanthin blocks preeclampsia progression by suppressing oxidative stress and inflammation, Mol. Med. Rep. 14 (2016) 2697–2704.

[162] H. Kang, H. Kim, Astaxanthin and beta-carotene in Helicobacter pylori-induced gastric inflammation: a mini-review on action mechanisms, J. Can. Prev. 22 (2017) 57–61.

[163] F.C. Wan, C. Zhang, Q. Jin, C. Wei, H.B. Zhao, X.L. Zhang, W. You, X.M. Liu, G.F. Liu, Y.F. Liu, X.W. Tan, Protective effects of astaxanthin on lipopolysaccharide-induced inflammation in bovine endometrial epithelial cells, Biol. Reprod. 102 (2019) 339–347.

[164] W. Xu, M. Wang, G. Cui, L. Li, D. Jiao, B. Yao, K. Xu, Y. Chen, M. Long, S. Yang, J. He, Astaxanthin protects OTA-induced lung injury in mice through the Nrf2/NF-kappaB pathway, Toxins 11 (2019) 1–15.

[165] L. Cui, F. Xu, M. Wang, L. Li, T. Qiao, H. Cui, Z. Li, C. Sun, Dietary natural astaxanthin at an early stage inhibits N-nitrosomethylbenzylamine-induced esophageal cancer oxidative stress and inflammation via downregulation of NFkappaB and COX2 in F344 rats, OncoTargets Ther. 12 (2019) 5087–5096.

[166] X. Cai, Y. Chen, X. Xie, D. Yao, C. Ding, M. Chen, Astaxanthin prevents against lipopolysaccharide-induced acute lung injury and sepsis via inhibiting activation of MAPK/NF-kappaB, Am. J. Transl. Res. 11 (2019) 1884–1894.

[167] G.L. Renju, G.M. Kurup, C.H.S. Kumari, Anti-inflammatory activity of lycopene isolated from Chlorella marina on Type II collagen induced arthritis in Sprague Dawley rats, Immunopharmacol. Immunotoxicol. 35 (2013) 282–291.

[168] F. von Nussbaum, Alkaloids. Nature's curse or blessing? By Manfred Hesse, Angew. Chem. Int. Ed. 42 (2003) 4852–4854.

[169] K.C. Güven, A. Percot, E. Sezik, Alkaloids in marine algae, Mar. Drugs 8 (2010) 269–284.

[170] M. Steiner, T. Hartmann, The occurence and distribution of volatile amines in marine algae, Planta 79 (1968) 113–121.

[171] A. Percot, A. Yalçin, V. Aysel, H. Erdugan, B. Dural, K.C. Güven, b-Phenylethylamine content in marine algae around Turkish coasts, Bot. Mar. 52 (2009) 87–90.

[172] I. Rolle, H.E. Hobucher, H. Kneifel, B. Paschold, W. Riepe, C.J. Soeder, Amines in unicellular green algae. 2. Amines in Scenedesmus acutus, Anal. Biochem. 77 (1977) 103–109.

[173] N. Barroso, M. Rodriguez, Action of beta-phenylethylamine and related amines on nigrostriatal dopamine neurotransmission, Eur. J. Pharmacol. 297 (1996) 195–203.

[174] K.C. Güven, B. Coban, E. Sezik, H. Erdugan, F. Kaleağasıoğlu, Alkaloids of marine macroalgae, in: K.G. Ramawat, J.-M. Mérillon (Eds.), Natural Products: Phytochemistry, Botany and Metabolism of Alkaloids, Phenolics and Terpenes, Springer, Berlin, Heidelberg, 2013, pp. 25–37.

[175] Y.X. Li, Y. Li, S.H. Lee, Z.J.I. Qian, S.E.K. Kim, Inhibitors of oxidation and matrix metalloproteinases, floridoside, and D-isofloridoside from marine red alga Laurencia undulata, J. Agric. Food Chem. 58 (2010) 578–586.

[176] A. Bruno, C. Rossi, G. Marcolongo, A. Di Lena, A. Venzo, C.P. Berrie, D. Corda, Selective in vivo anti-inflammatory action of the galactolipid monogalactosyldiacylglycerol, Eur. J. Pharmacol. 524 (2005) 159–168.

[177] C.E. Sansom, L. Larsen, N.B. Perry, M.V. Berridge, E.W. Chia, J.L. Harper, V.L. Webb, An antiproliferative bis-prenylated quinone from the New Zealand brown alga Perithalia capillaris, J. Nat. Prod. 70 (2007) 2042–2044.
[178] M.S. Bousquet, R. Ratnayake, J.L. Pope, Q.-Y. Chen, F. Zhu, S. Chen, T.J. Carney, R.Z. Gharaibeh, C. Jobin, V.J. Paul, H. Luesch, Seaweed natural products modify the host inflammatory response via Nrf2 signaling and alter colon microbiota composition and gene expression, Free Radic. Biol. Med. 146 (2019) 306–323.
[179] J.I. Choi, I.H. Yun, Y. Jung, E.H. Lee, T.J. Nam, Y.M. Kim, Effects of γ-Aminobutyric acid (GABA)-enriched sea tangle Laminaria japonica extract on lipopolysaccharide-induced inflammation in mouse macrophage (RAW 264.7) cells, Fish. Aquat. Sci. 15 (2012) 293–297.
[180] S.E.N. Ayyad, K.O. Al-Footy, W.M. Alarif, T.R. Sobahi, S.A. Bassaif, M.S. Makki, A.M. Asiri, A.Y. Al Halawani, A.F. Badria, F.A.A.R. Badria, Bioactive C15 acetogenins from the red alga Laurencia obtusa, Chem. Pharm. Bull. 59 (2011) 1294–1298.
[181] W.J. Yoon, S.J. Heo, S.C. Han, H.J. Lee, G.J. Kang, H.K. Kang, J.W. Hyun, Y.S. Koh, E.S. Yoo, Anti-inflammatory effect of sargachromanol G isolated from Sargassum siliquastrum in RAW 264.7 cells, Arch. Pharm. Res. 35 (2012) 1421–1430.
[182] M.N. Islam, I.J. Ishita, S.E. Jin, R.J. Choi, C.M. Lee, Y.S. Kim, H.A. Jung, J.S. Choi, Anti-inflammatory activity of edible brown alga Saccharina japonica and its constituents pheophorbide a and pheophytin a in LPS-stimulated RAW 264.7 macrophage cells, Food Chem. Toxicol. 55 (2013) 541–548.
[183] P. Balachandran, V. Parthasarathy, T.A. Kumar, Isolation of compounds from Sargassum wightii by GCMS and the molecular docking against anti-inflammatory marker COX2, Int. Lett. Chem. Phys. Astron. 63 (2016) 1–12.

10

Medicinal plants and their potential use in the treatment of rheumatic diseases

Diego P. de Oliveira[a], Fernão C. Braga[a], and Mauro M. Teixeira[b]

[a]Department of Pharmaceutical Products, Faculty of Pharmacy, Federal University of Minas Gerais (UFMG), Belo Horizonte, Brazil, [b]Department of Biochemistry and Immunology, Institute of Biological Sciences, Federal University of Minas Gerais (UFMG), Belo Horizonte, Brazil

10.1 Introduction

The term arthritis encompasses 100–150 rheumatic conditions associated with joint affections [1]. Generally, arthritis conditions can be classified as low-inflammatory disorders, such as osteoarthritis and post-traumatic arthritis, and as high-inflammatory disorders, like rheumatoid, infectious, and metabolic arthritis (gout and psoriatic arthritis) [2, 3].

Rheumatoid arthritis is among the most common joint inflammatory disorders. The disease has no known etiology, and progression is associated with joint degradation and functional disability in 10%–15% of the cases. Rheumatoid arthritis is estimated to affect 3.2% of the world population belonging to different age groups, sex, and races, but it is more prevalent in women (relation of 2.7:1) and tends to appear in advanced age [3–5]. A recent study with 1.5 million Australian active adult patients (with maximum age of 65 years old) revealed the incidence of arthritic diseases in 9.5% of them, including osteoarthritis, rheumatoid arthritis, polyarthritis, arthropathy, polyarthralgia, arthralgia, osteoarthrosis, and arthrosis. The same study pointed out that 1.6% and 0.9% of these patients are affected by gout and rheumatoid arthritis, respectively [6].

Molecular and cellular mediators play a crucial role in the physiopathological mechanism of arthritic diseases. Molecular patterns associated with microorganisms or tissue damage are capable to trigger the inflammatory response by modulating chemokines, cytokines, and leukocyte activation, in special due to the recruitment of neutrophils [7]. This response modifies the endothelial permeability, thus

Inflammation and Natural Products. https://doi.org/10.1016/B978-0-12-819218-4.00014-6
© 2021 Elsevier Inc. All rights reserved.

allowing the cell influx into the injured site and therefore initiating the cardinal signs of the inflammation (pain, heat, redness, and edema). A nonresolved inflammation process can evolve to loss of function of the injured tissue and chronic diseases [8–10]. The pharmacological treatment aims to induce clinical remission and, therefore, to inhibit the progression of structural and functional damage of the joint. Five classes of drugs are currently employed for arthritis treatment: analgesics; corticosteroids; nonsteroidal antiinflammatory drugs (NSAIDs); disease-modifying antirheumatoid drugs (DMARDs), such as methotrexate, sulfasalazine, cyclosporine, hydroxychloroquine, chloroquine, and leflunomide; and biological agents.

The aforementioned drugs are not effective to treat all types of arthritis and show several side effects [11, 12]. Biological agents like tumor necrosis factor (TNF) inhibitors (e.g., adalimumab, certolizumab, etanercept, golimumab, and infliximab) are currently indicated to treat moderate to severe active rheumatoid arthritis in adult subjects, which respond inadequately to DMARDs. TNF inhibitors may also be employed concomitantly with DMARDs, an effective, but costly, pharmacological approach. According to the Brazilian Ministry of Health, the costs with anti-TNF therapy can reach up to US$ 1200 per patient annually. In addition, monoclonal antibody–based therapies demand subcutaneous or intravenous administration, which decreases patient adherence to the treatment [13–17].

Due to the high cost of arthritis therapy with biological drugs and their side effects, the development of new effective low-cost antiarthritis agents is demanded, and natural products may represent a source of bioactive compounds for exploitation. Several studies have demonstrated that medicinal plants and compounds isolated thereof exhibit significant activity in preclinical models of arthritis, reducing the production of proinflammatory mediators, the infiltration of leukocytes into the joints, and the clinical arthritis score [18]. A total of 1328 new pharmacological entities were introduced into therapy in the US market from 1981 to 2014, of which 686 (51.6%) derive directly or indirectly form natural products. In the case of antiarthritic agents, 12 of the new 22 pharmaceutical products are natural products or derivates [19].

10.2 Herbal products currently used in antiarthritic therapy

Herein, we present some examples of plant species largely used for their alleged antiinflammatory properties, for which there is scientific evidence to support their benefits in arthritis therapy. Contradictory results will be pointed whenever possible. This section does not intend to make an exhaustive review of globally used antiarthritic plants. Several recently published reviews cover antiarthritic plants used worldwide [20–25]. Here, we will present studies on the following plants and their constituents: *Boswellia serrata* Roxb. (Shallaki), *Curcuma longa* Linn. (turmeric), *Cinnamomum zeylanicum* Blume (Cinnamon), *Echinodorus grandiflorus* (Cham. & Schltdl.) Micheli. (Chapéu-de-couro), *Colchicum luteum* Baker (Suranjan-Talkh), *Tripterygium wilfordii* Hook F. (Thunder God Vine), *Andrographis paniculata* (Burm. fil.) Nees (creat or green chiretta), and *Nigella sativa* L. (black caraway). Each section will follow and overall scheme where we present data on the effects of the plant extract or fractions in screening models in vitro and in vivo. The isolated compounds and their effects are then discussed and next the availability of mechanistic, toxicological, and clinical studies.

10.2.1 *Boswellia serrata* Roxb.

The species *B. serrata* Roxb. (Burseraceae), popularly known as Shallaki, is a plant species native of India and Saudi Arabia, distributed in the tropical parts of Asia and Africa. The species is traditionally used in the production of Indian frankincense. Its ethnomedicinal indications include the treatment of bronchitis, rheumatism, asthma, cough, intestinal problems, syphilis, jaundice, dysentery, and pulmonary diseases and use as diuretic and expectorant. *B. serrata* has been used for thousands of years to treat swelling and inflammation in ayurvedic medicine and traditional Chinese medicine [26, 27].

Singh and Atal [28] studied the antiinflammatory activity of a standardized hydroethanolic extract of the gum resin of *B. serrata* enriched in boswellic acids, having α-boswellic acid as major constituent (Fig. 10.1). The extract reduced the carrageenan-induced paw edema by 40% and 65%–73% in rats treated with the doses of 50–200 mg/kg (p.o.) and 50–100 mg/kg (i.p.), respectively. Phenylbutazone (50 mg/kg p.o.), employed as reference drug, promoted 47% of edema inhibition. The extract also showed antiinflammatory activity in adrenalectomized rats. In a model of adjuvant-induced arthritis in rats, the extract inhibited paw swelling at 50, 100, and 200 mg/kg (p.o.). The authors also reported antipyretic effect for the extract; it did not present ulcerogenic effect and was well tolerated up to 2 g/kg (p.o.) in rats.

In a rat model of collagen-induced arthritis, a commercial ethanolic extract from gum resin of *B. serrata* (100 and 200 mg/kg; p.o.) suppressed proinflammatory mediators and improved the antioxidant status, as reflected by the reduced levels of lactoperoxidase, myeloperoxidase, catalase, superoxide dismutase (SOD), glutathione (GSH), and nitric oxide (NO). Histological studies of the joints demonstrated positive changes on all evaluated parameters, including articular elastase, MPO, LPO, GSH, catalase, SOD, and NO. In addition, the extract

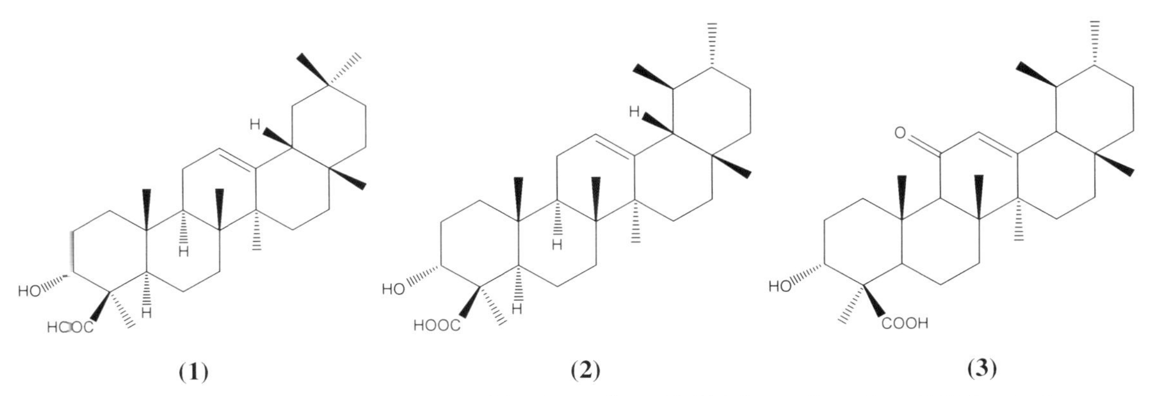

Fig. 10.1 Boswellic acids from *Boswellia serrata*. **(1)** α-Boswellic acid, **(2)** β-boswellic acid, and **(3)** 11-keto-β-boswellic acid.

significantly reduced the levels of proinflammatory mediators (IL-1, IL-6, TNF-α, IFN-α, and PGE2), whereas IL-10 was enhanced [29].

Kumar et al. [30] evaluated tablets containing *B. serrata* gum resin extract in a model of adjuvant-induced acute arthritis in rats by paw edema, at 180 mg/kg. Treatment with the tablets promoted significant improvement in body weight and decrease in ankle diameter and arthritic index; however, there was no significant reduction of paw volume. The activity was comparable with indomethacin. Histopathological analyses also demonstrated the reduction of inflammatory parameters, although the levels of TNF-α did not change significantly.

A mixture of boswellic acids (25, 50, and 100 mg/kg/day; p.o.) reduced the leukocyte population and inhibited its infiltration into the knee joint and in the pleural cavity of rabbits with bovine serum albumin (BSA)-induced arthritis. Treatment also changed the electrophoretic pattern of the synovial fluid proteins. The local injection of the boswellic acids mixture (5, 10, and 20 mg/Kg) into the knee joint 15 min prior to BSA challenge also reduced the accumulation of leukocytes into the knee significantly [31].

In another study the administration of boswellic acids conjugated with the active metabolite rhein at 15.73 mg/kg (p.o.) reduced the diameter of the knee and normalized the biochemical and hematological anomalies in a rat model of collagenase-induced osteoarthritis [32].

A randomized double-blind study was conducted to assess the efficacy, safety, and tolerability of a commercial extract of *B. serrata* (333 mg per capsule, per os, administered three times a day, for 8 weeks) in 30 patients with osteoarthritis of the knee. All patients treated with the extract reported decrease in the knee pain, increased knee flexion, and increased walking capacity. The frequency of swelling in the knee joint was also diminished. The extract was well tolerated by the subjects, except for minor gastrointestinal adverse effects. The authors did not report the chemical composition of the extract but correlated the activity with the presence of α-boswellic acid, β-boswellic acid, and 11-keto-β-boswellic acid, whose chemical structures are depicted in Fig. 10.1 [33].

The *n*-hexane extract of *B. serrata* gum resin administered in combination with a methanolic extract of *Glycyrrhiza glabra* rhizomes (1:1 at 100 mg/kg) exhibited antiarthritic activity in a model of adjuvant-induced arthritis in Wistar rats. The treatment prevented leukocyte migration into the inflamed area and decreased activity of membrane marker enzymes, such as alkaline phosphatase, serum glutamic oxaloacetic transaminase, and serum glutamate pyruvate transaminase. The combined administration of the extracts was shown to be more effective than treatment with any of them separately [34].

Singh et al. [35] evaluated the toxicity of a boswellic acid-rich fraction derived from *B. serrata* gum resin. The fraction did not cause any mortality in rats and mice when administered orally and intraperitoneally at doses up to 2 g/kg. Daily oral treatment with the fraction (250, 500, and 1000 mg/kg) to rats did not induce significant changes in general behavior, as well as in clinical, hematological, biochemical, and pathological parameters. A chronic toxicity study was carried out with 16 healthy monkeys treated with the fraction administered orally at 125, 250, and 500 mg/kg/day. The evaluation of biochemical hematological and histopathological parameters, in addition to other observations, did not indicate any toxicity.

Singh et al. [36] evaluated the toxicity of a commercial extract of *B. serrata* gum resin given by oral route (100, 500, and 1000 mg/kg/day) to rats for 90 days. Rats treated with the highest dose of the extract gained body weight at lower rate than the control group. The effect was shown to be reversible after a recovery period, and the metabolic activity was restored. According to the authors, *B. serrata* is relatively safe to rats up to 500 mg/kg, since no adverse effect was observed. An aqueous extract from *B. serrata* gum resin (50 mg/kg; p.o.) produced significant diuretic effects in albino rats. The extract did not show any acute toxic effect at 3000 mg/kg [37].

Therefore derivatives of the gum resin from *B. serrata* showed antiarthritic activity in preclinical and clinical trials. The studies suggest that the observed activity is attributed to boswellic acids and that these derivatives have no relevant toxicity, being apparently safe for therapeutic use.

10.2.2 *Curcuma longa* Linn.

C. longa Linn. (Scitaminaceae), commonly known as turmeric, is a perennial herb with measures up to 1 m high, with a short stem, showing two varieties: one with rich-colored oval rhizomes and other with larger and lighter-colored rhizomes. The species is spread in tropical and subtropical regions of the world, being widely cultivated in the Asian continent. *C. longa* is traditionally used to treat wound helminthic infections, fevers, skin eruption, conjunctivitis, cough, parasitic infections, and liver diseases. Its major chemical constituents are curcumin, methylcurcumin, demethoxycurcumin, sodium curcuminate, and (+)-(*S*)-ar-turmerone. In Fig. 10.2 are depicted some chemical structures of cucuminoids found in *C. longa* [38–40].

The *C. longa* oil was shown to be effective in preventing chemically induced inflammation and arthritis in rats [41]. In an animal arthritis model, a *C. longa* methanolic extract strongly suppressed joint inflammation and periarticular damage, in correlation with decreased activation of NF-κB and the ensuing cascade of events involving mediators of inflammation and injury, such as chemokines, cyclooxygenase 2, and receptor activator of nuclear factor kappa B ligand (RANKL). The administered dose of the extract was normalized to contain 46 mg/kg/day of curcuminoids [42].

A commercial extract of *C. longa* enriched in curcuminoids was evaluated in collagen-induced arthritis at 30, 60, and 110 mg/kg/day. Treatment with the highest dose showed significant improvement in erythrocyte sedimentation rate, arthritis, and radiographic scores 28 days after inflammation induction [43].

In a rat-model of adjuvant-induced arthritis, a combination of ginger and turmeric rhizomes (extracts at 200 or 400 mg/kg, given orally for 28 days) was shown to induce a more potent response than indomethacin, a potent NSAID. Treatment alleviated joint histopathological changes and extraarticular manifestations, including systemic inflammation (leukocytosis, thrombocytosis, and hyperglobulinemia),

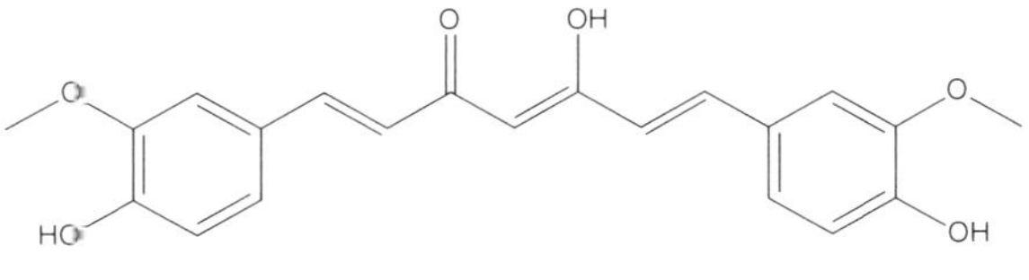

O OH O HO OH (1)

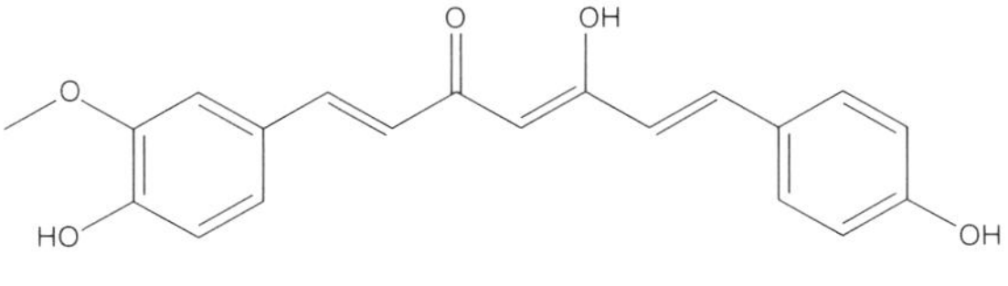

O OH O HO OH (2)

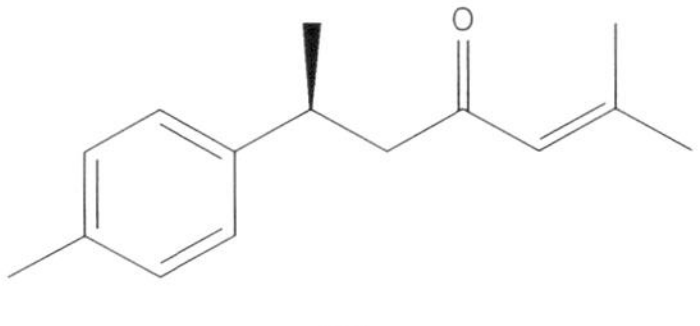

O (3)

Fig. 10.2 Curcuminoids from *Curcuma longa*. **(1)** Curcumin, **(2)** demethoxycurcumin, and **(3)** (+)-(*S*)-ar-turmerone.

decreased body weight gain, and hypoalbuminemia, in addition to iron deficiency anemia, with no prejudice to kidney function and reduced risk of cardiovascular disease [44].

Funk et al. [40] evaluated the antiarthritic activity of crude and refined essential oils of *C. longa* rhizomes in LPS-induced acute and chronic arthritis. Both oils dramatically inhibited joint swelling (90%–100% inhibition) in female rats when given by intraperitoneal route at 10 and 100 mg/kg. However, the antiarthritic effect was accompanied by significant morbidity and mortality. On the other hand, the oral administration of a 20-fold higher dose was nontoxic, but only mildly joint protective (20% inhibition of joint swelling).

A randomized double-blind study was conducted to evaluate the antiosteoarthritis activity of a *C. longa* extract (500 mg/day) standardized to contain 95% w/w of curcuminoids. The extract-treated group showed improvement in the osteoarthritis index and reduction in the levels of IL-1β and oxidative stress biomarkers in comparison with the placebo-treated group [45].

Treatment of healthy and arthritic animals (28 mg/kg/day; i.p.) with crude or refined *C. longa* essential oil resulted in a mortality rate of 20% (crude oil) or 36% (refined oil) by the end of the month-long experimental period [40]. The authors observed elevation in serum alanine aminotransferase (ALT) level, consistent with hepatocellular damage, and anemia in control animals. Necropsies performed 2–4 weeks after treatment indicated signs of mild to moderate peritonitis, including a small intestine perforation in one animal treated with the crude essential oil. There was no evidence of renal toxicity on necropsy. In contrast, no deaths were observed at 2.8 mg/kg/day [40]. A fraction derived from the methanolic extract of *C. longa* rhizomes had no signs of toxicity, as determined by measurement of ALT, creatinine, leukocyte counts, hematocrit, and daily weight gain in female rats [42].

In a human safety study, turmeric oil administered orally showed no clinical, hematologic, renal, or hepatic toxicities after 1 and 3 months of treatment [46]. A study conducted by Liju et al. [47] evaluated the acute and subchronic toxicities, as well as the mutagenic effect of turmeric essential oil. Results showed no mortality, adverse clinical signs, or changes in body weight and water and food consumption during acute and subchronic toxicity studies.

Studies on the safety profile of other *Curcuma* species have been also reported. Balaji et al. (2010) predicted the toxicity of 200 compounds found in *Curcuma* species by using chemoinformatics approaches. According to the authors, 184, 136, 153, and 64 compounds constituents of *Curcuma* species were, respectively, predicted to be potentially toxigenic, mutagenic, carcinogenic, and hepatotoxic [48].

In conclusion, methanolic extract of *C. longa* reduced the progression of arthritis by NF-κB inactivation. The crude and refined essential oils from *C. longa* rhizomes showed antiarthritic activity in preclinical and clinical trials with safety. The authors suggested that the observed activities are associated with curcuminoids, constituents of turmeric oil.

10.2.3 *Cinnamomum zeylanicum* Blume

The genus *Cinnamomum* comprises approximately 250 species, distributed in Asian and Australian continents. *C. zeylanicum* Blume (Lauraceae), popularly known as cinnamon, is a tropical tree that grows to a height of 7–10 m, cultivated in Sri Lanka, Myanmar, and southern coastal strips of India and Brazil [49, 50]. The traditional uses of the species for medicinal purposes include the treatment of vaginitis, inflammations, neuralgia, wounds, diabetes, leukorrhea, and rheumatism. The bark of cinnamon is one of the oldest herbal medicines mentioned in many traditional texts as antiinflammatory to treat pain, enteralgia, bronchitis, and rheumatism [51, 52].

The chemical composition of *C. zeylanicum* comprises A-type procyanidins, dimeric, trimeric and oligomeric proanthocyanidins, camphene, sabinene, myrcene, fenchone, nerol, bornyl and cinnamyl acetates, geranial, cinnamaldehyde, and eugenol. The chemical structures of some constituents of *C. zeylanicum* are shown in Fig. 10.3 [53–55]. The biologic activities already reported for the species include analgesic, antipyretic, antifungal, antiinflammatory, antimicrobial, antidiabetic, and antioxidant effects [54, 56].

A-type procyanidin polyphenols (TAPP), extracted from the bark of *C. zeylanicum*, were evaluated in rat models of carrageenan-induced paw edema and adjuvant-induced arthritis. TAPP showed significant antiinflammatory effect at 4, 8, and 25 mg/kg per os, promoting edema reduction. TAPP treatment (8 mg/kg, daily from day 12 to day 21)

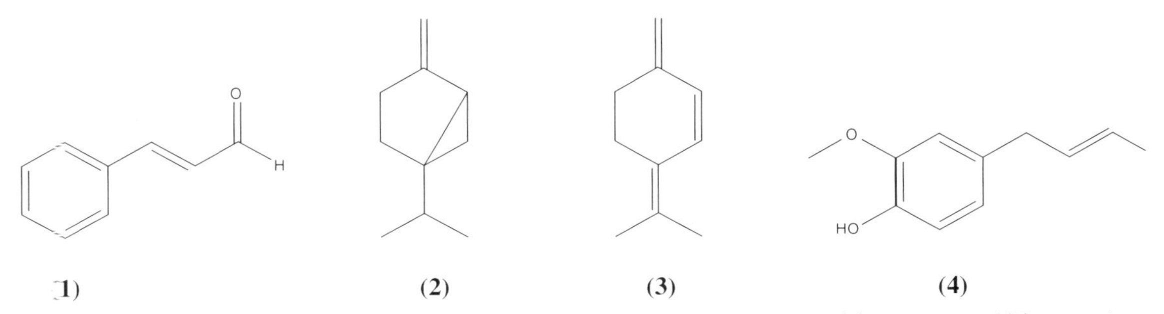

Fig 10.3 Constituents of *Cinnamomum zeylanicum*. **(1)** Cinnamaldehyde, **(2)** sabinene, **(3)** myrcene, and **(4)** eugenol.

significantly reversed changes observed in the arthritis model, including body weight loss, ankle diameter, arthritic score, and serum C-reactive protein levels [55].

Rathi et al. [56] reported the efficacy of a polyphenol-rich fraction from *C. zeylanicum* bark in inflammation models, such rat paw inflammation (edema), localized immune system distress (granuloma), and adjuvant-induced polyarthritis, at 50, 100, and 200 mg/kg. The fraction induced a marked dose-dependent reduction of paw volume, weight loss reversal effect against carrageenan-induced paw edema, and cotton pellet–induced granuloma models in rats, at all assayed doses. Treatment with the fraction (200 mg/kg p.o., for 10 days) promoted a significant reduction of serum TNF-α concentration without causing gastric ulcerogenicity in the arthritis model. It reduced cotton pellet granuloma and also presented mild analgesic effect during acute treatment, as evidenced by the reduction in paw withdrawal threshold of the inflamed paw, in the acetic acid–induced writhing model and Randall-Selitto test in rats.

A double-blind, randomized, controlled trial study was undertaken to compare the effect of an herbal ointment containing cinnamon and a salicylate ointment in patients suffering from knee osteoarthritis. The herbal ointment was shown to be clinically effective regarding pain relief, morning stiffness, and limited motion, producing an effect comparable with the salicylate ointment. However, the authors did not describe the assayed doses, posology, and quantitative composition of the herbal ointment [57].

A randomized double-blind clinical trial was performed with 36 women with rheumatoid arthritis, treated with 500 mg cinnamon powder or placebo daily, for 8 weeks. At the end of the study, there was a significant decrease of serum levels of C-reactive protein and TNF-α in the cinnamon-treated group as compared with the placebo group. Diastolic blood pressure was also significantly lower in the intervention group in comparison with the control group. Administration of cinnamon also significantly reduced the arthritis score in comparison with the placebo group. The authors did not specify which species was tested, but only the genus *Cinnamomum* [58].

The European Medicines Agency reported that extracts, oleoresins, essential oils, and other herbal derivates of cinnamon have no major safety concerns [59]. However, the type and concentration of solvents employed in the extraction process, which are controlled by specific regulations, should be considered [60].

Therefore the polyphenol rich fraction from *C. zeylanicum* bark showed antiarthritic activity in preclinical trials, while a cinnamon powder and an ointment containing cinnamon improved the arthritic score in patients with rheumatoid arthritis in clinical trials.

10.2.4 *Echinodorus grandiflorus* (Cham. & Schltdl.) Micheli.

E. grandiflorus (Cham. & Schltdl.) Micheli. (Alismataceae), popularly known as "Chapéu-de-couro," is a rooted aquatic plant widely distributed throughout the Central and South America, ranging from north Mexico and insular region of the West Indies to southern Argentina, widely found in the Brazilian territory [61–63]. The leaves are traditionally used as diuretic, tonic, antirheumatic, and antiarthritic and to treat oral, renal, hepatic, and dermatological disorders, as well as in cases of hyperuricemia and atherosclerosis [61, 64–66]. The scientific literature reports diuretic, antinociceptive, antiinflammatory, antimicrobial, antiparasitic, cytotoxic, hypocholesterolemic, antiarthritic, and antihypertensive activities for the species [67–71].

Its chemical composition comprises cembrane and clerodane diterpenes such as fitol [72] and hardwickiic acid [73] and flavone C-glycosides such as swertiajaponin and swertisine, in addition to tartaric acid derivatives like chicoric acid, along with *cis*- and *trans*-aconitic acids. The chemical structures of some compounds from *E. grandiflorus* associated with the antiinflammatory activity are depicted in Fig. 10.4 [74, 75].

Different extracts from *E. grandiflorus* leaves and derived fractions reduced TNF-α release by LPS-stimulated THP-1 cells, showing up to 100% inhibition. A positive correlation between chemical composition and biological response was observed for the concentrations

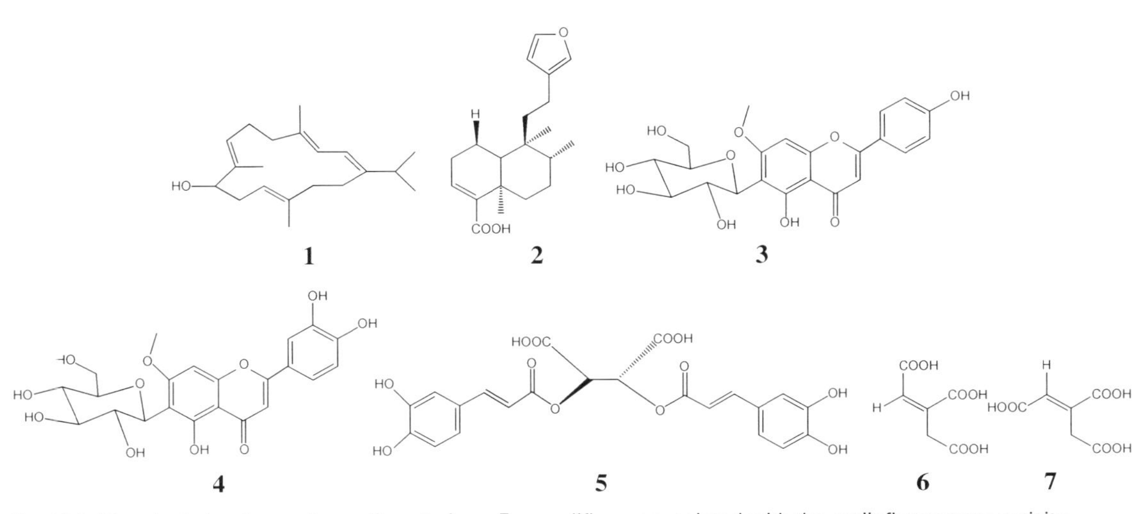

Fig. 10.4 Chemical structures of constituents from *E. grandiflorus* associated with the antiinflammatory activity of the species. **(1)** Fitol, **(2)** hardwickiic acid, **(3)** swertiajaponin, **(4)** swertisine, **(5)** chicoric acid, **(6)** *cis*-aconitic acid, and **(7)** *trans*-aconitic acid.

of swertiajaponin, swertisine, and chicoric acid. In turn the isolated compounds tested alone did not inhibited TNF-α release in the same extension of the extracts and fractions, suggesting a synergistic effect of the compounds present in these matrices [75].

A 70% ethanol extract from *E. grandiflorus* leaves and a flavonoid-rich fraction thereof were evaluated in an AIA model in mice. Both samples reduced neutrophil accumulation into the joint cavity and in the periarticular tissue. The levels of CXCL-1, TNF-α, and IL-1β were also reduced. Histological analyses confirmed that both the extract and the flavonoid-rich fraction suppressed joint inflammation [71].

On its turn, *trans*-aconitic acid, along with its ethyl, *n*-buthyl, and *n*-octyl aconitate esters tested at 172.3 μmol/kg p.o. (equivalent to 30 mg/kg) reduced neutrophil accumulation in the joint of LPS-induced arthritic mice. Interestingly, the diester derivatives induced a similar response at a nanomolar dose (17 nmol/kg) [76]. A formulation of mucoadhesive microspheres containing *trans*-aconitic acid (30 mg/kg) induced prolonged antiinflammatory response in LPS-induced acute arthritis by reducing total cell count and neutrophilic accumulation in the joint cavity when given 48 and 36 h before the stimulus, respectively, in comparison with free *trans*-aconitic acid, whose effect was observed up to 24 and 6 h after treatment, respectively [77].

Regarding the toxicological evaluation, a 70% hydroethanolic extract from *E. grandiflorus* leaves did not show cytotoxic and genotoxic effects in the micronucleus and comet assay in the range of 500–2000 mg/kg in Swiss mice [78]. On its turn the oral administration of an aqueous extract from *E. grandiflorus* leaves at 1000 mg/kg/day to pregnant rats suggested that the extract was able to cause anemia, leukocytosis, and increased cholesterol levels, with some possible changes in liver and kidneys [79]. More recently, Coelho et al. [80] described the genotoxic and antiproliferative potential of an aqueous extract from *E. grandiflorus* leaves. Moreover, extracts from dried leaves of *E. grandiflorus* prepared by infusion and steam sterilized have shown genotoxic activity by the bacterial lysogenic induction assay. According to the authors the genotoxic activity of the infusions implies in some risk of developing degenerative diseases in subjects using these preparations for a long term, without control [81].

The toxicity of some bioactive constituents of the species has been also evaluated. Hence, Misra et al. [82] described absence of acute toxicity for *trans*-aconitic acid in BALB mice up to 2 g/kg administered intraperitoneally. Andersen and Jensen [83] evaluated the mutagenic potential of aconitic acid by the AMES test in the concentration range of 32–20,000 μg per plate. The authors did not find any mutagenic potential for the compound.

In conclusion, preclinical trials with ethanolic extracts, fractions, constituents, and semisynthetic derivatives of *E. grandiflorus* reduced the TNF-α release in vitro and showed antiarthritic activity in vivo in mice. The effect is majorly attributed to flavone C-glycosides and *trans*-aconitic acid and its diester derivatives. The few toxicological reports described genotoxic, antiproliferative, hematological, and biochemical alterations induced by aqueous extracts from *E. grandiflorus* leaves, whereas the hydroethanolic extract and *trans*-aconitic acid did not show toxicological potential.

10.2.5 *Colchicum luteum* Baker

C. luteum Baker (Colchicaceae), commonly known as Suranjan-Talkh in Urdu language, is an annual herb widely distributed in the Western Himalayas (China, India, Pakistan, and Afghanistan), found at altitudes of 600–2700 m. Its corms and seeds are used in the traditional medicine for the treatment of gout, rheumatism, and diseases of the liver and spleen [84–86].

Besides the antiinflammatory and antiarthritic activity, the methanolic extract of this plant has been shown to inhibit lipoxygenase activity and to possess antimicrobial activity [87], antioxidant activity [88], and cytotoxic potential [89]. Phytochemically, it has described the isolation of alkaloids such as colchicine, β-lumicolchicine, 2-desmethylcolchicine, *N*-deacetyl-*N*-formylcolchicine, and luteidine [90, 91], in addition to phenolic compounds and flavonoids like chlorogenic acid and luteolin. Fig. 10.5 depicts the chemical structures of some constituents of *C. luteum* [88].

The 50% ethanolic extract of *C. luteum* corm reduced the joint diameter in a model of formaldehyde-induced arthritis at 34 and 68 mg/kg, in days 8 and 9 after arthritis induction [92]. The extract also reduced the joint diameter in complete Freund's adjuvant (CFA) arthritis at 17, 34, and 68 mg/kg, after 3–21 days of arthritis induction. According to the authors the extract was more effective than

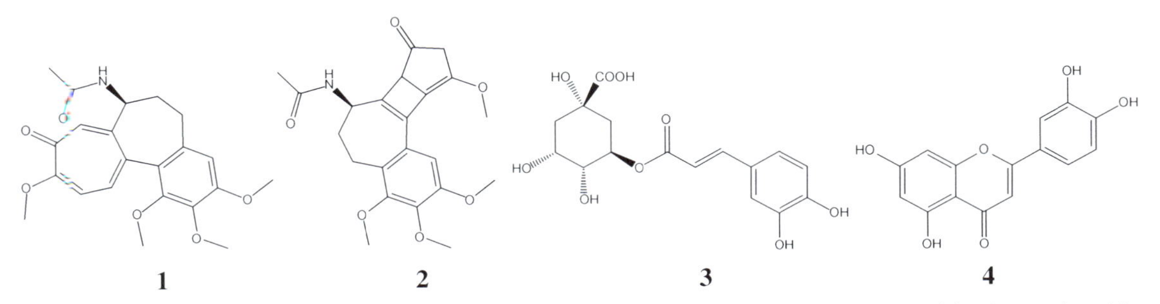

Fig. 10.5 Chemical structures of compounds from *C. luteum*. **(1)** Colchicine, **(2)** β-lumicolchicine, **(3)** chlorogenic acid, and **(4)** luteolin.

indomethacin in mitigating the joint swelling during the period of observation. The mechanism of action of the antiarthritic activity was ascribed to the inhibition of the production of proinflammatory cytokines such as TNF-α, IL-6, and IL-1β [92].

The same research group posteriorly reported that the *C. luteum* corm extract reduced the paw edema induced by carrageenan in rats at 17, 34, and 68 mg/kg, as well as granuloma formation by subcutaneous cotton pellet induction at 68 mg/kg, in addition to diminished serum levels of TNF-α, IL-6, and IL-1β. The authors also observed a reduction on the expression of TNF-R1 in peritoneal macrophage induced by CFA. Many scientific studies have demonstrated the involvement of TNF-R1 in pathophysiological effects of TNF-α, which consequently leads to arthritic conditions. The antiarthritic activities of the 50% ethanol extract was attributed to the presence of colchicines [93].

A few clinical studies in the field of alternative medicine have reported symptomatic improvement in patients with RA after treatment with polyherbal formulations containing *C. luteum* extract. However, the authors did not describe the vegetal drug derivative employed in the formulation, and the results are solely based on subjective evaluation by the patient and on the increased performance of articulation [94, 95].

Colchicine is considered to be the major antiinflammatory constituent of *C. luteum*, and its activity is attributed to the inhibition of microtubules in proinflammatory cells, including macrophages. This effect can reach healthy cells, thus justifying the compound toxicity [96]. Colchicine has been used to treat gout for centuries; however, relatively few controlled trials have assessed its efficacy. In a randomized study with 184 patients with acute gout, they were treated either with a low-dose colchicine regimen (1.2 mg followed by one single 0.6 mg dose 1 h later) or with a high-dose regimen (1.2 mg followed by 0.6 mg doses every hour up to 4.8 mg) 12 h of the onset of attack, in addition to a placebo-treated group [97]. The low- and high-dose regimens demonstrated similar efficacy (37.8% vs 32.7%, achieving $\geq$ 50% improvement within 24 h). It is noteworthy that adverse events in the low-dose regimen were similar to placebo.

In addition to its role in acute gout, colchicine is used prophylactically to reduce gout flare frequency, particularly when patients are initiating urate-lowering therapy. An analysis of three randomized controlled trials found that colchicine used for up to 6 months during initial urate lowering provided greater prophylaxis of flares than its use for only 8 weeks [98]. Colchicine has a narrow therapeutic window; it is a tolerated medication at 2 mg/day for children and 3 mg/day for adults [99]. Common side effects are cramping, abdominal pain, hyperperistalsis, diarrhea, and vomiting. These effects are usually mild and transient [100]. Colchicine doses of 0.5–0.8 mg/kg are highly toxic, and doses above 0.8 mg/kg are typically lethal [101].

Administration of a 50% ethanolic extract of *C. luteum* corm at 2000 mg/kg body weight did not produce any behavioral abnormalities in any of the tested animals. However, one out of five tested animals died. Therefore the LD_{50} in rats was found to be > 2000 mg/kg body weight. The mortality was observed 48 h after extract administration. Chronic administration of the extract at 340 mg/kg body weight for 28 days did not produce any significant physiological changes in the tested animals, as compared with the control group [92].

In conclusion, colchicine is the main active compound of *C. luteum* used for gout management due to its antiinflammatory properties. The ethanolic extract of *C. luteum* also exhibited antiinflammatory and antiarthritic activity in preclinical models. The toxicological studies suggest safety for use of *C. luteum* and colchicine.

10.2.6 *Tripterygium wilfordii* Hook F.

T. wilfordii Hook F. (Celastraceae), vulgarly called Lei Gong Teng (Thunder God Vine) or Mang Cao (rank grass) in China, is a perennial twining vine plant with 2–3 m high, originally found in Taiwan. The species is used in the traditional Chinese medicine (TCM) to treat joint pain, fever, chills, edema, and inflammatory and autoimmune diseases. The phytochemical composition of *T. wilfordii* comprises mainly diterpenes, triterpenes, glycosides, and alkaloids [102, 103]. The diterpenes triptolide, tripdiolide, triptonide, and celastrol (Fig. 10.6) are the most abundant ones and account for the immunosuppressive and antiinflammatory activity elicited by the root extracts [104].

The scientific literature encompasses several publications on preclinical and clinical trials with *T. wilfordii* and its constituents. We do not intend to make an exhaustive revision on the species, but to present some relevant examples of preclinical and clinical studies, as well as some toxicological data.

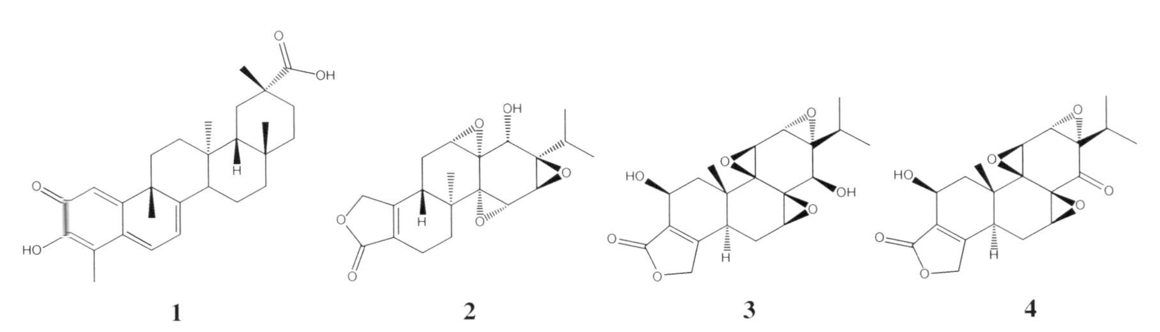

Fig. 10.6 Diterpenes from *T. wilfordii*. **(1)** Celastrol, **(2)** triptolide, **(3)** tripdiolide, and **(4)** triptonide.

A commercial chloroform/methanol extract of *T. wilfordii* reduced the incidence and severity of collagen-induced arthritis in mice (40, 78, and 140 mg/day; treatment for 21 days after collagen immunization), as well as the titers of anticollagen antibodies. Arthritic joint counts, arthritic severity scores, anticollagen antibody titers, and histopathological injuries decreased in all treated groups, as compared with control group [105].

An ethyl acetate fraction of *T. wilfordii* roots was evaluated in a double-blind, placebo-controlled study, in patients with long-standing rheumatoid arthritis. The subjects were randomly assigned to receive placebo, or low dose (180 mg/day), or high dose (360 mg/day) of the extract for 20 weeks, followed by an open-label extension period. A total of 35 patients were enrolled in the trial; 21 patients completed the 20-week study. One patient from each group withdrew because of side effects. The treatment reduced the duration of morning stiffness, the serum titers of rheumatoid factor, and the numbers of tender joints and swollen joints. In addition, the patient assessment of pain and physical function along with the patient and physician global assessments of disease activity improved significantly in the high-dose group from the first evaluation 4 weeks after the beginning of treatment and throughout the trial. No patients withdrew because of adverse events [106].

Asano et al. [107] evaluated a commercial chloroform/methanol extract of *T. wilfordii* in collagen type II induced arthritis in mice. Mice were treated with the extract orally once a day, for 14 days, at daily doses of 240, 320, and 400 μg/kg; treatment began on the day of the first immunization. Treatment with all doses suppressed the development of arthritis, as well as antibody production and delayed-type hypersensitivity. On the other hand, the extract did not affect the clinical course of the disease and the immune response to collagen type II.

A clinical study was conducted with a standardized extract of *T. wilfordii*, administered at 60 mg/patient, three times daily, for 24 weeks, in patients with rheumatoid arthritis. Sulfasalazine (1 g/patient) was administered twice daily as positive control. The standardized extract presented lower performance than sulfasalazine, but it showed relevant antiarthritic activity, improving biochemical parameters and inflammatory cytokines, such as IL-6 and rheumatoid factor levels, as well as the regression of rheumatoid arthritis, assessed by radiological joint evaluation [108].

Zhou et al. [109] conduced a randomized nonblinded clinical controlled study to evaluate the treatment with *T. wilfordii* extract and methotrexate for 2 years. One hundred and nine subjects out of 207 completed the 2 years of the protocol. The changes in total Sharp scores, joint erosion, and joint space narrowing during the 2 years were associated with changes in disease activity, measured by the 28-joint count disease activity score, and they were comparable among the 3 groups (extract and methotrexate in monotherapy and both associated). Adverse events were similar in the three treatment groups. During the 2-year therapy period, *T. wilfordii* monotherapy was similar to methotrexate monotherapy in controlling disease activity and retarding radiological progression in patients with active rheumatoid arthritis. The authors used a commercial extract of *T. wilfordii*, standardized to contain 1.2 μg/10 mg of triptolide and 36.6 μg/10 mg of wilforlide.

Triptolide seems to be important to the therapeutic effects of *T. wilfordii* extracts, but it is also implicated with the toxic effects. A clinical trial compared the effects of isolated triptolide (0.5–0.75 mg/day) and an ethanol extract (120 mg/day) in the treatment of RA. Both treatments induced significant improvements, but more frequent and more severe adverse effects were observed in the group treated with triptolide [110].

Administration of a *T. wilfordii* ethyl acetate fraction at 60 mg/kg for 60–80 days to mice did not affect body weight loss or the histology of most organs, except for the thymus. Long-term treatment of dogs with 5 mg/kg of the aforementioned fraction for 14.5 months did not affect body weight and hepatic or renal function. No pathological changes on the treated animals were observed. However, 4.2% of the treated female mice reduced the frequency of pregnancies and the number of fetuses in pregnant mice in the 5 months of the treatment. The treatment also decreased the activity of sperm of rats in a time-dependent manner. Administration of the fraction to dogs at 10 mg/kg, for 15 months caused significant testicular atrophy [111, 112].

Concluding, the species *T. wilfordii* is widely used in the traditional Chinese medicine for arthritic conditions. This fact reflects in the large number of available presentations on the market of the species and also in the number of the preclinical and clinical studies describing the antiarthritic activity of extracts and fractions of *T. wilfordii*. These studies attributed the observed activity to triptolide and suggest that it is also implicated with the toxic effects.

10.2.7 *Andrographis paniculata* (Burm. fil.) Nees

A. paniculata (Burm. fil.) Nees (Acanthaceae), commonly known as creat or green chiretta, is an annual herbaceous plant native to India and Sri Lanka. It is widely cultivated in Southern and Southeastern Asia, where it has been traditionally used to treat infections and metabolic and inflammatory diseases; mostly the leaves and roots are used for medicinal purposes [113–115].

The major phytoconstituents present in the herb are andrographolide, *bis*-andrographolide A, isoandrographolide, neoandrographolide, homoandrographolide, andrographiside, andrographin, andrographan, andropanoside, andrographosterin, 14-deoxy-11,12-didehydroandrographolide, 14-deoxyandrographolide, panicolin, chlorogenic and myristic acids, and various methoxyflavones [116–122]. Andrographolide is regarded as the main ingredient responsible for *A. paniculata* therapeutic effects. However, the effectiveness of this diterpene is hampered by its low aqueous solubility (3.29 ± 0.73 μg/mL), high lipophilicity (log P value of 2.632 ± 0.135), and low bioavailability. The chemical structures of some diterpenes found in *A. paniculata* are shown in Fig. 10.7 [123].

Andrographolide (100 mg/kg/day) administrated p.o. for 2 weeks decreased the severity of arthritis and joint destruction in a model of collagen-induced arthritis in mouse. Treatment significantly reduced the production of serum anti-CII, TNF-α, IL-1β, and IL-6 by inhibiting the MAPK signaling pathway, thus reducing the phosphorylation of p38 MAPK and ERK1/2 expression [124]. Complete Freund's adjuvant–induced paw edema in rats was also significantly inhibited by andrographolide (50 mg/kg; i.p.) treatment [125].

Yan et al. [126] investigated the effects of andrographolide on fibroblastoid synoviocytes isolated from RA human patients at 10, 20, and 30 μM. The authors observed inhibition in the proliferation of fibroblastoid synoviocytes in a concentration-dependent way. They suggested that this inhibition occurred by proapoptotic mechanism, disrupting the cell replication cycle from phase G0 to phase G1.

In a prospective, double-blind, randomized, placebo-controlled study carried out with 60 patients of rheumatoid arthritis (RA), tablets prepared with an extract of *A. paniculata* containing 30% w/w of

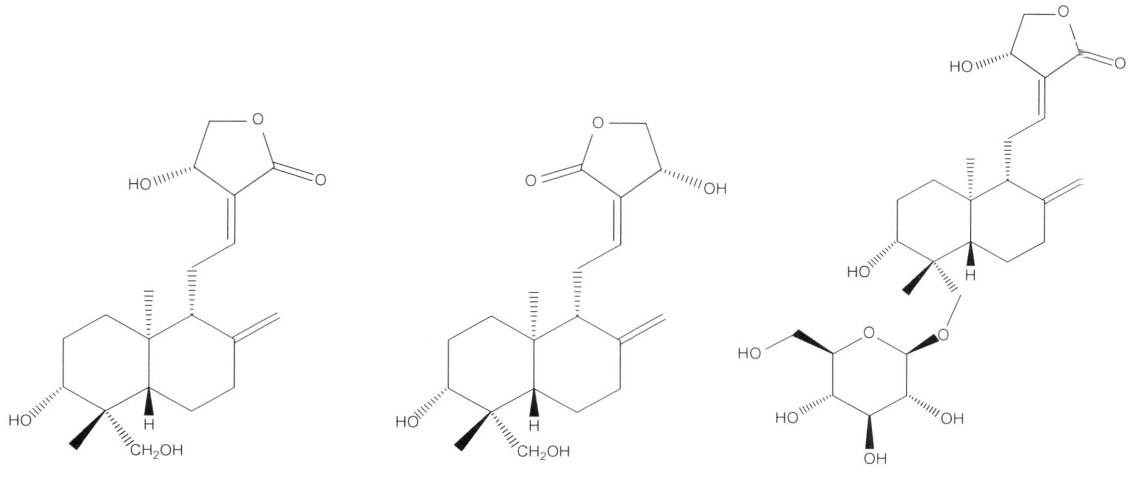

1 2 3

Fig. 10.7 Some diterpenes from *A. paniculata*. **(1)** Andrographolide, **(2)** isoandrographolide, and **(3)** andrographiside.

andrographolide were administered three times per day for 14 weeks, followed by a 2-week washout period. At the end of the treatment, the intensity of joint pain decreased in the test group, as compared with the placebo group, although the variance was not statistically significant. A reduction was also observed in the rheumatoid factor, C4, and IgA [127].

Hidalgo et al. [128] administered a standardized extract of *A. paniculata* (Paractin) containing 30 mg of andrographolide, three times daily, to rheumatoid arthritis patients. The authors reported a reduction in the production of proinflammatory parameters, such as iNOS, COX-2, and cytokines, in patients treated with andrographolide. According to the authors the mechanism of antiarthritic activity of the compound would be related to the reduction of activation of transcription factors like NF-κB, AP-1, STAT3, and NFAT, with inhibition of intracellular signaling pathways, directly interfering with cell proliferation.

A randomized, double-blind, placebo-controlled study was conducted to assess the efficacy of an andrographolide-containing supplement (Paractin; 300 and 600 mg daily) in patients with knee osteoarthritis [129]. A total of 103 male and female subjects with I–II osteoarthritis of the knee joint were assessed. Patients treated with the supplement showed a significant reduction in pain at days 28, 56, and 84 of treatment in comparison with the placebo group. WOMAC stiffness scores, physical function score, and fatigue score showed a significant improvement in both Paractin-treated groups compared with the placebo group. The main adverse events related to Paractin treatment were gastrointestinal symptoms such as acidity, constipation, and oral ulcers. Overall, Paractin at 300 and 600 mg/day dosages was found to be effective and safe in reducing pain in individuals suffering from mild to moderate knee osteoarthritis [129].

An ethanolic extract of *A. paniculata* leaves was evaluated in an acute toxicity assay at 300, 2000, and 5000 mg/kg. All treated animals survived, and no apparent adverse effects were observed during the duration of the study. Gross necropsy observation revealed no lesion in any organ. Although significant alterations in lymphocytes, neutrophils, hematocrit and hemoglobin were observed, these alterations were not treatment-related toxic effects [130].

No testicular toxicity was found by treatment with a 70% ethanol extract of *A. paniculata* at 20, 200, and 1000 mg/kg for 60 days. The authors evaluated the reproductive organ weight, testicular histology, ultrastructural analysis of Leydig cells, and testosterone levels after 60 days of treatment and did not find significant alterations [131].

Therefore in vitro studies demonstrate the reduction of different inflammatory mediators by andrographolide. *A. paniculata* showed antiarthritic activity in different clinical trials, including the commercial presentation Paractin. The ethanolic extract of *A. paniculata* did not exhibit relevant toxicological denotation.

10.2.8 *Nigella sativa* L.

N. sativa L. (Ranunculaceae), popularly known as kalonji or black caraway, grows in South Asia and Southwest Asia, where the seeds have been traditionally used by ancient physicians to treat several diseases. The seeds contain different terpenes and alkaloids such as thymoquinone, thymol, limonene, carvacrol, *p*-cymene, alpha-pinene, 4-terpineol, longifolene, *trans*-anethole, nigellimine N-oxide, nigellimine, and nigellicine, in addition to flavonoids and saponins. The chemical structures of some of these compounds are depicted in Fig. 10.8 [132–136].

A hydroethanolic extract from *N. sativa* seeds enriched in polyphenols was evaluated in different inflammatory models. In the acetic acid–induced writhing test, oral administration of *N. sativa* at 1.25–5.0 g/kg decreased the number of abdominal constrictions. Both oral and intraperitoneal administration of the extract significantly suppressed, in a dose-dependent manner, the nociceptive response in the early and late phases of the formalin test, being the effect on the late phase more pronounced. Oral administration of the extract did not produce a significant reduction in carrageenan-induced paw edema. However, when injected intraperitoneally, the extract inhibited paw edema in a dose-dependent manner [137].

An 80% methanolic extract of the seeds showed antiarthritic activity in a collagen-induced arthritis in rats, at 400 and 500 mg/daily, administered for 20 days. The extract decreased myeloperoxidase activity and associated elastase activity dose dependently and also decreased lipid levels. SOD activity increased significantly with a parallel increment in catalase activity; the articular nitrite content was reduced, which was further confirmed by the histological findings [138].

N. sativa oil at 1.82 mL/kg or 0.91 mL/kg was orally administered for 25 days, starting from the day of immunization with complete Freund's adjuvant in rats. The oil elicited antiinflammatory,

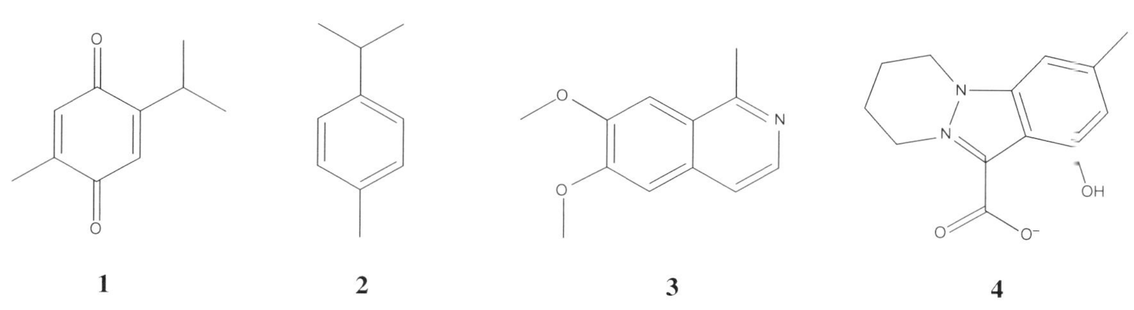

Fig. 10.8 Chemical structures of monoterpenes and alkaloids constituents from *N. sativa*. **(1)** Thymoquinone, **(2)** *p*-cymene, **(3)** nigellimine, and **(4)** nigellicine.

antiarthritic, and antinociceptive activity in a dose-dependent way, controlling the acute phase of the inflammation. The oil attenuated the paw edema in adjuvant-induced arthritis in rats, and the higher dose elicited an inhibition of 56% [139].

Gheita and Kenawy [140] reported a clinical study with 40 female patients with rheumatoid arthritis treated with *N. sativa* oil (capsules containing 500 mg twice/day). The patients showed disease activity score significantly decreased after treatment, as compared with the placebo group. Similarly the number of swollen joints and the duration of morning stiffness improved. The authors suggested that supplementation with *N. sativa* oil during DMARD therapy in rheumatoid arthritis may be considered an affordable potential adjuvant therapy.

A randomized double-blind placebo-controlled clinical trial was performed with 42 patients with rheumatoid arthritis to evaluate the effect of *N. sativa* oil on inflammatory mediators. Subjects from the intervention group received two capsules daily, each containing 500 mg of oil, for 8 weeks. The serum level of IL-10 was increased in the oil-treated group, in addition to a significant reduction of MDA and NO levels. On the other hand, there were no significant differences in TNF-α, superoxide dismutase, catalase, and total antioxidant capacity values between treated and control groups, before and after the intervention [141].

A randomized, double-blinded placebo-controlled, parallel-group clinical trial was conducted during 2 months with 43 female patients (20–50 years old) with rheumatoid arthritis. The treated group (23 patients) received 2 capsules containing *N. sativa* oil at 500 mg/day, and 20 subjects received placebo. Treatment led to significant reduction of serum high-sensitivity C-reactive protein and improved the number of swollen joints, as compared with baseline and placebo groups. A relatively comparable CD4 + T-cell percentage was observed in the treated and placebo groups either in baseline or at the end of study. The treatment also resulted in reduced counts of CD8 + and increased CD4 + CD25 + T-cell proportion, and CD4 +/CD8 + ratio, as compared with placebo and baseline values [142].

Several in vivo studies have demonstrated the activity of thymoquinone, a constituent of *N. sativa*, in inflammatory diseases. The antiarthritic effect was observed after its administration for consecutive 21 days in rats with collagen-induced arthritis. Thymoquinone produces antiarthritic effects by reducing joint elastase and myeloperoxidase activity. Thymoquinone is also responsible for reducing the expression of proinflammatory cytokines including IL-1β, TNF-α, IL-10, IFN-c, PGE-2, and IL-6 [143–145].

The administration of *N. sativa* seed extract (50 mg/kg) intraperitoneally to rats for 5 days did not affect significantly the activities of several enzymes and metabolites, which are indicative of hepatic and renal function [146]. Oral administration of the seed oil at doses up to

10 mL/kg in rats and mice did not cause any mortality or overt toxicity during the observation period of 48 h [147]. In an acute toxicity study carried out in mice with an ethanol extract from *N. sativa* seeds, toxicity signs were first noticed after 4–6 h of extract administration. The median lethal dose (LD_{50}) was 470 mg/kg body weight. The toxic signs observed included decreased locomotor activity, decreased sensitivity to touch, and jerking. After 10 h of administration, the mice exhibited tachypnea, prostration, and reduced food intake [148].

The LD_{50} value of thymoquinone was found to be 2.4 g/kg (range 1.52–3.77). The acute administration of high doses (2 g/kg or higher) caused hypoactivity and difficulty in respiration. Biochemically, these high doses depleted GSH concentrations in the liver, kidney, and heart, and damaged the liver and kidney, as evidenced by significant increases in plasma metabolites and enzymes [149]. Administration of thymoquinone in the drinking water of mice at concentrations up to 0.03% w/v for 90 days did not evidence any sign of toxicity, except for a significant decrease in fasting plasma glucose concentration [149].

N. sativa fixed oil was given to mice orally and intraperitoneally in an acute and chronic study. In the chronic toxicity study, rats treated daily for 3 months presented no changes in key hepatic enzyme levels, particularly aspartate aminotransferase, alanine aminotransferase, and gamma-glutamyltransferase. Moreover the histopathological results attested absence of tissue toxicity in the heart, liver, kidneys, and pancreas. LD_{50} values for single doses given orally and intraperitoneally to mice were reported to be 28.8 and 2.02 mL/kg, respectively [150]. The LD_{50} of thymoquinone in mice was found to be 104.7 mg/kg (i.p.) and 870.9 mg/kg (p.o.). On its turn the LD_{50} values reported for rats were 57.5 mg/kg (i.p.) and 794.3 mg/kg (p.o.) [151].

Administration of *N. sativa* oil (0.5 mL/kg/day) simultaneously with the insecticide acetamiprid mitigated and modulated the adverse effects induced by acetamiprid on reproductive organs weights, semen quality, and testosterone, thus protecting the reproductive system against the toxic effects promoted by the insecticide [152].

In conclusion, different extracts and oils of *N. sativa* showed preclinical and clinical improvement of arthritic conditions. The activity is associated with terpenes, mainly the monoterpene thymoquinone, and the toxicological reports suggest safety use of this compound or preparations containing it.

10.3 Conclusion and perspectives

Different medicinal plants are currently used in Eastern medicine to treat arthritic diseases, as evidenced by the number of publications reporting the effects of species like *T. wilfordii*, *B. serrata*, *C. longa*, *C. zeylanicum*, *A. paniculata*, and *N. sativa*. Clinical trials have been also described for some of these species aiming to attest their efficacy in arthritic diseases; however, in several of these studies, relevant data regarding the plant derivative (vegetal drug used, extract/fraction preparation, standardization of the extract/fraction), posologic schemes, and proper delineation of the study are missing. Additionally, some clinical trials carried out with these species in subjects with rheumatoid arthritis and osteoarthritis are based on subjective parameters such as pain intensity and improvement of joint conditioning and mobility. It is worth mentioning that toxicological assessments of some species are scarce and limited to evaluations in mice and rats, thus hindering the investigation of these species in clinical trials with a larger number of subjects.

On the other hand, numerous studies carried out with *T. wilfordii*, *A. paniculata*, and *B. serrata*, related to their traditional uses to treat osteoarthritis and rheumatoid arthritis, demonstrate the therapeutic potential of these species, whether in association with DMARDS, as reported for *T. wilfordii* and methotrexate, or individually like the commercial herbal Chinese medicine Paractin.

In conclusion, there is a high potential for some of the species briefly reviewed in this chapter to manage arthritic diseases, either employed individually or in association with currently used antiinflammatory drugs. To explore their therapeutic potential, there must be an accurate standardization of the plant preparations along with additional well-delineated clinical trials and toxicological assays to attest their efficacy and safety.

References

[1] J.M.W. Hazes, J.J. Luime, The epidemiology of early inflammatory arthritis, Nat. Rev. Rheumatol. 7 (2011) 381–390.

[2] L.G. Mercuri, Surgical management of TMJ arthritis, in: D.M. Laskin, C.S. Green, W.L. Hylander (Eds.), Temporomandibular Joint Disorders: An Evidence-Based Approach to Diagnosis and Treatment, Quintessence, Chicago, 2006, pp. 455–468.

[3] L.G. Mercuri, S. Abramowicz, Arthritic conditions affecting the temporomandibular joint, in: C. Farah, R. Balasubramaniam, M. McCullough (Eds.), Contemporary Oral Medicine, Springer, Cham, 2018.

[4] J.J. Rodrigo, M.E. Gershwin, Management of the arthritic joint, in: M.W. Chapman (Ed.), Chapman's Orthopaedic Surgery, third ed., Lippincott, Williams & Wilkins, Philadelphia, 2001, pp. 2551–2572.

[5] L.L. Xavier, P.R. Viacava, V.O.M. Teixeira, M.R. Munhoz, P.S. Lora, P.G. Oliveira, L.I. Filippin, R.M. Xavier, Desenvolvimento de artrite induzida por colágeno em camundongos DBA/1J entre os gêneros, Clin. Biomed. Res. 32 (2012) 436–442.

[6] D.A. González-Chica, S. Vanlint, E. Hoon, N. Stocks, Epidemiology of arthritis, chronic back pain, gout, osteoporosis, spondyloarthropathies and rheumatoid arthritis among 1.5 million patients in Australian general practice: NPS Medicine Wise Medicine Insight dataset, BMC Musculoskelet. Disord. 19 (2018) 20.

[7] Z. Chen, A. Bozec, A. Ramming, G. Schett, Anti-inflammatory and immune-regulatory cytokines in rheumatoid arthritis, Nat. Rev. Rheumatol. 15 (2019) 9–17.

[8] S.K. Chang, Z. Gu, M.B. Brenner, Fibroblast-like synoviocytes in inflammatory arthritis pathology: the emerging role of cadherin-11, Immunol. Rev. 233 (2010) 256–266.

[9] K.W. Frommer, B. Zimmermann, F.M. Meier, M.H. Shroder, A. Shaffler, C. Bucler, J. Steinmeyer, F. Brentano, F. Gay, U.M. Ladner, E. Neumann, Adiponectin-mediated changes in effector cells involved in the pathophysiology of rheumatoid arthritis, Arthritis Rheum. 62 (2010) 2886–2899.

[10] A.M.S. Carvalho, L. Heimfarth, K.A. Santos, A.G. Guimarães, L. Picot, J.R.G.S. Almeida, S.S. Jullyana, L.J. Quintans-Júnior, Terpenes as possible drugs for the mitigation of arthritic symptoms—a systematic review, Phytomedicine 57 (2019) 137–147.

[11] R. Venson, A. Wiens, C.J. Correr, M.F. Otuki, M.G. Grochocki, D.R.S. Pontarolli, R. Pontarolo, Economic evaluation of anticitokines adalimumab, etanercept and infliximab for treatment of rheumatoid arthritis in Paraná State, Brazil, Physis 21 (2011) 359–376.

[12] J.A. Singh, R. Christensen, G.A. Wells, M.E. Suarez-Almazor, R. Buchbinder, M.A. Lopez-Olivo, E.T. Ghogomu, P. Tugwell, Biologics for rheumatoid arthritis: an overview of Cochrane reviews, Cochrane Database Syst. Rev. 7 (2009) CD007848.

[13] M. Feldmann, Development of anti-TNF therapy for rheumatoid arthritis, Nat. Rev. Immunol. 2 (2002) 364–371.

[14] Brasil, Ministério da saúde. Relatório de Recomendação da Comissão Nacional de Incorporação de Tecnologias no SUS—CONITEC—12, 2012.

[15] R.J. Desai, J.K. Rao, R.A. Hansen, G. Fang, M.L. Maciejewski, J.F. Farley, Predictors of treatment initiation with tumor necrosis factor-alpha inhibitors in patients with rheumatoid arthritis, J. Manag. Care Pharm. 20 (2014) 1110–1120.

[16] S. Hopson, K. Saverno, L.Z. Liu, A. Al-Sabbagh, J. Orazem, M.E. Costantino, M.K. Pasquale, Impact of out-of-pocket costs on prescription fills among new initiators of biologic therapies for rheumatoid arthritis, J. Manag. Care Spec. Pharm. 22 (2016) 122–130.

[17] J. Laine, T.S. Jokiranta, K.K. Eklund, M. Väkeväinen, K. Puolakka, Cost-effectiveness of routine measuring of serum drug concentrations and anti-drug antibodies in treatment of rheumatoid arthritis patients with TNF-α blockers, Biologics 10 (2016) 67–73.

[18] N. Choudhary, L.K. Bhatt, K.S. Prabhavalkar, Experimental animal models for rheumatoid arthritis, Immunopharmacol. Immunotoxicol. 40 (2018) 193–200.

[19] D.J. Newman, G.M. Cragg, Natural products as sources of new drugs from 1981 to 2014, J. Nat. Prod. 79 (2016) 629–661.

[20] M.H. Farzaei, F. Farzaei, M. Abdollahi, Z. Abbasabadi, A.H. Abdolghaffari, B. Mehraban, A mechanistic review on medicinal plants used for rheumatoid arthritis in traditional Persian medicine, J. Pharm. Pharmacol. 68 (2016) 1233–1248.

[21] M. Ghasemian, S. Owlia, M.B. Owlia, Review of anti-inflammatory herbal medicines, Adv. Pharmacol. Sci. 2016 (2016), 9130979.

[22] S. Sadia, A. Tariq, S. Shaheen, K. Malik, F. Khan, M. Ahmad, H. Qureshi, B.G. Nayyar, Ethnopharmacological profile of anti-arthritic plants of Asia—a systematic review, J. Herb. Med. 13 (2018) 8–25.

[23] J.C. DeSalvo, M.B. Skiba, C.L. Howe, K.E. Haiber, J.L. Funk, Natural product dietary supplement use by individuals with rheumatoid arthritis: a scoping review, Arthritis Care Res. 71 (2019) 787–797.

[24] M.C. Lu, H. Livneh, L.M. Chiu, N.S. Lai, C.C. Yeh, T.Y. Tsai, A survey of traditional Chinese medicine use among rheumatoid arthritis patients: a claims data-based cohort study, Clin. Rheumatol. 38 (2019) 1393–1400.

[25] S. Saleem, R. Khan, I. Kazmi, M. Afzal, Medicinal plants in the treatment of arthritis, in: M. Ozturk, K. Hakeem (Eds.), Plant and Human Health, 3, Springer, Cham, 2019, pp. 101–137.

[26] A. Upaganlawar, B. Ghule, Pharmacological activities of *Boswellia serrata* Roxb-mini review, Ethnobot. Leaflets 13 (2009) 766–774.
[27] P. Pungle, M. Banavalikar, A. Suthar, M. Biyani, S. Mengi, Immunomodulatory activity of boswellic acids of *Boswellia serrata* Roxb, Indian J. Exp. Biol. 41 (2003) 1460–1462.
[28] G.B. Singh, C.K. Atal, Pharmacology of an extract of salai guggal ex-*Boswellia serrata*, a new non-steroidal anti-inflammatory agent, Agents Actions 18 (1986) 407–412.
[29] S. Umar, K. Umar, A.H.M.G. Sarwar, A. Khan, N. Ahmad, S. Ahmad, C.K. Katiyar, S.A. Husain, H.A. Khan, *Boswellia serrata* extract attenuates inflammatory mediators and oxidative stress in collagen induced arthritis, Phytomedicine 21 (2014) 847–856.
[30] R. Kumar, S. Singh, A.K. Saksena, R. Pal, R. Jaiswal, R. Kumar, Effect of *Boswellia serrata* extract on acute inflammatory parameters and tumor necrosis factor-α in complete Freund's adjuvant-induced animal model of rheumatoid arthritis, Int. J. Appl. Basic Med. Res. 9 (2019) 100–106.
[31] M.L. Sharma, S. Bani, G.B. Singh, Anti-arthritic activity of boswellic acids in bovine serum albumin (BSA)-induced arthritis, Int. J. Immunopharmacol. 11 (1989) 647–652.
[32] S. Dhaneshwar, P. Dipmala, H. Abhay, B. Prashant, Disease-modifying effect of anthraquinone prodrug with boswellic acid on collagenase-induced osteoarthritis in Wistar rats, Inflamm. Allergy Drug Targets 12 (2013) 288–295.
[33] N. Kimmatkar, V. Thawani, L. Hingorani, R. Khiyani, Efficacy and tolerability of *Boswellia serrata* extract in treatment of osteoarthritis of knee—a randomized double blind placebo controlled trial, Phytomedicine 10 (2003) 3–7.
[34] N.K. Mishra, S. Bstia, G. Mishra, K.A. Chowdary, S. Patra, Anti-arthritic activity of *Glycyrrhiza glabra Boswellia serrata* and their synergistic activity in combined formulation studied in Freund's adjuvant induced arthritic rats, J. Pharm. Educ. Res. 2 (2011) 92–98.
[35] G.B. Singh, S. Bani, S. Singh, Toxicity and safety evaluation of boswellic acids, Phytomedicine 3 (1996) 87–90.
[36] P. Singh, K.M. Chacko, M.L. Aggarwal, B. Binu, R.K. Khandal, S. Sutana, T. Binu, A-90 day gavage safety assessment of *Boswellia serrata* in rats, Toxicol. Int. 19 (2012) 273–278.
[37] M. Asif, Q. Jabeen, A.M.S.A. Majid, M. Atif, Diuretic activity of *Boswellia serrata* Roxb. oleo gum extract in albino rats, Pak. J. Pharm. Sci. 27 (2014) 1811–1817.
[38] R. Srimal, N. Khanna, B. Dhawan, A preliminary report on anti-inflammatory activity of curcumin, Int. J. Pharm. 3 (1971) 10–13.
[39] C.A.C. Araujo, L.L. Leon, Biological activities of *Curcuma longa* L, Mem. Inst. Oswaldo Cruz 96 (2001) 723–728.
[40] J.L. Funk, J.B. Frye, J.N. Oyarzo, H. Zhang, B.N. Timmermann, Anti-arthritic effects and toxicity of the essential oils of turmeric (*Curcuma longa* L.), J. Agric. Food Chem. 58 (2010) 842–849.
[41] D. Chandra, S.S. Gupta, Anti-inflammatory and anti-arthritic activity of volatile oil of *Curcuma longa* (Haldi), Indian J. Med. Res. 60 (1972) 138–142.
[42] J.L. Funk, J.B. Frye, J.N. Oyarzo, N. Kuscuoglu, J. Wilson, G. McCaffrey, G. Stafford, G. Chen, R.C. Lantz, S.D. Jolad, A.M. Sólyom, P.R. Kiela, B.N. Timmermann, Efficacy and mechanism of action of turmeric supplements in the treatment of experimental arthritis, Arthritis Rheum. 54 (2006) 3452–3464.
[43] A.F. Zahidah, O. Faizah, K. Nur Aqilah, K. Tatyanna, Curcumin as an anti-arthritic agent in collagen-induced arthritic Sprague-Dawley rats, Sains Malays. 41 (2012) 591–595.
[44] G. Ramadan, O. El-Menshawy, Protective effects of ginger-turmeric rhizomes mixture on joint inflammation, atherogenesis, kidney dysfunction and other complications in a rat model of human rheumatoid arthritis, Int. J. Rheum. Dis. 16 (2013) 219–229.

[45] S. Srivastava, A.K. Saksena, S. Khattri, S. Kumar, R.S. Dagur, *Curcuma longa* extract reduces inflammatory and oxidative stress biomarkers in osteoarthritis of knee: a four-month, double-blind, randomized, placebo-controlled trial, Inflammopharmacology 24 (2016) 377–388.
[46] J. Joshi, S. Ghaisas, A. Vaidya, R. Vaidya, D.V. Kamat, A.N. Bhagwat, S. Bhide, Early human safety study of turmeric oil (*Curcuma longa* oil) administered orally in healthy volunteers, J. Assoc. Physicians India 51 (2003) 1055–1060.
[47] V.B. Liju, K. Jeena, R. Kuttan, Acute and subchronic toxicity as well as mutagenic evaluation of essential oil from turmeric (*Curcuma longa* L), Food Chem. Toxicol. 53 (2013) 52–61.
[48] S. Balaji, B. Chempakam, Toxicity prediction of compounds from turmeric (*Curcuma longa* L), Food Chem. Toxicol. 48 (2010) 2951–2959.
[49] M. Das, S. Mandal, B. Mallick, J. Hazra, Ethnobotany, phytochemical and pharmacological aspects of *Cinnamomum zeylanicum* Blume, Int. Res. J. Pharm. 4 (2013) 58–63.
[50] P.V. Rao, S.H. Gan, Cinnamon: a multifaceted medicinal plant, Evid. Based Complement. Alternat. Med. 2014 (2014), 642942.
[51] P.K. Warriers, V.P. Nambiar, C. Ramankutty, P.S. Vaidhyarathnam, Indian Medicinal Plants: A Compendium of 500 Species, Orient Longman Ltd, Chennai, New-Delhi, 1993.
[52] M. Vangalapati, N.S. Satya, D.S. Prakash, S. Avanigadda, A review on pharmacological activities and clinical effects of cinnamon species, Res. J. Pharm., Biol. Chem. Sci. 3 (2012) 653–663.
[53] U.M. Senanayake, T.H. Lee, R.B.H. Wills, Volatile constituents of cinnamon (*Cinnamomum zeylanicum*) oils, J. Agric. Food Chem. 26 (1978) 822–824.
[54] G.K. Jayaprakasha, L.J.M. Rao, Chemistry, biogenesis, and biological activities of *Cinnamomum zeylanicum*, Crit. Rev. Food Sci. Nutr. 51 (2011) 547–562.
[55] V. Sachin, L.B. Subhash, M. Vishwaraman, P.A. Thakurdesaib, Anti-inflammatory and anti-arthritic activity of type-A procyanidine polyphenols from bark of *Cinnamomum zeylanicum* in rats, Food Sci. Human Wellness 2 (2013) 59–67.
[56] B. Rathi, S. Bodhankar, V. Mohan, P. Thakurdesai, Ameliorative effects of a polyphenolic fraction of *Cinnamomum zeylanicum* L. bark in animal models of inflammation and arthritis, Sci. Pharm. 81 (2013) 567–590.
[57] M. Zahmatkash, M.R. Vafaeenasab, Comparing analgesic effects of a topical herbal mixed medicine with salicylate in patients with knee osteoarthritis, Pak. J. Biol. Sci. 14 (2011) 715–719.
[58] F. Shishehbor, M. Rezaeyan Safar, E. Rajaei, M.H. Haghighizadeh, Cinnamon consumption improves clinical symptoms and inflammatory markers in women with rheumatoid arthritis, J. Am. Coll. Nutr. 37 (2018) 1–6.
[59] E. Panel, Opinion of the scientific panel on food additives, flavourings, processing aids and materials in contacts with food (AFC) on a request from the commission related to coumarin, EFSA J. 104 (2004) 1–36.
[60] D.R.A. Muhammad, K. Dewettinck, Cinnamon and its derivatives as potential ingredient in functional food—a review, Int. J. Food Prop. 20 (2017) 2237–2263.
[61] M.P. Correa, Dicionário de plantas úteis do Brasil e das exóticas cultivadas, vol. 2, Imprensa Nacional, Rio de Janeiro, 1931. 205, 214–215.
[62] H. Lorenzi, Plantas Daninhas do Brasil: terrestres, aquáticas, parasitas e tóxicas, third ed., Instituto Plantarum, Nova Odessa, SP, 2000, p. 43.
[63] A.M. Marques, D.W. Provance, M.A.C. Kaplan, M.R. Figueiredo, *Echinodorus grandiflorus*: ethnobotanical, phytochemical and pharmacological overview of a medicinal plant used in Brazil, Food Chem. Toxicol. 109 (2017) 1032–1047.
[64] G.L. Cruz, Livro Verde Das Plantas Medicinais E Industriais Do Brasil, first ed., Veloso S. A., Belo Horizonte, 1965, pp. 314–315.
[65] G.L. Cruz, Dicionário das plantas úteis do Brasil, third ed., Civilização Brasileira S.A., Rio de Janeiro, 1985, p. 229.

[66] S. Panizza, Plantas que curam: cheiro de mato, twenty eighth ed., IBRASA, São Paulo, 1997, pp. 79–80.

[67] M.G.R. Duarte, I.A.A. Soares, M. Brandão, R.L.R.P. Jácome, M.D. Ferreira, C.R.F. Silva, A.B. Oliveira, Perfil fitoquímico e atividade antibacteriana in vitro de plantas invasoras, Rev. Lecta Bragança Paulista. 20 (2002) 177–182.

[68] G.C. Souza, A.P.S. Haas, G.L. Von Poser, E.E.S. Schapoval, E. Elisabetsky, Ethnopharmacological studies of antimicrobial remedies in the south of Brazil, J. Ethnopharmacol. 90 (2004) 135–143.

[69] C.I. Wright, L. Van-Buren, C.I. Kroner, M.M.G. Koning, Herbal medicines as diuretics: a review of the scientific evidence, J. Ethnopharmacol. 114 (2007) 1–31.

[70] E. Tibiriça, A. Almeida, S. Cailleaux, D. Pimenta, M.A. Kaplan, M.A. Lessa, M.R. Figueiredo, Pharmacological mechanisms involved in the vasodilator effects of extracts from *Echinodorus grandiflorus*, J. Ethnopharmacol. 111 (2007) 50–55.

[71] E.F. Garcia, M.A. Oliveira, L.C. Candido, F.M. Coelho, V.V. Costa, C.M. Queiroz-Junior, D. Boff, F.A. Amaral, D.G. Souza, M.M. Teixeira, F.C. Braga, Effect of the hydroethanolic extract from *Echinodorus grandiflorus* leaves and a fraction enriched in flavone-C-glycosides on antigen-induced arthritis in mice, Planta Med. 82 (2016) 407–413.

[72] D. Manns, R. Hartmann, Echinodol: a new cembrene derivative from *Echinodorus grandiflorus*, Planta Med. 59 (1993) 465–466.

[73] C.M.A. Tanaka, Constituintes químicos de cinco espécies de *Echinodorus* e avaliação do beta-pineno como substrato para obtenção de quirons mais elaborados, Theses (Chemistry PhD), Instituto de Química, Universidade Estadual de Campinas, Campinas, 2000.

[74] M. Schnitzler, F. Petereit, A. Nahrstedt, *Trans*-aconitic acid, glucosylflavones and hydroxycinnamoyltartaric acids from the leaves of *Echinodorus grandiflorus* ssp. aureus, a Brazilian medicinal plant, Rev. Bras. Farm. 17 (2007) 149–154.

[75] E.F. Garcia, M.A. Oliveira, L.P.A. Dourado, D.G. Souza, M.M. Teixeira, F.C. Braga, In vitro TNF-α inhibition elicited by extracts from *Echinodorus grandiflorus* leaves and correlation with their phytochemical composition, Planta Med. 82 (2015) 337–343.

[76] D.P. Oliveira, T.V. Moreira, N.V. Batista, J.D. Souza Filho, F.A. Amaral, M.M. Teixeira, R.M. Padua, F.C. Braga, Esterification of *trans*-aconitic acid improves its anti-inflammatory activity in LPS-induced acute arthritis, Biomed. Pharmacother. 99 (2018) 87–95.

[77] D.P. Oliveira, G.G. Augusto, N.V. Batista, V.S. Louise, D.S. Ferreira, M.A.C. Souza, C. Fernandes, F.A. Amaral, M.M. Teixeira, R.M. Pádua, M.C. Oliveira, F.C. Braga, Encapsulation of *trans*-aconitic acid in mucoadhesive microspheres prolongs the anti-inflammatory effect in LPS-induced acute arthritis, Eur. J. Pharm. Sci. 119 (2018) 112–120.

[78] C.J. Da Silva, J.K. Bastos, C.S. Takahashi, Evaluation of the genotoxic and cytotoxic effects of crude extracts of *Cordia ecalyculata* and *Echinodorus grandiflorus*, J. Ethnopharmacol. 127 (2010) 445–450.

[79] S.S.S. Brugiolo, V.M. Peters, D.S. Pimenta, B.J.V. Aarestrup, A.S.S. Brugiolo, D.M. Ribeiro, M.O. Gerra, Reproductive toxicity of *Echinodorus grandiflorus* in pregnant rats, J. Toxicol. Sci. 35 (2010) 911–922.

[80] A.P.D. Coelho, H.D. Laughinghouse, A.W. Kuhn, A.A. Boligon, T.S. Canto-Dorow, A.C.F. Silva, S.B. Tedesco, Genotoxic and antiproliferative potential of extracts of *Echinodorus grandiflorus* and *Sagittaria montevidensis* (Alismataceae), Caryologia 70 (2017) 82–91.

[81] E.C. Lima-Dellamora, K.C.V. Waldhelm, A.M. Alves, C.A.S. Lage, A.A.C. Leitão, R.M. Kuster, Genotoxic Maillard Byproducts in Current Phytopharmaceutical Preparations of *Echinodorus grandiflorus*, vol. 86, Anais Da Academia Brasileira de Ciências, 2014, pp. 1385–1394.

[82] S. Misra, T. Sanyal, D. Sarkar, P.K. Bhattacharya, D.K. Ghosh, Evaluation of antileishmanial activity of *trans*-aconitic acid, Biochem. Med. Metab. Biol. 42 (1989) 171–178.

[83] P.H. Andersen, N.J. Jensen, Mutagenic investigation of flavourings: dimethyl succinate, ethyl pyruvate and aconitic acid are negative in the Salmonella/mammalian-microsome test, Food Addit. Contam. 1 (1984) 283–288.

[84] P.B. Singh, B.S. Aswal, Conservation and cultivation of medicinal plants in Himachal Pradesh, J. Econ. Taxon. Bot. 18 (1994) 715–722.

[85] S.K. Kapur, P. Singh, Traditionally important medicinal plants of Udhampur district (Jammu province) part-I, J. Econ. Taxon. Bot. 12 (1996) 75–81.

[86] Z.K. Shinwari, S.S. Gilani, Sustainable harvest of medicinal plants at Bulashbar Nullah. Astore (Northern Pakistan), J. Ethnopharmacol. 84 (2003) 289–298.

[87] B.H. Ahmad, S. Khan, M. Bashir, M. Nisar, M. Hassan, Inhibition activities of *Colchicum luteum* Baker on lipoxygenase and other enzymes, J. Enzyme Inhib. Med. Chem. 21 (2006) 449–452.

[88] M. Ahmad, M.A. Khan, M. Zafar, M. Arshad, S. Sultana, B.H. Abbasi, S.U. Din, Use of chemotaxonomic markers for misidentified medicinal plants used in traditional medicines, J. Med. Plant Res. 4 (2010) 1244–1252.

[89] H. Khan, S.A. Tariq, M.A. Khan, Biological and phytochemical studies on corms of *Colchicum luteum* Baker, J. Med. Plant Res. 5 (2011) 7031–7035.

[90] B. Chommadov, M.K. Yusupov, A.S. Sadykov, Structure of alkaloid L-5 from *Colchicum luteum*, Chem. Nat. Compd. 6 (1970) 82–88.

[91] S.K. Koul, R.S. Thakur, Alkaloids of *Colchicum luteum* Baker fresh corms, Indian J. Pharm. 39 (1977) 115–116.

[92] V. Nair, R. Kumar, S. Singh, Y.K. Gupta, Investigation into the anti-inflammatory and antigranuloma activity of *Colchicum luteum* Baker in experimental models, Inflammation 35 (2012) 881–888.

[93] V. Nair, S. Singh, Y.K. Gupta, Evaluation of the disease modifying activity of *Colchicum luteum* Baker in experimental arthritis, J. Ethnopharmacol. 133 (2011) 303–307.

[94] V.R. Joshi, R.D. Lele, R.D. Kulkarni, Treatment of rheumatoid arthritis with rumalaya, Probe 13 (1973) 22–24.

[95] M. Javed, J.A. Khan, M.M.H. Siddiqui, Effect of *Colchicum luteum* Baker in the management of rheumatoid arthritis, Indian J. Tradit. Knowl. 4 (2005) 421–423.

[96] P. Rao, L.A. Falk, S.F. Dougherty, S.F. Sawada, T. Pluznik, Colchicine down-regulates lipopolysaccharide-induced granulocyte-macrophage colony stimulating factor production in murine macrophage, J. Immunol. 159 (1997) 3531–3539.

[97] R.A. Terkeltaub, D.E. Furst, K. Bennett, K.A. Kook, R.S. Crockett, M.W. Davis, High versus low dosing of oral colchicine for early acute gout flare: twenty-four-hour outcome of the first multicenter, randomized, double-blind, placebo-controlled, parallel-group, dose-comparison colchicine study, Arthritis Rheum. 62 (2010) 1060–1068.

[98] R.L. Wortmann, P.A. MacDonald, B. Hunt, R.L. Jackson, Effect of prophylaxis on gout flares after the initiation of urate-lowering therapy: analysis of data from three phase III trials, Clin. Ther. 32 (2010) 2386–2397.

[99] S. Padeh, M. Gerstein, Y. Berkun, Colchicine is a safe drug in children with familial mediterranean fever, J. Pediatr. 161 (2012) 1142–1146.

[100] M. La Regina, E. Ben-Chetrit, A.Y. Gasparyan, A. Livneh, H. Ozdogan, R. Manna, Current trends in colchicine treatment in familial Mediterranean fever, Clin. Exp. Rheumatol. 31 (2013) 41–46.

[101] A. Slobodnick, B. Shah, M.H. Pillinger, S. Krasnokutsky, Colchicine: old and new, Am. J. Med. 128 (2015) 461–470.

[102] L. Zhang, Z.X. Zhang, D.K. An, The achievements in the studies of chemical components of *Tripterygium wilfordii* Hook family, Zhongguo Yao Ke Da Xue Xue Bao 21 (1990) 251–256.
[103] L. Zhou, Q.J. He, L.W. Lu, F. Zhao, Y. Zhang, X.X. Huang, B. Lin, S.J. Song, Tripterfordins A–O, dihydro-β-agarofuran sesquiterpenoids from the leaves of *Tripterygium wilfordii*, J. Nat. Prod. 82 (2019) 2696–2706.
[104] B.J. Chen, Triptolide, a novel immunosuppressive and anti-inflammatory agent purified from a Chinese herb *Tripterygium wilfordii* Hook F, Leuk. Lymphoma 42 (2001) 253–265.
[105] W.Z. Gu, S.R. Brandwein, S. Banerjee, Inhibition of type II collagen induced arthritis in mice by an immunosuppressive extract of *Tripterygium wilfordii* Hook f, J. Rheumatol. 19 (1992) 682–688.
[106] X. Tao, J. Younger, F.Z. Fan, B. Wang, P.E. Lipsky, Benefit of an extract of *Tripterygium wilfordii* Hook F. in patients with rheumatoid arthritis: a double-blind, placebo-controlled study, Arthritis Rheum. 46 (2002) 1735–1743.
[107] K. Asano, J. Matsuishi, Y. Yu, T. Kasahara, T. Hisamitsu, Suppressive effects of *Tripterygium wilfordii* Hook F., a traditional Chinese medicine, on collagen arthritis in mice, Immunopharmacology 39 (1998) 117–126.
[108] R.G. Mansk, M. Wilson, R. Fleishman, N. Olsen, J. Silverfild, P. Kempf, A. Kivitz, Y. Sherrer, F. Pucino, G. Csako, R. Costello, T.H. Pham, T.H. Snyder, D.V.D. Heijde, X. Tao, R. Welsley, P.E. Lipsky, Comparison of *Tripterygium wilfordii* Hook F versus sulfasalazine in the treatment of rheumatoid arthritis, Ann. Intern. Med. 151 (2009) 229–240.
[109] Y.Z. Zhou, L.D. Zhao, H. Chen, Y. Zhang, D.F. Wang, L.F. Huang, Q. Lv, B.L.Z. Li, W. Wei, H. Li, X. Liao, X. Liu, H. Jin, Y.F. Wang, Q. Wo, W. Zhang, Q. Shi, W. Zhenf, F. Zhang, F. Tang, P. Lipsky, X. Zhang, Comparison of the impact of *Tripterygium wilfordii* Hook F and methotrexate treatment on radiological progression in active rheumatoid arthritis: 2-year follow up of a randomized, non-blinded, controlled study, Arthritis Res. Ther. 20 (2018) 70.
[110] D.F. Su, R.L. Li, Y. Sun, Comparative study of triptolide and the ethyl acetate extract of *Tripterygium wilfordii* Hook F. in the treatment of rheumatoid arthritis, Zhong Cao Yao 10 (1990) 144–146.
[111] J.R. Zheng, J.H. Liu, L.F. Hsu, J.W. Gao, B.L. Jiang, Studies on toxicity of total glycosides in *Tripterygium wilfordii*, Acta Acad. Med. Sin. 5 (1963) 73–78 (Chem Abstr 1984, 100, 44921).
[112] J.R. Zheng, J.L. Fang, L.E. Xu, G. Jiwei, G. Hongzhi, L. Ziren, S. Huizhen, Effects of total glycosides of *Tripterygium wilfordii* on reproductive organs of experimental animals. II. Experiments in female rats, Acta Acad. Med. Sin. 7 (1985) 56–259.
[113] R.N. Chopra, S.L. Nayar, I.C. Chopra, L.V. Asolkar, K.K. Kakkar, Glossary of Indian Medicinal Plants, Council of Scientific & Industrial Research, New Delhi, 1956.
[114] Y.R. Chadha, The Wealth of India: Raw Materials, vol. 1A, CSIR, New Delhi, 1985.
[115] M.S. Hossain, Z. Urbi, A. Sule, K.M.H. Rahman, *Andrographis paniculata* (Burm. f.) Wall. Ex Nees: a review of ethnobotany, phytochemistry, and pharmacology, Sci. World. J. 2014 (2014), 274905.
[116] H.Y. Cheung, C.S. Cheung, C.K. Kong, Determination of bioactive diterpenoids from *Andrographis paniculata* by micellar electrokinetic chromatography, J. Chromatogr. A 930 (2001) 171–176.
[117] R.M.V. Bhaskar, P.H. Kishore, C.V. Rao, D. Gunasekar, C. Caux, B. Bodo, New 2′-oxygenated flavonoids from *Andrographis affinis*, J. Nat. Prod. 66 (2003) 295–297.
[118] N. Pholphana, N. Rangkadilok, S. Thongnest, S. Ruchirawat, M. Ruchirawat, J. Satayavivad, Determination and variation of three active diterpenoids in *Andrographis paniculata* (Burm.f.) Nees, Phytochem. Anal. 15 (2004) 365–371.

[119] P.L. Smith, K.N. Maloney, R.G. Pothen, J. Clardy, D.E. Clapham, Bisandrographolide from *Andrographis paniculata* activates TRPV4 channels, J. Biol. Chem. 281 (2006) 29897–29904.
[120] S. Pramanick, S. Banerjee, B. Achari, S. Mukhopadhyay, Phytochemicals from the genus Andrographis, in: J.N. Govil, V.K. Singh, R. Bhardwaj (Eds.), Recent Progress in Medicinal Plants: Phytomedicines, Studium Press LLC, Houston, 2007, pp. 339–387.
[121] W.W. Chao, B.F. Lin, Isolation and identification of bioactive compounds in *Andrographis paniculata* (Chuanxinlian), Chin. Med. 5 (2010) 17.
[122] P. Radhika, Y.R. Prasad, K.R. Lakshmi, Flavones from the stem of *Andrographis paniculata* Nees, Nat. Prod. Commun. 5 (2010) 59–60.
[123] R. Parveen, F.J. Ahmad, Z. Iqbal, M. Samim, S. Ahmad, Solid lipid nanoparticles of anticancer drug andrographolide: formulation, in vitro and in vivo studies, Drug Dev. Ind. Pharm. 40 (2014) 1206–1212.
[124] Z. Li, J. Tan, L. Wang, Q. Li, Andrographolide benefits rheumatoid arthritis via inhibiting MAPK pathways, Inflammation 40 (2017) 1599–1605.
[125] S. Gupta, K.P. Mishra, S.B. Singh, L. Ganju, Inhibitory effects of andrographolide on activated macrophages and adjuvant-induced arthritis, Inflammopharmacology 26 (2017) 447–456.
[126] J. Yan, Y. Chen, C. He, Z. Yang, C. Lu, X. Chen, Andrographolide induces cell cycle arrest and apoptosis in human rheumatoid arthritis fibroblast-like synoviocytes, Cell Biol. Toxicol. 28 (2012) 47–56.
[127] R.A. Burgos, J.L. Hancke, J.C. Bertoglio, V. Aguirre, S. Arriagada, M. Calvo, D.D. Caceres, Efficacy of an *Andrographis paniculata* composition for the relief of rheumatoid arthritis symptoms: a prospective randomized placebo-controlled trial, Clin. Rheumatol. 28 (2009) 931–946.
[128] M.A. Hidalgo, L.J. Hancke, J.C. Bertoglio, R.A. Burgos, Andrographolide a new potential drug for the long term treatment of rheumatoid arthritis disease, in: Innovative Rheumatology, Intech, 2013, https://doi.org/10.5772/55642.
[129] J.L. Hancke, S. Srivastav, D.D. Cáceres, R.A. Burgos, A double-blind, randomized, placebo-controlled study to assess the efficacy of *Andrographis paniculata* standardized extract (ParActin®) on pain reduction in subjects with knee osteoarthritis, Phytother. Res. 33 (2019) 1469–1479.
[130] L. Worasuttayangkurn, W. Nakareangrit, J. Kwangjai, P. Sritangos, N. Pholphana, P. Watcharasit, J. Satayavivad, Acute oral toxicity evaluation of *Andrographis paniculata*-standardized first true leaf ethanolic extract, Toxicol. Rep. 6 (2019) 426–430.
[131] R.A. Burgos, E.E. Caballero, N.S. Sánchez, R.A. Schroeder, G.K. Wikman, J.L. Hancke, Testicular toxicity assessment of *Andrographis paniculata* dried extract in rats, J. Ethnopharmacol. 58 (1997) 219–224.
[132] S. Malik, H. Cun-Heng, J. Clardy, Isolation and structure determination of nigellicine, a novel alkaloid from the seeds of *Nigella sativa*, Tetrahedron Lett. 26 (1985) 2759–2762.
[133] A.S. Malik, S.S. Hasan, M.I. Choudhary, C.Z. Ni, J. Clardy, Nigellidine—a new indazole alkaloid from the seeds of *Nigella sativa*, Tetrahedron Lett. 36 (1995) 1993–1996.
[134] C.C. Toma, G.M. Simu, D. Hanganu, N. Olah, F.M.G. Vata, M. Hammami, Chemical composition of the Tunisian *Nigella sativa*. Note I. Profile on essential oil, Farmacia 58 (2010) 458–464.
[135] S. Gharby, H. Harhar, D. Guillaume, A. Roudani, S. Boulbaroud, M. Ibrahimi, M. Ahmad, S. Sultana, T.B. Hadda, I. Chafchaouni-Mossaoui, Z. Charrouf, Chemical investigation of *Nigella sativa* L. seed oil produced in Morocco, J. Saudi Soc. Agric. Sci. 14 (2015) 172–177.

[136] W. Kooti, Z. Hasanzadeh-Noohi, N. Sharafi-Ahvazi, M. Asadi-Samani, D. Ashtary-Larky, Phytochemistry, pharmacology, and therapeutic uses of black seed (*Nigella sativa*), Chin. J. Nat. Med. 14 (2016) 732–745.

[137] A. Ghannadi, V. Hajhashemi, H. Jafarabadi, An investigation of the analgesic and anti-inflammatory effects of *Nigella sativa* seed polyphenols, J. Med. Food 8 (2005) 488–493.

[138] M. Sajad, M. Asif, S. Umar, J. Zargan, M. Rizwan, S.H. Ansari, M. Ahmad, H.A. Khan, Amelioration of inflammation induced oxidative stress and tissue damage by aqueous methanolic extract of *Nigella sativa* Linn. in arthritic rats, J. Complement. Integr. Med. 7 (2010) 51, https://doi.org/10.2202/1553-3840.1454.

[139] C. Nasuti, D. Fedeli, L. Bordoni, M. Piangerelli, M. Servili, R. Selvaggini, R. Gabbianelli, Anti-inflammatory, anti-arthritic and anti-nociceptive activities of *Nigella sativa* oil in a rat model of arthritis, Antioxidants 8 (2019) 342.

[140] T.A. Gheita, S.A. Kenawy, Effectiveness of *Nigella sativa* oil in the management of rheumatoid arthritis patients: a placebo controlled study, Phytother. Res. 26 (2011) 1246–1248.

[141] V. Hadi, S. Kheirouri, M. Alizadeh, A. Khabbazi, H. Hosseini, Effects of *Nigella sativa* oil extract on inflammatory cytokine response and oxidative stress status in patients with rheumatoid arthritis: a randomized, double-blind, placebo-controlled clinical trial, Avicenna J. Phytomed. 6 (2016) 34–43.

[142] S. Kheirouri, V. Hadi, M. Alizadeh, Immunomodulatory effect of *Nigella sativa* oil on T lymphocytes in patients with rheumatoid arthritis, Immunol. Invest. 45 (2016) 271–283.

[143] M.J. Laughton, Inhibition of mammalian 5-lipoxygenase and cyclooxygenase by flavonoids and phenolic dietary additives relationship to antioxidant activity and to iron ion-reducing ability, Biochem. Pharmacol. 42 (1991) 1673–1681.

[144] F. Vaillancourt, Elucidation of molecular mechanisms underlying the protective effects of thymoquinone against rheumatoid arthritis, J. Cell. Biochem. 112 (2011) 107–117.

[145] S. Umar, Modulation of the oxidative stress and inflammatory cytokine response by thymoquinone in the collagen induced arthritis in Wistar rats, Chem. Biol. Interact. 197 (2012) 40–46.

[146] E.S. El Daly, Protective effect of cysteine and vitamin E, *Crocus sativus* and *Nigella sativa* extracts on cisplatin induced toxicity in rats, J. Pharm. Belg. 53 (1998) 87–95.

[147] T. Khanna, F.A. Zaidi, P.C. Dandiya, CNS and analgesic studies on *Nigella sativa*, Fitoterapia 5 (1993) 407–410.

[148] Y. Tanko, A. Mohammed, M.A. Okasha, A. Shuaibu, M.G. Magaji, A.H. Yaro, Analgesic and anti-inflammatory activities of ethanol seed extract of *Nigella sativa* (black cumin) in mice and rats, Eur. J. Sci. Res. 18 (2007) 277–281.

[149] O.A. Badary, O.A. Al-Shabanah, M.N. Nagi, A.M. Al-Bekairi, M.M.A. Almazar, Acute and subchronic toxicity of thymoquinone in mice, Drug Dev. Res. 44 (1998) 56–61.

[150] A. Zaoui, Y. Cherrah, N. Mahassini, K. Alaoui, H. Amarouch, M. Hassar, Acute and chronic toxicity of *Nigella sativa* fixed oil, Phytomedicine 9 (2002) 69–74.

[151] M. Khader, N. Bresgen, P.M. Eckl, In vitro toxicological properties of thymoquinone, Food Chem. Toxicol. 47 (2009) 129–133.

[152] R. Mosbah, Z. Djerrou, A. Mantovani, Protective effect of *Nigella sativa* oil against acetamiprid induced reproductive toxicity in male rats, Drug Chem. Toxicol. 41 (2017) 206–212.

11

Natural product–derived drugs for the treatment of inflammatory bowel diseases (IBD)

Cristina C. Salibay[a], Tooba Mahboob[b], Ajoy Kumar Verma[c], Jonnacar S. San Sebastian[a], Hazel Anne Tabo[a], Chandramathi Samudi Raju[b], and Veeranoot Nissapatorn[d]

[a]College of Science and Computer Studies, De La Salle University-Dasmariñas, Dasmariñas, Cavite, Philippines, [b]Department of Medical Microbiology, Faculty of Medicine, University of Malaya, Kuala Lumpur, Malaysia, [c]National Institute of Tuberculosis and Respiratory Diseases (NITRD), New Delhi, India, [d]School of Allied Health Sciences, Southeast Asia Water Team (SEA Water Team) and World Union for Herbal Drug Discovery (WUHeDD), Walailak University, Nakhon Si Thammarat, Thailand

11.1 Introduction

Inflammatory bowel disease (IBD) is a general term given to the condition of chronic inflammation occurring in the gastrointestinal tract (GIT). The two main types of IBD often seen are ulcerative colitis and Crohn's disease whereby the former affects the lining of the colon (large intestine), while the latter affects any part of the digestive tract from the mouth to the perianal area (usually small or large intestines) and causes inflammation in both the lining and deeper layers of intestines. Apart from this, in Crohn's disease (CD), the affected or inflamed areas appear in patches, which are next to areas of healthy tissue. Ulcerative colitis (UC) involves continuous injured parts, not patchy, and usually starts at the rectum and spreads further into the colon. Despite these pathological differences, the common symptoms exhibited by patients diagnosed with IBD include abdominal pain, rectal bleeding or bloody stools, persistent diarrhea, weight loss, and fatigue [1]. Therefore diagnosis of IBD, especially Crohn's diseases, involves combination of endoscopy and radiography such as contrast radiography, magnetic resonance imaging (MRI), or computed tomography (CT). Similarly, for ulcerative

Inflammation and Natural Products. https://doi.org/10.1016/B978-0-12-819218-4.00017-1
© 2021 Elsevier Inc. All rights reserved.

colitis, imaging studies need to be done together with colonoscopy to confirm the diagnosis. In addition to this, physicians usually carry out microbiological examination on patients' stool samples along with full blood profile to rule out infections.

11.2 Epidemiology

Generally, IBD seems to be highly prevalent in developed countries than developing regions. Nevertheless, based on numerous epidemiological studies in the past decade, the disease has developed as a global challenge of public health. In Western countries, namely, North America and Europe, more than 1.5 million and 2 million people have been diagnosed with CD and UC, respectively. Also the prevalence of IBD in North America, Australia, and many countries in Europe has increased to more than 0.3% of the population [2]. A prevalence study carried out in 2015 had estimated that 3.1 million (1.3%) of US adults have been diagnosed with IBD, which is either Crohn's disease or ulcerative colitis [3]. Most of these patients were classified as adults aged more than 45 years. On the other hand, newly industrialized countries outside Western world such as Africa, Asia, and South America whose cultures are gradually but becoming Westernized and urbanized are also exhibiting the emergence of IBD with rapid increasing in disease incidence [2]. In Southeastern Asian countries including Malaysia, the overall prevalence of IBD reported since year 1990 was approximately 9 per 100,000 people, whereby the prevalence of ulcerative colitis was approximately 6.67 and the prevalence of Crohn's disease was approximately 2.17 per 100,000 people. Collectively the high prevalence of IBD reported in both Western countries and elsewhere is challenging clinicians and health policy makers to deliver eminence and cost-efficient care to IBD patients.

11.3 Pathogenesis

Crohn's disease and ulcerative colitis have different pathologic and clinical characteristics, but with considerable similarities, the disease pathogenesis is still poorly understood. Numerous scientific reports have recognized roles of both host and microbial agents in the disease pathogenesis, for example, continuous immune or inflammatory response toward the microbial community in the GIT. In other words, while the exact cause or obvious mechanism of IBD remains unclear, past studies implicate that it encompasses a complex interaction between the genetic, environmental, or microbial factors and the host immune responses.

In term of genetic factors, scientists have claimed that the disease can be closely related to familial history and identified several candidate genes such as CDH1 and LAMB1 (functions in regulation of intestinal epithelial barrier) especially in patients with ulcerative colitis [4]. More genetic investigations with the application of whole genome association study (GWAS) and single nucleotide polymorphisms (SNPs) have revealed the inflammatory pathway-related genes, namely, IL23R, IL12B, JAK2, and STAT3, in both ulcerative colitis and Crohn's disease [5]. With current next-generation sequencing (NGS) technology, researchers are continuing in identifying the role of more significant genes that can be used either as biomarker for diagnostic purposes or therapeutic target.

Besides this, environmental factors such as smoking, diet, social stress, and drugs are well-known risk factors of IBD. Of these, drug plays a major role in IBD. Two main drug examples are nonsteroidal antiinflammatory drugs (NSAIDs) and antibiotics as they have the ability to cause dysbiosis of gut microbiome. This has been proven via animal models and human cohort studies. Dysbiosis of gut microbiome will eventually disrupt the mucosal immunity, resulting in overactivation of inflammation, and thus prolonged use of these drugs will lead to IBD [6, 7]. Based on metagenomic analysis, human GIT is known to be colonized by more than 10^{14} microorganisms with each individual host having roughly 160 species. Following this, microbiome analysis has been carried out in CD and UC in both inflamed and noninflamed segments and showed that there is a significantly reduced biodiversity of fecal microbiome in IBD patients compared with healthy controls [8].

The role of host immune response, precisely mucosal immunity, is very well established in gut-related diseases including IBD. Mucosal immunity is defined as the immune response that occurs at mucosal membranes of the intestines, the urogenital tract, and the respiratory system, of which the GIT is known to have the largest surface area of mucosal membrane. In general, mucosal immunity can be divided into (a) innate and (b) adaptive immunity. Innate immunity of this region comprises epithelia, macrophages, monocytes, neutrophils, eosinophils, basophils, dendritic cells (DCs), and natural killer cells. In normal circumstances, intraluminal microbes will constantly communicate with the aforementioned cells via receptors such as Toll-like receptors (TLRs) and lectin receptors. TLR signaling leads to tolerance or unresponsiveness toward luminal pathogens or microbes as the pattern recognition receptors (PRRs) will be downregulated, thus preventing mucosal inflammation. In contrast, in IBD patients, impaired TLR signaling often leads to increased intestinal permeability and subsequently results in inflammatory responses and damage of the intestinal epithelial layer. Antigen-presenting cells (APCs) of innate immunity interact with adaptive immunity by presenting antigens to T lymphocytes via the major histocompatibility complex. In non-IBD conditions, as part of the immune homeostasis, the dendritic cells will trigger upregulation of T regulatory cells or T_{reg} (suppress inflammatory response) instead of T helper cells, which promote recruitment of leukocytes to the site of infection leading to inflammation. This mechanism is also orchestrated by pro- and antiinflammatory cytokines such as IL-12 (Th1) and IL-10 (Th2), respectively. In contrast to this, IBD patients will have overregulation of T helper cells that leads to inflammation. Apart from Th1 and Th2 response, another T-cell subset, Th17, which is characterized by its signature cytokine, IL-17, has also been studied in IBD. This is because IL-17 is a key player in triggering inflammation at any site of injury or infection. Increased transcript levels of IL-17A have been detected in IBD patient's mucosa of both CD and UC in contrast to normal gut [9].

As natural products or compounds have been commonly known to have properties of immune booster, antioxidant, and antimicrobial, it is noteworthy that one should not rule out their potentials as treatment agent for IBD including UC and/or CD. Having this in mind the following topics will focus on various types of natural products that have been studied so far as treating agent for IBD.

11.3.1 Crohn's disease

CD is characterized as chronic relapsing inflammation and skipping noncaseating granulomatous lesions that affect any portion of the gastrointestinal tract (from the mouth to the anus) more commonly on the terminal ileum and colon but sparing the rectum [10–12]. In advance stage of the disease, CD patients are at risk of other complications locally such as gallstones, renal stones, and bacterial overgrowth extending from small intestines and gastrointestinal tract [11]. Another type of complication may involve systemically, hence affecting extraintestinal parts or any part of the body outside the intestine. The signs and symptoms associated with CD can range from mild to severe, which usually develop in gradual manner. However, sudden signs of the disease may be felt without warning, or a remission may happen when there are times that a CD patient may not show any signs or symptoms. Because of the different localizations of the affected area, some specific symptoms are likely to be associated with the type of CD (Table 11.1).

Symptoms can be variable considering that there can be an overlap between types of CD; hence more than one area of the digestive tract can be affected. However, in severe cases, symptoms do not only emanate from the gastrointestinal tract but systemically such as inflammation of the skin, eyes, and joints, as well as liver or bile ducts. In children, severe CD can lead to delayed growth or sexual development.

11.3.2 Ulcerative colitis

Ulcerative colitis is a chronic, recurrent disease characterized by diffuse mucosal inflammation involving only the colon. It invariably involves the rectum and may exceed proximally in a continuous fashion to involve part or the entire colon. Clinical symptoms are highly variable in ulcerative colitis patients. The major symptoms are bloody diarrhea. Other symptoms are rectal bleeding, abdominal cramps, abdominal pain, fecal urgency, and tenesmus. On abdominal examination the patients will have abdominal tenderness, evidence of peritoneal tenderness, and

Table 11.1 The different types of Crohn's disease with brief description and signs and symptoms.

Types of Crohn's disease[a]	Description	Signs and symptoms
Ileocolitis	It affects the small intestine, particularly the ileum, and the colon. This is the most common among the types	Diarrhea, cramping, weight loss, pain felt in the middle or lower right part of your abdomen
Ileitis	This type affects the ileum	Considerable weight loss, diarrhea, cramping, pain felt in the middle or lower right part of your abdomen. Fistulas may form in the lower right part of the abdomen
Gastroduodenal	This type affects the stomach and duodenum (the first part of the small intestine)	Nausea, weight loss, loss of appetite, vomiting (if narrow segments of bowel are obstructed)
Jejunoileitis	This type is characterized by the inflammation in the jejunum (the middle part of the small intestine)	Cramps after meals, fistulas, diarrhea, abdominal pain that can become intense
Crohn's (granulomatous) colitis	This type affects only the colon	Skin lesions, joint pain, diarrhea, rectal bleeding ulcers, fistulas, and abscesses around the anus

[a] Sources: Ref. [13] and https://www.mayoclinic.org/diseases-conditions/crohns.../symptoms-causes/syc-2035330.

presence of red blood on digital rectal examination. Symptoms such as weight loss, tachycardia, fever, anemia, and bowel distension have been observed in severe cases. Prior to start of any medical treatment, other etiologies of colitis/enteritis, particularly infections, toxic reactions (e.g., antibiotics and NSAID colitis), mesenteric ischemia, or intestinal malignancies, should be precluded. In patients under immunosuppressive treatment with a corticosteroid-refractory course, opportunistic infection should be excluded before medical treatment. UC is characterized as mild, moderate, and severe on the basis of clinical and laboratory parameters. The intensity of disease can be classified as follows:

Remission: 3 or less stools per day without any presence of blood or increased urgency of defecation;

Mild: up to 4 stools per day, possibly bloody having normal pulse rate, temperature, and hemoglobin concentration;

Moderate: 4–6 bloody stools daily with no signs of systemic involvement;

Severe: more than 6 bloody stools daily with signs of systemic involvement such as temperature above 37.5°C, heart rate above 90/min, and hemoglobin concentration below 10.5 g/dL.

Distribution patterns depend on the part of the colon involved and are designated according to the Montreal classification such as proctitis, left-sided colitis, extensive colitis, or severe colitis [14], as shown in Table 11.2.

Table 11.2 Clinical and laboratory parameters for assessment of disease activity.

Clinical and laboratory parameters	Disease activity		
	Mild	**Moderate**	**Severe**
Stool frequency/day	< 4	4–6	> 6 (bloody)
Pulse rate/min	< 9	90–100	> 100
Hematocrit (%)	Normal	30–40	< 30
Weight loss (%)	None	1–10	> 10
Temperature (°F)	Normal	99–100	> 100
ESR (mm/h)	< 20	0–30	> 30
Albumin (g/dL)	Normal	3–35	< 3

ESR, erythrocyte sedimentation rate; *°F*, degrees Fahrenheit.

11.4 The role of natural products on IBD

Natural products are compounds with biological activities that are derived from sources such as terrestrial plants, vertebrates and invertebrates, microorganisms, and marine organisms [15, 16]. These sources are considered as reservoir of numerous types of bioactive compounds including alkaloids, glycosides, flavonoids, phenolics and polyphenols, saponins, tannins, terpenes, anthraquinones, essential oils, and steroids (Table 11.3) that manifest vast array of therapeutic and medicinal properties.

The earliest records of natural products were written on hundreds of clay tablets in cuneiform from Mesopotamia (2600 BC) describing approximately 1000 plants and plant-derived substances from *Cedrus* species (cedar), resin of *Commiphora myrrha* (myrrh), and juice of the poppy seed, *Papaver somniferum* [16]. Likewise, the Ebers Papyrus (2900 BC) in Egypt documented over 700 plant-based drugs such as *Aloe vera* (aloe), *Boswellia carterii* (frankincense), and *Ricinus communis* (castor), while the Chinese recorded over 11,000 herbal remedies in their Chinese Materia Medica, Shennong Herbal, and Tang Herbal [24–26] that formed the base of commercially available modern drugs at present [27, 28]. According to World Health Organization (2017), about 3.4 billion people in the developing countries depend on plant-based traditional medicines. This represents approximately 88% of the world's inhabitants who rely on traditional medicine for their primary health care.

The development of traditional medicines continues, from the principle of decoction to the introduction of techniques on herbal extraction; natural products were recognized to exhibit antiinflammatory [29, 30], analgesic [31], immunomodulatory [32], antibacterial [33, 34],

Table 11.3 List of bioactive compounds in plants, animals, and microorganisms with medicinal properties.

Bioactive compounds	Pharmacological properties	Reference
Terpenoids (carotenoids, monoterpenoids, diterpenoids, triterpenes, triterpenoid saponin, sesquiterpenoids, sesquiterpene lactones, and polyterpenoids)	Antiinflammatory	[17]
	Antimicrobial	[18]
	Antifungal	[19]
	Antiviral	[20]
	Antiallergenic	
	Antiparasitic	
	Antispasmodic	
	Antihyperglycemic	
	Anticancer	
Phenolic acids (flavonoids, phenolic acids, stilbenoids, tannins, lignans, xanthones, quinones, coumarins, phenylpropanoids, and benzofurans)	Antioxidant	[21]
	Anticarcinogenic	
	Antimutagenic	
	Antiinflammatory	
Alkaloids and other nitrogen-containing metabolites (glucosinolates, betalain, indole, isoquinolone, pyrroloindole, piperidine, aporphine, pyridine, methylxanthine derivatives, vinca, lycopodium, and *Erythrina* spp.)	Analgesic	[22]
	Antiinflammatory	[23]
	Antidepressant	
	Muscle relaxant	
	Antihypertensive	
	Antiviral	
	Antiulcer	
	Antiproliferative	

antifungal [35], antiviral, and antitumor/proliferative activities [36, 37]. Thus it is considered indispensable source of both preventive and curative lead compounds that inexorably cradle pharmaceutical drug discovery and industry since ancient civilization across all cultures [38, 39]. With these traditional medicine practices, subsequently clinical and pharmacologic studies lead to the synthesis of commercially available drugs as presented in Table 11.4.

Despite the remarkable discovery of these naturally derived drugs, there is a constant demand to develop effective and affordable medicines that could cope with the emerging and growing incidences of infectious diseases, lipid disorder, neurological diseases, cardiovascular and metabolic diseases, and immunological, inflammatory, and oncologic diseases and related diseases due to lifestyle changes such consumption of tobacco, alcohol, and diets high in fat and low in fiber. Among the most prevalent diseases, those require a promising medicine from natural products are Crohn's diseases and ulcerative colitis.

Table 11.4 Commercially available drugs extracted from plants, animals, and microorganisms.

Commercial drugs	Natural source	Medicinal use	Reference
Morphine (Merck in 1826)	*Papaver somniferum* L. (opium poppy)	For pain	[40] [41]
Acetylsalicylic acid (aspirin) (Bayer in 1899)	*Salix alba* L. (willow tree)	Antiinflammatory	
Digitoxin and digoxin	*Digitalis purpurea* L. (foxglove)	Management of congestive heart failure	
Penicillin	*Penicillium notatum*		[42]
Norcardicin	*Nocardia uniformis*		[43]
Imipenem (more stable product of thienamycin)	*Streptomyces cattleya*		[44] [45]
Aztreonam	*Chromobacterium violaceum* *Acetobacter* sp.	Antibacterial	[46]
Vancomycin	*Amycolatopsis orientalis*		
Erythromycin	*Saccharopolyspora erythraea*		
Fumagillin (Flisint)	*Aspergillus fumigatus*	Antiparasitic	[47]
Exenatide (Byetta)	*Heloderma suspectum* (Gila monster lizard saliva)	For type 2 diabetes mellitus	[48]
Galantamine (Reminyl)	*Galanthus caucasicus* (Caucasian snowdrop) *Galanthus woronowii* (Woronov's snowdrop) *Narcissus* (daffodil) *Leucojum aestivum* (snowflake) *Lycoris radiata* (red spider lily)	For treatment of cognitive decline in mild to moderate Alzheimer's disease and other various memory impairments	[49]
Paclitaxel (Taxol)	*Taxus brevifolia* (Pacific yew tree)	Lung, ovarian, breast, and pancreatic cancer and AIDS-related Kaposi sarcoma	[50]
Artemisinin	*Artemisia annua* (sweet wormwood)	Multidrug resistant malaria	[51, 52] [53–55]

11.5 Natural products for ulcerative colitis and Crohn's disease

The term "herb" is derived from the Latin word herba, meaning "grass." The term has been implied to plants of which the leaves, stems, or fruits are being used by human for food and medicines and as a scent or flavor. Herbal medicines are commonly referred to primeval or traditional medicinal put into practice based on the use of plants and plant extracts for the treatment of medical conditions. The use of herbs to take care of diseases is almost universal among native people. A number of civilization have come to direct the carry out of herbal medicine around the world. Herbal remedy is one of the most common traditional Chinese medicines (TCM) used by Chinese people. It has been estimated that 28.9% of US adults regularly use one or more herbal products for treating diseases, around 10% of which are in the form of herbal products [56]. Recent studies have indicated that the percentage of adults using herbal therapies for their gastrointestinal symptoms ranges from 20% to 26%, mainly in chronic GI conditions [57, 58].

The use of complementary medicine among patients with ulcerative colitis, particularly in the form of herbal therapies, is widespread in the Western world and in many Asian countries including China and India [57]. There are limited restricted evidences indicating the efficacy of herbal medicine such as *A. vera* gel, wheatgrass juice, *Boswellia serrata*, and bovine colostrum enemas in the management of patients with ulcerative colitis. Currently, herbal medicine is widely used in the treatment of ulcerative colitis in many parts of the world and is used for the treatment of other inflammatory bowel diseases (IBD). The herbal medicines used are slippery elm, fenugreek, devil's claw, Mexican yam, tormentil, and wei tong ning (a TCM). Slippery elm, fenugreek, devil's claw, tormentil, and wei tong ning are novel drugs in the management of inflammatory bowel diseases (IBD). The facts indicate that treatment of herbal medicine is superior to that by simple allopathic or ayurvedic medicine, such as prednisone, hydrocortisone, mesalamine, sulfasalazine, balsalazide, and methylprednisolone, and that it is also safe and effective in maintaining remission of ulcerative colitis. Following are herbal medicines being used for the treatment of ulcerative colitis [59].

11.5.1 *Aloe vera*

A. vera is a tropical plant used in traditional medicine throughout the world. It has been studied for its ability to relieve ulcerative colitis. *A. vera* gel is the mucilaginous aqueous extract of the leaf pulp of *A. barbadensis* Miller. *A. vera* juice has antiinflammatory activity and has been used by some doctors for patients with UC. It was the single most widely used herbal therapy [60].

Mechanism of action: In vitro studies on human colonic mucosa have demonstrated that *A. vera* gel could inhibit prostaglandin E2 and IL-8 secretion, indicating its role in antimicrobial and antiinflammatory responses [59].

Dose: 100 mL of a twice daily for 4 weeks.

Outcome of treatment: Clinical remission, improvement, and response occurred in 2 weeks.

11.5.2 *Artemisia absinthium* (Compositae)

The plant is commonly known as the wormwood that is widely distributed all over the world, which contains dimeric guaianolides absinthins [61] that is beneficial in treating Crohn's disease [62].

Mechanism of action: Dimeric guaianolides absinthins act as inhibitor of TNF-α that plays a key role in pathogenesis of Crohn's disease [63].

Dose: 3×500 mg/day for 10 weeks [64]; 3×750 mg/day for 6 weeks [62].

Outcome: Almost complete remission in 65% of the patients, whereas no beneficial effect was observed in those receiving the placebo [64].

11.5.3 *Boswellia* spp.

Boswellia or Indian frankincense is an ayurvedic herb that is found from the resin of the plant and is used to treat ulcerative colitis and Crohn's diseases. Boswellic acid, the major constituent of *Boswellia*, is thought to contribute to most of the herbal pharmacologic activities.

Mechanism of action: In vitro studies and animal models have shown that boswellic acid could inhibit 5-lipoxygenase selectively with antiinflammatory and antiarthritic effects. Since the inflammatory process in IBD is associated with increased function of leukotrienes, the benefits of *Boswellia* in the treatment of ulcerative colitis have proven a positive result. Moreover, it has also been found to directly inhibit intestinal motility with a mechanism involving L-type Ca^{2+} channels. *Boswellia* has been found to reduce chemically induced edema and inflammation in the intestine in rodents. Other studies suggest that it has cytotoxic properties [65, 66].

Doses: 900 mg daily divided into 3 doses for 6 weeks.

Outcome of treatment: This herbal medicine has very good response in ulcerative colitis patients. Study has showed that 70%–90% improvement seen among chronic ulcerative colitis patients [67].

11.5.4 Butyrate

Butyrate is an important energy source for intestinal epithelial cells and plays a role in the maintenance of colonic homeostasis. Butyrate enemas have been studied for use in treating chronic ulcerative colitis patients. It may help in decreasing the inflammation in the colon. Mechanism of action: Possible mechanism is to decrease oxidation in ulcerative colitis patients who showed that butyrate oxidation could be reduced by TNF-α at concentrations found in inflamed human mucosa. This antiinflammatory effect of butyrate via NF-κB inhibition, contributing, for example, to decreased concentrations of myeloperoxidase, cyclooxygenase-2, adhesion molecules, and different cytokine levels, has been confirmed in several in vitro and in vivo studies [56, 68, 69].

Doses: Administration of 4 g of butyrate daily via enteric-coated tablets in combination with mesalazine in patients with mild to moderate UC.

11.5.5 Licorice

Licorice, which is derived from the root of the plant, is used extensively in chronic ulcerative colitis patients. *Licorice* has also got immunomodulatory and adaptogenic property, which is required for the pathogenesis of chronic ulcerative colitis.

Mechanism of action: A number of active chemicals, including glycyrrhizin, are thought to account for its biologic activity. Diammonium glycyrrhizinate is a substance that is extracted and purified from licorice and may be useful in the treatment of UC. Evidence has also reported that diammonium glycyrrhizinate could improve intestinal mucosal inflammation in rats and, importantly, reduce expression of NF-κB, tumor necrosis factor (TNF-α), and intercellular adhesion molecule (ICAM-1) in inflamed mucosa. Clinical studies on licorice have also been performed in combination with other herbs and demonstrated to be effective in the management of UC. The antiestrogenic action documented for licorice at high concentration has been associated with licorice-binding estrogen receptors. However, estrogenic activity has also been reported for licorice and is attributed to its isoflavone constituents. It has been suggested that licorice may exert its mineralocorticoid effect via an inhibition of 11β-hydroxysteroid dehydrogenase. Evidences have proven that licorice could also suppress both plasma renin activity and aldosterone secretion. In addition, licorice has been shown to have chemopreventive effects through influencing Bcl-2/Bax and inhibiting carcinogenesis [70, 71].

11.5.6 Slippery elm (*Ulmus fulva*)

Slippery elm is a supplement that is made from the powdered bark of the slippery elm tree. It has long been used by Native Americans to treat cough, diarrhea, and other GI complaints. Recently, slippery elm has been studied for use as a supplement for ulcerative colitis. A study has confirmed the antioxidant effects of slippery elm when used in patients with IBD. The research has so far been promising, but there is not enough to warrant the widespread use of slippery elm in the treatment of chronic ulcerative colitis patients [72].

11.5.7 *Tormentil* extracts

Tormentil extracts have antioxidative properties and are used as a complementary therapy for chronic IBD. In individual patients with UC, positive effects have been observed.

Doses: *Tormentil* extracts used in escalating doses of 1200, 1800, 2400, and 3000 mg/day for 3 weeks.

11.5.8 *Triticum aestivum* (wheatgrass)

The wheatgrass juice has been used for the treatment of ulcerative colitis and gastrointestinal conditions. Wheatgrass juice for 1 month results in clinical improvement in 78% of people with ulcerative colitis [73].

Doses: The amount of wheatgrass used is 20 mL per day initially, and this is increased by 20 mL/day to a maximum of 100 mL per day (approximately 3.5 oz.). No serious side effects are noticed. Wheatgrass juice appears to be effective and safe as a single or adjuvant treatment of active distal UC.

11.5.9 *Cannabis sativa* (Cannabaceae)

This is also known as the medical marijuana that has been recognized for the treatment of different diseases such as chronic pain and neurological conditions [74]. It contains over 70 various cannabinoid compounds, but the two major active compounds include cannabinol and 8,9-tetrahydrocannabinol [75, 76].

Mechanism of action: The intestinal antiinflammatory effects of cannabis can be related to the capacity of cannabinoids to downregulate the production and release of different proinflammatory mediators including TNFα, IL-1β, and nitric oxide, thus restoring the altered immune response that occurs in IBD [77]. Most probably, these effects would be related to cannabinoid receptor type 1 (CB1) activation that mediates essential protective signals and counteracts proinflammatory pathways, since it has been reported that the severity of two different experimental models of colitis, induced by the intrarectal infusion of 2,4-dinitrobenzenesulfonic acid (DNBS) or by oral administration of DSS, is higher in CB1-deficient mice (CB1(−/−)) than in wild type [78].

Dose: Two cigarettes containing 115 mg of THC/day [79].

Outcome: Complete remission in 50% of the patients given with the plant treatment and 10% for the placebo group patients. Significant amelioration of the CD activity index in majority the patients treated with *Cannabis*.

11.5.10 *Curcuma longa* (Zingiberaceae)

Curcumin is a compound in turmeric (*Curcuma longa*) that has been reported to have antiinflammatory activity. It has been found to induce the flow of bile, which helps break down fats. Additionally, it could reduce the secretion of acid from the stomach and protect against injuries such as inflammation along the stomach (gastritis) or intestinal walls and ulcers from certain medications, stress, or alcohol. In a preliminary trial, five of five people with chronic ulcerative proctitis had an improvement in their disease after supplementing with curcumin.

Curcumin is an organically active phytochemical stuff showing antioxidant, antiinflammatory, anticarcinogenic, hypocholesterolemic, antibacterial, wound healing, antispasmodic, anticoagulant, antitumor, and hepatoprotective activities. It inhibits many cytokine pathways including interleukin (IL)-6, concurrently having a favorable safety profile. It has antiinflammatory and antioxidant effect.

Doses: Curcumin given as 1 g in two divided dose for months. Remission rate is 50%. Curcumin seems to be promising and safe medication for maintaining remission in patients with quiescent UC.

Mechanism of action: Curcumin inhibits the activation of NF-κB. NF-κB promotes the synthesis of many antioxidant enzymes. Curcumin directly binds to thioredoxin reductase and irreversibly changes its activity from an antioxidant to a strong prooxidant.

Doses: The amount of curcumin used was 550 mg twice a day for 1 month, followed by 550 mg three times a day for 1 month [59].

Hanai and colleagues published the results of the first randomized, multicenter, double-blind, placebo-controlled trial from Japan to study curcumin's effect on UC maintenance.

11.5.11 Germinated barley foodstuff

Two open-label Japanese trials have shown the efficacy of germinated barley foodstuff (GBF) in the treatment of UC, consisting mainly of dietary fibers and glutamine-rich protein that function as a probiotic.

Mechanism of action: The potency of GBF on modulating microflora, as well as the high water holding capacity, may play an important role in the treatment and prolongation of remission in UC [80].

11.5.12 Bromelain

Bromelain is an antiinflammatory and has been used as a digestive aid and a blood thinner, as well as to treat sports injuries, sinusitis, arthritis, and swelling. It has been studied for use as a supplement for IBD, especially UC. Emerging research on pineapple suggests that pineapple's "active" component, bromelain, may help relieve the inflammation associated with UC. The mechanisms that are primarily responsible for its antiinflammatory effects are still unclear. However, proteolytic activity is required for the antiinflammatory effect of bromelain on T-cell activation and cytokine secretion in vitro and in murine models of IBD in vivo.

Mechanism of action: The major action of bromelain appears to be proteolytic in nature, although evidence also suggests an immunomodulatory and hormone-like activity acting via intracellular signaling pathways. Bromelain has been shown to reduce cell surface receptors, such as hyaluronan receptor CD44, which is associated with leukocyte migration and induction of proinflammatory mediators. Additionally, it is also reported to significantly reduce CD4 + T-cell infiltrations, which are primary effectors in animal models of inflammation in the gut. It has been found to be effective in improvement of clinical and histologic severity of colonic inflammation in a murine colitis model of IL-10-deficient mice.

11.5.13 Psyllium

Psyllium comes from a shrub-like herb called *Plantago ovata* and is classified as a mucilaginous fiber due to its gel-forming properties in water. It has a long history of use as a laxative as it absorbs water and expands as it travels through the digestive tract. The seeds isolated from psyllium is also used as an effective drug in the treatment of IBD. The psyllium husk contains a largely insoluble fiber (hemicellulose), which helps to retain water within the bowel and effectively increases stool moisture content and weight. Soluble fibers (including psyllium) are noted for their effect on the stomach and small intestine, whereas insoluble fibers are noted for their effect on the large intestine, although some carbohydrates (such as psyllium) have an effect on both. Psyllium also has hypocholesterolemic effects, although the exact mechanism by which psyllium husk brings about a reduction of cholesterol is not totally clear.

Mechanism of action: Animal studies have shown that psyllium increases the activity of cholesterol 7α-hydroxylase (a rate-limiting enzyme in bile acid synthesis, also referred to as cytochrome 7A) more than twice that of cellulose or oat bran, but less than cholestyramine. In animals fed a high-fat diet, psyllium could increase the activity of cholesterol 7α-hydroxylase and HMG-CoA reductase.

Doses: 20 g of ground psyllium seeds twice daily with water for 4 weeks [81, 82].

11.5.14 *Withania somnifera*

A member of the family Solanaceae, *Withania somnifera* has good response in antiinflammatory activity. Immunomodulatory role of *W. somnifera* roots and antiinflammatory activity using adjuvant-induced arthritic rat models were also demonstrated. Considering the various biological activities, roots of *W. somnifera* can potentially be utilized for the effective treatment of various inflammatory conditions.

Doses: Use as a rectal gel applied at 1000 mg of *W. somnifera* root extract that showed significant mucorestorative efficacy in the IBD-induced rats [83].

11.5.15 Plant tannins

It helps in decreasing the inflammation of UC and CD patients. Patients with UC don't have the protective benefit of normal mucin production, which can also leave them vulnerable to oxidized molecules that increase the inflammation and mucosal injury seen in UC.

Mechanism of action: The tannins appear to exert a protective effect against oxidative stress–induced cell death. Condensed tannins can also help return the GI flora to a state of balance. Patients with UC have GI flora that favor pathogenic bacteria. Current research with flavonoids and UC demonstrate a protective effect in mice treated with a colitis-inducing agent, dextran sulfate sodium, so as to prevent the occurrence of colitis. Green tea polyphenols have shown similar benefits in mice by attenuating colonic injury induced by experimental colitis [84].

11.5.16 Guggulsterone

It is a plant steroid found in the resin of the guggul plant and is an antiinflammatory compound with the capacity to prevent and ameliorate T-cell-induced colitis. These data ground the use of GS, a natural cholesterol-lowering agent, in the treatment of chronic inflammatory diseases.

Mechanism of action: Guggulsterone inhibits LPS- or IL-1b-induced ICAM-1 gene expression, NF-κB transcriptional activity, IκB phosphorylation/degradation, and NF-κB DNA-binding activity in IEC and strongly blocked IKK activity. It significantly reduced the severity of DSS-induced murine colitis as assessed by clinical disease activity score, colon length, and histology. Furthermore, tissue upregulation of IκB and IKK phosphorylation induced by DSS was attenuated in guggulsterone-treated mice [62]. The guggulsterone derivative GG-52 has both protective and therapeutic effects on inflammation in the colon, indicating that it has a potential clinical value for the treatment of IBD [85].

11.5.17 *Agaricus subrufescens* (*Agaricus blazei*)

It has an antiinflammatory effect [64]. This is being used for the treatment of chronic nonspecific ulcerative colitis, which is better than that of the oral administration of sulfasalazine with less adverse reaction. A randomly controlled trial study showed that 70% were successfully treated with *Agaricus subrufescens*, whereas 25% were cured by allopathic mode of treatment [86].

11.5.18 Fufangkushen colon-coated capsule

It is an effective and safe in the treatment of active ulcerative colitis. In the double-blind multicenter, randomized, and controlled study, it was observed that 72% of patients are successfully managed by Fufangkushen colon-coated (FCC) capsule [87].

11.5.19 *Andrographis paniculata* (Acanthaceae)

Andrographis paniculata, a plant belonging to the family of Acanthaceae, grows mainly in India and Sri Lanka, as well as in South and Southeastern Asia. A recent randomized, double-blind, placebo-controlled study compared the extract of *A. paniculata* (HMPL-004) with placebo in 224 adult patients with mild to moderately active UC.

Doses: *A. paniculata* is given in the dose of 1800 mg per day.

Outcome: Clinical remission was obtained from patients with highest doses of plant extract, although occurrence was observed in 8% of patients receiving *A. paniculata* and 1% of patients receiving placebo. The rashes were mostly mild (with the rest moderate) and reversible and did not cause treatment discontinuation [88, 89].

11.5.20 Jian Pi Ling

Jian Pi Ling (JPL) is considered as one of the current plant treatments in patients with ulcerative colitis. It consists of nine components and is available in the form of tablets containing 0.75 g of dry herbal. The rate of remission by using this herbal medicine is higher in comparison with other herbal drugs. The low rate of remission has been reported and raises questions about the real value of this herbal product.

11.5.21 Xilei-san

Xilei-san is a mixture of herbs of Chinese medicine that harbors significant antiinflammatory properties. It seems to be effective in a number of inflammatory conditions of gastrointestinal tract such as ulcerative colitis and other digestive disorders such as esophagitis.

Doses: In a randomized control trial of Xilei-san herbal medicine for 8 weeks, a good response is seen in man with UC diseases and other inflammatory bowel diseases with no significant side effect.

11.5.22 Anthocyanin-rich bilberry preparation

Anthocyanins, which can be found in large quantities in bilberries (*Vaccinium myrtillus*), were shown to have antioxidative and antiinflammatory effects. They have very good therapeutic potential value. In randomized controlled trial study, 90% response is seen in patients having mild to moderate ulcerative colitis diseases. It is being observed that at the end of the sixth week, 63.4% of patients achieved remission and 90.9% showed a response.

11.5.23 *Plantago ovata* (Plantaginaceae)

P. ovata is a tiny plant with typical flowers. The juice derived from the plant leaves has been used in the treatment of IBD. The plant has antiinflammatory and antioxidative properties. It inhibits the protein kinase C, downregulates the expression of intercellular adhesion molecule-1, and inhibits the inflammation produced from 5-hydroxy-6, 8,11,14-eicosatetraenoic acid and leukotriene B4. The enzymatic dissolution of the seeds of *P. ovata* results in the production of short-chain fatty acids that have favorable effects in patients with patients with UC. It has also been used to treat patients of peptic ulcer.

Doses: It is given in 1g in two divided dose daily for months. Remission rate with this herbal medicine is 40%. There were few side effects, mainly constipation and abdominal bloating.

11.5.24 *Oenothera biennis*

Oenothera biennis belongs to the group of *Oenothera*, which can be found in North America and other tropical and subtropical countries. The evening primrose oil is the main product of the plant. The main constituent of *O. biennis* seeds is the γ-linolenic acid. The plant has been used as maintenance treatment in patients with ulcerative colitis with reasonable results.

11.5.25 *Bunium persicum*

The fruit of *Bunium persicum*, known as "Zirehkermani," is another natural product used for the treatment of IBD in traditional medicine. The essential oil of *B. persicum* has strong antibacterial effects. This property could be the result of relatively high amounts of terpinenes and cumin aldehyde in the essential oil. In addition, this essential oil has shown antioxidant properties. It was able to reduce the oxidation rate of soybean oil in the accelerated condition [90].

Doses: *B. persicum* at concentration of 0.88mg/mL was found to exhibit strong antioxidant potential.

11.5.26 *Cassia fistula*

The fruit from *Cassia fistula*, known as "Flous," is another drug for the treatment of IBD. The only known mechanism related to the beneficial effect of this plant is its antimicrobial properties. Crude extract of *C. fistula* exhibited significant antimicrobial activity [91].

Doses: Concentrations ranging from 50 to 200 μg/mL showed promising antibacterial effects.

11.5.27 *Cydonia oblonga*

The fruit from *Cydonia oblonga*, known as "Beh," is also used for the treatment of IBD. This fruit has shown radical scavenging and antimicrobial activities. The phenolic extract exhibited the strongest antioxidant activity among other extracts. The antioxidant functions of its phenolic extracts were superior to that of chlorogenic acid and ascorbic acid as standard antioxidants [92, 93].

Doses: The concentrations ranging from 6.5 μg/mL, 7.4 μg g/mL, and 8.4 μg/mL were found to be effective as strong antioxidant agent.

11.5.28 *Solanum nigrum*

The fruit of *Solanum nigrum*, known as "Tajrizi," is another natural product for the treatment of IBD (4,5). A glycoprotein isolated from this fruit (*S. nigrum* L. (SNL) glycoprotein) has demonstrated a dose-dependent inhibitory effect on NO production and free radical formation in DSS-induced colitis in mice. It exhibited a suppressive effect on the activities of NF-κB and regulated the expression of iNOS and Cox-2 in the downstream signaling pathway [94]. *S. nigrum* fruits showed effective free radical scavenging activities. Treatment with *S. nigrum* extract significantly inhibited the gastric lesions induced by cold restraint stress (76.6%), indomethacin (73.8%), pyloric ligation (80.1%), and ethanol (70.6%) with equal or higher potency than omeprazole in experimental ulcer models. It also showed concomitant attenuation of gastric secretory volume, acidity, and pepsin secretion in ulcerated rats. In addition, it accelerated the healing of acetic acid-induced ulcers after 7 days of treatment. Furthermore, it significantly inhibited H + K + ATPase activity and decreased gastrin secretion in the ethanol-induced ulcer model. The severity of the reaction of the ulcerogen and the reduction in ulcer size by *S. nigrum* extract was evident from histological findings [95, 96]. Doses: 200 and 400 mg/kg of *S. nigrum* extract accelerated the healing of acetic acid-induced ulcers after the treatment for 7 days.

11.5.29 *Juglans regia*

The kernel of *Juglans regia*, known as "gerdou," has been used for the treatment of IBD in traditional Iranian medicine. Polyphenol compounds isolated from *n*-butanol extract of *J. regia* demonstrated a significant decrease in lipid peroxidation and a remarkable increase in antioxidant potential [97].

Doses: *J. regia* at EC 50 21.4–190 μM exhibited superoxide dismutase and a remarkable radical scavenging effect against 1,1-diphenyl-2-picrylhydrazyl (DPPH) at EC50 0.34–4.72 μM.

11.5.30 *Althaea* spp.

The flower and seed of various species of *Althaea*, known as "Khatmi," have been claimed to be efficacious in IBD. The ethanol extract of *Althaea officinalis* demonstrated significant antibacterial activity against *E. coli* [39]. The ethanol extract of the flower of *Althaea rosea* showed antiinflammatory and analgesic effects in carrageenan- or dextran-induced rat paw edema.

Doses: The ratio of dry herb/extractant (w/v) was used in the preparation, the final ethanol/water concentrations (v/v), and their suppliers were 1:4 of *Althea* spp. [98].

11.6 Conclusion remarks and future perspective

Natural products are compounds with biological activities that are derived from sources such as terrestrial plants, vertebrates and invertebrates, microorganisms, and marine organisms [15, 16]. Based on the published studies, there is variety of plants likely to be more effective in the treatment of current IBD cases. There is no clear relationship between the class of plants investigated and their efficacy, which evidenced hypothesis of a complicated pathogenesis of IBD. However, no potential adverse events have been reported as result of these herbal remedies. The main mechanism of the anti-IBD effect of natural products involves modulation of cytokine activity. Cytokines can upregulate and/or downregulate several genes and their associated transcription factors, which may lead to a reduction in IBD symptoms. The anti-IBD effects of these natural products are majorly mediated by TNF-α, NF-κB, IL-6, or iNOS, whereas IL-8, IL-2, and IL-4 have less of an influence on IBD. The main active constituents of anti-IBD reservoir include numerous types of bioactive compounds including alkaloids, glycosides, flavonoids, phenolics and polyphenols, saponins, tannins, terpenes, anthraquinones, essential oils, and steroids that manifest vast array of therapeutic and medicinal properties. However, only a few of these substances have been used as anti-IBD drugs over the last decade. Further studies are warranted targeting the exact mechanisms behind the anti-IBD effects of these natural products and their effects and to investigate which specific factors are related to improvement IBD in humans. The available data concerning the administration of herbal medicine derived from plants and thyme are clearly indicating a beneficial alternative to the patients. It helps in concurring the chronic diseases and gives the gastroenterologist an arm to treat and explain the benefit of herbal treatment. Herbal medicines are economical, easy to produce, and simple in administration. A strong recommendation and social awareness toward the herbal medicine can make a significant improvement in the disease population. Thus natural products are considered indispensable source of both preventive and curative lead compounds that inexorably cradle pharmaceutical drug discovery and industry since ancient civilization across all cultures [25, 38, 39]. Botanical drugs have been preferred by patients due to their efficacy (as shown in multitude clinical trials of managing ulcerative colitis and Crohn's disease), acceptable safety (no major adverse effects), and comparatively low cost. However, evidences are still partial, intricate, and unquestionably related to both benefits and side effects.

References

[1] Crohn's and Colitis Foundation of America, Understanding IBD Medications and Side Effects, CCFA, 2018. Reviewers: Balzora S, Iskandar H, Keyashian K, Kinnucan J and Taleban S. Contributors: DeBourcy T and Klapman G. November 2018. Access at: www.crohncolitisfoundation.org.

[2] S.C. Ng, H.Y. Shi, N. Hamidi, F.E. Underwood, W. Tang, E.I. Benchimol, R. Panaccione, S. Ghosh, J.C.Y. Sung, G.G. Kaplan, Worldwide incidence and prevalence of inflammatory bowel disease in the 21st century: a systematic review of population-based studies, Lancet 390 (2017) 2769–2778.

[3] J.M. Dahlhamer, E.P. Zammitti, B.W. Ward, A.G. Wheaton, J.B. Croft, Prevalence of inflammatory bowel disease among adults aged ≥ 18 years—United States, 2015, MMWR Morb. Mortal. Wkly Rep. 65 (2016) 1166–1169.

[4] UK IBD Genetics Consortium, J.C. Barrett, J.C. Lee, C.W. Lees, N.J. Prescott, C.A. Anderson, A. Phillips, E. Wesley, K. Parnell, H. Zhang, H. Drummond, E.R. Nimmo, D. Massey, K. Blaszczyk, T. Elliott, L. Cotterill, H. Dallal, A.J. Lobo, C. Mowat, J.D. Sanderson, D.P. Jewell, W.G. Newman, C. Edwards, T. Ahmad, J.C. Mansfield, J. Satsangi, M. Parkes, C.G. Mathew, Wellcome Trust Case Control Consortium 2, P. Donnelly, L. Peltonen, J.M. Blackwell, E. Bramon, M.A. Brown, J.P. Casas, A. Corvin, N. Craddock, P. Deloukas, A. Duncanson, J. Jankowski, H.S. Markus, C.G. Mathew, M.I. McCarthy, C.N. Palmer, R. Plomin, A. Rautanen, S.J. Sawcer, N. Samani, R.C. Trembath, A.C. Viswanathan, N. Wood, C.C. Spencer, J.C. Barrett, C. Bellenguez, D. Davison, C. Freeman, A. Strange, P. Donnelly, C. Langford, S.E. Hunt, S. Edkins, R. Gwilliam, H. Blackburn, S.J. Bumpstead, S. Dronov, M. Gillman, E. Gray, N. Hammond, A. Jayakumar, O.T. McCann, J. Liddle, M.L. Perez, S.C. Potter, R. Ravindrarajah, M. Ricketts, M. Waller, P. Weston, S. Widaa, P. Whittaker, P. Deloukas, L. Peltonen, C.G. Mathew, J.M. Blackwell, M.A. Brown, A. Corvin, M.I. McCarthy, C.C. Spencer, A.P. Attwood, J. Stephens, J. Sambrook, W.H. Ouwehand, W.L. McArdle, S.M. Ring, D.P. Strachan, Genome-wide association study of ulcerative colitis identifies three new susceptibility loci, including the HNF4A region, Nat. Genet. 41 (2009) 1330–1334.

[5] S. Brand, Crohn's disease: Th1, Th17 or both? The change of a paradigm: new immunological and genetic insights implicate Th17 cells in the pathogenesis of Crohn's disease, Gut 58 (2009) 1152–1167.

[6] S.Y. Shaw, J.F. Blanchard, C.N. Bernstein, Association between the use of antibiotics in the first year of life and pediatric inflammatory bowel disease, Am. J. Gastroenterol. 105 (2010) 2687–2692.
[7] A.N. Ananthakrishnan, L.M. Higuchi, E.S. Huang, H. Khalili, J.M. Richter, C.S. Fuchs, A.T. Chan, Aspirin, nonsteroidal anti-inflammatory drug use, and risk for Crohn disease and ulcerative colitis: a cohort study, Ann. Intern. Med. 156 (2012) 350–359.
[8] M. Joossens, G. Huys, M. Cnockaert, V. De Preter, K. Verbeke, P. Rutgeerts, P. Vandamme, S. Vermeire, Dysbiosis of the faecal microbiota in patients with Crohn's disease and their unaffected relatives, Gut 60 (2011) 631–637.
[9] T. Sugihara, A. Kobori, H. Imaeda, T. Tsujikawa, K. Amagase, K. Takeuchi, Y. Fujiyama, A. Andoh, The increased mucosal mRNA expressions of complement C3 and interleukin-17 in inflammatory bowel disease, Clin. Exp. Immunol. 160 (2010) 386–393.
[10] M. Sans, Crohn's disease-prognosis, and long-term complications: what to expect? Gastroenterol. Hepatol. 4 (2008) 47–50.
[11] A.S. Cheifetz, Management of active crohn disease, JAMA 309 (2013) 2150–2158.
[12] D.J. Wong, E.M. Roth, J.D. Feuerstein, V.Y. Poylin, Surgery in the age of biologics, Gastroenterol. Rep. (Oxf.) 7 (2019) 77–90.
[13] J. Cosnes, S. Cattan, A. Blain, L. Beaugerie, F. Carbonnel, R. Parc, J.P. Gendre, Long-term evolution of disease behavior of Crohn's disease, Inflamm. Bowel Dis. 8 (2002) 244–250.
[14] J. Meier, A. Sturm, Current treatment of ulcerative colitis, World J. Gastroenterol. 17 (2011) 3204–3212.
[15] D.D. Baker, M. Chu, U. Oza, V. Rajgarhia, The value of natural products to future pharmaceutical discovery, Nat. Prod. Rep. 24 (2007) 1225–1244.
[16] D.J. Newman, G.M. Cragg, K.M. Snader, The influence of natural products upon drug discovery, Nat. Prod. Rep. 17 (2000) 215–234.
[17] T. Rabi, A. Bishayee, Terpenoids and breast cancer chemoprevention, Breast Cancer Res. Treat. 115 (2009) 223–239.
[18] N. Sultana, A. Ata, Oleanolic acid and related derivatives as medicinally important compounds, J. Enzyme Inhib. Med. Chem. 23 (2008) 739–756.
[19] G. Wang, W. Tang, R. Bidigare, Terpenoids as therapeutic drugs and pharmaceutical agents, in: Zhang, Demain (Eds.), Natural Products: Drug Discovery and Therapeutic Medicine, Humana Press Inc., Totowa, NJ, 2005, pp. 197–227.
[20] K.H. Wagner, I. Elmadfa, Biological relevance of terpenoids. Overview focusing on mono-, di- and tetraterpenes, Ann. Nutr. Metab. 47 (2003) 95–106.
[21] W.Y. Huang, Y.Z. Cai, Y. Zhang, Natural phenolic compounds from medicinal herbs and dietary plants: potential use for cancer prevention, Nutr. Cancer 62 (2010) 1–20.
[22] N. Bribi, Pharmacological activity of alkaloids: a review, Asian J. Bot. 1 (2018) 1–6.
[23] H. de Sousa Falcão, J.A. Leite, J.M. Barbosa-Filho, P.F. de Athayde-Filho, M.C. de Oliveira Chaves, M.D. Moura, A.L. Ferreira, A.B. de Almeida, A.R. Souza-Brito, M. De Fátima Formiga Melo Diniz, L.M. Batista, Gastric and duodenal antiulcer activity of alkaloids: a review, Molecules 13 (2008) 3198–3223.
[24] D.A. Dias, S. Urban, U. Roessner, A historical overview of natural products in drug discovery, Metabolites 2 (2012) 303–336.
[25] G.M. Cragg, D.J. Newman, Biodiversity: a continuing source of novel drug leads, Pure Appl. Chem. 77 (2005) 7–24.
[26] G.S. Zhong, F. Wan, An outline on the early pharmaceutical development before Galen, China J. Med. Hist. 29 (1999) 178–182.
[27] R. Dhandapani, B. Sabna, Phytochemical constituents of some Indian medicinal plants, Anc. Sci. Life 4 (2008) 1–8.

[28] H.O. Edeoga, D.E. Okwu, B.O. Mbaebie, Phytochemical constituents of some Nigerian medicinal plants, Afr. J. Biotechnol. 4 (2005) 685–688.

[29] R.C.F. Cheung, T.B. Ng, J.H. Wong, Y. Chen, W.Y. Chan, Marine natural products with anti-inflammatory activity, Appl. Microbiol. Biotechnol. 100 (2016) 1645–1666.

[30] S.A. Adebayo, J.P. Dzoyem, L.J. Shai, J.N. Eloff, The anti-inflammatory and antioxidant activity of 25 plant species used traditionally to treat pain in Southern African, BMC Complement. Altern. Med. 15 (2015) 1–10.

[31] A. Dellai, H.B. Mansour, A. Clary-Laroche, M. Deghrigue, A. Bouraoui, Anticonvulsant and analgesic activities of crude extract and its fractions of the defensive secretion from the Mediterranean sponge, *Spongia officinalis*, Cancer Cell Int. 12 (2012) 15.

[32] X. Song, T. Wang, Z. Zhang, H. Jiang, W. Wang, Y. Cao, N. Zhang, Leonurine exerts anti-inflammatory effect by regulating inflammatory signalling pathways and cytokines in lps-induced mouse mastitis, Inflammation 38 (2015) 79–88.

[33] L.E. Ishaku, F.S. Botha, L.J. McGaw, J.N. Eloff, The antibacterial activity of extracts of nine plant species with good activity against *Escherichia coli* against five other bacteria and cytotoxicity of extracts, BMC Complement. Altern. Med. 17 (2017) 133.

[34] S.A. Rhoden, A. Garcia, V.A. Bongiorno, J.L. Azevedo, J.A. Pamphile, Antimicrobial activity of crude extracts of endophytic fungi isolated from medicinal plant *Trichilia elegans* a. Juss, J. Appl. Pharm. Sci. 2 (2012) 57–59.

[35] D.A. Putri, D. Pringgenies, Effectiveness of marine fungal symbiont isolated from soft coral *Sinularia* sp. from Panjang Island as antifungal, Procedia Environ. Sci. 23 (2015) 351–357.

[36] M.F. Visintini Jaime, F. Redko, L.V. Muschietti, R.H. Campos, V.S. Matino, L.V. Cavallaro, *In vitro* antiviral activity of plant extracts from Asteraceae medicinal plants, Virol. J. 10 (2013) 245.

[37] B. Uzair, Z. Mahmood, S. Tabassum, Antiviral activity of natural products extracted from marine organisms, Bioimpacts 1 (2011) 203–211.

[38] B.B. Mishra, V.K. Tiwari, Natural products: an evolving role in future drug discovery, Eur. J. Med. Chem. 46 (2011) 4769–4807.

[39] J. Rey-Ladino, A.G. Ross, A.W. Cripps, D.P. McManus, R. Quinn, Natural products and the search for novel vaccine adjuvants, Vaccine 29 (2011) 6464–6471.

[40] T. Tarver, in: A. DerMarderosian, A. John, J. Beutler (Eds.), The Review of Natural Products, eighth ed., vol. 18, Consumer Health Internet, 2014, pp. 291–292.

[41] H.P. Albrecht, K.H. Geiss, Cardiac glycosides and synthetic cardiotonic drugs, in: Ullmann's Encyclopedia of Industrial Chemistry, Wiley-VCH Verlag GmbH & Co. KGaA, Weinheim, 2000, pp. 1–18.

[42] A. Fabbretti, C.O. Gualerzi, L. Brandi, How to cope with the quest for new antibiotics, FEBS Lett. 585 (2011) 1673–1681.

[43] P.M. Dewick, Medicinal Natural products: A Biosynthetic Approach, second ed., John Wiley and Son, West Sussex, 2009, p. 520.

[44] J.D. Williams, B-lactamases and B-lactamase inhibitor, Int. J. Antimicrob. Agents 12 (1999) S2–S7.

[45] A.D. Buss, R.D. Waigh, Antiparasitic drugs, in: M.E. Wolff (Ed.), Burger's Medicinal Chemistry and Drug Discovery, fifth ed., vol. 1, Wiley-Interscience, New York, 1995, pp. 1021–1028.

[46] J. Mann, Murder, Magic and Medicine, Oxford University Press, New York, 1994, pp. 164–170.

[47] J.P. Van den Heever, T.S. Thompson, J.M. Curtis, A. Ibrahim, Fumagillin: an overview of recent scientific advances and their significance for apiculture, J. Agric. Food Chem. 62 (2014) 2728–2737.

[48] A. Bond, *Exenatide* (byetta) as a novel treatment option for type 2 diabetes mellitus, Proc. (Baylor Univ. Med. Cent.) 19 (2006) 281–284.
[49] T. Ohnishi, Y. Sakiyama, Y. Okuri, Y. Kimura, N. Sugiyama, T. Saito, M. Takahashi, T. Kobayashi, The prediction of response to galantamine treatment in patients with mild to moderate Alzheimer's disease, Curr. Alzheimer Res. 11 (2014) 110–118.
[50] B.E. Lee, B.Y. Choi, D.K. Hong, J.H. Kim, S.H. Lee, A.R. Kho, H. Kim, H.C. Choi, S.W. Suh, The cancer chemotherapeutic agent paclitaxel (Taxol) reduces hippocampal neurogenesis via down-regulation of vesicular zinc, Sci. Rep. 7 (2017) 11667.
[51] M. Butler, Natural products to drugs: natural product derived compounds in clinical trials, Nat. Prod. Rep. 22 (2005) 162–195.
[52] M. Butler, Natural products to drugs: natural product-derived compounds in clinical trials, Nat. Prod. Rep. 25 (2008) 475–516.
[53] A. Ganesan, The impact of natural products upon modern drug discovery, Curr. Opin. Chem. Biol. 2 (2008) 306–317.
[54] A. Harvey, Natural products in drug discovery, Drug Discov. Today 13 (2008) 894–901.
[55] Y. Chin, M. Balunas, H. Chai, A. Kinghorn, Drug discovery from natural sources, AAPS J. 8 (2006) E239–E253.
[56] J.K. Triantafillidis, C. Stanciu (Eds.), Inflammatory Bowel Disease: Etiopathogenesis, Diagnosis, Treatment, fourth ed., Technogramma, Athens, 2012.
[57] K.M. Comar, D.F. Kirby, Herbal remedies in gastroenterology, J. Clin. Gastroenterol. 39 (2005) 457–468.
[58] K. Tillisch, Complementary and alternative medicine for gastrointestinal disorders, Clin. Med. 7 (2007) 224–227.
[59] L. Langmead, D.S. Rampton, Complementary and alternative therapies for inflammatory bowel disease, Aliment. Pharmacol. Ther. 23 (2006) 341–349.
[60] Q. Chen, H. Zhang, Clinical study on 118 cases of ulcerative colitis treated by integration of traditional Chinese and Western medicine, J. Tradit. Chin. Med. 19 (1999) 163–165.
[61] A. Turak, S.P. Shi, Y. Jiang, P.F. Tu, Dimeric guaianolides from *Artemisia absinthium*, Phytochemistry 105 (2014) 109–114.
[62] S. Krebs, T.N. Omer, B. Omer, Wormwood (*Artemisia absinthium*) suppresses tumour necrosis factor alpha and accelerates healing in patients with Crohn's disease—a controlled clinical trial, Phytomedicine 17 (2010) 305–309.
[63] R. Altwegg, T. Vincent, TNF blocking therapies and immunomonitoring in patients with inflammatory bowel disease, Mediators Inflamm. 2014 (2014) 172821.
[64] B. Omer, S. Krebs, H. Omer, T.O. Noor, Steroid-sparing effect of wormwood (*Artemisia absinthium*) in Crohn's disease: a double-blind placebo-controlled study, Phytomedicine 14 (2007) 87–95.
[65] L. Langmead, M. Chitnis, D.S. Rampton, Use of complementary therapies by patients with IBD may indicate psychosocial distress, Inflamm. Bowel Dis. 8 (2002) 174–179.
[66] L. Langmead, R.J. Makins, D.S. Rampton, Anti-inflammatory effects of *Aloe vera* gel in human colorectal mucosa *in vitro*, Aliment. Pharmacol. Ther. 19 (2004) 521–527.
[67] U. Dahmen, Y.L. Gu, O. Dirsch, L.M. Fan, J. Li, K. Shen, C.E. Broelsch, Boswellic acid, a potent antiinflammatory drug, inhibits rejection to the same extent as high dose steroids, Transplant. Proc. 33 (2001) 539–541.
[68] I. Gupta, A. Parihar, P. Malhotra, S. Gupta, R. Lüdtke, H. Safayhi, H.P.T. Ammon, Effects of gum resin of *Boswellia serrata* in patients with chronic colitis, Planta Med. 67 (2001) 391–395.

[69] S. Nancey, D. Moussata, I. Graber, S. Claudel, J.C. Saurin, B. Flourié, Tumor necrosis factor α reduces butyrate oxidation *in vitro* in human colonic mucosa: a link from inflammatory process to mucosal damage? Inflamm. Bowel Dis. 11 (2005) 559–566.

[70] J.P. Segain, D. Raingeard de la Blétière, A. Bourreille, V. Leray, N. Gervois, C. Rosales, L. Ferrier, C. Bonnet, H.M. Blottière, J.P. Galmiche, Butyrate inhibits inflammatory responses through NFkappaB inhibition: implications for Crohn's disease, Gut 47 (2000) 397–403.

[71] M. Song, B. Xia, J. Li, Effects of topical treatment of sodium butyrate and 5-aminosalicylic acid on expression of trefoil factor 3, interleukin 1β, and nuclear factor κB in trinitrobenzenesulphonic acid induced colitis in rats, Postgrad. Med. J. 82 (2006) 130–135.

[72] T. Kudo, S. Okamura, Y. Zhang, T. Masuo, M. Mori, Topical application of glycyrrhizin preparation ameliorates experimentally induced colitis in rats, World J. Gastroenterol. 17 (2011) 2223.

[73] H. Yuan, W.S. Ji, K.X. Wu, J.X. Jiao, L.H. Sun, Y.T. Feng, Anti-inflammatory effect of *diammonium glycyrrhizinate* in a rat model of ulcerative colitis, World J. Gastroenterol. 12 (2006) 4578.

[74] C.D. Schubart, I.E. Sommer, P. Fusar-Poli, L. de Witte, R.S. Kahn, M.P. Boks, Cannabidiol as a potential treatment for psychosis, Eur. Neuropsychopharmacol. 24 (2014) 51–64.

[75] T. Naftali, R. Mechulam, L.B. Lev, F.M. Konikoff, *Cannabis* for inflammatory bowel disease, Dig. Dis. 32 (2014) 468–474.

[76] C. Hasenoehrl, M. Storr, R. Schicho, Cannabinoids for treating inflammatory bowel diseases: where are we and where do we go? Expert Rev. Gastroenterol. Hepatol. 11 (2017) 329–337.

[77] S.H. Burstein, R.B. Zurier, Cannabinoids, endocannabinoids, and related analogs in inflammation, AAPS J. 11 (2009) 109–119.

[78] F. Massa, G. Marsicano, H. Hermann, A. Cannich, K. Monory, B.F. Cravatt, G.L. Ferri, A. Sibaev, M. Storr, B. Lutz, The endogenous cannabinoid system protects against colonic inflammation, J. Clin. Invest. 113 (2004) 1202–1209.

[79] T. Naftali, L. Bar-Lev Schleider, I. Dotan, E.P. Lansky, F. SklerovskyBenjaminov, F.M. Konikoff, *Cannabis* induces a clinical response in patients with Crohn's disease: a prospective placebo-controlled study, Clin. Gastroenterol. Hepatol. 11 (2013) 1276–1280.

[80] E. Ben-Arye, E. Goldin, D. Wengrower, A. Stamper, R. Kohn, E. Berry, Wheat grass juice in the treatment of active distal ulcerative colitis: a randomized double-blind placebo-controlled trial, Scand. J. Gastroenterol. 37 (2002) 444–449.

[81] S. Kane, M.J. Goldberg, Use of bromelain for mild ulcerative colitis, Ann. Intern. Med. 132 (2000) 680.

[82] L.P. Hale, P.K. Greer, C.T. Trinh, M.R. Gottfried, Treatment with oral bromelain decreases colonic inflammation in the IL-10-deficient murine model of inflammatory bowel disease, Clin. Immunol. 116 (2005) 135–142.

[83] M.J. Shale, S.A. Riley, Studies of compliance with delayed-release mesalazine therapy in patients with inflammatory bowel disease, Aliment. Pharmacol. Ther. 18 (2003) 191–198.

[84] F. Fernandez-Banares, J. Hinojosa, J.L. Sanchez-Lombrana, E. Navarro, J.F. Martınez-Salmerón, A. Garcıa-Pugés, F. González-Huix, J. Riera, V. González-Lara, F. Domínguez-Abascal, J.J. Giné, J. Moles, F. Gomollón, M.A. Gassull, Randomized clinical trial of *Plantago ovata* seeds (dietary fiber) as compared with mesalamine in maintaining remission in ulcerative colitis, Am. J. Gastroenterol. 94 (1999) 427–433.

[85] P. Pawar, S. Gilda, S. Sharma, S. Jagtap, A. Paradkar, K. Mahadik, P. Ranjekar, A. Harsulkar, Rectal gel application of *Withania somnifera* root extract expounds anti-inflammatory and muco-restorative activity in TNBS-induced inflammatory bowel disease, BMC Complement. Altern. Med. 11 (2011) 34.

[86] J.P. Spencer, H. Schroeter, A.R. Rechner, C. Rice-Evans, Bioavailability of flavan-3-ols and procyanidins: gastrointestinal tract influences and their relevance to bioactive forms *in vivo*, Antioxid. Redox Signal. 3 (2001) 1023–1039.

[87] J.H. Cheon, J.S. Kim, J.M. Kim, N. Kim, H.C. Jung, I.S. Song, Plant sterol guggulsterone inhibits nuclear factor-κB signaling in intestinal epithelial cells by blocking IκB kinase and ameliorates acute murine colitis, Inflamm. Bowel Dis. 12 (2006) 1152–1161.

[88] T. Tang, S.R. Targan, Z.S. Li, C. Xu, V.S. Byers, W.J. Sandborn, Randomised clinical trial: herbal extract HMPL-004 in active ulcerative colitis—a double-blind comparison with sustained release mesalazine, Aliment. Pharmacol. Ther. 33 (2011) 194–202.

[89] W.J. Sandborn, S.R. Targan, V.S. Byers, D.A. Rutty, H. Mu, X. Zhang, T. Tang, *Andrographis paniculata* extract (HMPL-004) for active ulcerative colitis, Am. J. Gastroenterol. 108 (2013) 90–98.

[90] N. Shahsavari, M. Barzegar, M.A. Sahari, H. Naghdibadi, Antioxidant activity and chemical characterization of essential oil of *Bunium persicum*, Plant Foods Hum. Nutr. 63 (2008) 183–188.

[91] V.P. Kumar, N.S. Chauhan, H. Padh, M. Rajani, Search for antibacterial and antifungal agents from selected Indian medicinal plants, J. Ethnopharmacol. 107 (2006) 182–188.

[92] B.M. Silva, P.B. Andrade, P. Valentão, F. Ferreres, R.M. Seabra, M.A. Ferreira, Quince (*Cydonia oblonga* Miller) fruit (pulp, peel, and seed) and jam: antioxidant activity, J. Agric. Food Chem. 52 (2004) 4705–4712.

[93] Y. Hamauzu, H. Yasui, T. Inno, C. Kume, M. Omanyuda, Phenolic profile, antioxidant property, and anti-influenza viral activity of Chinese quince (*Pseudocydonia sinensis* Schneid), quince (*Cydonia oblonga* Mill.), and apple (*Malus domestica* Mill.) fruits, J. Agric. Food Chem. 53 (2005) 928–934.

[94] H.Y. Joo, K. Lim, K.T. Lim, Phytoglycoprotein (150 kDa) isolated from *Solanum nigrum* Linne has a preventive effect on dextran sodium sulfate-induced colitis in A/J mouse, J. Appl. Toxicol. 29 (2009) 207–213.

[95] M.S. Akhtar, M. Munir, Evaluation of the gastric antiulcerogenic effects of *Solanum nigrum*, *Brassica oleracea* and *Ocimum basilicum* in rats, J. Ethnopharmacol. 27 (1989) 163–176.

[96] M. Jainu, C.S. Devi, Antiulcerogenic and ulcer healing effects of *Solanum nigrum* (L.) on experimental ulcer models: possible mechanism for the inhibition of acid formation, J. Ethnopharmacol. 104 (2006) 156–163.

[97] T. Fukuda, H. Ito, T. Yoshida, Antioxidative polyphenols from walnuts (*Juglans regia* L.), Phytochemistry 63 (2003) 795–801.

[98] K. Watt, N. Christofi, R. Young, The detection of antibacterial actions of whole herb tinctures using luminescent *Escherichia coli*, Phytother. Res. 21 (2007) 1193–1199.

12

Smart drug delivery systems of natural products for inflammation: From fundamentals to the clinic

Akhila Nair, Bincicil Annie Varghese, Sreeraj Gopi, and Joby Jacob
R&D Centre, Aurea Biolabs (P) Ltd, Kolenchery, Cochin, Kerala, India

12.1 Introduction

Drug delivery can be described as the method or route by which an active pharmaceutical ingredient is administered to promote its desired pharmacological effect and/or convenience and/or to reduce adverse effects. The drug delivery system is roughly defined as a device or formulation that delivers the pharmaceutical ingredient in site-directed and timely release (sustained or immediate) of the ingredient. This system is primarily inactive, but improves the safety and/or efficacy of the pharmaceutical ingredient that it carries [1]

Currently, nano-derived drug delivery system provides remarkable contribution to precision medicine. However, there are certain limitations surrounding drug delivery systems such as multiple reactions with repeated purifications, drug leakage, and low yield [2]. Therefore research is now moving forward toward smart drug delivery system to provide superior controllability (reduce adverse off-target effects), maximize the efficiency, and control the release of the drug [2, 3]. These drug delivery systems enhance drug specificity and reduce systemic toxicity. For the treatment of various inflammation-related diseases such as cancer, cardiovascular, and neurodegenerative, these smart drug delivery systems provide enhanced permeability and retention effect to deliver the drugs at the desired site [4]. There are various types of smart drug delivery systems such as pH-, temperature-, redox-, magnet-, and enzyme-responsive drug delivery systems, which are discussed further in the chapter. These systems utilize smart polymers that are capable of reversible and irreversible changes in their chemical structure and/or physical properties with slight changes in

Inflammation and Natural Products. https://doi.org/10.1016/B978-0-12-819218-4.00013-4
© 2021 Elsevier Inc. All rights reserved.

the external environment including temperature, light, mechanical force, magnetism, ion concentration, pH electricity, and bioactive molecules [5]. These factors take advantage of the external and internal stimuli depending upon their route of administration and time of arrival at the site of action, which consequently effect the rupture of the carrier and drug release [4]. Hence smart drug delivery systems are intelligent systems with self-regulation capability, integrated sensing, monitoring, and activation by the stimuli and the environment [6]

12.2 Stimuli-responsive drug delivery system

Stimuli-responsive drug delivery systems are designed to release the drug at specified site in a controlled manner. These systems are of two types, external and internal stimuli. The internal stimuli approaches are pH, temperature, redox, enzymes, glucose, urea morphine, or diseased body conditions, and the external stimuli are electric, magnetic, thermal, and ultrasonic [6] (Table 12.1). The stimuli-responsive drug delivery systems from normal to tumor tissues utilizing these approaches are depicted in Fig 12.1 [7]. These approaches have been explored for the development of organosilica nanoparticles [8] (Fig. 12.2). Natural products utilized in stimuli responsive drug delivery systems for various inflammatory diseases are tabulated in Table 12.2.

Table 12.1 Different smart drug delivery systems.

S. no.	Smart drug delivery systems	Features	References
1.	**Internally responsive drug delivery system**		
1.a	Redox-responsive drug delivery system	**(a)** Rely on $-H_2O_2$ and $-OH$ radicals (reactive oxygen species) **(b)** Polymeric nanomaterials possessing disulfide bonds oxidize the extracellular media and reduce to thiol groups in reducing environment **(c)** The polymeric nanocarriers will rupture to release the bioactive agents in excess of glutathione inside the tumor cell	[6]
1.b	Enzyme-responsive drug delivery system	**(a)** Biocatalytic activity **(b)** Enzymes involved are lipase, protease, phospholipase, and glycosidase **(c)** Specific moieties are inserted either in the side groups or the main chain to produce self-assembled structures that are bonded by selected enzyme	[6]

Table 12.1 Different smart drug delivery systems.—cont'd

S. no.	Smart drug delivery systems	Features	References
1.c	pH-responsive drug delivery system	**(a)** The polymers with functional groups are utilized to act as proton donors or acceptors in response to the variation in the environmental pH **(b)** Introduction of cleavable acid-responsive bonds in the structure of nanocarriers **(c)** The polymer-drug hybrids link drug to the polymer backbone, which play an important responsive part	[7]
1.d	Thermo-responsive drug delivery system	**(a)** The phase transition occurs via temperature known as the critical solution temperature (CST) where the polymers become soluble or insoluble **(b)** Lower critical solution temperature (LCST) becomes higher, and the polymeric chain gets hydrated, which results in more hydrophobic state to further rupture the polymeric matrix and release of the drug	[7]
1.e	Light-responsive drug delivery system	**(a)** Drug carriers are cleaved or destroyed in accordance with the specific wavelength of light (UV light, visible light, and NIR) to release the drug **(b)** Photoinduced transition of hydrophobicity/hydrophilicity **(c)** Photocleavage reaction **(d)** Photoinduced heating UV sensitive and visible light sensitive	
2.	**Externally responsive drug delivery system**		
2.a	Ultrasonically responsive drug delivery system	**(a)** Follows captivation by pressure-dependent alternating sizing of gas-filled microbubbles that are produced by ultrasonic energy **(b)** Carrier destabilization and accelerated vessel permeability result in cellular uptake and drug release	[9]
2.b	Magnetically responsive drug delivery system	**(a)** Utilizes magnetite (Fe_3O_4) or maghemite (γ-Fe_2O_3) to form either magnetoliposomes or porous metallic nanocapsules **(b)** This system apply magnetic current to bring about changes in the size or physical shape of the formulation	[9]

12.2.1 Internally responsive drug delivery system

Internally responsive drug delivery systems mainly rely on the internal environment of the diseased person. These systems have increased reproducibility in vivo, predictability, and feasibility during large-scale production. These include redox-responsive, enzyme-responsive, different physiological pH, thermoresponsive (physiological temperature), and biological factors including glucose, inflammation responsive, and cancer responsive [9].

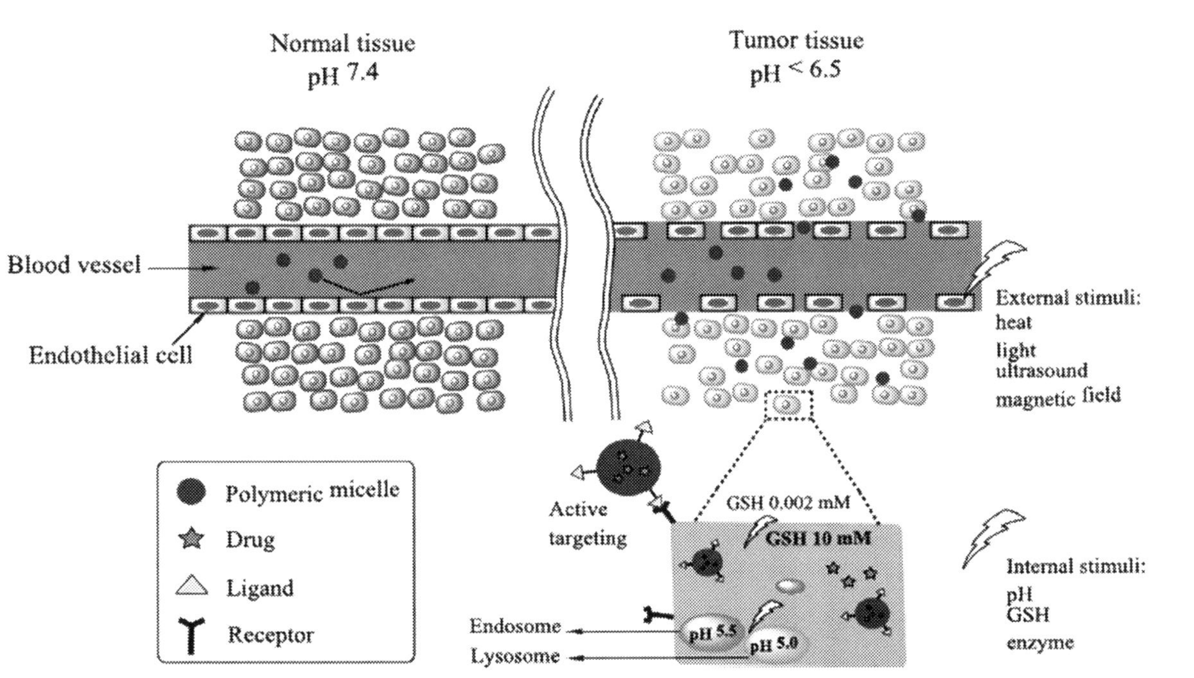

Fig. 12.1 Transport of stimuli-responsive polymeric nanocarriers through normal and tumor tissues via several stimuli-responsive drug delivery strategies. From M. Alsehli, Polymeric nanocarriers as stimuli-responsive systems for targeted tumor (cancer) therapy: recent advances in drug delivery, Saudi Pharm. J. 28 (2020) 255–265. Copyright Elsevier.

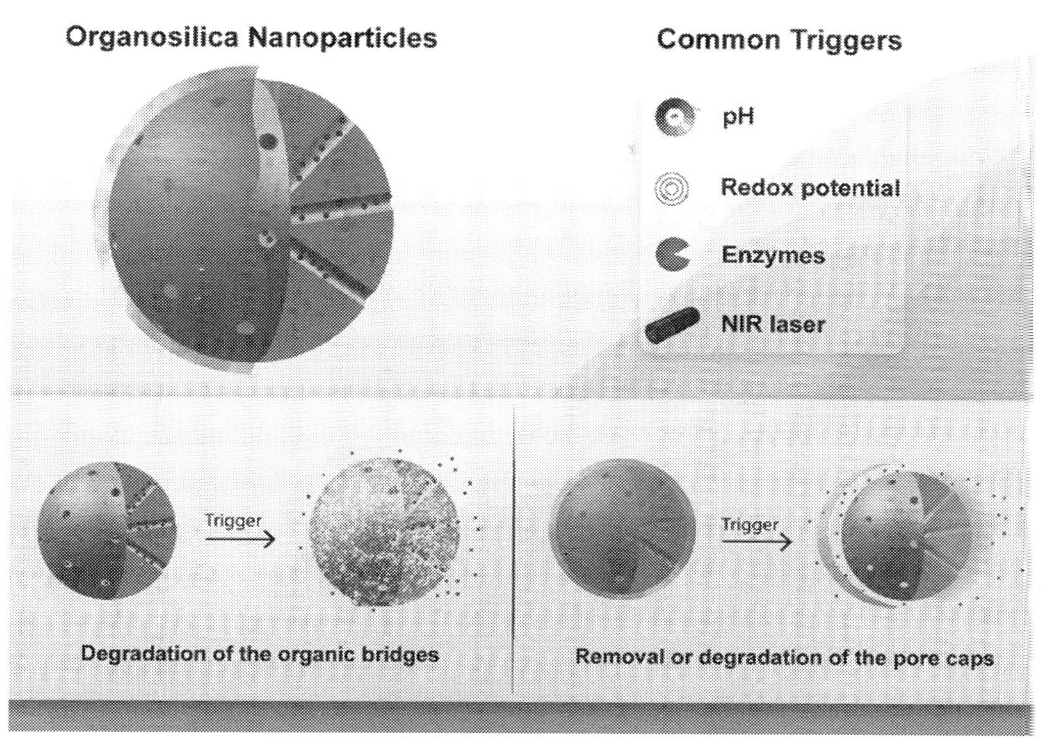

Fig. 12.2 Representation of the approaches explored for developing stimuli-responsive organosilica nanoparticles. The stimuli-responsive drug release can be mediated by the degradation of the organic bridges present on the nanoparticle matrix or through the utilization of pore capping agents. From R.S. Guimarães, C.F. Rodrigues, A.F. Moreira, I.J. Correi, Overview of stimuli-responsive mesoporous organosilica nanocarriers for drug delivery, Pharmacol. Res. 155 (2020) 104742. Copyright Elsevier.

12.2.1.1 Redox-responsive drug delivery system

Redox-sensitive drug delivery system is an internal stimuli-responsive drug delivery system that has gained lot of acclamation to release drug intracellularly. This system takes advantage of the redox status of the tissues. They mainly rely on reactive oxygen species (ROS), specially $-H_2O_2$ and $-OH$ radicals. ROS are omnipresent in tissues and are linked to numerous pathological conditions like inflammation, heart, nerve injuries, and arteriosclerosis [6]. The redox potential difference between the reducing intracellular space and the oxidizing extracellular space acts as a potential stimulus for the timely release of the drug at desired site. The cytosol possesses a low redox potential because of the reduced abundance of glutathione (GSH) as its average extracellular concentration is approximately 2 μm and intracellular concentration is about 10 nm. The GSH concentration is lower in healthy people than in tumor tissues. The design of redox-responsive polymeric

Table 12.2 Smart drug delivery systems for natural products for various inflammatory diseases.

S. no.	Smart drug delivery formulation	Drug	Features	Inflammatory activities	Reference
1.	**Redox-responsive drug delivery system**				
a.	Glutathione-responsive drug delivery system	Curcumin (CUR)	**(a)** Tannic acid and iron chloride on silica nanoparticles **(b)** Lowering the pH from 7.4 to 6.0 or 4.5 **(c)** Adjusting the GSH level	Biomedicines	[10]
b.	Redox-responsive polymersomes	CUR and doxorubicin (DOX)	**(a)** MMP-7 peptide linker (GPMG-IAGQ) masked domain (EEEE) and NLS peptide (PKKKRKV) where the latter was activated at high concentration of MMP-7 isozyme within the tumors **(b)** More toxic toward cancer cells (AsPC-1 and BxPC-3 cells) than normal cells	Pancreatic cancer	[11]
2.	**Enzyme-responsive drug delivery system**				
a.	Glucose oxidase-catalase-chitosan nanogel	Insulin	**(a)** Swelled in hyperglycemic conditions through enzymatic conversion of glucose **(b)** Protonation of the pH-responsive chitosan matrix **(c)** Release drug as a self-regulating nanovalve system	Antidiabetic	[12]
3.	**pH-responsive drug delivery system**				
a.	*N*-Succinyl chitosan-ZnO nanoparticles	CUR	**(a)** Drug release from the system showed a pH-sensitive release profile **(b)** Neutral pH 60% of drug was released in 16 h, and in acidic pH 45% drug released in 1 h as per drug release study	Antibacterial, anticancer	[13]
b.	Chitosan, graphene, and cellulose nanowhisker hydrogel	CUR, DOX	**(a)** Sensitive to external stimuli including pH and amino acid cysteine **(b)** In vitro drug release study showed a pH-dependent release of drugs in PBS solution	Anticancer, antibacterial	[14]
c.	Lactosylated nanoparticles (LAC-NPs)	CUR, sorafenib	**(a)** Drug release of these nanoparticles was more efficient in pH 5.5 than pH 7.4 **(b)** The inhibition rate was found to be 77.4%	Anticancer (hepatocellular carcinoma)	[15]
d.	Chitosan microcapsules	CUR, catechin	**(a)** Sequential drug release		[16]

4.	**Thermo-responsive drug delivery system**				
a.	Temperature-responsive amphiphilic triblock copolymers poly[2-(2-methoxyethoxy) ethyl methacrylate-*co*-oligo(ethylene glycol) methacrylate]-*b*-poly(L-lactide)-*b*-poly[2-(2-methoxyethoxy) ethyl methacrylate-co-oligo(ethylene glycol) methacrylate] [P(MEO2MA-*co*-OEGMA)-*b*-PLLA-*b*-P(MEO2MA-*co*-OEGMA)]	CUR	**(a)** Local temperature around the nanocarriers slightly higher than the LCST **(b)** The polymeric chain becomes dehydrated and will be more hydrophobic **(c)** Collapse, which triggers the release of the encapsulated drug	Anticancer	[7]
b.	κ-Carrageenan polysaccharide thermosensitive nanogels	Methylene blue	**(a)** Increase in drug release when the temperature increased from 25°C to 37°C and 45°C, as the nanogel swelled	Biomedicines	[12]
5.	**Light-responsive drug delivery system**				
	Nanogels of hyaluronic acid-*g*-7-*N*,*N*-diethylamino-4-hydroxymethylcoumarin (HA-CM)	DOX	**(a)** The coumarin moiety has a very high two-photon absorption cross section that promotes the release of drug by either UV (1-photon) or NIR (2-photon) triggers **(b)** In vitro studies demonstrated that drug-loaded nanogels irradiated by UV had a higher drug release than near infrared because of the cleavage of the urethane bonds under UV irradiation **(c)** Effective uptake by CD44 + MCF-7 cells via a receptor-mediated pathway and intracellular drug release under near-infrared radiation	Anticancer	[17]

Continued

Table 12.2 Smart drug delivery systems for natural products for various inflammatory diseases—cont'd

S. no.	Smart drug delivery formulation	Drug	Features	Inflammatory activities	Reference
6.	**Ultrasonically responsive drug delivery system**				
a.	Starch nanocomposite	Zolpidem	**(a)** Improved drug release and loading capacity **(b)** Hydrophilic molecule into the multiwalled carbon nanotube (MWCNT) and sonochemical method could prevent aggregation of nanotubes in starch	Antiinsomnia	[18]
b.	Nanoemulsion	Protein	**(a)** Stabilize proteins in nanoemulsion formulations and for the safety of excipients, various additives such as glycerin, cholesterol, and chloroform and soybean lecithin **(b)** Charged in oil phase to increase its solubility and stability **(c)** Improved protein delivery	Topical	[19]
7.	**Magnetically responsive drug delivery system**				
a.	Magnetic reduction–responsive alginate-based microcapsules (MRAMCs) Thiolated alginate and oleic acid modified Fe_3O_4 nanoparticles (OA-Fe_3O_4)	Coumarin 6	**(a)** MRAMCs exhibited good magnetic targeted ability originated from the superparamagnetism of OA-Fe_3O_4 NPs **(b)** Reductive triggered drug release capacity owing to the response of disulfide bonds on the shell to reduction of GSH	Anticancer	[20]
b.	Alginate-gelatin and Fe_3O_4 magnetic nanoparticles	DOX	**(a)** Alginates were partially oxidized to form hydrogels through Schiff-base condensation reaction **(b)** Fe_3O_4 NPs impregnated in situ chemical coprecipitation **(c)** Saturation magnetization (δs) value of the alginate-gelatin-Fe_3O_4 found to be 31 emu g^{-1}, which represented proper magnetic property	Anticancer	[21]
c.	Folate receptor–targeted hybrid protein inorganic nanoparticles	CUR	**(a)** Hybridization with superparamagnetic calcium ferrite **(b)** The drug release rate was higher in acidic conditions, increased drug concentrations was due to the influence of magnetic field		[22]

nanocarriers depends upon each redox-responsive section such as disulfide bonds that oxidizes extracellular media with high stability on the one hand and reduces to thiol groups in reducing environment on the other hand.

Therefore, in the presence of excess GSH inside the cell, the polymeric nanocarriers will rupture to release the bioactive agents. A study demonstrated that the release of curcumin (CUR) could be effectively controlled by adjusting the GSH level. In this study a pH and GSH-responsive drug delivery system was prepared by tannic acid-iron (III) complex was formed on the mesoporous silica nanoparticles (MSN), which was successfully accomplished by the incorporation of tannic acid and iron chloride ($FeCl_3$) dispersed aqueous MCM-41. By lowering the pH from 7.4 to 6.0 or 4.5 and adjusting the GSH level (competitive liganding accelerated the decomposition of TA-Fe (III) complex), pH- and GSH-responsive drug delivery system was formed, which could be effectively used as biomedicines [10]. The encapsulation or conjugation of drug into a polymeric nanocarrier depends upon the disulfide bond. Numerous polymeric nanocarriers such as dendritic polymers, block copolymers, and redox-responsive biodegradable polymers are discovered till now. Among them, biodegradable nanocarriers for drug delivery help avoid poor in vivo metabolism and elimination of other nanocarrier formulations. Hence, polymeric nanocarriers having biodegradability and triggering signals can be designed efficiently with the help of disulfide bonds. These disulfide bonds can be employed as crosslinking agents either in the form of the shell or the core in the polymeric micelles that help trigger the nanosystem to rupture and release the drug intracellularly [7] (Fig. 12.3). Redox-responsive polymersomes were prepared for the delivery of CUR and doxorubicin to the cell nucleus of pancreatic cancer. MMP-7 peptide linker (GPMG-IAGQ) was utilized to join the masking domain (EEEE) and NLS peptide (PKKKRKV) where the latter was activated at high concentration of MMP-7 isozyme within the tumors. The nanoparticulate system prepared were more toxic toward cancer cells (AsPC-1 and BxPC-3 cells) than normal cells (Fig. 12.4). Thus redox-responsive drug delivery systems synthesized from natural ingredients are potential carriers that might be promising for biomedical applications.

12.2.1.2 Enzyme-responsive drug delivery system

Enzyme-activated drug delivery system is another emerging class of smart drug delivery system that releases drug in a controlled manner via enzymatic reactions. As major reactions in the body involve enzymes including lipase, protease, phospholipase, and glycosidase, this system provides high degree of specificity. Numerous enzymes

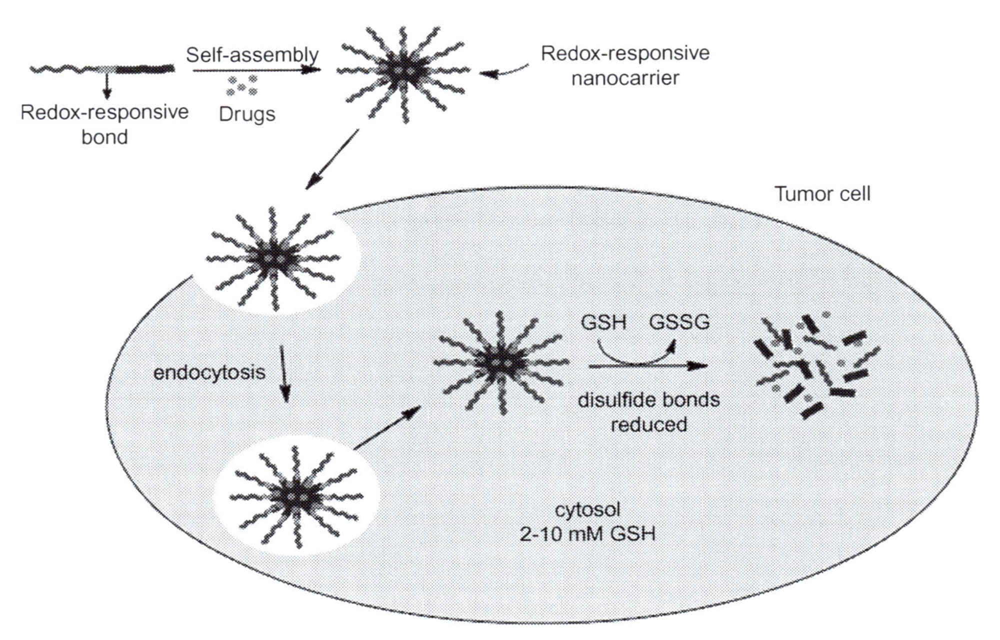

Fig. 12.3 Schematic illustration of the mechanism of action of redox-responsive micelles. The drug-loaded redox-responsive nanocarrier is taken up into the cancer cell by endocytosis and then actively releases the biologically active agent into the cytosol inside the cells owing to GSH-triggered disassembly. From M. Alsehli, Polymeric nanocarriers as stimuli-responsive systems for targeted tumor (cancer) therapy: recent advances in drug delivery, Saudi Pharm. J. 28 (2020) 255–265. Copyright Elsevier.

are utilized for carrying the drug, which are fabricated with the help of covalent bonding or encapsulation. The enzyme-mediated drug is released through biocatalytic activity at malignant or inflammatory sites. Proteases are useful in tumor invasion, wound healing, and tissue remodeling [11].

Different types of proteases such as matrix metalloproteinases (MMPs), protein kinase C alpha, and urokinase-type plasminogen activator (uPA) are mainly used for controlled drug release; however, precise control of the initial response time is required. Another important group of enzymes is phospholipases such as phospholipase A2 (PLA2) enzyme, which release drug via small unilamellar vesicle (SUV) or liposome-mediated drug delivery. These are enzymes that are involved in malignancies, inflammation, neurodegeneration, and infection. In addition, glucose oxidase of oxidoreductase family that is involved in glucose metabolism is helpful in drug delivery and diagnosis via glucose level response [6]. The enzyme-responsive nanocarriers are fabricated by inserting specific moieties either in the side groups or in the main chain to produce self-assembled structures that are bonded by selected enzyme. Lysosomes are novel

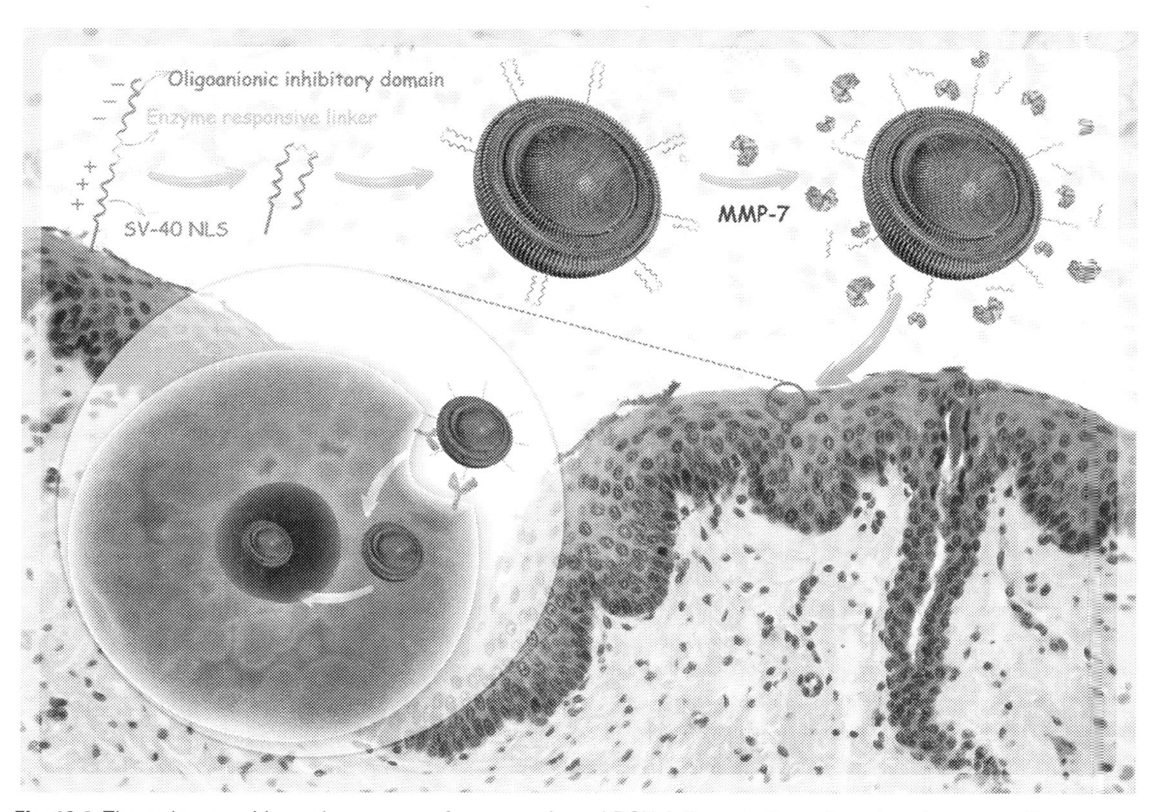

Fig. 12.4 The redox-sensitive polymersomes for curcumin and DOX delivery to the cell nucleus in pancreatic cancer was developed and conjugated to the antennae consisted of MMP-7 peptide linker (GPMG-IAGQ) coupled to the NLS peptide (PKKKRKV) and the masking domain (EEEE). From S. Indermun, M. Govender, P. Kumar, Y.E. Choonara, V. Pillay, Stimuli-responsive polymers as smart drug delivery systems: classifications based on carrier type and triggered-release mechanism, in: A.S.H. Makhlouf, N.Y. Abu-Thabit (Eds.). Stimuli Responsive Polymeric Nanocarriers for Drug Delivery Applications, Types and Triggers, Woodhead Publishing Series in Biomaterials, 2018, pp. 43–58. Copyright Elsevier.

membrane-bound vesicles that carry numerous enzymes such as sulfatases and glycosidases. These nanocarriers are stable during blood circulation but release the drug under enzymatic environment [7]. Glucose oxidase-catalase-chitosan nanogel was successfully fabricated for the encapsulation of insulin, an antidiabetic drug, as this nanoparticulate swelled in hyperglycemic conditions through enzymatic conversion of glucose and protonation occurred to the pH-responsive chitosan matrix to release the drug [12]. Thus enzyme-responsive drug delivery systems prepared from natural materials are promising carriers providing control and effective release of drug the inflammatory sites of any diseases.

12.2.1.3 pH-responsive drug delivery system

Among the various smart drug delivery systems, pH is one of the important stimuli that are used to the release of drug in controlled manner. It is a smart tool to construct stimuli-responsive drug delivery system. In the research of precision medicine, it may occur in the body during the treatment process that effectively releases drugs in the target cells. Normally, pH-responsive carriers are based on significant variation of pH values in organs. In different organs, the pH varies, including stomach (pH=2) and intestinal tract (pH=7). The designed carriers can sensitively differentiate delicate pH changes in specific disease sites, such as inflammatory, ischemic, and tumor tissues, even in different organelles, like endosomes and lysosomes. The pH of normal tissues and the blood circulation system is about 7.4, while the pH of some diseased tissues, such as inflammatory tissues or tumor cells, is about 6.5. Through drug release study, it is reported that the drug release rate at pH 5 was better than that of 7.4 [23]. There are two main strategies adopted for pH-responsive drug delivery system to facilitate a response toward acidic environment. Firstly, the polymers with functional groups are utilized that act as proton donors or acceptors in response to the variation in the environmental pH. Under acidic environment the polymers become deprotonated to cause structural damage and reverse the hydrophobicity to the polymers to release the drug, whereas at physiological pH deprotonation occurs (Fig. 12.5A). The polymers having ionizable groups, namely, weak bases (amines) and weak acids (carboxylic acids), are utilized to induce pH-responsive nanocarriers to induce the disruption of polymeric micelles in acidic media at the interior and/or exterior of tumor cells. Secondly, introduction of cleavable acid-responsive bonds in the structure of nanocarriers. These bonds in between the polymer and drug or inside the amphiphilic block can be easily broken at acidic pH to release the drug at the intended site (Fig. 12.5B and C). Generally, cleavable acid-responsive bonds impregnated into the polymeric nanocarriers include imine, hydrazine, oxime, orthoester, hydrazide, and vinyl ether bonds. Particularly the pH-responsive chemical bonds facilitate the self-assembly of block polymers forming micelles. In acidic media the acid-responsive bonds on these polymers, which are usually stable at pH 7.4, become hydrolyzed, thus disrupting the core-corona micelle structure and releasing the encapsulated drugs (Fig. 12.5B). Thirdly, the polymers-drug hybrids link drug to the polymer backbone, which play an important responsive part (Fig. 12.5C) [7]

Inflammation is the main reason for the development of chronic wounds to heal rapidly, due to pH changes in inflammatory tissues. Therefore changes in pH values can be used as a trigger point for the drug delivery to reduce the inflammation in the wound cells [13]. The drug carried in acidic conditions is better than under neutral

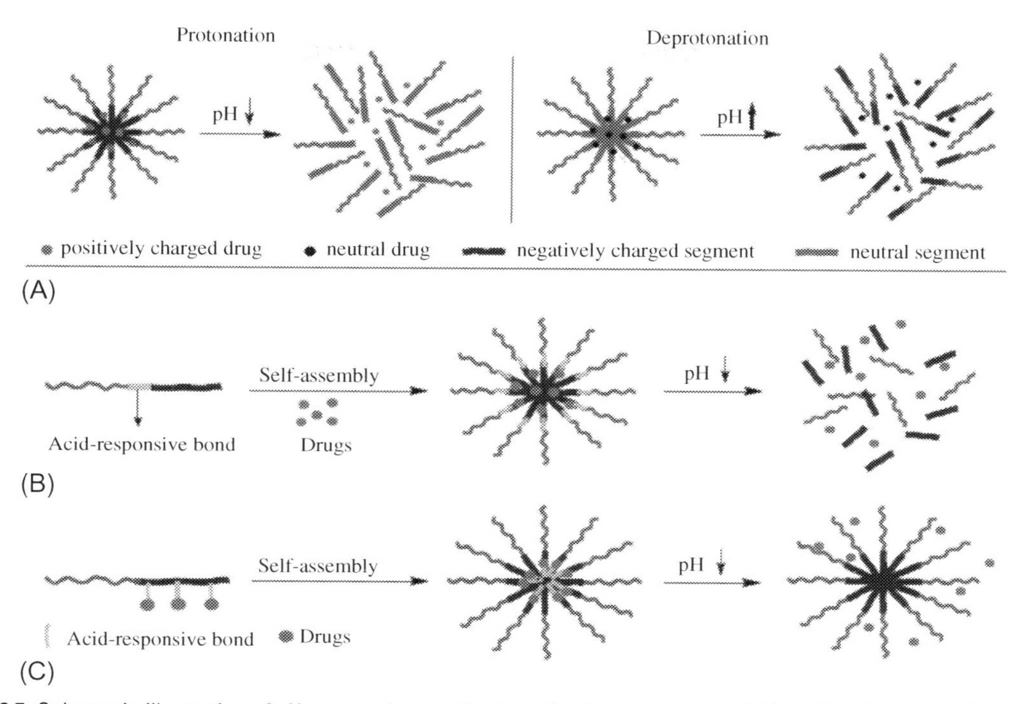

Fig. 12.5 Schematic illustration of pH-responsive mechanisms for the drug release initiated by changes in the microenvironment. (A) Protonation or deprotonation of polymers leads to structural damage of nanocarriers. (B) Breakage of acid-responsive bonds within the polymer at acidic pH causes damage to the amphiphilic blocks. (C) Breakage of the acid-responsive bond between the anticancer drug and polymer. From M. Alsehli, Polymeric nanocarriers as stimuli-responsive systems for targeted tumor (cancer) therapy: recent advances in drug delivery, Saudi Pharm. J. 28 (2020) 255–265. Copyright Elsevier.

conditions (pH = 4.5 > 5.8 > 7.4). Many research studies have been found that the amount of drug released is higher at acidic pH, which consequently suppresses cancer cells.

CUR is an active ingredient of turmeric, which has significant pharmacological activities; mainly, they are related to antioxidant and antiinflammatory properties. The therapeutic effect of CUR is well utilized in anticancer studies. Ghaffari et al. [14] demonstrated that a pH-sensitive delivery system based on *N*-succinyl chitosan-zinc oxide (ZnO) nanoparticles for improving antibacterial and anticancer activities of CUR. Drug release from the system showed a pH-sensitive release profile. In neutral pH, 60% of drug was released in 16 h and in acidic pH, 45% drug released in 1 h as per drug release study. 3-[4,5-Dimethylthiazole-2-yl]-2,5-diphenyltetrazolium bromide (MTT) and Annexin V-FITC/PI assays revealed the superior anticancer activity of CUR-CS-ZnO compared with free CUR against breast cancer cells (MDA-MB-231) by inducing the apoptotic response with no cytotoxic effects on HEK293 normal cells. Moreover, CUR conjugation to the system notably dropped the MIC (25- to 50-fold) and MBC values (10- to 40-fold) against *Staphylococcus aureus* and *Escherichia coli*. These features proved promising for anticancer and antimicrobial applications [14]. One of the study was synthesized to hydrogel using chitosan, graphene, and cellulose nanowhisker (CGW). The hydrogel was found pH sensitive and injectable. The hydrogel showed a pH-sensitive release of doxorubicin and CUR. The synthesized hydrogel was sensitive to various external stimuli including pH and amino acids. CGW hydrogel showed a pH-responsive release behavior for anticancer drugs. In vivo test indicated the fast gelation of the hydrogel in the rat's skin by the subcutaneous injections. Antimicrobial investigations confirmed that CGW showed strong antibacterial activity against gram-positive bacteria [15]. pH-sensitive lactosylated nanoparticles (LAC-NPs) for the treatment of hepatocellular carcinoma treatment for the codelivery of sorafenib and CUR. It was observed that this lactosylated sorafenib and CUR nanoparticles were spherical, the size was 115.5 + 3.6 nm, and zeta potential was − 34.6 + 2.4. The drug release of these nanoparticles was more efficient in pH 5.5 than pH 7.4, and the inhibition rate was found to be 77.4%, which proved that these pH-sensitive NPs were promising system for hepatocellular carcinoma [16]. The sequential drug release of CUR and catechin from pH-responsive chitosan microcapsules is depicted in Fig. 12.6.

12.2.1.4 Thermo-responsive drug delivery system

Thermosensitive polymers undergo changes in solubility due to small changes in temperature. This temperature-induced solubility change effectively controls the rate of drug release while maintaining physicochemical stability and biological activity within the human body. The swelling balance of the polymers is controlled by the hydrophilic/hydrophobic balance of the nanocarrier material. In this approach the phase transition occurs via temperature known as the critical solution temperature (CST) where the polymers become soluble or insoluble. Therefore solubility transformation is attributed to the presence of hydrophobic (alkyl) groups that are important in establishing critical solution temperatures (CSTs) or upper and lower critical solution temperatures (UCST and LCST, respectively). The lower critical solution temperature (LCST) is the thermal transition that occurs from more soluble state to less soluble state. If this LCST becomes higher, the polymeric chain gets hydrated, which results in more hydrophobic state to further rupture the polymeric matrix and release of the drug [7] (Fig. 12.7).

Natural polymers go through various changes due to its thermosensitive nature. Among them, one example is their temperature-induced solubility change that effectively controls the

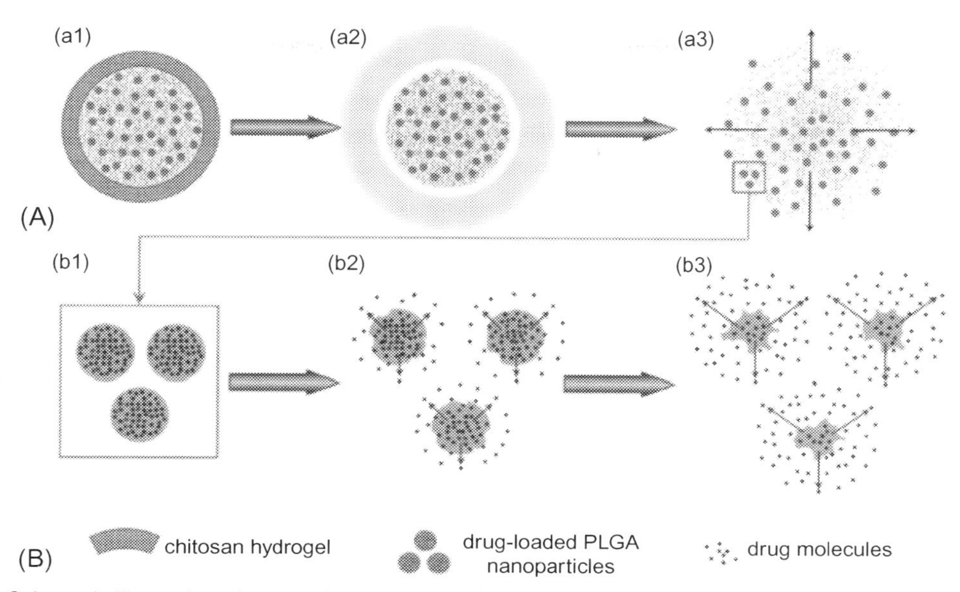

Fig. 12.6 Schematic illustration of sequential drug release (Cur and catechin) from pH-sensitive core-shell chitosan microcapsules. (A) Decomposition of chitosan shell with burst release of Cur and catechin in acidic condition. (B) Destruction of PLGA NPs and sustained release of Cur and catechin in 2 days' degradation of PLGA core in acidic conditions. From Y. Bian, D. Guo, Targeted therapy for hepatocellular carcinoma: co-delivery of sorafenib and curcumin using lactosylated pH-responsive nanoparticles, Drug Des. Devel. Ther. 14 (2020) 647–659. Copyright Elsevier.

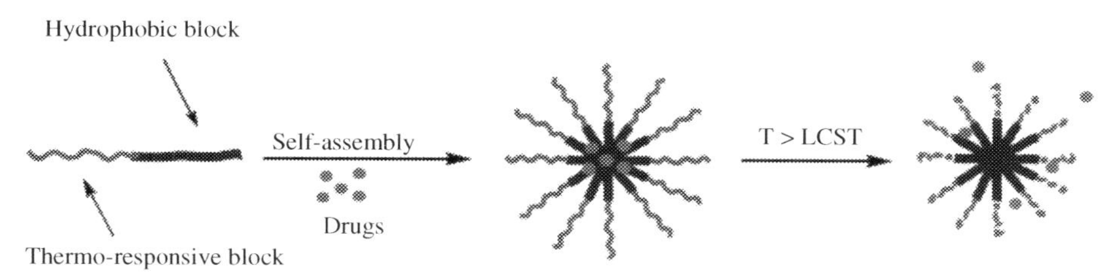

Fig. 12.7 Schematic illustration of temperature-responsive amphiphilic polymer mechanisms for drug release initiated by a variation in the surrounding temperature. Below the LCST the temperature-responsive shell is hydrated and is hydrophilic. Once the temperature (*T*) is slightly above the LCST, the hydrophilic corona collapses, which triggers the drug release. From M. Alsehli, Polymeric nanocarriers as stimuli-responsive systems for targeted tumor (cancer) therapy: recent advances in drug delivery, Saudi Pharm. J. 28 (2020) 255–265. Copyright Elsevier.

release of drug at the same that it controls the physicochemical stability and biological activity within the human body. Thermally triggered drug delivery is the most traditional and easily implemented stimulus, and the literature using this trigger is vast [24]. Typically, thermogels rely on the existence of a phase transition between room temperature and physiological temperature, for example, a lower critical solution temperature, such as poly(*N*-isopropylacrylamide)

(PNIPAM) or poly(oligoethylene glycol methacrylate), or a volume phase transition temperature. This transition is usually actuated by the change in temperature occurring from contact of the material with the body, for example, after injection, the situ gelation of the material was thus used as a depot formulation. The field of thermogelling polymers is thriving for a range of biomedical applications and not limited to drug delivery, especially in particular tissue engineering and cell therapy. The challenge in the design of thermoresponsive nanodevices lies in the use of materials that are both safe and sensitive enough to respond to slight temperature changes around the physiological temperature of 37°C [24]. Major studies are adopting thermally responsive drug delivery system. In this case the drug is released in the diseases site over temperature changes. The variation in the temperature changes around the physiological temperature of 37°C can be used to test the design of thermoresponsive nanodevices that provide materials that are both safe and sensitive. Normally, body temperature is 37°C, and that for tumor is 40–42°C. In cancer therapy, the drug loaded in thermo-responsive nanodevices at normal body temperature of 37°C, release drug from such devices, in tumors cell at 40–42°C. Another example is a temperature-responsive amphiphilic triblock copolymers poly[2-(2-methoxyethoxy)ethyl methacrylate-*co*-oligo (ethyleneglycol)methacrylate]-*b*-poly(L-lactide)-*b*-poly[2-(2-methoxyethoxy) ethyl methacrylate-*co*-oligo(ethylene glycol)methacrylate] [P(MEO2MA-*co*-OEGMA)-*b*-PLLA-*b*-P(MEO2MA-*co*-OEGMA)], which was capable of efficiently delivering the hydrophobic drug curcumin. The local temperature around the nanocarriers is slightly higher than the LCST, which facilitated the polymeric chain to become dehydrated and more hydrophobic. Hence, these nanocarriers collapse, to trigger the release of the encapsulated drug [7]. Karimi et al. reviewed that methylene blue drug, which was loaded into κ-carrageenan polysaccharide-thermosensitive nanogels, successfully increased the drug release when the temperature increased from 25°C to 37°C and 45°C, as the nanogel swelled [12].

12.2.1.5 Light-responsive drug delivery system

As light can be controlled in space or by time, it can be applied as a special external stimulus. So far, various photosensitive drug carrier materials have been successively developed. These drug carriers are cleaved or destroyed in accordance with the specific wavelength of light (UV light, visible light, and NIR) so that the successful release of the drug can achieve the corresponding therapeutic purpose [9].

In biomaterial research, the smart and stimulus-responsive drug delivery systems are gaining immense popularity. Mainly, stimulus responsive drug delivery systems such as pH, temperature, light and redox, result in the release of drug in controlled manner. One of the most important external stimuli was light. Nowadays, many photosensitive drugs have been developed. To a corresponding therapeutic purpose, the drug carrier can be cleaved with the specific wavelength of light; thus a successful release of the drug has been achieved [7]. Light offers many advantages as the trigger; among them are low cost, ease of tenability of wavelength, and intensity. Light ranges from UV and visible allow low penetration and can damages tissue at much lower than NR [6]. Hyaluronic acid is a natural polysaccharide formed from disaccharide units of D-glucuronic acid with β(1,4)- and β(1,3)-glucosidic bonds and *N*-acetyl-D-glucosamine. This natural moiety has the ability to recognize the hyaluronic acid receptor also known as CD44, a high-affinity receptor that is overexpressed on numerous cancer cells [25]. The UV-responsive degradable nanogels fabricated by hyaluronic acid-*g*-7-*N*,*N*-diethylamino-4-hydroxymethylcoumarin (HA-CM) allowed CD44-targeted delivery of DOX, which was triggered by near-infrared radiation. The coumarin moiety has a very high two-photon absorption cross section that promotes the release of drug by either UV (1-photon) or NIR (2-photon) triggers. In vitro studies demonstrated that DOX-loaded nanogels irradiated by UV had a higher drug release than near infrared because of the cleavage of the urethane bonds under UV irradiation. They also showed effective uptake by CD44+MCF-7 cells via a receptor-mediated pathway and intracellular DOX release under near-infrared radiation [17]. When drug reaches the target cells (inflammatory cells), the light sources help release the drug. The light responsive drug delivery systems, which utilises natural polymers for the delivery of drug, are biocompatible, water soluble, biodegradable and safe. The light field–responsive polymers can be broken into three subclasses: (1) photoinduced transition of hydrophobicity/hydrophilicity, (2) photocleavage reaction, and (3) photoinduced heating UV sensitive and visible light sensitive [18]. These light-responsive drug delivery systems are promising for the encapsulation of natural bioactive components, proteins, and DNA for wide drug delivery applications including wound healing, anticancer therapy, and tissue engineering [24].

12.2.2 Externally responsive drug delivery system

Externally responsive stimuli focus on targeted and responsive drug delivery system. These systems can be modified according to the patient health conditions and therapy and are independent of the physiological properties. These systems mainly consist of magnetically responsive and ultrasonically responsive drug delivery system [9].

12.2.2.1 Ultrasonic-responsive drug delivery system

Ultrasonic-responsive drug delivery system utilizes the mechanical or thermal effect to release drug at the desired site. This system is absolutely harmful because it encourages tissue penetration depth by tuning the frequency in the absence of ionizing energy and exposure time. This external stimuli-responsive drug delivery system utilizes nonerodible polymers for enhanced permeation. The main mechanism of action is captivation, which is achieved by pressure-dependent alternating sizing of gas-filled microbubbles that are produced by ultrasonic energy. This system allows the release of drug from the microbubbles by carrier destabilization and accelerated vessel permeability, resulting in the cellular uptake of the drug [9]. The sonochemical method used to prepare starch nanocomposite containing multiwalled carbon nanotube proved to be economical, eco-friendly, effective, and fast method. The nanoparticles formed by this method proved promising for the release of the intended drug and improved the loading capacity as well [18]. The nanoemulsion formed by sonication or ultrasonic method for the delivery of protein drug proved to be efficient and stable. For this study, other parameters were also taken into account such as the concentration of various oil types such as olive oil and sesame oil. The size of the nanoparticles was dependent upon the concentration of the oil because as the oil concentration increased, the particle size decreased. Twenty-five percent olive oil was used as optimized formulation with droplet size 143.1 nm, higher zeta potential (− 33.3 mV), and lower polydispersity index. Hence the nanoemulsion formed by sonication method was spherical as analyzed by transmission electron microscopy (TEM), which is physically stable and proved promising in delivering protein drugs [19]

12.2.2.2 Magnetically responsive drug delivery system

Magnetically responsive drug delivery system is widely used in cancer therapy, biomimetic actuators, separation medias, switches, sensors, artificial muscles, and drug delivery. They utilize magnetite (Fe_3O_4) or maghemite (γ-Fe_2O_3) to form either magnetoliposomes or porous metallic nanocapsules impregnated into the desired formulation. These types of drug delivery systems apply magnetic current to bring about changes in the size or physical shape of the formulation, which are usually in the form of gels [9]. He et al. [21] demonstrated the magnetic reduction-responsive alginate microcapsules developed by thiolated alginate and oleic acid modified Fe_3O_4 nanoparticles via sonochemical method. This preparation proved efficient in delivering hydrophobic drugs intended for targeted drug delivery and was biocompatible and immunogenetic. The encapsulation of super-paramagnetic OA-FE3O4 NPs was confirmed by TEM and vibrating sample magnetometer. These nanoparticles improved and controlled the release of hydrophobic drug, which was helpful in biomedical applications [20]. Another study illustrated that hydrogel formed by alginate-gelatin and Fe_3O_4 magnetic nanoparticles was efficient against inflammatory diseases, namely, cancer. Alginates were partially oxidized to form hydrogels through Schiff-base condensation reaction, and Fe_3O_4 NPs were impregnated in situ chemical coprecipitation approach. The porous microstructure was confirmed by scanning electron microscopy (SEM), and TEM confirmed the formation of Fe_3O_4 NPs throughout the alginate gel with size 25 + 10 nm. The saturation magnetization (δs) value of the alginate-gelatin-Fe_3O_4 found to be 31 emu g^{-1}, which represented proper magnetic property. The obtained alginate-gelatin-Fe_3O_4 loaded the anticancer drug (doxorubicin hydrochloride) efficiently with efficient drug loading and encapsulation efficiencies against Hela cells. The natural matrix formulated from alginate-gelatin-Fe_3O_4 exhibited pH-dependent drug release behavior due to the presence of carboxylic acid groups in the DDS. According to the results, this magnetic hydrogel can be considered as an efficient and "smart" drug delivery for cancer therapy and diagnosis [21]. CUR, an herbal bioactive with plethora of biological activity, specially anti-inflammatory and anticancer activity, was controlled and effectively released when encapsulated in folate receptor targeted hybrid protein inorganic carrier via hybridization with superparamagnetic calcium ferrite. The folate was conjugated to facilitate receptor-mediated endocytosis. The synthesized materials were characterized by using SEM, VSM, XRD, FTIR, DLS, and drug loading, which were confirmed by Taguchi technique. CUR release was thus influenced by magnetic field effect, pH, and concentration adopting magnetic-responsive technique. The drug release rate was higher in acidic conditions, the increased drug concentrations was due to the influence of magnetic field. Biocompatibility and cytotoxicity tests were performed for the synthesized drug carrier systems using L929 murine fibroblast and MCF-7 breast cancer cells, respectively (Fig. 12.8). The IC50 value reduced nearly sixfold for MCF-7 cells and for casein-CUR with folate conjugation in comparison with the carrier without folate. Hence, magnetically synthesized casein-CUR-folate formulation proved potential for targeted drug delivery [26]. Ahmadi et al. reviewed that the permeability of liposomes is increased by alternating magnetic field (AMF) heating to facilitate the sequential release of the drug [27]

12.3 Conclusion

In a nutshell, smart drug delivery systems are very important emerging nanodrug delivery systems, which make use of the internal environment of the diseased site to release drug in a controlled

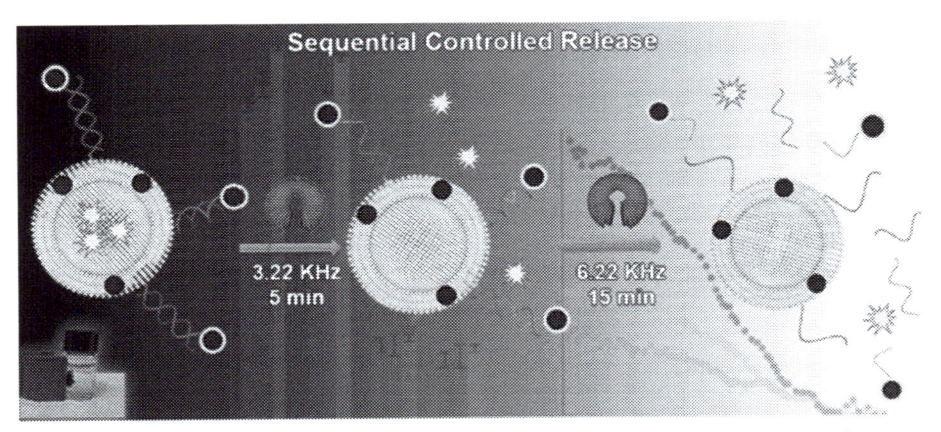

Fig. 12.8 Schematic illustration of multifunctional magnetoliposomes and the sequential release of carboxyfluorescein and therapeutic zipper ON. Carboxyfluorescein was first released after exposure to a 3.22-kHz AMF over a short period of time; subsequently, after applying 6.22-kHz AMF, zipper ON (double-stranded DNA hybridized with zipper) was released. From S. Omidi, M. Pirhayati, A. Kakanejadifard, Co-delivery of doxorubicin and curcumin by a pH-sensitive, injectable, and in situ hydrogel composed of chitosan, graphene, and cellulose nanowhisker, Carbohydr. Polym. 231 (2020) 115745. Copyright Elsevier.

manner. Under the smart drug delivery system, the endogenous stimuli that are commonly studied are redox, enzymes, pH, and temperature, and exogenous stimuli are magnetic- and ultrasonic-responsive drug delivery system. However, exogenous stimuli require adequate force to trigger more release of the drug; hence, coordination with drug accumulation becomes a challenge. For this reason, exogenous stimuli–responsive drug delivery systems are well suited for imaging purposes to monitor the drug distribution and also help to design patient-based or personalized regimens for particular diseases [4]. The smart drug delivery systems facilitate site-specific and timed drug delivery, which becomes a promising tool for natural bioactive molecules to portray their maximum beneficial effects in any inflammatory-related diseased state. More in vivo studies or excessive clinical studies are still to be conducted to establish their safety and efficacy parameters. However, these types of naturally based drug delivery systems are promising candidates for various inflammatory diseases.

References

[1] Y.H. Bae, K. Park, Advanced drug delivery 2020 and beyond: perspectives on the future, Adv. Drug Deliv. Rev. (2020). (in press) S0169-409X(20)30064-8.

[2] N. Zhao, B. Ding, Y. Zhang, J.L. Klockow, K. Laua, F.T. Chin, Z. Cheng, H. Liu, Reactive oxygen species and enzyme dual-responsive biocompatible drug delivery system for targeted tumor therapy, J. Control. Release 324 (2020) 330–340.

[3] B. Tian, Y. Liu, J. Liu, Smart stimuli-responsive drug delivery systems based on cyclodextrin: a review, Carbohydr. Polym. (2020), 116871.

[4] A. Duro-Castano, M. Talelli, G. Rodríguez-Escalona, M.J. Vicent, Smart polymeric nanocarriers for drug delivery, in: M.R. Aguilar, J.S. Román (Eds.), Smart Polymers and Their Applications, Woodhead Publishing in Materials, 2019, pp. 439–479.
[5] M. Mu, M. Ebara, Smart polymers, in: R. Narain (Ed.), Polymer Science and Nanotechnology, Elsevier, 2020.
[6] P. Darvin, A. Chandrasekharan, T.R. Santhosh Kumar, Introduction to smart drug delivery systems, in: A.R. Unnithan, A.R. Kurup Sasikala, C.S. Kim (Eds.), Biomimetic Nanoengineered Materials for Advanced Drug Delivery, Elsevier, Amsterdam, 2019, pp. 1–9.
[7] M. Alsehli, Polymeric nanocarriers as stimuli-responsive systems for targeted tumor (cancer) therapy: recent advances in drug delivery, Saudi Pharm. J. 28 (2020) 255–265.
[8] R.S. Guimarães, C.F. Rodrigues, A.F. Moreira, I.J. Correi, Overview of stimuli-responsive mesoporous organosilica nanocarriers for drug delivery, Pharmacol. Res. 155 (2020), 104742.
[9] S. Indermun, M. Govender, P. Kumar, Y.E. Choonara, V. Pillay, Stimuli-responsive polymers as smart drug delivery systems: Classifications based on carrier type and triggered-release mechanism, in: A.S.H. Makhlouf, N.Y. Abu-Thabit (Eds.), Stimuli Responsive Polymeric Nanocarriers for Drug Delivery Applications, Types and Triggers, Woodhead Publishing Series in Biomaterials, 2018, pp. 43–58.
[10] S. Kim, S. Philippot, S. Fontanay, R. Duval, E. Lamouroux, N. Canilho, A. Pasc, pH- and glutathione-responsive release of curcumin from mesoporous silica nanoparticles coated using tannic acid-Fe(III) complex, RSC Adv. (2015) 1–3.
[11] M. Shahriari, M. Zahiri, K. Abnous, S.M. Taghdisidi, M. Ramezani, M. Alibolandi, Enzyme responsive drug delivery systems in cancer treatment, J. Control. Release 308 (2019) 172–189.
[12] M. Karimi, A. Ghasemi, P.S. Zangabad, R. Rahighi, S.M.M. Basri, H. Mirshekari, M. Amiri, Z.S. Pishabad, A. Aslani, M. Bozorgomid, D. Ghosh, A. Beyzavi, A. Vaseghi, A.R. Aref, L. Haghani, S. Bahramia, M.R. Hamblin, Smart micro/nanoparticles in stimulus-responsive drug/gene delivery systems, Chem. Soc. Rev. 45 (2016) 1457–1501.
[13] S. Moradi, A. Barati, A.E. Tonelli, H. Hamedi, Chitosan-based hydrogels loading with thyme oil cyclodextrin inclusion compounds: from preparation to characterization, Eur. Polym. 122 (2020). 109303S.-B.
[14] Ghaffari, M.-H. Sarrafzadeh, M. Salami, M.R. Khorramizadeh, A pH-sensitive delivery system based on N-succinyl chitosan-ZnO nanoparticles for improving antibacterial and anticancer activities of curcumin, Int. J. Biol. Macromol. 151 (2020) 428–440.
[15] S. Omidi, M. Pirhayati, A. Kakanejadifard, Co-delivery of doxorubicin and curcumin by a pH-sensitive, injectable, and in situ hydrogel composed of chitosan, graphene, and cellulose nanowhisker, Carbohydr. Polym. 231 (2020), 115745.
[16] Y. Bian, D. Guo, Targeted therapy for hepatocellular carcinoma: co-delivery of sorafenib and curcumin using lactosylated pH-responsive nanoparticles, Drug Des. Devel. Ther. 14 (2020) 647–659.
[17] C. Hang, Y. Zou, Y. Zhong, Z. Zhong, F. Meng, NIR and UV-responsive degradable hyaluronic acid nanogels for CD44-targeted and remotely triggered intracellular doxorubicin delivery, Colloids Surf. B: Biointerfaces 158 (2017) 547–555.
[18] H.J. Cho, M. Chung, M.S. Shim, Engineered photo-responsive materials for near infrared-triggered drug delivery—review, J. Ind. Eng. Chem. 31 (2015) 15–25.
[19] S. Mallakpour, L. Khodadadzadeh, Ultrasonic-assisted fabrication of starch/MWCNT-glucose nanocomposites for drug delivery, Ultrason. Sonochem. 40 (2018) 402–409.

[20] S. Mohamadi Saani, J. Abdolalizadeh, S.Z. Heris, Ultrasonic/sonochemical synthesis and evaluation of nanostructured oil in water emulsions for topical delivery of protein drugs, Ultrason. Sonochem. 55 (2019) 86–95.
[21] S. He, S. Zhong, L. Xua, Y. Dou, Z. Li, F. Qiao, Y. Gao, X. Cui, Sonochemical fabrication of magnetic reduction-responsive alginate based microcapsules for drug delivery, Int. J. Biol. Macromol. 155 (2020) 42–49.
[22] R. Jahanban-Esfahlan, H. Derakhshankhah, B. Haghshenas, B. Massoumi, M. Abbasian, M. Jaymand, A bio-inspired magnetic natural hydrogel containing gelatin and alginate as a drug delivery system for cancer chemotherapy, Int. J. Biol. Macromol. 156 (2020) 438–445.
[23] Y. Kato, S. Ozawa, C. Miyamoto, Y. Maehata, A. Suzuki, T. Maeda, Acidic extracellular microenvironment and cancer, Cancer Cell Int. 13 (2013) 1–8.
[24] H.P. James, R. John, A. Alex, K.R. Anoop, Smart polymers for the controlled delivery of drugs – a concise overview, Acta Pharm. Sin. B 4 (2014) 120–127.
[25] S. Pedrosa, C. Gonc, L. David, M. Gama, A novel crosslinked hyaluronic acid nanogel for drug delivery, Macromol. Biosci. 14 (2014) 1–13.
[26] B.K. Purushothaman, M. Harsha S, P. Uma Maheswari, K.M.M.S. Begum, Magnetic assisted curcumin drug delivery using folate receptor targeted hybrid casein-calcium ferrite nanocarrier, J. Drug Deliv. Sci Technol. 52 (2019) 509–520.
[27] S. Ahmadi, N. Rabiee, M. Bagherzadeh, F. Elmi, Y. Fatahi, F. Farjadian, N. Baheiraei, B. Nasseri, M. Rabiee, N.T. Dastjerd, A. Valibeik, M. Karimi, M.R. Hamblin, Stimulus-responsive sequential release systems for drug and gene delivery, Nano Today 34 (2020) 100914.

13

Systems pharmacology and molecular docking strategies prioritize natural molecules as antiinflammatory agents

Anjana S. Nair and Anupam Paliwal
R&D Centre, Aurea Biolabs (P) Ltd, Kolenchery, Cochin, Kerala, India

13.1 Introduction

Inflammation is the body defense mechanism in response to pathogens and injuries. Various inflammatory mediators and many cellular effects are generated during every inflammation process. The large number of chemotactic, vasoactive, and proliferative factors generated at different stages and antiinflammatory action has different targets of action [1, 2]. In the event of inflammation, metabolism of arachidonic acid plays a substantial role. When a cell is activated, phospholipase A2 cleaves arachidonic acid from cell membrane phospholipids and donates to lipoxygenases (LOX) by LOX-activating protein, which then metabolizes arachidonic acids in a series of reactions to leukotrienes. Leukotrienes act as phagocyte chemoattractant and recruiting cells to sites of inflammation of the innate immune system. The pain associated with inflammation is resulted by prostaglandins metabolized by the cyclooxygenase (COX) pathway to prostaglandins and thromboxane A2 or to hydroperoxyeicosatetraenoic acid (HPETEs) and leukotrienes (LTs) by lipoxygenase (LOX) pathway which plays vital roles in variety of inflammatory events as active mediators. Thus through the 5-lipoxygenase (5-LOX) or cyclooxygenase (COX) pathways, leukotrienes and prostaglandins are produced by arachidonic acid cleavage from membrane phospholipids during appropriate inflammatory stimulation [3].

The study of inflammatory associated events has been one of the most rapidly advancing and expanding areas of research in recent years. Inflammatory mediators are crucially involved in the genesis, persistence, and severity of pain following trauma, infection, or nerve

Inflammation and Natural Products. https://doi.org/10.1016/B978-0-12-819218-4.00016-X
© 2021 Elsevier Inc. All rights reserved.

injury. Inflammatory events have also become important due to their strong connection with the neoplastic transformation and cancer progression. This connection is due to one of the most important links, that is, the transcription factor nuclear factor-kappa B (NF-κB). NF-κB is activated by various types of stress, such as proinflammatory cytokine tumor necrosis factor (TNF), viruses, (gamma) radiation, bacterial cell wall components, lipopolysaccharide (LPS), or chemotherapeutic agents. It is seen that NF-κB DNA binding leads to the activation of over 400 genes, many of which lead to a variety of diseases such as Alzheimer's disease and arthritis besides cancer [4]. Other important links between inflammation and cancer are TNF, interleukins (IL-6), and chemokines (IL-8 or CXCL8) [5]. Similarly, LOX pathway plays important role in leukocytes and many immune-competent cells including mast cells, monocytes, neutrophils, eosinophils, and basophils. When inflammation becomes beyond control, it leads to several chronic diseases like cardiovascular diseases, cancer, arthritis, asthma, and type 2 diabetes mellitus [1].

Both nonsteroidal and steroidal antiinflammatory drugs, to relieve inflammatory responses, were effective antiinflammatory agents providing analgesic, antiinflammatory, and antipyretic effects and were widely recommended for long time. The most prominent members of nonsteroidal group are aspirin, naproxen, diclofenac, ibuprofen, and indomethacin. Most of these drugs inhibit the activity of COX-1 and COX-2 and therefore the synthesis of prostaglandins and thromboxanes. Inhibiting COX-2 leads to the desirable analgesic, antiinflammatory, and antipyretic activities, whereas the inhibition of COX-1 leads to undesirable side effects like kidney problems, gastrointestinal bleeding, and central nervous system (CNS) effects [33]. Nonselective inhibition of COX enzyme remains a major adverse effect associated with the available common nonsteroidal antiinflammatory drugs, which has lead a comeback of natural plant products to cure inflammations and its outcomes. Similarly, steroidal class of common antiinflammatory drugs includes various glucocorticoids, which increase the transcription of antiinflammatory cytokines and decrease the transcription of proinflammatory cytokines [6]. Their benefits are limited by variety of systemic side effects and the development of resistance after chronic use. In this context, it is important to find an alternative to these drugs, and there comes the significance of developing new drugs from natural products.

Many plants and herbs such as ginger, turmeric, rosemary, boswellia, and galangal have been shown to exhibit potent antiinflammatory effect. These plant metabolites (bioactives) show excellent antioxidant, antibacterial, antiinflammatory, antimicrobial, and anticancer activities (Table 13.1) [34]. The mechanisms of action of the bioactive compounds from these plants (such as flavonoids and tannins,

Table 13.1 Structure and properties of plant metabolites with potent antiinflammatory effects.

Plant species	Phytochemical	Structure	Properties	Reference
Curcuma longa L.	Curcumin	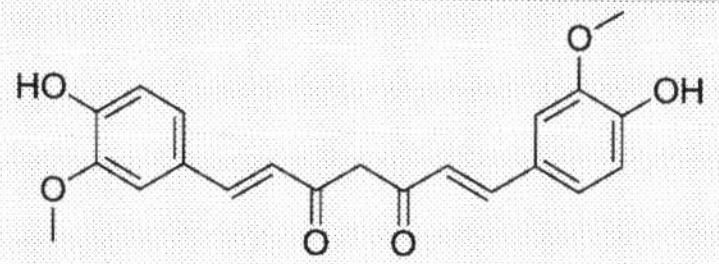	Enzyme inhibitors; coloring agents; antiinflammatory; antineoplastic, and antioxidant activity	[1, 6–9]
	Bisdemethoxycurcumin	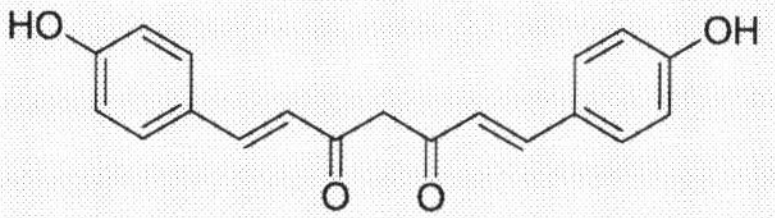	Antiinflammatory; antiproliferative response; antitumor, antioxidant, and antidiabetic activity	[9–12]
	Tetrahydrocurcumin	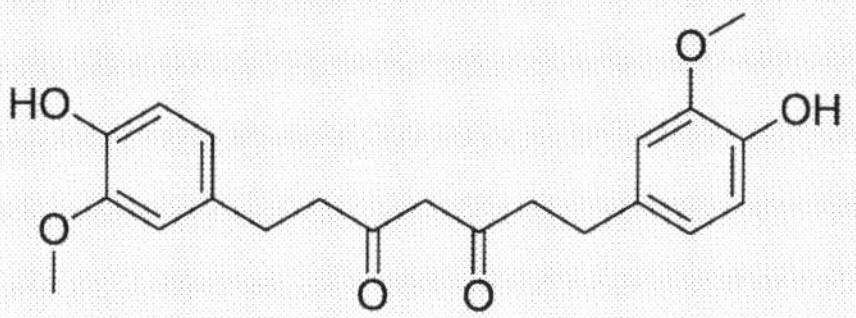	Antiinflammatory; antioxidant; neuroprotection; antihypertensive properties; anticarcinogenic attributes; antidiabetic action and reduction of nephrotoxicity	[10]
	Demethoxycurcumin	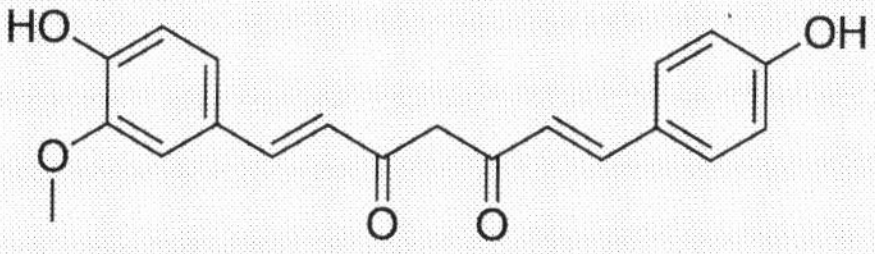	Antiinflammatory, antioxidative, and anticancerous activity	[11]
Zingiber officinale	6-Gingerol	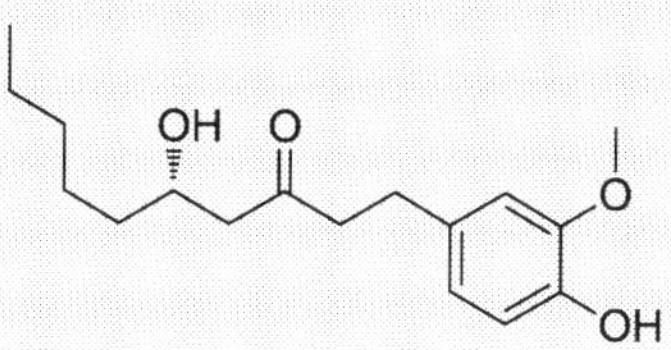	Antiinflammatory, antioxidant, antitumor, antianalgesic, antimicrobial, and hepatoprotective activity	[13–17]
	Zerumbone	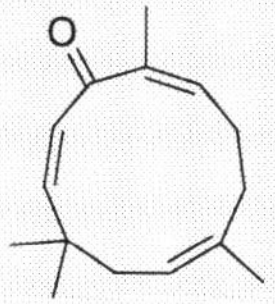	Antiinflammatory, antitumor, and antimicrobial activity	[18]

Continued

Table 13.1 Structure and properties of plant metabolites with potent antiinflammatory effects—cont'd

Plant species	Phytochemical	Structure	Properties	Reference
	Zingiberene	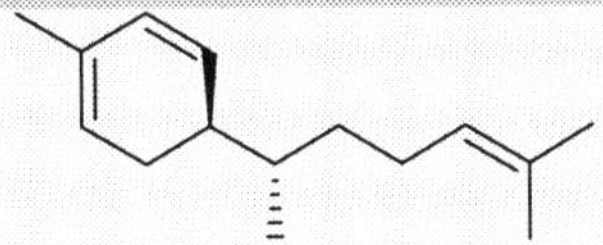	Antiinflammatory, antioxidant, and antinociceptive activity	[18]
	Zingerone	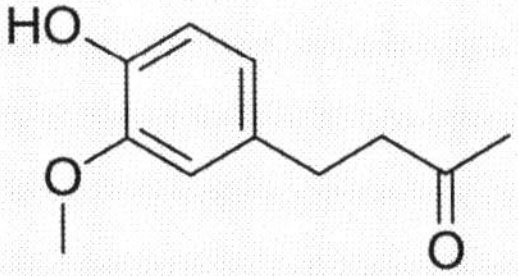	Anticancer, antioxidative, antiemetic, antidiarrheal, and lipolytic activity	[18]
	Gingerdiol	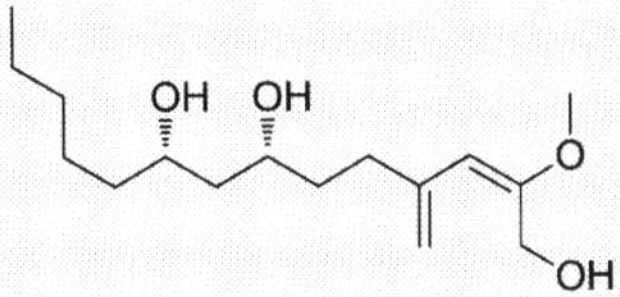	Antimimetic, antithrombotic, and antiinflammatory	[18]
	6-Shogaol	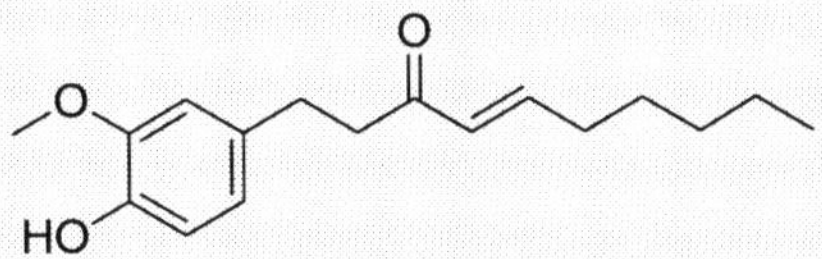	Antioxidant, antiinflammatory, anticancer, antiproliferation, and antiinvasion activity	[14, 15, 18–23]
Boswellia serrata	Boswellic acid	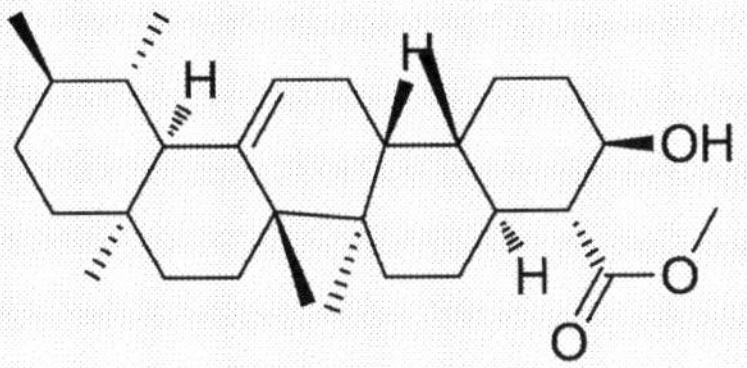	Antiinflammatory and antitumor activity	[24, 25]

	Acetyl-11-keto-β-boswellic acid	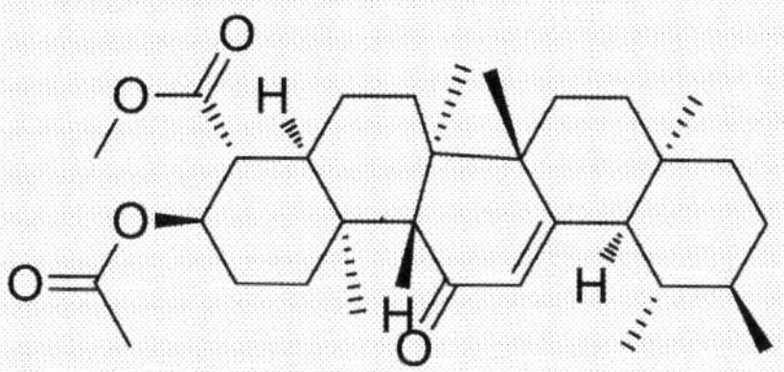	Antiinflammatory and antitumor activity	[94, 118, 119]
	Incensole acetate	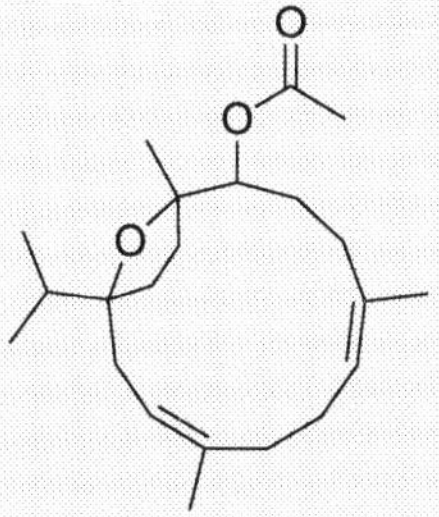	Antiinflammatory activity and antioxidant activity	[26]
Alpinia galanga	Galangin	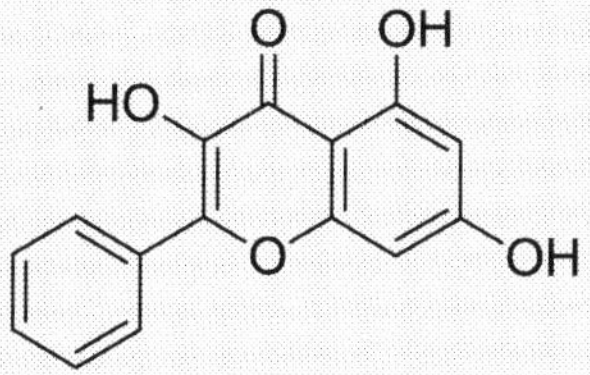	Antiinflammatory, antiproliferative, antimicrobial, anticancer, antioxidant, and enzyme-modulating activity	[27]
Rosmarinus officinalis	Limonene	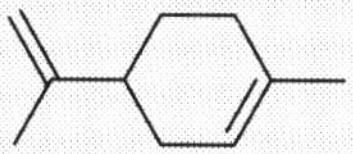	Antiinflammatory, anticancerous, antimicrobial, antioxidant, antiseptic, antinociceptive, insecticidal, and anticatarrhal activity	[28]
	1,8-Cineole	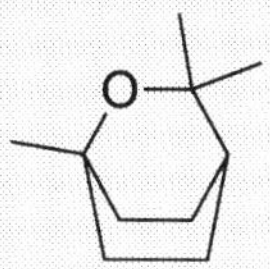	Antiinflammatory; antimicrobial; insect repellents; antitussive and antiinfective activity	[28]

Continued

Table 13.1 Structure and properties of plant metabolites with potent antiinflammatory effects—cont'd

Plant species	Phytochemical	Structure	Properties	Reference
	Camphor	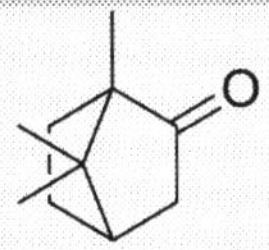	Antiinflammatory, anticancerous, anthelmintic, and antibacterial activity	[28]
	Carnosol	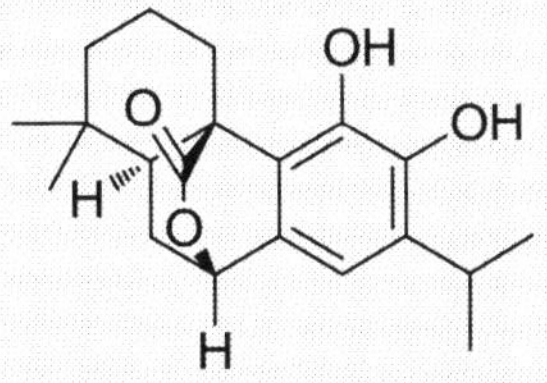 OH, OH, O, O, H, H	Antiinflammatory, anticancer, antiproliferative, and antioxidant activity	[29–31]
	Carnosic acid	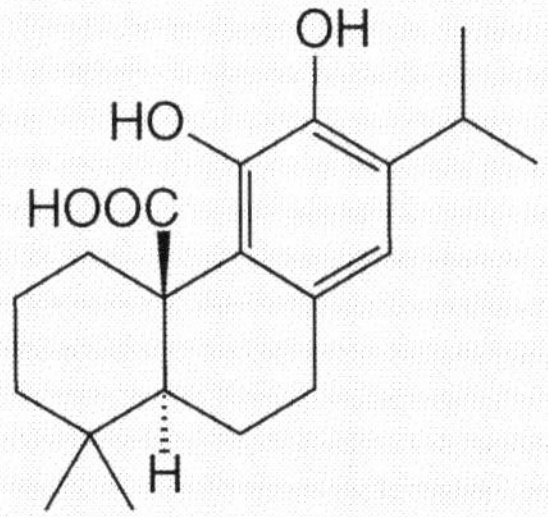 OH, HO, HOOC, H	Antiinflammatory, anticancer, antiproliferative, and antioxidant activity	[31, 32]

saponins, terpenoids, and alkaloids) are thought to be via their free radical scavenging activities or by the inhibition of COX and LOX in the inflammatory cascades [35, 36]. Natural products and phytochemicals have been described and used since centuries in traditional medicine and home remedies. Nowadays, natural products are considered as an important source of lead compounds for drug discovery, and more than 50% of FDA-approved pharmacological classes of drugs are natural products or natural product derivatives [37, 38]. Aspirin, atropine, artemisinin, colchicine, ephedrine, physostigmine, pilocarpine, quinine, quinidine, reserpine, taxol, vincristine, and vinblastine are a few examples of important molecules that medicinal plants have given us in the past. Also, there are many historical examples in which the natural product has not only been the medicinal product but has also helped in revealing novel aspects of pharmacology and physiology [39–41]. In recent research works, more and more efforts are toward construction of structure-activity relationship for unraveling the mechanism of action and molecular target of various natural compounds. In these efforts, the systems-level approach and molecular docking studies are proving instrumental to delineate the different interactions of these natural compounds with different molecular targets in the body. Such structure-activity relationships have effectively been used in the inflammatory field to develop new derivative natural compounds with higher antiinflammatory activity [6].

This chapter highlights on the recent study evidences prioritizing natural molecules (curcumin, bisdemethoxycurcumin, tetrahydrocurcumin, demethoxycurcumin, 6-gingerol, zerumbone, zingiberene, zingerone, gingerdiol, 6-shogaol, boswellic acid, acetyl-11-keto-β-boswellic acid, incensole acetate, galangin, limonene, 1,8-cineole, camphor, carnosol, and carnosic acid) from the commonly available and widely used botanical sources as safe and effective natural antiinflammatory agents.

13.2 Systems pharmacology and antiinflammatory agents

Systems pharmacology is a field of study that uses experimental and computational approaches and provides us with a broad view of drug action rooted in molecular interactions between the target and its drug in the context of such a target, which is interacting with and regulating other cellular components [38]. Systems pharmacology provides powerful new tools and approaches for natural product lead discovery.

In the past, through classical pharmacology approach, a single transduction pathway, connecting the processes on the causal path between drug administration and response, was considered as the basis of drug action [42]. Such one-to-one pathway analysis has yielded many useful drugs that are often taken chronically to control symptoms of a disease. In many instances however, these drugs do not modify the disease process. The focus of ancient-pharmacology approaches, on a single transduction pathway, as the basis of drug action was also reflected in the structure of physiology-based pharmacokinetic-pharmacodynamic (PB-PKPD) models, which were increasingly applied for prediction of drug effects in drug discovery and development [37, 38, 43–46]. Two primary pillars of classic pharmacology are the study and quantification of drug behavior in the body comprising pharmacokinetics (PK; what the body does to the drug), pharmacodynamics (PD; what the drug does to the body), and the receptor hypothesis, the idea that drug action is mediated through binding to specific target molecules (which are usually proteins).

While considering the extremely high cost of new drug development, the high drug attrition rate, and complexity of human body, it became evident that to deep understand the interplay between drugs and complex disease, there is a pressing need to understand the pathophysiology at a systems level [47]. Systems pharmacology emerged as a comprehensive pharmacology approach incorporating systems biology principles in pharmacology to give a clear picture of network of interactions of different drugs in human body. In other words, to investigate biological systems and design therapeutic interventions, systems pharmacology sets out to use quantitative concepts rooted in the synergy between modeling and simulation and large-scale data collection and analysis [43]. The scientific basis of modeling pharmacodynamic interactions and mathematical expressions remain in accordance with receptor theory, which has been developed to account for different mechanisms of interactions like interdependence, summation, allosterism, bliss, modulation, and competition. These play important role in modeling of the interactions in cascading turnover models that are increasingly used to describe complex patterns of pharmacological responses and disease progression [38].

Research in systems pharmacology mainly focuses on the structure or molecular level of organization of the biological network. A network is defined as a series of nodes connected to one another on a defined criterion, wherein the nodes represent different types of objects such as proteins, genes, drugs, and diseases and are also used to specify the state of a system. The biological phenomena are further described as a dynamic process across widely different time scales for the prediction of drug effects in vivo. In systems pharmacology, simultaneous measurements on a large number of variables in response to a perturbation are required for the data sets needed to build networks, such as a drug treatment. Omics experiments (genomics, proteomics, or metabolomics) are used to measure a large number of output variables in response to one or more perturbations [48–50]. For developing a drug, it is important to accompany systems pharmacology by a computational method to accelerate and economize drug discovery and development process.

13.3 Computational methodologies and molecular docking studies with natural compounds

Use of computational methods is an effective strategy in systems pharmacology, leading for rapid and economized drug discovery and development. In recent years, computational drug discovery methods such as molecular docking, pharmacophore modeling and mapping, de novo design, molecular similarity calculation, and sequence-based virtual screening have been greatly improved [51]. These computational approaches are broadly categorized into structure-based, ligand-based, and sequence-based approaches and are mostly used in combinational and hierarchical manner. The structure-based methods rely on the knowledge of the structure of the target macromolecule composed of molecular docking and de novo drug design approaches, which are mainly obtained from crystal structures, NMR data, and homology models. In ligand-based approaches the structure-activity relationship, molecular field analysis, pharmacophore modeling, and 2D or 3D similarity assessment are used to provide crucial insights into the nature of the interactions between drug targets and ligands. However, in sequence-based approaches, bioinformatics methods are applied to analyze and compare multiple sequences and homology within to identify potential targets [52].

Availability of structural data on molecules/targets identity, interaction network and associated detailed information on inflammatory pathways and natural molecules has facilitated many structure-based molecular docking studies focusing on prioritizing natural molecules as antiinflammatory agents. In docking, it involves molecular biology and computer-aided drug design (CADD) strategy to investigate and predict the candidate drug (ligand) interaction at the molecular level by binding to the target protein or receptor, analyzing the energies and interactions involved between them [53]. It is a promising strategy to mimic intermolecular binding modes and interactions. Mainly, molecular docking relies on intermolecular affinity, binding site topology, and interaction of key residues with the ligand [23]. For putative binding modes and affinities of ligands for macromolecules, molecular docking is very useful and reasonably reliable [54]. Over the years the speed and accuracy of molecular docking methods has improved, and these methods now play vital role in structure-based drug design [55–57].

In a typical docking study, thus a network is created by using docking data obtained from a series of ligands tested on multiple targets of a specific disease, and this helps to understand mechanism by providing information about the number and quality of the compound's interactions. The importance of nodes in a network is characterized through degree and centrality parameters. Information concerning all of the bindings between a ligand and a receptor is revealed by a score-weighted docking prediction model. QSAR predictions are also validated by docking studies [58]. Molecular docking is a promising strategy to mimic intermolecular binding modes and interactions. There are many molecular docking softwares available for studying different molecular interactions, namely, AutoDock software, Glide module of Schrodinger Maestro 9.1 software, CDOCKER software, and LibDock software. Docking relies on the concept of target-ligand interactions such as binding energy, electron distribution, geometry complementarity, hydrogen bond donor acceptor properties, polarizability, and hydrophobicity. The final configuration also should be of low energy similar to the molecules in nature, which have a tendency to be found in their low energy form. There is a secured mode of predictions concerning both energies and conformations of target-ligand complexes. Conformers are found ranked according to energy values, and along with structural interaction hypothesis, various ligands are explored by this method [59–63].

13.4 Inflammatory pathway network and key node targets of antiinflammatory agents

13.4.1 Cyclooxygenase-2 (COX-2)

As described before, the COX enzymes (COX-1 and COX-2) are enzymes that catalyze the biosynthesis of prostaglandins, prostacyclins, and thromboxanes from arachidonic acid. The COX-2 selective inhibitors are designed to inhibit COX-2 over COX-1 to obtain desired antiinflammatory activity with minimal gastric toxicity side effect. COX-1 and COX-2 were almost identical, despite the key catalytic site residues of Ile434, His513, and Ile523 in COX-1, are substituted with Val434, Arg513, and Val523 in COX2, which results in increased volume of the COX-2 active site and additional side pocket off the main channel. Several lines of evidence showed that COX-2 is overexpressed in a wide variety of human cancers, such as colon, liver, pancreas, breast, lung, bladder, skin, stomach, head, and neck cancers [64]. The expression and activity of

COX-2 can be downregulated by curcumin both in vitro and in vivo. A study shows that in TPA-treated mouse skin, curcumin potently inhibited COX-2, as well as TPA-stimulated NF-κB activation [6].

13.4.2 Phospholipase A2 (PLA2)

Phospholipase A2 (PLA2) catalyzes the hydrolysis of the ester bond at the sn-2 position of the glycerol backbone of phospholipids to liberate arachidonic acid (AA), a precursor of eicosanoids including prostaglandins (PGs) and leukotrienes (LTs) [65, 66]. They are required to increase the level of arachidonic acid for metabolism and biosynthesis of eicosanoid under physiological condition and in inflammatory cell activation. The catalysis reaction also produces lysophospholipids, another class of lipid mediators [65]. PLA2 superfamily is classified on the basis of structural relationships into several families such as Ca^{2+}-independent PLA2 (iPLA2, also called PNPLA [patatin-like phospholipase]), cytosolic PLA2 (cPLA2), secreted PLA2 (sPLA2), platelet-activating factor acetylhydrolase (PAF-AH), lysosomal PLA2 (LPLA2), PLA/acyltransferase (PLAAT), and α/β-hydrolase (ABHD) families. The secretory PLA2 (sPLA2) family, in which 10 isozymes have been identified, consists of low-molecular-weight, Ca^{2+}-requiring secretory enzymes that have been implicated in a number of biological processes, such as modification of eicosanoid generation, inflammation, host defense, and atherosclerosis [66]. But the platelet-activating factor (PAF) acetyl hydrolase (PAF-AH) family represents a unique group of PLA that contains 4 enzymes exhibiting unusual substrate specificity toward PAF and oxidized phospholipids. It is estimated that the human genome encodes more than 30 (even 50) PLA2s or related enzymes [6].

13.4.3 NF-κB-inducing kinase (NIK)

Nuclear factor (NF)-κB is a group of eukaryotic transcription factors that regulates the expression of gene important for immune responses. There are around five NF-κB families in mammals, namely, p65 (RelA), RelB, p100/p52, p105/p50, and c-Rel. It is regulated using two distinct pathways. The well-studied canonical pathway is mediated by κB kinase complex (IKKα/β/γ) inhibitor leading to the phosphorylation and degradation of inhibitor of κBs (IκBs) [6]. To control transcriptional expression of certain proteins such as cytokines, chemokines, and NF-κB signaling molecules, they activate the canonical and noncanonical NF-κB pathways. By promoting proteolytic processing, NIK activates NF-κB2 and the generation of NF-κB transcription of the targeted gene. NIK is also required in the signaling pathways elicited by other cytokines. Both inflammation-induced and tumor-associated angiogeneses are regulated by NIK [67].

13.4.4 Interleukin-1 receptor-associated kinase-4 (IRAK-4)

IRAK-4 (EC: 2.7.11.1) is a protein kinase involved in signaling innate immune responses from Toll-like receptors. It is also known as renal carcinoma antigen NYREN-64, a proinflammatory cytokine that propagates and amplifies signals. The signaling pathways mediated by IL-1 and other cytokine receptors communicate in various cross-talk mechanisms, and thus inhibition of IL-1 receptor would have profound effects on overall inflammatory responses. Its catalyzes the transfer of a phosphate moiety from ATP molecule to substrate peptide, resulting in a phosphoprotein and ADP molecule. The RNA expression of human IRAK-4 protein is ubiquitous in the cytoplasm. The crucial role of IRAK-4 in Toll/interleukin-1 (IL-1) receptor (TIR) signaling and its ubiquitous expression in various cells play a pivotal role in the development of novel therapeutics to trigger innate immune actions and combat various related pathological conditions [6].

13.4.5 Dihydrofolate reductase (DHFR)

DHFR is a member of the reductase enzyme family, which is ubiquitously expressed in all organisms and crucial for cell growth and cell proliferation. DHFR reduces dihydrofolate to tetrahydrofolate using NADPH as electron donor. Tetrahydrofolate and its derivatives serve as 1-C donors in purine synthesis and nucleic acid synthesis, essential for cell growth and cell proliferation. Since then, DHFR has been a target for antiinflammatory and anticancer therapeutic agents, which have been developed to target this key enzyme. A classic example is methotrexate, which is a folate analogue and used to inhibit DHFR to finally inhibit proliferation of the lymphocytes and other cells responsible for inflammation in the joint. The amino acid residues Ile-7, Leu-22, Phe-31, Phe-34, Arg-70, Val-115, and Tyr121 comprise the active site of DHFR. The NMR structure of DHFR shows an eight-stranded β-pleated sheet in the center of the molecule. DHFR inhibition is essential to the action of antifolate medications used to treat cancer and some inflammatory diseases [68].

13.4.6 5-Lipooxygenase (5-LOX)

5-LOX enzymes are enzymes predominant to convert endogenous arachidonic acid to 5-hydroxyeicosatetraenoic acid (5-HETE) and leukotrienes. They are responsible for vasoconstriction, bronchospasm, increased permeability, and chemotaxis. Leukotrienes are potent mediators of inflammation and allergy, which play roles in bronchial asthma, rheumatoid arthritis, and psoriasis and are also likely to be involved in the pathogenesis of inflammatory bowel diseases. Calcium stimulates both the oxygenase and the LTA4 synthase activities, whereby 5-LOX catalyzes the oxidation of arachidonic acid at the 5-position to yield 5-hydroperoxyeicosatetraenoic acid (5-HPETE) and then the conversion of 5-HPETE to leukotriene (LT) A4. LTA4 is then metabolized to various other leukotriene cleavage products (LTB4, LTC4, LTD4, and LTE4), which exert generally proinflammatory effects [69].

13.5 Prioritization strategy and systems pharmacology approach for screening of potential therapeutic agents

Prioritization through utilization of systems pharmacological techniques and molecular docking remains the most economic and rapid process in drug candidate discovery today [70]. The focus here shall be to highlight in brief the entire prioritization process, which includes the selection of targets, systems-level large-scale "omics" data analysis for target prioritization, computational approaches of molecular docking, database-mediated compound target/activity prioritization, toxicity risk analysis, and drug score assessment.

13.5.1 Target prediction and pathway enrichment analysis

For prioritizing a potential drug molecule, it is important to predict the target and have a detailed pathway enrichment analysis. There are different softwares used for searching target molecules such as PharmMapper, which is designed to identify potential target candidates for the given molecules (natural molecules, drugs, or other newly discovered compounds with unidentified targets) via a reverse pharmacophore mapping approach. DAVID (Database for Annotation, Visualization, and Integrated Discovery) or alike bioinformatics online tools are further used to dissect associated pathways. The smaller densely connected modules, nodes, or seed proteins within the module are likely to work collectively in a biological pathway. The possible targets of compounds can also be predicted via structure likeness. Among them, two potential web servers that use structural similarity scores to find the possible targets are SwissTargetPrediction (http://www.swisstargetprediction.ch) and

SuperPred (http://prediction.charite.de.) [37]. SwissTargetPrediction web server predicts the target of bioactive molecules based on a combination of 2D and 3D similarity measures with known ligands, and SuperPred web server connects chemical similarity (2D structure) of drug-like compounds with molecular targets [71, 72].

The compound targets and relevant pharmacological information of natural molecules are obtained via database mining of important databases such as Protein Data Bank (PDB), Herbal Ingredients' Targets Database (HIT) (http://lifecenter.sgst.cn/hit/), Traditional Chinese Medicine Integrated Database (TCMID) (http://www.megabionet.org/tcmid/), and Traditional Chinese Medicine Systems Pharmacology Database and Analysis Platform (TCMSP) (http://sm.nwsuaf.edu.cn/lsp/tcmsp.php), and their result shows that all the compounds we are focusing can be prioritized as antiinflammatory agents [73–75]. In general the action mechanism of natural molecules is described as multiple target interaction. Unlike other synthetic drugs a single natural molecule deals with different molecular targets of inflammation. The network-based computational approach for drug target identification is not restricted to the single target protein structure. TCMID is an integrative database that contains data of herbal ingredients, herbal targets, disease-related gene or proteins, drugs, and their targets, many of which were collected through text mining.

13.5.2 Systems biology approach for target prioritization and molecular docking

The key step for unraveling the pharmacological efficacy of a natural molecule is the prediction of interaction or inhibitory potential of compounds with valid or experimentally proven systems-level potential targets. Reanalysis of large-scale data sets (generated from microarray, proteomics, or metabolomics studies) is performed to explore the potential target components of important pathways involved in inflammatory remodeling events apart from conventional targets. To explore the putative binding site of compounds on prioritized target structure, specific docking approaches are used. Among them, AutoDock Vina in PyRx 0.8 is an easy-to-use user interface that significantly improves the speed and accuracy of docking with a new scoring function, efficient optimization, and multithreading [76]. LigPlot plus version 1.4 (http://www.ebi.ac.uk/thornton srv/software/LigPlus) is used for the 2D visualization of ligand-receptor interaction [77]. The Catalytic Site Atlas (CSA) (http://www.ebi.ac.uk/thornton-srv/databases/CSA) [78] is a software used to retrieve the information of enzyme catalytic site residues to explore the possible orthosteric (binding at the active site) and allosteric (binding elsewhere) binding mode of ligands with receptor. Reverse docking is a viable direction in the network exploration of herbal medicine and in systematic understanding of drug pharmacology and toxicology [79].

13.5.3 Database-mediated compound target/activity prioritization

The compound targets and relevant pharmacological information of natural molecules thus obtained via previously described approaches provide compounds that can be prioritized as antiinflammatory agents. In general the action mechanism of natural molecules is described as multiple target interaction. Unlike other synthetic drugs a single natural molecule may deal with different molecular targets of inflammation. Such virtual screening method is essential as a complement to experimental techniques for rapid screening and predicting the location of functional binding pockets of ligands on targets [80–84].

13.6 Natural compounds and their interactions in the inflammatory pathway network and prioritization as antiinflammatory agents

There are multiple interaction studies using natural molecules, highlighting their significant antiinflammatory activities and interactions with the inflammatory pathway molecules. The ability of certain potent/common natural molecules to act against multiple targets of inflammation has been studied in detail including systems pharmacology approaches, enabling on prioritization of such natural molecules as antiinflammatory agents.

13.6.1 Turmeric bioactives targeting the inflammatory pathway network

Curcumin is a highly pleiotropic molecule in turmeric that interacts with multiple molecular targets in inflammatory pathway. There are many studies indicating that curcumin selectively inhibits COX-2 in a dose- and time-dependent manner [85]. Curcumin exerts this effect by directly targeting COX-2 and prostaglandin (PG) production and by upregulating 5′-adenosine monophosphate-activated protein kinase (AMPK) that further leads to a suppression of COX-2 production [86]. Moreover, curcumin also prevents biosynthesis of prostaglandin E2 (PGE2) from prostaglandin H2 (PGH2) [87].

Zhang et al. investigated whether curcumin inhibited chenodeoxycholate (CD) or phorbol ester (phorbol 12-myristate 13-acetate, PMA) mediated induction of COX-2 in several gastrointestinal cell lines (IEC-18 SK-GT-4, SCC450, and HCA7) [88]. Treatment with curcumin suppressed CD- and PMA-mediated induction of COX-2 protein, induction of COX-2 mRNA, and synthesis of prostaglandin E2. Kim et al. demonstrated that the inhibitory action of curcumin on Janus kinase (JAK)-STAT signaling could contribute to its antiinflammatory activity in the brain. The study was performed using primary microglia and murine BV2 microglial cells from rat brain, wherein both the LPS and interferon (IFN-γ)-stimulated induction of COX-2 was effectively suppressed by curcumin. Curcumin attenuated the inflammatory response of brain microglial cells by inhibiting the phosphorylation of STAT1 and STAT 3 and JAK1 and JAK 2 in microglia activated with gangliosides, LPS, or IFN-gamma [88–90]. Interestingly the COX inhibitory activity of curcumin remains concentration dependent. In inhibitory activity studies, using COX-1-functionalized magnetic nanoparticles and curcumin, at 30 uM concentration, curcumin inhibited 59% of COX-1 activity and 20% of COX-2. These inhibitory activities were further confirmed using computational structural data with the docking studies. Molecular docking studies were done to understand the ligand-protein interactions and COX-2 selectivity in detail. The crystal structures of COX-2 enzyme complexes with curcumin were used for docking. The most favorable conformation resulted from the docking of curcumin into the active site of COX-2 is similar to that experimentally found for the COX-2 substrate arachidonic acid [88]. The docking protocol predicted the same conformation as was present in the crystal structure. It further suggested that a better activity and stability is seen when pharmacophore modification of the dienone functional group into monoketone and side chain of aromatic rings with symmetrical or asymmetrical substituents is done [91].

Antiinflammatory activity of curcumin is also related to its ability to inhibit cellular gene expression regulated by transcription factors NF-κB, activator protein-1 (AP-1), and early growth response protein 1 (EGR-1). Curcumin blocks cytokine-mediated NF-κB activation and proinflammatory gene expression by inhibiting inhibitory factor I-κB kinase (IKK) activity. Curcumin was also shown to inhibit allergic encephalomyelitis by blocking IL-12 signaling through JAK-STAT pathway in T lymphocytes [92, 93]. Gupta et al. demonstrated that curcumin inhibited TNF-α-induced expression of adhesion molecules such as vascular cell adhesion molecule 1 (VCAM-1), intercellular adhesion molecule 1 (ICAM-1), and endothelial leukocyte adhesion molecule 1 (E-selectin) on human umbilical vein endothelial cells [94]. Since diferuloylmethane significantly blocks the cytokine-induced transcript levels for the leukocyte adhesion molecules, at an early stage itself, it may be interfering the signaling event induced by TNF-α. Curcumin produced significant inhibition of interleukin-1β (IL-1β) and IL-8 but minimal inhibition of TNF-α expression at 20 μM concentrations. It significantly inhibited adult peripheral blood mononuclear cell expression of IL-8 at 20 μM concentrations [95]. Bisdemethoxycurcumin A (BDMC-A), an analogue of curcumin, inhibits markers of invasion, angiogenesis, and metastasis in breast cancer cells via NF-κB pathway [7, 8]. Curcumin potentially inhibits LPS-induced inflammation mediated through reactive oxygen species/Toll-like receptor 4-mitogen-activated protein kinase (ROS/TLR4-MAPK)/NF-κB pathway in rat vascular smooth muscle cells, RAW 264.7 cells. It suppressed the induction of proinflammatory transcription factors NF-κB and AP-1, signal transducer, activators of signal transducer and activator of transcription (STAT) proteins, and Wnt/β-catenin and also downregulated the expression of NF-κB-regulated gene products such as TNF-α, monocyte chemoattractant protein-1 (MCP1), inducible nitric oxide synthase (iNOS), IL-1, IL-6, IL-8, matrix metalloproteinase-2 (MMP-2), and matrix metalloproteinase-9 (MMP-9) in in vivo model [8]. Docking analysis studies further proved the efficiency of curcumin in inhibiting the molecular targets of inflammation.

Single ligand docking of curcumin at the active pocket of DHFR, IL-6, PLA2, and COX-2 is shown in Fig. 13.1. Docking study details and the identified binding score or target/affinity details are given in Table 13.2. The antiinflammatory activity of curcumin was found associated with the downregulation/inhibition of key inflammatory network candidates' mRNA expression and protein expression. Docking receptor was prioritized as the factor deregulating expression of few important genes responsible for the enhancement and progression of LPS-induced inflammatory pathways. The outcome explicated the interacting mode of curcumin with the active amino acid residues at the binding site in the active groove of target protein molecules: COX-2, IL-6, and transforming growth factor beta (TGF-β) [8, 100]. Docking studies of curcumin on bovine α-lactalbumin indicated that Trp-118 has the most contribution to the process of curcumin binding. The strong interactions of curcumin with bovine α-lactalbumin suggested that this protein can act as a suitable carrier of these bioactive compounds. Protein-ligand molecular docking to evaluate binding mode and interaction energy of the curcumin toward six major enzymatic targets of skin disorders also highlighted its potential to downregulate the inflammatory reactions [88]. Curcumin was docked in the active site of six enzymes, namely, phosphodiesterase type 1 (PDE1), protein kinase B (AKT), protein kinase C delta (PKCδ), phosphorylase kinase (PhK), COX-2, and phosphoinositide 3-kinase (PI3K). The curcumin molecule docked in the active sites remained fully stabilized by several hydrophobic contacts and hydrogen bonds established with the active site residues [101].

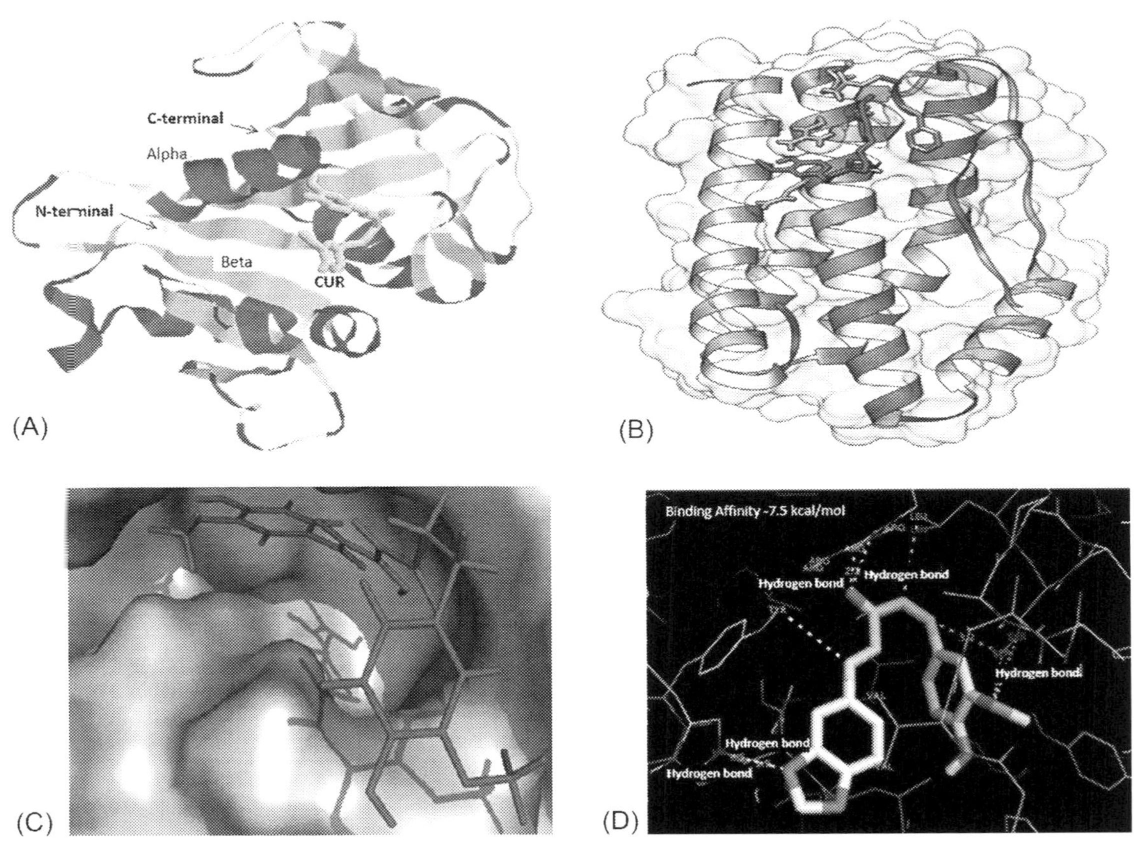

Fig. 13.1 Docking studies indicating curcumin's (natural bioactive principle from turmeric root with antiinflammatory activity) interaction with key target molecules in the inflammatory pathway: interaction of curcumin with DHFR, IL-6, PLA2, and COX-2. (A) A structural rendering of the docked CUR-DHFRA complex showing CUR (cyan, grey in print version) in the active site of DHFRA. Secondary structural features of DHFRA are shown in standard color scheme (alpha, red (dark grey in print version); beta, yellow (grey in print version)) [96]. (B) Docking of curcumin at the active pocket of IL-6 [8]. (C) Binding mode of the curcumin at the active site of PLA2 [97]. (D) Docking of curcumin with COX-2 [6].

Curcumin also docked with dihydrofolate reductase (enzyme crucial for cell proliferation and cell growth). Several significant interactions are revealed in the docking study of complex of curcumin with DHFRA. It was found that curcumin binds to DHFR with an affinity comparable with that of methotrexate, an established anticancer drug targeting DHFR. It was concluded to be due to the flexibility in its structure, which allows curcumin to bind in a bent conformation in the active site of DHFR, thus optimizing interactions on either side of the active site pocket [96].

Table 13.2 Summary of in silico docking studies, indicating on interaction of the prioritized natural molecules with molecular targets in the inflammatory pathway network.

Plant species	Phytochemical	Molecular targets	Molecular docking software	Docking score/binding energy (kcal/mol)	Reference
Curcuma longa L.	Curcumin	DHFR	AutoDock 4 0.2	– 9.02	[96]
		COX-2	AutoDock Vina	– 8.9	[6]
		PLA2	Schrödinger Maestro 9.1 software package	– 4.32	[97]
			AutoDock Vina	– 8.4	[6]
		NIK	AutoDock Vina	– 9.0	[6]
		IRAK-4	AutoDock Vina	– 9.3	[6]
		NF-κB	Discovery Studio v3.1.	95.696	[7]
		IL6	AutoDock 4.2	– 4.49	[8]
		TGF-β	AutoDock 4.0	– 5.61	[8]
	Bisdemethoxycurcumin	MMPs	AutoDock 3.0	– 11.46	[9]
		1AAP	Auto Dock 4.0	– 13.11	[10]
		HPV 16 E6	AutoDock 4.0.	– 4.09	[11]
		TTR	AutoDock 4.0.	– 4.76	[12]
	Tetrahydrocurcumin	1AAP	Auto Dock 4.0	– 12.03	[10]
	Demethoxycurcumin	HPV 16 E6	AutoDock 4.0.	– 3.81	[11]
Zingiber officinale	6-Gingerol	NF-κB	C-DOCKER	– 76.0004	[15]
		AP-1	C-DOCKER	– 138.2092	[15]
		EGFR	C-DOCKER	– 107.9914	[15]
		C-Met	C-DOCKER	– 66.7825	[15]
		PI3K	C-DOCKER	– 83.9303	[15]
		COX-2	C-DOCKER	– 61.3134	[15]
		TRPV1	AutoDock Vina	– 6.44	[16]
		DNA	AutoDock 4.2	– 21.91	[13]
		ERα	AutoDock Tools (ADT)	– 6.59	[14]
		SdiA protein-4Y13-S	Dock Ligands tool	– 61.34	[17]

Continued

Table 13.2 Summary of in silico docking studies, indicating on interaction of the prioritized natural molecules with molecular targets in the inflammatory pathway network—cont'd

Plant species	Phytochemical	Molecular targets	Molecular docking software	Docking score/binding energy (kcal/mol)	Reference
	Zerumbone	Bax	YASARA	6.631	[18]
		Bcl2	YASARA	6	[18]
		IL-6	YASARA	6.38	[18]
		TNF-α	YASARA	5.852	[18]
	Zingiberene	Bax	YASARA	6.619	[18]
	NF-κB		YASARA	5.574	[18]
	Zingiberol	COX	YASARA	6.297	[18]
		Bcl2	YASARA	5.803	[18]
		Vegf	YASARA	4.897	[18]
	Gingerdiol	NF-κB	YASARA	5.591	[18]
	6-Shogaol	COX	YASARA	4.515	[18]
		Vegf	YASARA	6.99	[18]
		ERα	AutoDock Tools	– 5.7	[14]
		TRPV1	AutoDock Vina	– 7.1	[19]
		EGFR	C-DOCKER	– 107.9644	[15]
		C-Met	C-DOCKER	– 107.9644	[15]
		PI3K	C-DOCKER	– 42.7721	[15]
		COX-2	C-DOCKER	– 38.7325	[15]
		NF-κB	C-DOCKER	– 40.7234	[15]
			LigandFit docking	41.27	[20]
		AP-1	C-DOCKER	– 117.683	[18]
		HSA	AutoDock 4.2	– 23.27	[21]
		bcl-2	Surflex-Dock	6.29	[22]
		SaHPPK	GOLD	55.48	[23]

Boswellia serrata	Boswellic acid	5-LOX	Gromacs software version 5.1.2	− 2.37	[24]
		COX-1	GOLD 3.1.1	37.3	[25]
	Acetyl-11-keto-β-boswellic acid	Pin1	Discovery Studio 3.0		[98]
		CatG	GOLD 3.1.1	44.1	[99]
		COX-1	GOLD 3.1.1	14.5	[25]
	Incensole acetate	EcAspTA	MolegroVirtualDockerv. 6.0.1	− 106.9	[26]
		BcPhzA/B	MolegroVirtualDockerv. 6.0.2	− 103.7	[26]
Alpinia galanga	Galangin	COX-2	Schrödinger molecular modeling software	− 11.33	[27]
Rosmarinus officinalis	Limonene	COX-1	GOLD	50.14	[28]
		COX-2	GOLD	45.85	[28]
		PGI-2	GOLD	38.21	[28]
	1,8-Cineole	COX-1	GOLD	36.3	[28]
		COX-2	GOLD	37.76	[28]
		PGI-2	GOLD	33.71	[28]
	Camphor	COX-1	GOLD	33.49	[28]
		COX-2	GOLD	34.38	[28]
		PGI-2	GOLD	36.13	[28]
	Carnosol	Nrf2-Keap1 PPI	C-DOCKER	–	[29]
		RSK2	Schrödinger Suite	–	[30]
		DPP-4	MOE software	− 4.569	[31]
	Carnosic acid	DPP-4	MOE software	− 4.633	[31]
		α-Amylase	Discovery Studio 3.5	− 48.53	[32]
		α-Glucosidase	Discovery Studio 3.5	− 40.2	[32]

There are many studies showing the ability of BDMC to target MMPs, Alzheimer's disease amyloid a4 protein (1AAP), human papillomavirus (HPV 16 E6), and transthyretin (TTR) [9–12] (Table 13.2). For the treatment of inflammation-related diseases, compounds inhibiting PLA2 have been implicated as potential therapeutic agents. Dileep et al. in their study found that curcumin analogues, namely, rosmarinic acid, tetrahydrocurcumin, dihydrocurcumin, and hexahydrocurcumin, can effectively block the interaction of original substrate with the target molecule catalytic site residues and thus explain the antiinflammatory activity of curcumin and its analogues [10, 11, 97].

13.6.2 Ginger bioactives targeting the inflammatory pathway network

The antiinflammatory property of ginger has been known for centuries, but a good deal of recent scientific evidence has further confirmed its antiinflammatory property in humans. Gingerols are the major antiinflammatory bioactives present in the ginger tuber. The antiinflammatory properties of 6-gingerols were evidenced by their inhibitory effects on prostaglandin and leukotriene synthesis and by mimicking dual-acting nonsteroidal antiinflammatory drugs [102]. Different docking studies reveal that 6-gingerols have the potential for protecting human body from severity of inflammation by acting against NF-κB, AP-1, epidermal growth factor receptor (EGFR), tyrosine-protein kinase Met (C-Met), PI3K, COX-2, transient receptor potential cation channel subfamily V member 1 (TRPV1), deoxyribonucleic acid (DNA), estrogen receptor α (ERα), and SdiA protein-4Y13-S (Table 13.2) [13–17]. The mechanism of the antiinflammatory action of ginger is also due to dual inhibition of COX and 5-lipoxygenase, enzymes essential for arachidonate metabolism, and then extended to downregulation of the induction of inflammatory genes [103]. It has been reported that ginger extract inhibits the production of nitric oxide (NO) and proinflammatory cytokines in LPS-stimulated BV-2 microglial cells via the NF-κB pathway. Ginger can modulate pathophysiological pathways activated in chronic inflammation. The major in vitro inhibitors of 5-LOX are due to the two labdanum-diterpene-like dialdehydes isolated from ginger extracts. Various experimental reports have evidenced that ginger extract is capable of inhibiting targets such as COX-1, COX-2, AChE, BuChE, JNK, and NOS [104].

Ginger extract is used for treatment of Alzheimer's disease. Its molecular complexity leads to the concept of multitargeted therapies, which are becoming increasingly important for long term, since they maximize the therapeutic effect and overcome the adverse effects associated with combination therapy [105]. The natural active compounds gingerols and zerumbone were found to be potent inhibitors for NF-κB and proinflammatory cytokine TNF-α. Ginger may block any one or more steps in the NF-κB signaling pathway, such as DNA binding of dimers or interactions with the basal transcriptional machinery, the signals that activate the NF-κB signaling cascade or translocation of NF-κB into the nucleus. The characterization of the pharmacological properties of ginger entered a new phase with the discovery of a ginger extract derived from *Zingiber officinale*. It inhibits the induction of several genes involved in the inflammatory response such as genes encoding for cytokines, chemokines, and the inducible enzyme COX-2. Ginger extract significantly reduced the elevated expression of NF-κB and TNF-α in rats with liver cancer. Through the suppression of the proinflammatory TNF-α, ginger acts as an anticancer and antiinflammatory agent by inactivating NF-κB [5].

Topically applied 6-gingerol renders NF-κB transcriptionally inactive by inhibiting 12-O-tetradecanoylphorbol-13-acetate (TPA)-induced phosphorylation of p65 at Ser 536 and its interaction with the coactivator cAMP response element binding protein-binding protein (CBP/p300) in mouse skin. This effect of 6-gingerol appears to be associated with inhibition of p38 MAPK. Topical application of 6-gingerol or 6-paradol before TPA attenuated 7,12-dimethylbenz[a]anthracene initiated skin papillomagenesis in female ICR mice [5, 106]. It also inhibited anchorage-independent growth of mouse epidermal JB-6 cells stimulated with epidermal growth factor and has been shown to inhibit UVB-induced activation of NF-κB and COX-2 expression in hairless mouse skin and also in an immortalized human keratinocyte cell line [107].

In another study by Lee et al., 6-gingerol suppresses iNOS expression by inhibiting IkBα phosphorylation in TLR4 agonist LPS-activated macrophages. It also inhibits COX-2 expression by blocking p38 MAPK-mediated phosphorylation of IkBα or the NF-κB p65 subunit in phorbol ester-treated mouse skin. Recently, 6-gingerol has been shown to inhibit UVB-induced activation of NF-κB and COX-2 expression in hairless mouse skin and also in an immortalized human keratinocyte cell line [103]. Oyagbemi et al. studied the activity of 6-gingerol and concluded that it decreases inducible nitric oxide synthase and TNF-α expression through suppression of IkBα phosphorylation, NF-κB nuclear activation, and PKC-alpha translocation, which in LPS-stimulated macrophages inhibits Ca2þ mobilization and disruption of mitochondrial membrane potential. It also inhibited the TNF-α production, tumor promoter-stimulated inflammation, and activation of epidermal ornithine decarboxylase in mice and the superoxide production stimulated by TPA in differentiated HL-60 cells. Additionally, 6-shogaol from ginger also suppresses NF-κB-regulated gene expression of iNOS and COX-2 by activation of PI3K, p44/42 MAPK, and IKKβ. 6-Shogaol inhibits LPS-induced dimerization of

TLR4, resulting in the inhibition of NF-κB activation and expression of COX-2 [5]. Owing these antiinflammatory effects, 6-shogaol also significantly proved as neuroprotective agent in vivo in transient global ischemia via the inhibition of microglia. Microglial activation induced by LPS is suppressed by both in primary cortical neuron-glia culture and in an in vivo neuroinflammatory model [104]. These activities of 6-shogaol is supported by different docking studies, where it efficiently docks COX, vascular endothelial growth factor (VEGF), ERα, TRPV1, EGFR, C-Met, PI3K, COX-2, NF-κB, AP-1, B-cell lymphoma 2 (bcl-2), and 6-hydroxymethyl-7,8-dihydropterin pyrophosphokinase (SaHPPK) (Table 13.2) [14, 15, 18–23].

Interestingly, 1-dehydro-10-gingerdione (D10G) in ginger appears to be a more effective compound than 6-shogaol and other pungent constituents that inhibit the production of NO in LPS-activated macrophages. D10G inhibited the catalytic activity of IKKβ by interacting directly with Cys179 in the activation loop of IKKβ. Its ability to inhibit IKKβ-catalyzed IkBα phosphorylation became a converging step in the NF-κB activating pathways in macrophages mediated by TLR2/6, TLR4, or TLR5. It sequentially suppresses NF-κB-regulated expression of inflammatory genes such as iNOS, COX-2, or IL-6 in the cells. D10G interaction with cysteine 179 (Cys179) in the activation loop of IKKβ was a good evidence of its selective inhibition of in vitro kinase activities of wild-type IKKβ proteins [109].

In a study by Rashmi et al., major constituents of ginger are studied for its antiinflammatory activity. Docking study revealed that compounds gingerol, shogaol, zerumbone, zingiberol, zingiberene, zingerone, and zingerdiol were found to have strong binding affinity

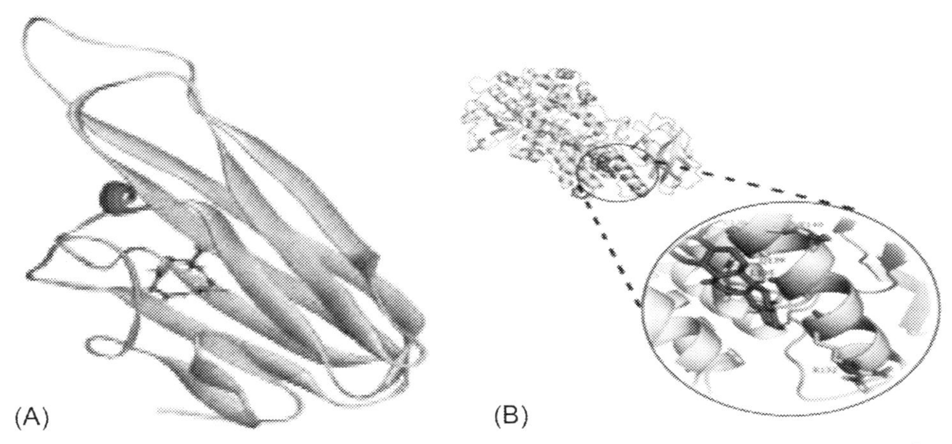

Fig. 13.2 Docking studies indicating on interactions of ginger and galangal bioactives with key target molecules in the inflammatory pathway: 3D representation of the best pose of interaction of zerumbone, 6-shogaol, and AKBA. (A) Docking of zerumbone at the active pocket of IL-6 [18]. (B) Molecular interaction of AKBA and 5-LOX [108].

toward major molecular targets and these compounds can be further investigated in vitro and in vivo to develop novel chemical scaffold on which further derivatization can be done for further optimization of its antiinflammatory activity. Docking of Zerumbone in active pocket of IL-6 is depicted in Fig. 13.2A [18]. Another study by Haris et al. deals with molecular docking and molecular dynamics (MD) studies that provide a detailed understanding on the interaction of 6-gingerol binding in the minor groove of DNA [13]. 6-Gingerol has high level of interaction with ERα active site in terms of hydrogen bonding, whereas hydrophobic interactions are observed with both 6-gingerol and 6-shogaol. Both ginger bioactive compounds pose low potential as substitute in comparison with tamoxifen against ERα [14].

13.6.3 Boswellia bioactives targeting the inflammatory pathway network

Boswellia species contain more than 200 bioactive compounds [110]. Among them, boswellic acid (BA) is shown to have appreciable antiinflammatory activity. The major targets of BA are the members of the arachidonic acid cascade, proteolytic enzymes, oxygen radicals, and the complement system. BA exerts antiinflammatory activity by inhibition of NF-κB, TNF-α, IL-6, IL-2, interferon gamma (IFN-γ), 5-LOX, and COX-1 [24, 25, 98, 111]. Boswellic acids, especially 3-acetyl-11-keto-β-boswellic acid (AKBA), inhibited peptidyl-prolyl cis-trans isomerase (Pin1) and cathepsin G (CatG) (Table 13.2) [25, 98] and COX-1 product formation in intact human platelets. They also inhibited the activity of isolated COX-1 enzyme in cell-free assays. This reversible effect is impaired by increased levels of arachidonic acid as substrate of COX-1 [25]. Molecular docking of BA into x-ray structures of COX-1 yielded positive ChemScore values for BA, indicating favorable binding to the active site of the enzyme. But BA exhibited weak binding with COX-2 and inhibited its activity less efficiently [110]. AKBA and acetyl-β-boswellic acid (AβBA) also inhibited constitutively activated NF-κB signaling by attenuating the activity of the major regulatory enzyme in the NF-κB pathway, IKK. In LPS-stimulated human peripheral monocytes, acetyl-α-boswellic acid (AαBA) and AKBA downregulate the TNF-α expression [112].

AKBA also inhibit 5-LOX, which is the producer of leukotrienes, which are responsible for plasma exudation, edema production, chemotaxis, and activation of white blood cells that release proteolytic enzymes and oxygen radicals (Fig. 13.2B). When compared with COX, 5-LOX is more sensitive to AKBA. AKBA is a novel specific nonredox inhibitor of 5-LOX as it neither impairs the cyclooxygenase and 12-lipoxygenase enzyme properties nor inhibits the peroxidation of arachidonic acid [108]. BAs directly inhibit proteolytic enzymes and oxygen radical formation in polymorphonuclear neutrophil leukocyte (PMNL) and also the complement system, which is an important link between immune and inflammatory reactions. In contrast to the widely used nonsteroidal antiinflammatory drugs, BAs exhibit their antiinflammatory action simultaneously affecting a variety of parameters that are involved in inflammatory processes. Moreover, BAs significantly reduce the glycosaminoglycan degradation, thus avoiding ulcer creation and articular damage in arthritic conditions [113]. BAs also act against CatG secreted from cytochalasin B/fMLP-stimulated human PMNL [99].

Incensole acetate, a *Boswellia* resin constituent, inhibited transforming growth factor β-activated kinase (TAK) and transforming growth factor β-activated kinase-binding protein (TAB)-mediated IKK activation loop phosphorylation, resulting in the inhibition of cytokine and LPS-mediated NF-κB activation. Incensole acetate had no effect on IKK activity in vitro while inhibiting IKK activity in TNF-stimulated cells, indicating that the kinase inhibition is indirect. It also inhibited the formation of interleukin (IL)-1β, IL-6, and TNF-α, as well as prostaglandin E2, in LPS-stimulated human peripheral monocytes. It significantly inhibits COX-2, *Escherichia coli* aspartate transaminase (EcAspTA), and *Burkholderia cepacia* phenazine biosynthesis protein A/B (BcPhzA/B) [26, 110].

13.6.4 Galangal bioactives targeting the inflammatory pathway network

Alpinia officinarum Hance (*A. officinarum*) contain several antiinflammatory phytochemicals, especially linear diarylheptanoids and flavonols. Diarylheptanoids are class of compounds with limited distribution in plant kingdom, found mainly in families Zingiberaceae and Betulaceae. LPS is known to exert their inflammatory effect via activation of MAPKs. As COX-2 expression regulation depends on MAPK activation of the NF-κB pathways, this sheds the light on the importance of galangal bioactives as therapeutic target for treating inflammatory diseases. Nearly 48 diarylheptanoid compounds had been isolated from galangal rhizome, including 43 linear diarylheptanoids, three cyclic diarylheptanoids, and a diarylheptanoid bearing flavonol moiety. They are known to inhibit biosynthesis of prostaglandin, proinflammatory mediators, S-phase arrest, and differentiation of human neuroblastoma cells, cytotoxic activities, and leukotrienes [114]. Elgazar et al. studied on five compounds isolated from *A. officinarum* rhizomes, and their antiinflammatory effect for prevention and management of inflammatory conditions of liver were evaluated. These compounds successively reduced the expression of inflammatory cytokines in human hepatocellular carcinoma (HepG2) cells induced for inflammation by LPS. In silico studies suggest that their antiinflammatory effect is attributed to their ability to interact with p38 alpha MAPK [115].

Ding et al. in their study describe that *A. officinarum* extract and galangin remarkably reverse the inhibition of indomethacin causing alleviation of COX-1 levels while having little influence on COX-2 levels, thus reducing the drug-associated side effects (gastric injury), while preserving the antiinflammatory activity of indomethacin. The COX-2-specific effect was further confirmed by in silico approaches, wherein bioactives from *A. officinarum*, such as galangal, kaempferide, 5-hydroxy-7-(4-hydroxy-3-methoxy phenyl)-1-phenyl-3-heptanone (DPHA), and 1,7-diphenyl-5-hydroxy-3-heptanone (DPHC), proved to be potent COX-2 inhibitors in molecular docking studies [27, 116]. Galangin inhibited β-amyloid production and acetylcholinesterase activity and has the potential to treat Alzheimer's disease and cerebral ischemia by a mechanism related to the modulation of PI3K, NF-κB, and proliferator-activated receptor gamma (PPAR-γ) signaling pathways [115]. Kim et al. further investigated and proved that galangal extracts significantly suppressed degranulation of phorbol myristate acetate (PMA)/A23187, stimulated rat basophilic leukemia cells (RBL)/2H3 cells, and inhibited MAPK signaling pathway, especially p38 activation, and MAPK and JNK were not affected [117].

13.6.5 Rosemary bioactives targeting the inflammatory pathway network

Carnosic acid (CA), a major phenolic compound isolated from the leaves of rosemary (*Rosmarinus officinalis* L.), has been reported to have remarkable antioxidative, antiinflammatory, and antimicrobial properties (Table 13.1). CA has the ability to act against α-amylase and α-glucosidase (Table 13.2) [32]. Liu et al. studied the effect of CA in rheumatoid arthritis (RA). CA suppressed the expression of proinflammatory cytokines including TNF-α, IL-1β, IL-6, IL-8, IL-17, and MMP-3 and downregulated the production of receptor activator of nuclear factor kappa-B ligand (RANKL). CA inhibited osteoclastogenesis and bone resorption in vitro and exerted in vivo therapeutic protection against joint destruction. Further biochemical analysis demonstrated that CA suppressed RANKL-induced activations of NF-κB and MAPKs (JNK and p38) leading to the downregulation of nuclear factor of activated T cells, cytoplasmic 1 (NFATc1). Taken together, they provide the convincing evidence that rosemary-derived CA is a promising natural compound for the treatment of RA [118].

Antiinflammatory activity of CA is nuclear factor erythroid 2-related factor 2 (Nrf2)-dependent. de Oliveira et al. investigated whether CA would modulate inflammation-related parameters in SH-SY5Y cells treated with paraquat (PQ), an agrochemical that is able to activate NF-κB in mammalian cells. By upregulating the transcription factor Nrf2, CA exerts antioxidant effect, which controls the expression of antioxidant and phase II detoxification enzymes. Nrf2 modulates heme oxygenase-1 (HO-1) expression and has been demonstrated as part of the mechanism underlying the CA-induced cytoprotection. Through a mechanism involving the activation of the Nrf2/HO-1 axis, CA suppressed the PQ-induced alterations on the levels of IL-1β, TNF-α, and COX-2. A crosstalk between the Nrf2/HO-1 signaling pathway and the activation of the NF-κB transcription factor is taking place since administration of ZnPP IX (specific inhibitor of HO-1) or Nrf2 knockdown using small interfering ribonucleic acid (RNA) (siRNA) abolished the antiinflammatory effects induced by CA. Moreover, administration of SN50 (specific inhibitor of NF-κB) inhibited the PQ-induced inflammation-related effects in SH-SY5Y cells [119].

The antiinflammatory activity of yet another bioactive from rosemary, that is, carnosol (CS), along with carnosic acid targets 5-lipoxygenase. It is seen that CA and CS activate the peroxisome proliferator-activated receptor gamma, implying an antiinflammatory potential on the level of gene regulation. Short-term effects of CA and CS on typical functions of PMNL were found to be by inhibition of the formation of proinflammatory leukotrienes in intact PMNL (IC50 = 15–20 mM [CA] and 7 mM [CS], respectively) and purified recombinant 5-LOX (EC number 1.13.11.34, IC50 = 1 mM [CA] and 0.1 mM [CS], respectively). Both CA and CS potently antagonize intracellular Ca^{2+} mobilization induced by a chemotactic stimulus, and they attenuated formation of reactive oxygen species and the secretion of human leukocyte elastase (EC number 3.4.21.37) [28]. CS also acts against Nrf2-Kelch-like ECH-associated protein 1 (Keap1), protein phosphatase 1 (PP1), ribosomal protein S6 kinase (RSK2), and DPP-4 (Table 13.2) [29–31].

Concerning the biological activities of essential oil of *R. officinalis* L., its potential as an antiinflammatory agent has also been reported in several pharmacological experiments (Table 13.2) [120]. The major components possessing antiinflammatory activity in rosemary essential oil are 1,8-cineol, camphor, eucalyptol, and α-pinene. The presence of these terpenes and their potential to inhibit NF-κB transcription are major contributors for rosemary essential oil's antiinflammatory potential. Among these terpenes, 1,8-cineol has been the most associated with the antiinflammatory activity. Santos and Rao studied its ability to inhibit in vivo the vascular permeability increase and granuloma formation in carrageenan-induced rat paw edema assay as demonstrated. In vitro production of cytokines and arachidonic acid metabolites in human blood monocytes is inhibited by 1,8-cineol by inhibiting NF-κB transcription [120]. Besides 1,8-cineol, other terpenes are also known to act against molecular targets of inflammation. In amyloid beta-induced inflammation of PC-12 cells, cineol-treated group showed decreased concentration of inflammatory cytokines TNF-α, IL-1β, and IL-6. Similarly, in vitro assays using eucalyptol done by Juergens et al. showed significant inhibition of leukotriene and prostaglandin production in monocytes from patients with bronchial asthma. It was able to decrease production of TNF-α, IL-1β, IL-4, and IL-5 in lymphocytes and TNF-α, IL-1β, IL-6, and IL8 in monocytes [121]. A molecular docking study confirmed the same and further identified that among different bioactive terpenes, camphor has the highest number of interactions with therapeutic targets related to inflammation, such as prostaglandin I2 (PGI2) and COX-2 [122].

13.7 Conclusion

Studies on inflammation are a broad area and studies are going on to explore the complex reactions and systems. With the progress in the field of systems biology, several novel insights in the mechanisms of inflammatory diseases and identification of secondary drug targets for the existing antiinflammatory drugs have become more and more evident, leading to the development of novel systems pharmacology approach to therapeutic interventions (precision treatments). This differs in many ways from traditional drugs and can help overcome the associated toxic side effects. Computational data analysis strategies identify the drug effects/interactions at systems level, followed by in silico docking studies to further predict and prove the possibility of a drug-target interaction serves as an excellent, rapid, and economic approach. They help to constitute a basis for the modeling of the interactions in cascading turnover models, which are increasingly used to describe complex patterns of pharmacological responses and disease progression.

Interestingly, in this scenario, the traditionally used plant-origin natural cure remedies (known for their antiinflammatory effects) turned out to be more interesting not only due to their comparable or equal efficacy but also because of less or nontoxic side effects. With more and more published studies and evidences through multilevel data integration, it is highlighted that natural molecules directly interact with multiple cellular target proteins involved in the disease/pathology occurrence and progression and thus can be prioritized as drug leads. Additionally, there is a proven case for prioritization of natural compounds that have antiinflammatory effects and protect human body from pathological conditions. This chapter highlights on the available evidences to conclude on prioritization of curcumin, bisdemethoxycurcumin, tetrahydrocurcumin, demethoxycurcumin, 6-gingerol, zerumbone, zingiberene, zingerone, gingerdiol, 6-shogaol, boswellic acid, acetyl-11-keto-β-boswellic acid, incensole acetate, galangin, limonene, 1,8-cineole, camphor, carnosol, and carnosic acid as effective natural antiinflammatory agents. Further studies may help better to develop a complete understanding of interactions of these natural molecules in specific cell type, at specific concentration, time space, age, metabolic state, and physiological condition for an individual. However, it is already encouraging to see that most of these natural molecules are experimentally proven to avoid interaction with off-target cell receptors/enzyme proteins involved in inflammatory network pathway. Years of usage of these molecules directly in food (spices/herbs) or traditional medicines (Ayurveda and Chinese traditional medicine) further support their efficacy and safe-for-humans claim.

References

[1] R. Thirumalaisamy, S. Ammashi, G. Muthusamy, Screening of anti-inflammatory phytocompounds from Crateva adansonii leaf extracts and its validation by in silico modelling, J. Genet. Eng. Biotechnol. 16 (2018) 711–719.

[2] K.D. Tripathi, Essentials of medical pharmacology, Vth ed., Jaypee Brothers Medical Publishers (P) Ltd., New Delhi, 2004, pp. 167–181.

[3] H. Bouriche, E.A. Miles, L. Selloum, P.C. Calder, Effect of Cleome arabica leaf extract, Rutin and Quercetin on soybean lipoxygenase activity and on generation of inflammatory eicosanoids by human neutrophils, Prostaglandins Leukot. Essent. Fatty Acids 72 (2005) 195–201.

[4] D. Weininger, SMILES, a chemical language and information system. Introduction to methodology and encoding rules, J. Chem. Inf. Model. 28 (1988) 31–36.

[5] A.A. Oyagbemi, A.B. Saba, O.I. Azeez, Molecular targets of [6]-gingerol: its potential roles in cancer chemoprevention, Biofactors 36 (3) (2010) 169–178.

[6] R. Herowati, G.P. Widodo, Molecular docking analysis: interaction studies of natural compounds to anti-inflammatory targets, in: Quantitative Structure-activity Relationship, 2017, p. 63.

[7] K. Mohankumar, S. Sridharan, S. Pajaniradje, V.K. Singh, L. Ronsard, A.C. Banerjea, R. Rajagopalan, BDMC-A, an analog of curcumin, inhibits markers of invasion, angiogenesis, and metastasis in breast cancer cells via NF-κB pathway—A comparative study with curcumin, Biomed. Pharmacother. 74 (2015) 178–186.

[8] T. Vasanthkumar, M. Hanumanthappa, R. Lakshminarayana, Curcumin and capsaicin modulates LPS induced expression of COX-2, IL-6 and TGF-β in human peripheral blood mononuclear cells, Cytotechnology 71 (2019) 963–976.

[9] C.R. Girija, P. Karunakar, C.S. Poojari, N.S. Begum, A.A. Syed, Molecular docking studies of curcumin derivatives with multiple protein targets for procarcinogen activating enzyme inhibition, J. Proteom. Bioinform. 3 (2010) 200–203.

[10] J.P. Ganugapati, P.R. Babu, S.J. Ahuja, M. Mukundan, S.S. Vutukuru, Screening and molecular docking studies of curcumin and its derivatives as inhibitors of amyloid-Î² protein-a key protein in Alzheimer's disease, Asian J. Pharm. Clin. Res. (2015) 98–101.

[11] A.K. Singh, K. Misra, Human papilloma virus 16 E6 protein as a target for curcuminoids, curcumin conjugates and congeners for chemoprevention of oral and cervical cancers, Interdiscipl. Sci. Comput. Life Sci. 5 (2013) 112–118.
[12] J. Jacob, A. Amalraj, K.J. Raj, C. Divya, A.B. Kunnumakkara, S. Gopi, A novel bioavailable hydrogenated curcuminoids formulation (CuroWhite™) improves symptoms and diagnostic indicators in rheumatoid arthritis patients-A randomized, double blind and placebo controlled study, J. Tradit. Complement. Med. 9 (2019) 346–352.
[13] P. Haris, V. Mary, C. Sudarsanakumar, Probing the interaction of the phytochemical 6-gingerol from the spice ginger with DNA, Int. J. Biol. Macromol. 113 (2018) 124–131.
[14] F. Sharif, A.M. Yunus, R.R. Saedudin, A.A.A. Hamid, S. Kasim, Molecular docking and dynamics (MD) simulation of 6-gingerol and 6-shogaol against human estrogen receptor alpha (ERα), Int. J. Integr. Eng. 10 (2018) 119–127.
[15] J. Jacob, G. Peter, S. Thomas, J.T. Haponiuk, S. Gopi, Chitosan and polyvinyl alcohol nanocomposites with cellulose nanofibers from ginger rhizomes and its antimicrobial activities, Int. J. Biol. Macromol. 129 (2019) 370–376.
[16] U. Muzaffer, V.I. Paul, N.R. Prasad, Molecular docking of selected phytoconstituents with signaling molecules of Ultraviolet-B induced oxidative damage, In Silico Pharmacol. 5 (2017) 17.
[17] F.A. de Almeida, E.L.G. Vargas, D.G. Carneiro, U.M. Pinto, M.C.D. Vanetti, Virtual screening of plant compounds and nonsteroidal anti-inflammatory drugs for inhibition of quorum sensing and biofilm formation in Salmonella, Microb. Pathog. 121 (2018) 369–388.
[18] K. Rashmi, S.P. Tripathi, R.M. Rawal, Exploring inhibitory potential of ginger against numerous targets of diverse forms of cancer, Res. J. Life Sci. Bioinform. Pharm. Chem. Sci. 4 (2018) 792–802.
[19] F.A. Fajrin, A.E. Nugroho, R. Susilowati, A. Nurrochmad, Molecular docking analysis of ginger active compound on transient receptor potential cation channel subfamily V Member 1 (TRPV1), Indones. J. Chem. 18 (2018) 179–185.
[20] J.C. Wang, L.H. Zhou, H.J. Zhao, S.X. Cai, Examination of the protective effect of 6-shogaol against LPS-induced acute lung injury in mice via NF-κB attenuation, Arch. Biol. Sci. 68 (2016) 633–639.
[21] S.R. Feroz, S.B. Mohamad, G.S. Lee, S.N.A. Malek, S. Tayyab, Supramolecular interaction of 6-shogaol, a therapeutic agent of Zingiber officinale with human serum albumin as elucidated by spectroscopic, calorimetric and molecular docking methods, Phytomedicine 22 (2015) 621–630.
[22] L.W. Qi, Z. Zhang, C.F. Zhang, S. Anderson, Q. Liu, C.S. Yuan, C.Z. Wang, Anticolon cancer effects of 6-shogaol through G2/M cell cycle arrest by p53/p21-cdc2/cdc25A crosstalk, Am. J. Chin. Med. 43 (2015) 743–756.
[23] S. Rampogu, A. Baek, R.G. Gajula, A. Zeb, R.S. Bavi, R. Kumar, Y. Kim, Y.J. Kwon, K.W. Lee, Ginger (Zingiber officinale) phytochemicals-Gingerenone-A and shogaol inhibit SaHPPK: molecular docking, molecular dynamics simulations and in vitro approaches, Ann. Clin. Microbiol. Antimicrob. 17 (2018) 16.
[24] M. Bishnoi, C.S. Patil, A. Kumar, S.K. Kulkarni, Analgesic activity of acetyl-11-keto-beta-boswellic acid, a 5-lipoxygenase-enzyme inhibitor, Indian J. Pharm. 37 (2005) 255.
[25] U. Siemoneit, B. Hofmann, N. Kather, T. Lamkemeyer, J. Madlung, L. Franke, G. Schneider, J. Jauch, D. Poeckel, O. Werz, Identification and functional analysis of cyclooxygenase-1 as a molecular target of boswellic acids, Biochem. Pharmacol. 75 (2008) 503–513.
[26] K. Byler, W. Setzer, Protein targets of frankincense: a reverse docking analysis of terpenoids from Boswellia oleo-gum resins, Medicines 5 (2018) 96.

[27] V.S. Honmore, A.D. Kandhare, P.P. Kadam, V.M. Khedkar, D. Sarkar, S.L. Bodhankar, A.A. Zanwar, S.R. Rojatkar, A.D. Natu, Isolates of Alpinia officinarum Hance as COX-2 inhibitors: evidence from anti-inflammatory, antioxidant and molecular docking studies, Int. Immunopharmacol. 33 (2016) 8–17.

[28] D. Poeckel, C. Greiner, M. Verhoff, O. Rau, L. Tausch, C. Hörnig, D. Steinhilber, S.M. Zsilavecz, O. Werz, Carnosic acid and carnosol potently inhibit human 5-lipoxygenase and suppress pro-inflammatory responses of stimulated human polymorphonuclear leukocytes, Biochem. Pharmacol. 76 (2008) 91–97.

[29] X. Li, Q. Zhang, N. Hou, J. Li, M. Liu, S. Peng, Y. Zhang, Y. Luo, B. Zhao, S. Wang, Y. Zhang, Carnosol as a Nrf2 Activator Improves Endothelial Barrier Function Through Antioxidative Mechanisms, Int. J. Mol. Sci. 20 (2019) 880.

[30] L. Wang, Y. Zhang, K. Liu, H. Chen, R. Yang, X. Ma, H.G. Kim, A.M. Bode, D.J. Kim, Z. Dong, Carnosol suppresses patient-derived gastric tumor growth by targeting RSK2, Oncotarget 9 (2018) 34200–34212.

[31] B. Salim, A. Hocine, G. Said, First study on anti-diabetic effect of rosemary and salvia by using molecular docking, J. Pharm. Res. Int. 19 (2017) 1–12.

[32] H. Wang, J. Wang, Y. Liu, Y. Ji, Y. Guo, J. Zhao, Interaction mechanism of carnosic acid against glycosidase (α-amylase and α-glucosidase), Int. J. Biol. Macromol. 138 (2019) 846–853.

[33] I.L. Meek, M.A. Van de Laar, H.E. Vonkeman, Non-steroidal anti-inflammatory drugs: an overview of cardiovascular risks, Pharmaceuticals 3 (7) (2010) 2146–2162.

[34] Z. Hodzic, H. Pasalic, A. Memisevic, M. Srabovic, M. Saletovic, M. Poljakovic, The influence of total phenols content on antioxidant capacity in the whole grain extract, Eur. J. Sci. Res. 28 (2009) 471–477.

[35] C.H.D. Sadik, H. Sies, T. Schewe, Inhibition of 15-lipoxygenase by flavonoids: structure activity relation and mode of action, Biochem. Pharmacol. 65 (2003) 773–781.

[36] S. Lee, I. Lee, W. Mar, Inhibition of inducible nitric oxide synthase and cycloxygenase-2 activity by 1,2,3,4,6-penta-O-galloyl-beta-Dglucose in murine macrophage cells, Arch. Pharm. Res. 26 (2003) 832–839.

[37] J. Manivannan, T. Silambarasan, R. Kadarkarairaj, B. Raja, Systems pharmacology and molecular docking strategies prioritize natural molecules as cardioprotective agents, RSC Adv. 5 (2015) 77042–77055.

[38] M. Danhof, Systems pharmacology—towards the modeling of network interactions, Eur. J. Pharm. Sci. 94 (2016) 4–14.

[39] M.J. Balunas, A. Kinghorn, Drug discovery from medicinal plants, Life Sci. 78 (2005) 431–441.

[40] A. Gurib-Fakim, Medicinal plants: traditions of yesterday and drugs of tomorrow, Mol. Aspects Med. 27 (2006) 1–93.

[41] D. Khanna, G. Sethi, A. Kwang Seok, M. Pandey, A.B. Kunnumakkara, B. Sung, A. Aggarwal, B. Aggarwal, Natural products as a gold mine for arthritis treatment, Curr. Opin. Pharmacol. 7 (2007) 344–351.

[42] M. Danhof, E.C. de Lange, O.E. Della Pasqua, B.A. Ploeger, R.A. Voskuyl, Mechanism-based pharmacokinetic-pharmacodynamic (PK-PD) modeling in translational drug research, Trends Pharmacol. Sci. 29 (2008) 186–191.

[43] P. Vicini, P.H. Van Der Graaf, Systems pharmacology for drug discovery and development: paradigm shift or flash in the pan? Clin. Pharmacol. Ther. 93 (2013) 379–381.

[44] M. Garrido, J. Gubbens-Stibbe, E. Tukker, E. Cox, J. von Frijtag, D. Kunzel, M. Danhof, P.H. van der Graaf, Pharmacokinetic-pharmacodynamic analysis of the EEG effect of alfentanil in rats following beta-funaltrexamine-induced mu-opioid receptor “knockdown” in vivo, Pharm. Res. 17 (2000) 653–659.

[45] P.H. Van der Graaf, E.A. Van Schaick, S.A. Visser, H.J. De Greef, A.P. Ijzerman, M. Danhof, Mechanism-based pharmacokinetic-pharmacodynamic modeling of antilipolytic effects of adenosine A (1) receptor agonists in rats: prediction of tissue-dependent efficacy in vivo, J. Pharmacol. Exp. Ther. 290 (1999) 702–709.
[46] A. Yassen, E. Olofsen, J. Kan, A. Dahan, M. Danhof, Pharmacokinetic pharmacodynamic modeling of the effectiveness and safety of buprenorphine a9nd fentanyl in rats, Pharm. Res. 25 (2008) 183–193.
[47] A.D. Boran, R. Iyengar, Systems pharmacology, Mt. Sinai J. Med. 77 (2010) 333–344.
[48] M.W. Mumtaz, A.A. Hamid, M.T. Akhtar, F. Anwar, U. Rashid, M.H. AL-Zuaidy, An overview of recent developments in metabolomics and proteomics-phytotherapic research perspectives, Front. Life Sci. 10 (2017) 1–37.
[49] A. Hashiguchi, J. Tian, S. Komatsu, Proteomic contributions to medicinal plant research: from plant metabolism to pharmacological action, Proteomes 5 (2017) 35.
[50] P.K. Mukherjee, R.K. Harwansh, S. Bahadur, S. Biswas, L.N. Kuchibhatla, S.D. Tetali, A.S. Raghavendra, Metabolomics of medicinal plants–a versatile tool for standardization of herbal products and quality evaluation of ayurvedic formulations, Curr. Sci. 111 (2016) 1624–1630.
[51] S.S. Ou-Yang, J.Y. Lu, X.Q. Kong, Z.J. Liang, C. Luo, H. Jiang, Computational drug discovery, Acta Pharmacol. Sin. 33 (2012) 1131.
[52] L. Chen, J.K. Morrow, H.T. Tran, S.S. Phatak, L. Du-Cuny, S. Zhang, From laptop to benchtop to bedside: structure-based drug design on protein targets, Curr. Pharm. Des. 18 (2012) 1217–1239.
[53] L. Scotti, F.J.B. Mendonca Junior, H.M. Ishiki, F.F. Ribeiro, R.K. Singla, J.M.B. Filho, M.T. Scotti, Docking studies for multi-target drugs, Curr. Drug Targets 18 (2017) 592–604.
[54] Y. Hobani, A. Jerah, A. Bidwai, A comparative molecular docking study of curcumin and methotrexate to dihydrofolate reductase, Bioinformation 13 (2017) 63.
[55] N. Brooijmans, I.D. Kuntz, Molecular recognition and docking algorithms, Annu. Rev. Biophys. Biomol. Struct. 32 (2003) 335–373.
[56] I. Halperin, B. Ma, H. Wolfson, R. Nussinov, Principles of docking: an overview of search algorithms and a guide to scoring functions, Proteins: Struct. Funct. Bioinf. 47 (2002) 409–443.
[57] B.K. Shoichet, S.L. McGovern, B. Wei, J.J. Irwin, Lead discovery using molecular docking, Curr. Opin. Chem. Biol. 6 (2002) 439–446.
[58] M.C. Guimaraes, D.G. Silva, E.G. da Mota, E.F.F. da Cunha, M.P. Freitas, Computer-assisted design of dual-target anti-HIV-1 compounds, Med. Chem. Res. 23 (2014) 1548–1558.
[59] D. Hecht, G.B. Fogel, Computational intelligence methods for docking scores, Curr. Comput-Aid. Drug Des. 5 (2009) 56–68.
[60] A. Lavecchia, C. Di Giovanni, Virtual screening strategies in drug discovery: a critical review, Curr. Med. Chem. 20 (2013) 2839–2860.
[61] W.F. De Azevedo, MolDock Applied to structure-based virtual screening, Curr. Drug Targets 11 (2010) 327–334.
[62] I.V. Ogungbe, W.N. Setzer, Comparative Molecular Docking of Antitrypanosomal natural products into multiple trypanosoma brucei drug targets, Molecules 14 (2009) 1513–1536.
[63] J. Xu, A. Hagler, Chemoinformatics and drug discovery, Molecules 7 (2002) 566–600.
[64] L. Minghetti, Cyclooxygenase-2 (COX-2) in inflammatory and degenerative brain diseases, J. Neuropathol. Exp. Neurol. 63 (2004) 901–910.
[65] M. Murakami, I. Kudo, Phospholipase A2, J. Biochem. 131 (2002) 285–292.

[66] M. Murakami, Novel functions of phospholipase A2s: overview, Biochim. Biophys. Acta Mol. Cell Biol. Lipid. 1864 (2019) 763.
[67] T. Lawrence, The nuclear factor NF-κB pathway in inflammation, Cold Spring Harb. Perspect. Biol. 1 (2009) a001651.
[68] M.V. Raimondi, O. Randazzo, M. La Franca, G. Barone, E. Vignoni, D. Rossi, S. Collina, DHFR inhibitors: reading the past for discovering novel anticancer agents, Molecules 24 (2019) 1140.
[69] C. Pergola, O. Werz, 5-Lipoxygenase inhibitors: a review of recent developments and patents, Expert Opin. Ther. Pat. 20 (2010) 355–375.
[70] T. Katsila, G.A. Spyroulias, G.P. Patrinos, M.T. Matsoukas, Computational approaches in target identification and drug discovery, Comput. Struct. Biotechnol. J. 14 (2016) 177–184.
[71] J. Nickel, B.O. Gohlke, J. Erehman, P. Banerjee, W.W. Rong, A. Goede, R. Preissner, SuperPred: update on drug classification and target prediction, Nucleic Acids Res. 42 (2014) W26–W31.
[72] D. Gfeller, A. Grosdidier, M. Wirth, A. Daina, O. Michielin, V. Zoete, SwissTargetPrediction: a web server for target prediction of bioactive small molecules, Nucleic Acids Res. 42 (2014) W32–W38.
[73] J. Zhang, Y. Li, X. Chen, Y. Pan, S. Zhang, Y. Wang, Systems pharmacology dissection of multi-scale mechanisms of action for herbal medicines in stroke treatment and prevention, PLoS One 9 (2014), e102506.
[74] X. Zeng, P. Zhang, W. He, C. Qin, S. Chen, L. Tao, Z. Chen, NPASS: natural product activity and species source database for natural product research, discovery and tool development, Nucleic Acids Res. 46 (2017) D1217–D1222.
[75] L. Huang, D. Xie, Y. Yu, H. Liu, Y. Shi, T. Shi, C. Wen, TCMID 2.0: a comprehensive resource for TCM, Nucleic Acids Res. 46 (2017) D1117–D1120.
[76] O. Trott, A.J. Olson, AutoDock Vina: improving the speed and accuracy of docking with a new scoring function, efficient optimization, and multithreading, J. Comput. Chem. 31 (2010) 455–461.
[77] R.A. Laskowski, M.B. Swindells, LigPlot+: multiple ligand-protein interaction diagrams for drug discovery, J. Chem. Inf. Model. 51 (2011) 2778–2786.
[78] N. Furnham, G.L. Holliday, T.A. de Beer, J.O. Jacobsen, W.R. Pearson, J.M. Thornton, The Catalytic Site Atlas 2.0: cataloging catalytic sites and residues identified in enzymes, Nucleic Acids Res. 42 (2014) D485–D489.
[79] H. Zhang, J. Pan, X. Wu, A.R. Zuo, Y. Wei, Z.L. Ji, Large-scale target identification of herbal medicine using a reverse docking approach, ACS Omega 4 (2019) 9710–9719.
[80] W.P. Walters, Virtual chemical libraries: miniperspective, J. Med. Chem. 62 (2018) 1116–1124.
[81] D.B. Kitchen, H. Decornez, J.R. Furr, J. Bajorath, Docking and scoring in virtual screening for drug discovery: methods and applications, Nat. Rev. Drug Discov. 3 (2004) 935–949.
[82] G. Klebe, Virtual ligand screening: strategies, perspectives and limitations, Drug Discov. Today 11 (2006) 580–594.
[83] A. Kolodzik, N. Schneider, M. Rarey, Structure-based virtual screening: achievements and future opportunities, in: T. Engel, J. Gasteiger (Eds.), Applied chemoinformatics, Wiley-VCH Verlag GmbH & Co. KGaA, 2018, pp. 313–331.
[84] K.E. Hevener, Computational toxicology methods in chemical library design and high-throughput screening hit validation, in: Computational Toxicology, Humana Press, New York, NY, 2018, pp. 275–285.
[85] L. Vollono, M. Falconi, R. Gaziano, F. Iacovelli, E. Dika, C. Terracciano, E. Campione, Potential of curcumin in skin disorders, Nutrients 11 (2019) 2169.
[86] H.S. Lee, Y.D. Kim, B.R. Na, H.R. Kim, E.J. Choi, W.C. Han, H.K. Choid, S.H. Leed, C. Juna, Phytocomponent p-Hydroxycinnamic acid inhibits T-cell activation by

modulation of protein kinase C-dependent pathway, Int. Immunopharmacol. 12 (2012) 131–138.
[87] A. Goel, C.R. Boland, D.P. Chauhan, Specific inhibition of cyclooxygenase-2 (COX-2) expression by dietary curcumin in HT-29 human colon cancer cells, Cancer Lett. 172 (2001) 111–118.
[88] F. Zhang, N.K. Altorki, J.R. Mestre, K. Subbaramaiah, A.J. Dannenberg, Curcumin inhibits cyclooxygenase-2 transcription in bile acid-and phorbol ester-treated human gastrointestinal epithelial cells, Carcinogenesis 20 (1999) 445–451.
[89] H.Y. Kim, E.J. Park, E.H. Joe, I. Jou, Curcumin suppresses Janus kinase-STAT inflammatory signaling through activation of Src homology 2 domain-containing tyrosine phosphatase 2 in brain microglia, J. Immunol. 171 (2003) 6072–6079.
[90] A. Amalraj, A. Pius, S. Gopi, S. Gopi, Biological activities of curcuminoids, other biomolecules from turmeric and their derivatives–A review, J. Tradit. Complement. Med. 7 (2017) 205–233.
[91] M.R. Sohilait, H.D. Pranowo, W. Haryadi, Molecular docking analysis of curcumin analogues with COX-2, Bioinformation 13 (2017) 356–359.
[92] S. Jude, A. Amalraj, A. Kunnumakkara, C. Divya, B.M. Löffler, S. Gopi, Development of validated methods and quantification of curcuminoids and curcumin metabolites and their pharmacokinetic study of oral administration of complete natural turmeric formulation (Cureit™) in human plasma via UPLC/ESI-Q-TOF-MS spectrometry, Molecules 23 (2018) 2415.
[93] C. Jobin, C.A. Bradham, M.P. Russo, B. Juma, A.S. Narula, D.A. Brenner, R.B. Sartor, Curcumin blocks cytokine-mediated NF-κB activation and proinflammatory gene expression by inhibiting inhibitory factor I-κB kinase activity, J. Immunol. 163 (1999) 3474–3483.
[94] B. Gupta, B. Ghosh, Curcuma longa inhibits TNF-α induced expression of adhesion molecules on human umbilical vein endothelial cells, Int. J. Immunopharmacol. 21 (1999) 745–757.
[95] S. Singh, B.B. Aggarwal, Activation of transcription factor NF-kappa B is suppressed by curcumin (diferuloylmethane), J. Biol. Chem. 270 (1995) 24995–25000.
[96] M.M. Thunnissen, P.A. Franken, G.H. de Haas, J. Drenth, K.H. Kalk, H.M. Verheij, B.W. Dijkstra, Crystal structure of a porcine pancreatic phospholipase A2 mutant: a large conformational change caused by the F63V point mutation, J. Mol. Biol. 232 (1993) 839–855.
[97] K.V. Dileep, I. Tintu, C. Sadasivan, Molecular docking studies of curcumin analogs with phospholipase A2, Interdiscipl. Sci. Comput. Life Sci. 3 (2011) 189–197.
[98] K. Li, L. Li, S. Wang, X. Li, T. Ma, D. Liu, Y. Jing, L. Zhao, Design and synthesis of novel 2-substituted 11-keto-boswellic acid heterocyclic derivatives as anti-prostate cancer agents with Pin1 inhibition ability, Eur. J. Med. Chem. 126 (2017) 910–919.
[99] L. Tausch, Novel anti-inflammatory targets and mechanisms of boswellic acids and celecoxib, Goethe Universität Frankfurt, Institut für Pharmazeutische Chemie. Frankfurt, 2008.
[100] S. Radaev, Z. Zou, T. Huang, E.M. Lafer, A.P. Hinck, P.D. Sun, Ternary complex of transforming growth factor-b1 reveals isoform-specific ligand recognition and receptor recruitment in the superfamily, J. Biol. Chem. 285 (2010) 14806–14814.
[101] F. Mohammadi, M. Moeeni, Study on the interactions of trans-resveratrol and curcumin with bovine α-lactalbumin by spectroscopic analysis and molecular docking, Mater. Sci. Eng. C 50 (2015) 358–366.
[102] F. Zehsaz, N. Farhangi, L. Mirheidari, The effect of Zingiber officinale R. rhizomes (ginger) on plasma pro-inflammatory cytokine levels in well-trained male endurance runners, Cent-Eur. J. Immunol. 39 (2014) 174–180.
[103] Y.J. Surh, H.K. Na, NF-κB and Nrf2 as prime molecular targets for chemoprevention and cytoprotection with anti-inflammatory and antioxidant phytochemicals, Genes Nutr. 2 (2008) 313–317.

[104] F. Azam, A.M. Amer, A.R. Abulifa, M.M. Elzwawi, Ginger components as new leads for the design and development of novel multi-targeted anti-Alzheimer's drugs: a computational investigation, Drug Des. Devel. Ther. 8 (2014) 2045–2059.
[105] S. Kumar, K. Saxena, U.N. Singh, R. Saxena, Anti-inflammatory action of ginger: a critical review in anemia of inflammation and its future aspects, Int. J. Herb. Med. 1 (2013) 16–20.
[106] S.H.M. Habib, S. Makpol, N.A.A. Hamid, S. Das, W.Z.W. Ngah, Y.A.M. Yusof, Ginger extract (Zingiber officinale) has anti-cancer and anti-inflammatory effects on ethionine-induced hepatoma rats, Clinics 63 (2008) 807–813.
[107] J.K. Kim, Y. Kim, K.M. Na, Y.J. Surh, T.Y. Kim, [6]-Gingerol prevents UVB-induced ROS production and COX-2 expression in vitro and in vivo, Free Radic. Res. 41 (2007) 603–614.
[108] F. Iram, S.A. Khan, A. Husain, Phytochemistry and potential therapeutic actions of Boswellic acids: a mini-review, Asian Pac. J. Trop. Biomed. 7 (2017) 513–523.
[109] H.Y. Lee, S.H. Park, M. Lee, H.J. Kim, S.Y. Ryu, N.D. Kim, B.Y. Hwang, J.T. Hong, S.B. Han, Y. Kim, 1-Dehydro- [10] -gingerdione from ginger inhibits IKKβ activity for NF-κB activation and suppresses NF-κB-regulated expression of inflammatory genes, Br. J. Pharmacol. 167 (2012) 128–140.
[110] A. Moussaieff, R. Mechoulam, Boswellia resin: from religious ceremonies to medical uses; a review of in-vitro, in-vivo and clinical trials, J. Pharm. Pharmacol. 61 (2009) 1281–1293.
[111] H.P.T. Ammon, Boswellic acids and their role in chronic inflammatory diseases, in: S.C. Gupta, S. Prasad, B.B. Aggarwal (Eds.), Anti-inflammatory Nutraceuticals and Chronic Diseases, Springer, Cham, 2016, pp. 291–327.
[112] T. Syrovets, B. Büchele, C. Krauss, Y. Laumonnier, T. Simmet, Acetyl-boswellic acids inhibit lipopolysaccharide-mediated TNF-α induction in monocytes by direct interaction with IκB kinases, J. Immunol. 174 (2005) 498–506.
[113] R. Satpathy, R.K. Guru, R. Behera, B. Nayak, Prediction of anticancer property of boswellic acid derivatives by quantitative structure activity relationship analysis and molecular docking study, J. Pharm. Bioallied Sci. 7 (2015) 21–25.
[114] P. Ding, L. Yang, C. Feng, J.C. Xian, Research and application of Alpinia officinarum in medicinal field, Chin. Herb. Med. 11 (2019) 132–140.
[115] A.A. Elgazar, N.M. Selim, N.M. Abdel-Hamid, M.A. El-Magd, H.M. El Hefnawy, Isolates from *Alpinia officinarum* Hance attenuate LPS-induced inflammation in HepG2: evidence from in silico and in vitro studies, Phytother. Res. 32 (2018) 1273–1288.
[116] J. Gong, Z. Zhang, X. Zhang, F. Chen, Y. Tan, H. Li, J. Jiang, J. Zhang, Effects and possible mechanisms of Alpinia officinarum ethanol extract on indomethacin-induced gastric injury in rats, Pharm. Biol. 56 (2018) 294–301.
[117] H.G. Kim, B. Shrestha, S.Y. Lim, D.H. Yoon, W.C. Chang, D.J. Shin, S.K. Han, S.M. Park, J.H. Park, H.I. Park, J.M. Sung, Cordycepin inhibits lipopolysaccharide-induced inflammation by the suppression of NF-κB through Akt and p38 inhibition in RAW 264.7 macrophage cells, Eur. J. Pharmacol. 545 (2006) 192–199.
[118] M. Liu, X. Zhou, L. Zhou, Z. Liu, J. Yuan, J. Cheng, J. Zhao, L. Wu, H. Li, H. Qiu, J. Xu, Carnosic acid inhibits inflammation response and joint destruction on osteoclasts, fibroblast-like synoviocytes, and collagen-induced arthritis rats, J. Cell. Physiol. 233 (2018) 6291–6303.
[119] M.R. de Oliveira, I.C.C. de Souza, C.R. Fürstenau, Carnosic acid induces anti-inflammatory effects in paraquat-treated SH-SY5Y cells through a mechanism involving a crosstalk between the Nrf2/HO-1 axis and NF-κB, Mol. Neurobiol. 55 (2018) 890–897.
[120] A.C. Dorni, A. Amalraj, S. Gopi, K. Varma, S.N. Anjana, Novel cosmeceuticals from plants—An industry guided review, J. Appl. Res. Med. Aromat. Plants 7 (2017) 1–26.

[121] U.R. Juergens, Anti-inflammatory Properties of the Monoterpene 1.8-cineole: current evidence for co-medication in inflammatory airway diseases, Drug Res. 64 (2014) 638–646.
[122] R.S. Borges, E.S. Lima, H. Keita, I.M. Ferreira, C.P. Fernandes, R.A.S. Cruz, J.L. Duarte, J. Velázquez-Moyado, B.L.S. Ortiz, N.A. Castro, J.V. Ferreira, L.I.S. Hage-Melim, J.C.T. Carvalho, Anti-inflammatory and antialgic actions of a nanoemulsion of Rosmarinus officinalis L. essential oil and a molecular docking study of its major chemical constituents, Inflammopharmacology 26 (2018) 183–195.

14

Bioavailability, pharmacokinetic, pharmacodynamic, and clinical studies of natural products on their antiinflammatory activities

Akhila Nair, Sreeraj Gopi, and Joby Jacob
R&D Centre, Aurea Biolabs (P) Ltd, Kolenchery, Cochin, Kerala, India

14.1 Introduction

Inflammation is a biological response or a protective effect developed in reply to the harmful stimuli such as damaged cells, pathogens, or irradiation to minus injurious stimuli and instigates the recovery process and involves numerous proinflammatory expressions. Prolonged inflammation is associated with various chronic diseases. Therefore targeting inflammation could overcome such diseases, but modern antiinflammatory drugs are associated with high adverse effect and costly [1]. Due to these shortcomings, nutraceuticals (natural medicines or green medicines) have gained a wider currency for their multifarious benefits that ameliorate the body functions and regulate chronic diseases [2]. These products have wide chemical diversity, extensive dispersals, low cost, minimal side effects, and biologically accepted potential. The term nutraceuticals was coined around the 20th century, which stated a link with medicine and nutrition. The European Nutraceutical Association defines nutraceuticals as nutritional products that have effects that are applicable to health. These products consist of bioactive agents with multitudinous functions, which are provided in a specific dose, intended to rectify the underlying cause of numerous diseases to enrich human health with minimal side effects and toxicity. This caused an inclination of the current medicinal wave toward herbal remedies, and the concept of "return to nature" came into existence. Besides, herbal therapies have been successfully utilized in traditional medicine systems, namely, Siddha, Unani, Ayurveda, Kampo, and traditional Chinese medicine. The consumption and demand of phytomedicines as

Inflammation and Natural Products. https://doi.org/10.1016/B978-0-12-819218-4.00006-7
© 2021 Elsevier Inc. All rights reserved.

health products, plant-based medicines, food supplements, and cosmetics has accelerated around the globe. A great deal of clinical studies on certain natural remedies has been successfully accomplished, whereas some are still ongoing. This phenomenon escalation of nutraceuticals has given inquisitiveness about their pharmacological aspects [3]. The pharmacological behavior covers the study of their pharmacokinetics, pharmacodynamics, and bioavailability of these natural products. In the case of natural products, the basic principle revolving around the achievement of pharmacological effects is when the intended metabolites reach and acquire a sustained state at the target site. The concentration that is achieved at the site of action after the administration of a natural product is determined by their potential and dose. Therefore pharmacokinetic studies have turned out to be an integral part for an optimal herbal drug development. Natural compounds witness phase I and/or phase II metabolism with uridine diphosphate glucuronosyltransferases (UGTs) and cytochrome P450s (CYPs) in vivo and are substrates of P-glycoprotein that are majorly found in the intestine, brain, liver, and kidney. Their mode of action is governed by the drug transporters and metabolizing enzymes and states their in vivo bioavailability, distribution, and disposition. Although rigorous research has been carried out to optimize the usage of natural products by conducting various clinical studies, there are many herbal products whose complete biological fate and data are unavailable [4]. This chapter deals with the prominent bioavailability, pharmacokinetic, and pharmacodynamic studies carried out on natural antiinflammatory medicines such as milk thistle, ginseng, ginger, curcumin, and *Ginkgo biloba.*

14.2 Antiinflammatory activities of natural products

Around 5,00,000 plant species around the globe are studied for their antiinflammatory and other therapeutic activities. Inflammation is triggered as a protective measure in response to the injurious stimuli; however, antiinflammatory drugs can help control prolonged inflammation. Modern antiinflammatory medications have found to be expensive and interfere with the inflammatory mechanisms, with immense adverse effects that accelerated the scope of natural products in the treatment of inflammation and inflammatory diseases [5]. Currently, scientific research focuses on the approach of "one molecule multiple targets several diseases"; therefore numerous herbal products in synergism with this approach are considered the drug of choice. Herbal drugs such as angelica root, astragalus, curcumin, danshen, echinacea, garlic, ginger, *Ginkgo biloba*, ginseng, and milk thistle are successfully studied for various diseases, especially inflammatory diseases, both clinically and nonclinically including pharmacodynamics bioavailability, and pharmacokinetics (Tables 14.1 and 14.2) [6].

Table 14.1 Active ingredients and general findings on pharmacodynamics of natura antiinflammatory agents.

S. no.	Active ingredients	Pharmacodynamics	References
1.	*Angelica sinensis* (Oliv.) Diels (angelica root)		
	Ferulic acids, *Z*-ligustilide, angelicide, senkyunolide A, butylphthalide, butylidenephthalide, phthalide dimers, organic acids, polyacetylenes, coniferyl ferulate, amino acids, and vitamins	**(a)** Inhibit the macrophage inflammatory protein-2 (MIP-2) production by murine macrophage in RAW 264.7 cells **(b)** Suppress the NF-κB luciferase activity and lower the PGE2 and NO production of lipopolysaccharide (LPS)/IFN-g-stimulated murine primary peritoneal macrophages **(c)** Decrease in the neutrophil count and IL-6 mRNA and TNF-α mRNA level in the pouch membrane and concentration of IL-6 and PGE2 in the pouch fluid **(d)** Increase the PGD2 concentration in pouch fluid **(e)** n-Butylidenephthalide decrease the secretion of IL-6 and TNF-α during LPS-stimulated activation of murine dendritic cells 2.4	[7–12]
2.	*Astragalus membranaceus* (astragalus)		
	(a) Saponins and (b) polysaccharides Astragaloside I, II, and IV and isoastragaloside II	**(a)** Inhibit TNF-α and IL-1β and suppress nuclear factor-κB (NF-κB) activation **(b)** Regulate cytokines and subside Th2 cytokine expression and TNF-α level in 1-chloro-2,4-dinitrobenzene-induced mice **(c)** Inhibit of NF-κB channeled transcription MKP-1-dependent Erk1/2 and p38 inactivation **(d)** Astragalus polysaccharides improve palmitate-induced proinflammatory responses via AMPK activity in RAW 264.7 and exhibit structure defensive properties in Caco2 cells that are infected by lipopolysaccharide **(e)** Their active constituents obstruct lipopolysaccharide-induced nitric oxide development in RAW 264.7 macrophages	[13–16]

Continued

Table 14.1 Active ingredients and general findings on pharmacodynamics of natural antiinflammatory agents—cont'd

S. no.	Active ingredients	Pharmacodynamics	References
3.	*Curcuma longa* L. (curcumin)		
	(a) Volatile portion—turmerone, zingiberone, and atlantone **(b)** Nonvolatile portion constitutes curcuminoids—curcumin, demethoxycurcumin, bisdemethoxycurcumin	**(a)** Suppress NF-κB via binding lipoxygenase (5-LOX) and cyclooxygenase-2 (COX-2) to inhibit the activity in vivo and suppress inducible nitric oxide synthase (iNOS) **(b)** Suppress the tumor necrosis factor alpha (TNF-α); macrophage inflammatory protein-1α; C-reactive protein (CRP); monocyte chemoattractant protein (MCP-1); chemokines interleukin (IL)-1β, IL-2, IL-6, IL-8, and IL-12; chemokine receptor CXCR-4; and so on **(c)** Downregulate epidermal growth factor receptor and the activity of extracellular signal-regulated kinase 1/2 (also called mitogen-activated protein kinase [MAPK]) and c-Jun N-terminal kinase (JNK) and inhibit the phosphatidylinositol-3 kinase/AKT pathway **(d)** Suppress the activity of intercellular signaling protein kinase C, protamine kinase, autophosphorylation-activated protein kinase, and pp60c-src tyrosine kinase **(e)** Downregulate ROS-mediated inflammation through the nuclear factor (erythroid-derived 2)- related factor 2 (Nrf2) pathway	[17–21]
4.	*Salvia miltiorrhiza* (danshen)		
	Danshensu and tanshinone IIA	**(a)** Subdue antiinflammatory cytokine induction such as IL-4, TGF-β, IL-10, and IL-1Ra and proinflammatory cytokines like NO, IL-6, TNF-α, and IL-1β **(b)** Tanshinone IIA reduces ICAM-1 and VCAM-1 expression to reduced TNF-α-induced neutrophil adhesion to BMVECs in dose-dependent manner via inhibition of ROS generation and NF-κB activation **(c)** Ailanthoidol subdues NO generation, prostaglandin E2, cyclooxygenase-2, and inducible NOS and inhibits IL-6 and IL-1β in RAW 264.7 cells **(d)** Cryptotanshinone inhibits the phosphorylation of mitogen-activated protein kinase (MAPKs) inclusive of p38MAPK, ERK1/2, and JNK and fully removes LPS-triggered nuclear factor-κB activation	[22–24]

Table 14.1 Active ingredients and general findings on pharmacodynamics of natural antiinflammatory agents—cont'd

S. no.	Active ingredients	Pharmacodynamics	References
5.	*Echinacea purpurea* (echinacea)		
		(a) Activate macrophages and control in vitro cytokine secretions **(b)** Alkamide-rich echinacea inhibit 5-lipoxygenase and cyclooxygenase **(c)** Reduce prostaglandin E2 production and arachidonic acid metabolism	[25]
6.	*Allium sativum* L. (garlic)		
	Allicin; ajoene; thiacremonone; S-allylmercaptocysteine; proteins; diallyl sulfide; inorganic elements like iron, copper, and selenium; polyphenols; and amino acids	**(a)** S-allylmercaptocysteine, diallyl sulfide, and ajoene suppress inflammatory factor NF-κB and *Z*- and *E*-ajoene downregulated nitric oxide, tumor necrosis factor alpha (TNF-α), prostaglandin E2, interleukin-6, and IL-1β **(b)** Decrease activity of protease in chondrocyte like cells and inhibit effect of matrix metalloproteinase and the tissue inhibitor of metalloproteinase-1 **(c)** Decrease development of IL-6 and TNF-α and reduction in IL-12 **(d)** In IL-3-dependent murine pro-B-cells Ba/F3, suppress NF-κB activation,COX-2 expression, and inducible nitric oxide synthase via Toll-like receptor-dependent pathway **(e)** Attenuate TNF-α-induced VCAM-1 expression through NF-κB-dependent pathway in human umbilical vein endothelial cells (HUVEC) **(f)** Thiacremonone inhibits 12-O-tetradecanoylphorbol-13-acetate-induced ear edema in ICR mice, carrageenan, and *Mycobacterium butyricum*-induced inflammatory as well as arthritic response **(g)** Suppress 1-chloro-2,4-dinitrobenzene-induced contact hypersensitivity measured by ear swelling	[26–28]
7.	*Zingiber officinale* Roscoe (ginger)		
	(a) 4-, 8-, 10-, and 12-gingerols **(b)** 6-, 8-, and 10-shogaols	**(a)** Inhibit leukotriene biosynthesis and prostaglandin via repression of prostaglandin synthetase or 5-lipoxygenase and inhibit proinflammatory cytokines synthesis such as IL-1, IL-8, and TNF-α **(b)** Shogaols downregulate inflammatory COX-2 and iNOS expression in macrophages **(c)** Production of NO, TNF-α, PGE, and IL-1β **(d)** Inhibit PGE2 production by inhibiting the LPS-induced COX-2 expression	[29,30]

Continued

Table 14.1 Active ingredients and general findings on pharmacodynamics of natural antiinflammatory agents—cont'd

S. no.	Active ingredients	Pharmacodynamics	References
8.	*Ginkgo biloba* L. (Ginkgo biloba)		
	(a) Flavonoids—quercetin, isorhamnetin glycosides, and kaempferol **(b)** 6% terpenes (ginkgolides, bilobalides)	**(a)** Suppress NF-κB factor and downregulate nitric oxide (NO) and PGE2 production inclusive of mRNA expression of iNOS, COX-2, and proinflammatory cytokinins (IL-6,IL-1β, and TNF-α) **(b)** Leaf extract suppresses the activation of macrophages and downregulates inflammatory stress markers like p53-phospho-serine-15 and p53; inflammatory expressions like iNOS, COX-2, TNF-α, and T-cell numbers like CD4 + and CD25-/FOX p3 were reduced in the colon of mice **(c)** Inhibit mRNA expression of TNF-α, IL-6, and NF-κBp65 in colitis of rats **(d)** Exhibit upregulation of inflammatory markers and downregulate mRNA expression of TNF-α and IL-1β in rat model	[31,32]
9.	*Panax ginseng* Meyer (ginseng)		
	(a) Protopanaxatriol (PPT)—ginsenoside Rb1, Rb2, Rb3, Rd., and Rc **(b)** Protopanaxadiol (PPD)—ginsenoside Rg1 and Re	**(a)** Ginsenosides Rg1 and Rb1 resulted about 40% upraise in NO levels in serum after 2 h of administration in isoproterenol (ISO)-induced myocardial ischemia rats **(b)** NO inducing effect is drug concentration dependent	[33]
10.	*Silybum marianum* (L.) Gaertn. (milk thistle)		
	Silybin, silymarin	**(a)** Silybin inhibits synthesis of leukotriene B4 (IC50 15 μmoL/L); no effect on the prostaglandin E2 formation at concentration up to 100 μmoL/L in isolated Kupffer cells of rats **(b)** Silymarin inhibits NF-κB DNA binding activity-dependent gene expression induced by okadaic acid **(c)** Silymarin inhibits leukotriene synthesis via 5-lipoxygenase pathway and formation of prostaglandins	[34,35]

Table 14.2 Pharmacokinetics and bioavailability of natural products.

S. no.	Natural products	Pharmacokinetics/bioavailability	Reference
1	Angelica root	Low solubility, low bioavailability, fast metabolism	[36]
2	Astragalus	**(a)** Low absorption due to high molecular weight, low lipophilicity, poor intestinal permeability, and their paracellular transport in Caco-2 cells **(b)** Astragalus IV combines with plasma protein was found to be 83%–90% and eliminated slowly by hepatic clearance of 0.004 L/kg/min **(c)** Maternal toxicity for astragalus IV—1.0 mg/kg and fetal toxicity—higher than 0.5 mg/kg (with no teratogenic effects in rabbits and rats) **(d)** Bioavailability in rats 3.66% and beagle was 7.4% for Astragalus IV after oral administration	[37–40]
3	Curcumin	**(a)** A complete plasma clearance results within 1 h with low intravenous doses of curcumin (40 mg/kg) and 1.8 ng/mL peak plasma concentration with 500 mg/kg oral dose **(b)** The mean plasma levels after 1 h was 11.1–0.6 nmol/L, which remained constant for a month in 15 colorectal cancer patients given 0.4–3.6 g of curcumin **(c)** Plasma concentration in micromolar level show poor absorption and low systemic bioavailability with 4–8 g of oral curcumin **(d)** Curcumin via reductases is converted into dihydrocurcumin and tetrahydrocurcumin transforms via β-glucuronide enzymes into dihydrocurcumin glucuronide and tetrahyrocurcumin glucuronide and then reduced to hexahydrocurcumin, tetrahydrocurcumin, octahydrocurcumin, and hexahydrocurcuminol **(e)** 75% in faces and negligible amount in urine with oral dose of 1 g/kg of curcumin, 11% in urine when administered intraperitoneally **(f)** Poor absorption due to low solubility of almost 11 ng/mL at 5.0 pH results in low dissolution rate, poor intestinal permeability in caco-2 cells of 0.07 × 10–6 cm/s, and extensive first-pass metabolism—hence low bioavailability	[41–43]
4	Danshen	**(a)** Danshensu and tanshinone IIA were noticed to absorb rapidly after oral administration in various animal studies **(b)** For danshensu, half-life of the drug, 32 min; steady-state volume of distribution, 149 mL/kg; mean residence time, 48 min; and total clearance, 3.13 mL/min/kg **(c)** Degradation of lithospermic acid B resulted in protocatechuic aldehyde, danshensu, lithospermic acid, and their isomers **(d)** Lithospermic acid B in Caco-2 cell monolayer system of rats resulted in low permeability and low bioavailability because of poor absorption and substantial metabolism **(e)** Excreted in urine; active phenolic compounds underwent metabolic transformation in liver and colon. 87% of the urine metabolite fraction constituted of caffeic acid, danshensu, and protocatechuic aldehyde	[44,45

Continued

Table 14.2 Pharmacokinetics and bioavailability of natural products—cont'd

S. no.	Natural products	Pharmacokinetics/bioavailability	Reference
5	Echinacea	**(a)** Absorption of alkylamides were fast with time taken 20–45 min for attaining maximum serum alkylamide concentration **(b)** Their glycosides have absorption is extremely rapid, Tmax of 15.0 min after intragastric administration in rats (100 mg/kg).The serum concentration was very low Cmax of 612.2 $\pm$ 320.4 ng/mL, elimination T1/2 was 74.4 min, and after 6-h postdosing, the concentration decreased to 36.3 ng/mL and distribution and elimination of echinacoside were extremely fast in rats about 12.4 min and 41.0 min, respectively, and after intravenous administration (5 mg/kg), and the mean concentrations decreased from 15,598.8 ng/mL in 2 min to 43.6 ng/mL in 4 h **(c)** Greater permeability	[46,47]
6	Garlic	**(a)** Human urinary samples showcased *N*-acetylcysteine S-conjugates (mercapturic acids) **(b)** Urinary concentration also detected sulfur compounds, namely, diallyldisulfide, diallylsulfide, and dimethyldisulfide **(c)** Urinary concentration detected *N*-acetyl-S-allyl-L-cysteine (allylmercapturic acid)	[48,49]
7	Ginger	**(a)** Phenolics were existed as phase II metabolites in plasma (glucuronide conjugates) **(b)** Absorb at 2 g dose with $t_{1/2}$: 75–120 min and Tmax: 55–65.6 min **(c)** 6-Gingerols: the area under the curve (AUC), 2.8 μg min/ml (250 mg) and 5.3 μg min/ml (500 mg); Cmax: 100 and 250 mg were 0.3 μg/mL (100) and 0.4 μg/mL (250 mg); their sulfates at doses 1.0–2.0 g	[50,51]
8	Gingko biloba	**(a)** Terminal half-life of around 4.5 h **(b)** Dose linear half-life of bilobalide (2.2 h) and ginkgolides A (1.7 h), ginkgolides B (2.0 h), clearance: 24.2–37.6 mL/min/kg **(c)** Cmax of bilobalides and total ginkgolides were two to three times more than phospholipid complex formulation, and mean elimination half-life of individual terpene was in between 120 and 180 min in healthy volunteers after oral administration of 160 mg extract **(d)** Half-lives of ginkgolides A (4.5 h), ginkgolides B (10.6 h), and bilobalide (3.2 h) **(e)** AUC for reference product was found to be 3 times greater for ginkgolide B, 1.5 times greater for ginkgolide A, and 1.2 times higher for bilobalide and obvious differences between Cmax in healthy patients **(f)** Peak plasma concentration—flavonol (2 – 3h), after acid hydrolysis, elimination complete in 24 h **(g)** Maximum plasma concentrations of flavonol conjugates within 30 min **(h)** Absolute bioavailability—ginkgolides A and B were greater than 80% and ginkgolide C very low after the oral intake of 80 mg of EGb 761 **(i)** Bioavailability—70% bilobalide after oral intake of 120 mg of EGb761 **(j)** Mean bioavailabilities—ginkgolide A (80%), ginkgolide B (88%), and bilobalide (79%)	[52–58]

Table 14.2 Pharmacokinetics and bioavailability of natural products—cont'd

S. no.	Natural products	Pharmacokinetics/bioavailability	Reference
9	Ginseng	**(a)** Ginsenosides with a short half-life of about 0.2–18 h and low bioavailability **(b)** Red ginseng—poorly absorbed from the gastrointestinal tract **(c)** Ginsenosides—rapid GIT absorption with bioavailability of 7.06% **(d)** Intravenous dose in diabetic rat exhibited rapid and voluminous biliary excretion of 43%–100% **(e)** Ginsenosides—CLR value of 624 mL/h/kg by glomerular filtration in Sprague-Dawley rats **(f)** Low bioavailability	[[illegible],60]
10	Milk thistle	**(a)** In 200 mg/kg of silybin phosphatidyline complex (IdB 1016), the conjugated form of silybin was about 94% cummulative in plasma. Cumulative biliary excretion was about 3.7% and urinary excretion was about 3.3% for IdB 1016 dose, biliary excretion was about 0.001% and urinary excretion was about 0.032% for silybin dose **(b)** Cumulative biliary and urinary excretion of IdB 1016: 13% and silymarin: 2% **(c)** AUC linear for 102, 153, 203, and 254 mg of silybin diastereomers on 6 healthy male volunteers; 10% of silybin was present as conjugates in plasma, and only 5% of the dose was excreted in urine in approximately 6 h **(d)** 70% or 80% of silymarin should be released in vitro within 15 min or 30 min for desirable plasma **(e)** The free and conjugated silybin concentration elevated at 2.4 and 3.8 h, respectively, and half-life was found to be 1.6 and 3.4 h, respectively, for 80 mg silybin equivalent dose, increase of Cmax and AUC by two- and threefold, respectively **(f)** Amounts of free and conjugated form of silybin recovered in bile were 11% and 3%, respectively, after 48 h, and silydianin and silychristin were found to be 1% of the total dose after oral administration of both formulation of silymarin and silipide equivalent to 120 mg silybin by patients requiring T-tube biliary drainage; AUC of total silybin was 40-fold more than free silybin when 120 mg of silybin was orally administered in patients with extrahepatic biliary obstruction **(g)** Bioavailability of silymarin is about 23%–47%, which indicated a low bioavailability profile due to poor solubility **(h)** Silybin in silipide had bioavailability 4.2 higher than silymarin	[61–66]

14.2.1 Angelica sinensis (Oliv.) Diels (Angelica root)

Over past 2000 years, *Angelica sinensis* Radix or roots of *A. sinensis* is used as traditional Chinese medicine as a painkiller and intestine softener and to invigorate and replenish blood. According to Chinese Pharmacopeia the roots of *Angelica* are obtained from *Angelica sinensis* (Oliv.) Diels from the family Umbellaceae. The main chemical constituents are ferulic acids, *Z*-ligustilide, angelicide, senkyunolide A, butylphthalide, butylidenephthalide, phthalide dimers, organic acids, polyacetylenes, coniferyl ferulate, amino acids, and vitamins (Fig. 14.1) [7]. Ferulic acid mainly contributes to the therapeutic activities [67].

14.2.1.1 Pharmacokinetics and bioavailability

A. sinensis are reported to have low solubility, low bioavailability, and fast metabolism. Therefore many attempts have been made to improve these parameters by involving nanodrug delivery system; especially encapsulation with poly(lacto-*co*-glycolic) acid has improved their characteristics and widen the scope of applications particularly in vaccine delivery and improve immune responses [36].

14.2.1.2 Pharmacodynamics

Antiinflammatory activity of angelica root is reflected through its components ferulic acid and isoferulic acid, which inhibit the macrophage inflammatory protein-2 (MIP-2) production by murine macrophage in RAW 264.7 cells [8,9]. A prescreening NF-κB dependent transactivation tool was used to study the antiinflammatory effect of angelica root (ethyl acetate fraction) by suppressing the NF-κB luciferase activity and lowering the PGE2 and NO production of lipopolysaccharide (LPS)/IFN-g-stimulated murine primary peritoneal macrophages [10]. Jung et al. utilized a murine air pouch model to demonstrate that the leukocyte count in the pouch exudate lowered in the BALB/c mice fed with 100 mg/kg body weight of angelica root

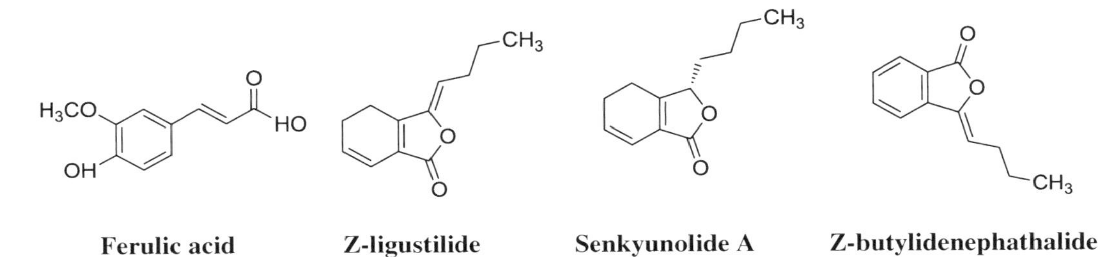

Fig. 14.1 Main chemical constituents of *Angelica sinensis* (Oliv) Diels.

extract. This was followed by a decrease in the neutrophil count, IL-6 mRNA level, and TNF-α mRNA in the pouch membrane and decreased concentration of IL-6 and PGE2 in the pouch fluid [11]. Also, PGD2 concentration in pouch fluid increased, exhibiting the anti-inflammatory activity of angelica root. Further, it was reported that n-butylidenephthalide decreased the secretion of IL-6 and TNF-α during LPS-stimulated activation of murine dendritic cells 2.4 via the suppression of the NF-κB-dependent pathways [7,12].

14.2.2 Astragalus membranaceus (Astragalus)

Astragalus membranaceus (AM) is a traditional Chinese medicine (TCM) that belongs to a leguminous family. The active ingredient of astragalus is astragalus IV (Fig. 14.2). Their dry roots known as *Astragali Radix* are used as tonic in TCM. This plant possesses a plethora of activities such as antihyperglycemic, antioxidant, immunomodulatory, antiviral, and antiinflammatory [68].

14.2.2.1 Pharmacokinetics and bioavailability

As pharmacodynamics focuses on the function of drug on the biological organism portraying a strong relation of drug effect-concentration-time. Astragalus is reported to have low absorption in Caco-2 cells particularly because of high molecular weight, low lipophilicity, poor intestinal permeability, and their paracellular transport [37]. They are found in selected tissues such as heart muscle, brain, kidney, lung, spleen, liver, skin, stomach, duodenum, and ovary

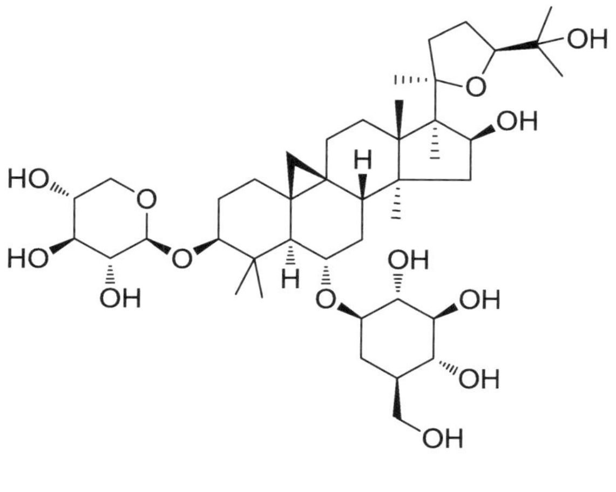

Astragalus IV

Fig. 14.2 Active ingredient of astragalus.

following intravenous administration of 1.5 mg/kg of astragalus IV in rats with highest concentration in the liver, kidney, and lung. The rate at which astragalus IV combines with plasma protein was found to be 83%–90%, and they are slowly eliminated by hepatic clearance of 0.004 L/kg/min [38]. The preclinical maternal and fetal toxicity of astragalus IV by Zhu et al. was reported at 1.0 mg/kg and higher than 0.5 mg/kg (with no teratogenic effects in rabbits and rats), respectively [39]. It is reported that bioavailability of astragalus IV is usually low. A study reported that after oral administration the bioavailability in rats and beagle was found to be around 3.66% and 7.4%, respectively [40]. Therefore the absolute bioavailability through oral administration poses tremendous restriction in its application.

14.2.2.2 Pharmacodynamics

Astragalus exhibits antiinflammatory mechanism through the inhibition of TNF-α and IL-1β along with the suppression of nuclear factor-κB (NF-κB) activation [13]. The antiinflammatory activity both in vivo and in vitro is registered for astragalus extract and their saponins and polysaccharides. Their extract demonstrated amelioration of preserving nuclear factor-κB expression, enhanced atopic dermatitis skin lesions in 1-chloro-2,4-dinitrobenzene-induced mice by regulation of cytokines, and subside Th2 cytokine expression and TNF-α level [14]. Besides, Astragali Radix were reported to exert antiinflammatory effect by inhibition of NF-κB channeled transcription and by MKP-1-dependent Erk1/2 and p38 inactivation [15]. In addition, their major constituents are studied to be effective against inflammation. The polysaccharides of astragalus efficiently improve the palmitate-induced proinflammatory responses via AMPK activity in RAW 264.7 and portrayed structure defensive properties in Caco2 cells that are infected by lipopolysaccharide. Moreover, astragaloside I, II, and IV and isoastragaloside II have demonstrated that the capability of obstructing lipopolysaccharide-induced nitric oxide development in RAW 264.7 macrophages, especially astragaloside IV, is more responsive as anti-inflammatory agent and lessens diabetic nephropathy by inhibiting NF-KB channeled inflammatory gene expression in rats [16].

14.2.3 *Curcuma longa* L. (Curcumin)

Since time immemorial, Ayurveda medicine has enlisted turmeric (*Curcuma longa*) as valuable for treating inflammatory conditions. Turmeric is broadly classified into volatile and nonvolatile constituents along with resins, proteins, sugars, and fibers. The volatile portion constitutes oils like turmerone, zingiberone, and atlantone, and the nonvolatile portion constitutes mainly of three curcuminoids, namely, curcumin, demethoxycurcumin, and

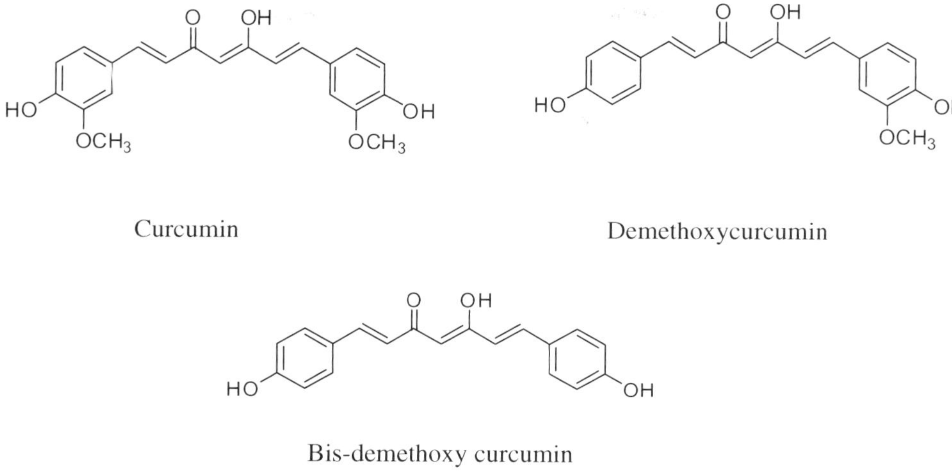

Curcumin

Demethoxycurcumin

Bis-demethoxy curcumin

Fig. 14.3 Main chemical constituents of *Curcuma longa* L.

bisdemethoxycurcumin (Fig. 14.3). Numerous pharmacological activities are linked to curcumin such as antibacterial, antioxidant, antifungal, and antiinflammation. Curcumin suppressed the inflammatory response via inhibition of the induction of iNOS and COX-2 and the production of cytokines like interferon-γ due to the suppression of JAK(Janus kinase)/STAT signaling cascade by exerting effect on the SHP-2 (Src homology 2 domain) having protein tyrosine phosphatase [41]. In vitro and in vivo studies have proven that it is well founded for various inflammation-stimulated conditions such as chronic anterior uveitis, pancreatitis, arthritis, inflammatory bowel diseases, and cancer [69].

14.2.3.1 Pharmacokinetics and bioavailability

Numerous clinical studies on curcumin depict a clear pattern on the pharmacokinetics (absorption, distribution, metabolism, and excretion). Curcumin is poorly absorbed but rapidly metabolized, which leaves only certain distributed portion in the kidney and spleen, which decreases with time and rapidly excreted in feces that is indicative of low systemic bioavailability [41]. There are numerous studies in vitro and in vivo that revealed that dose and route of administration may be factors crucial for attaining beneficial results. On the one hand, some studies on rats revealed that with low intravenous doses of curcumin (40 mg/kg), a complete plasma clearance results within 1 h and with 500 mg/kg oral dose results in 1.8 ng/mL peak plasma concentration. On the other hand, phase I clinical trials for 4-month duration involving 15 colorectal cancer patients who were given 0.4–3.6 g of curcumin demonstrated that at higher doses the mean plasma levels after 1 h was 11.1–0.6 nmol/L, which remained constant for a month. However, another study suggested that 4–8 g of oral curcumin resulted plasma concentration in micromolar level showcasing a case of poor absorption and low systemic bioavailability. Interestingly a study showed that addition of other bioactive alkaloid compounds such as *Piper longum* and *P. nigrum* to curcumin accelerated its intake in brain by 50% [42]. During metabolism, absorption from the bowel depicts a significant biotransformation into metabolites is witnessed by curcumin. Sulfate derivative and glucuronide are the two main metabolites after sulfate conjugation and glucuronide O-conjugation of curcumin, respectively. Chiefly, curcumin via reductases is converted into dihydrocurcumin and tetrahydrocurcumin, which then transforms via β-glucuronide enzymes into dihydrocurcumin glucuronide and tetrahyrocurcumin glucuronide, which can be further reduced to hexahydrocurcumin, tetrahydrocurcumin, octahydrocurcumin, and hexahydrocurcuminol. Further, excretion of curcumin metabolites also depends upon the route of administration and vehicle. An oral dose of 1 g/kg of curcumin in rats demonstrated 75% of it in faces and negligible amount in urine; however, a similar amount was seen in feces and 11% in urine when administered intraperitoneally.

Although curcumin has a plethora of therapeutic activities, low bioavailability is the major challenge faced by this green drug. This poor bioavailability reflects low absorption when administered orally because of various factors such as low solubility of almost 11 ng/mL at 5.0 pH, which consequently results in low dissolution rate; instability in intestinal pH; poor intestinal permeability in caco-2 cells of about 0.07×10^{-6} cm/s even after exclusion of ABCB1 efflux transporter involvement in curcumin uptake; and extensive first-pass metabolism (intestinal and hepatic). A possible hypothesis that BCS class II drugs (low solubility and high permeability) that are effluxed by intestinal transporters could be compared to the reasons associated with the low solubility of curcumin. A curcumin is available at low concentration to enterocytes eventually leading to minimal saturation of metabolizing enyzmes. Therefore many factors that are interlinked with low bioavailability can be resolved by adopting technological advancement in the form of encapsulated formulation [43]. Researchers have learned through various studies that structural modifications, complex formation, and biological transporters using formulation can help to ameliorate the bioavailability such as liposomes, cyclodextrin formulation, polymeric nanoparticles, lipid complexes and so on. A phosphatidylcholine curcumin complex (phytosomes) has demonstrated better bioavailability than unbound curcumin. In vitro rat models that were investigated observed five times higher area under the curve and peak plasma concentration in complex than in unbound drug.

14.2.3.2 Pharmacodynamics

The mechanism of action through which curcumin acts against various diseases is reported in several studies. The therapeutic activities of curcumin are linked to the performance of various inflammatory components. The downregulation/upregulation of inflammatory cascades has proven beneficial against numerous chronic diseases such as cardiovascular, neurodegenerative, autoimmune, and endocrine. The proinflammatory transcription factors are important targets of curcumin via which it exerts antiinflammatory activity. The inflammatory pathways of curcumin are denoted in Fig. 14.4 [17]. The nuclear factor (NF-κB) is an ubiquitous eukaryotic proinflammatory transcription factor that is responsible for the regulation of inflammation, transformation, cellular proliferation, and tumorigenesis suppressed by curcumin (Fig. 14.5) [18]. Studies on either alveolar macrophages or human peripheral blood monocytes have demonstrated that curcumin suppresses the tumor necrosis factor alpha

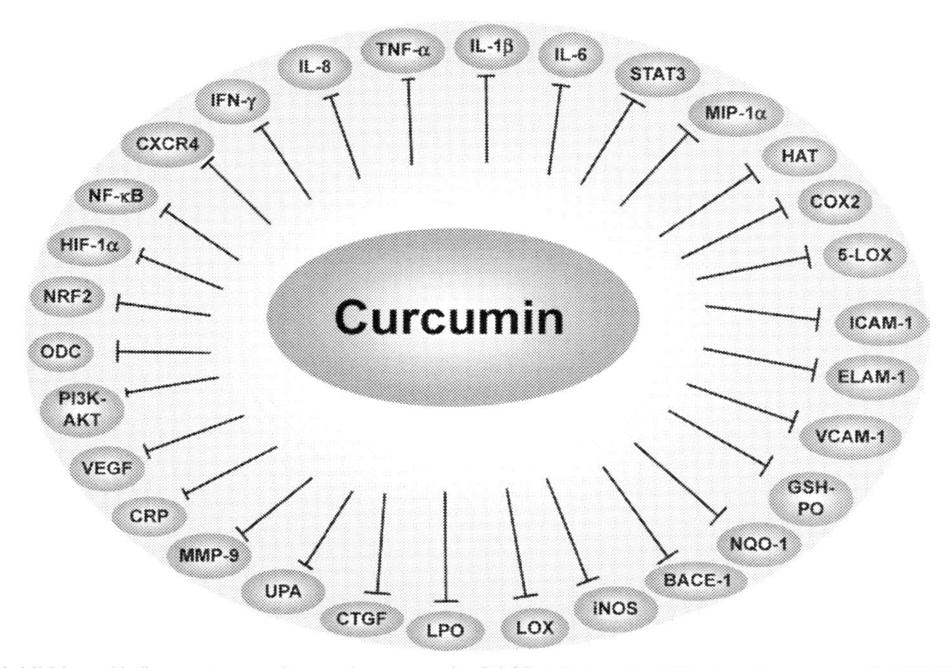

Fig. 14.4 Inhibition of inflammatory pathways by curcumin. BACE-1, beta-site APP-cleaving enzyme 1; CRP, C-reactive protein; CTGF, connective tissue growth factor; ELAM-1, endothelial leukocyte adhesion molecule-1; HAT, histone acetyltransferase; HIF, hypoxia inducible factor; ICAM-1, intracellular adhesion molecule-1; LPO, lipid peroxidation; MMP, matrix metalloprotease; NF-B, nuclear factor kappa B; ODC, ornithine decarboxylase; STAT, signal transducers and activator of transcription protein; TNF, tumor necrosis factor; VCAM, vascular cell adhesion molecule-1; VEGF, vascular endothelial growth factor [17].

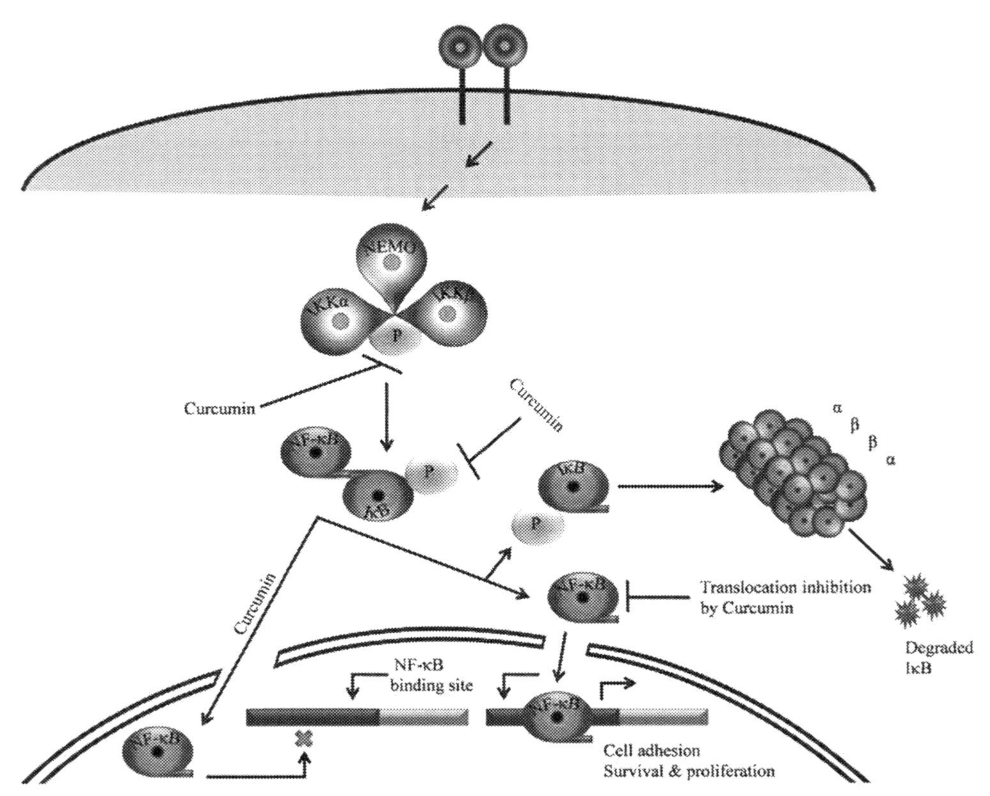

Fig. 14.5 Modulation of NF-κB pathway by curcumin. Curcumin inhibits phosphorylation of inhibitors of IκBs; as a result, nuclear translocation of NF-κB is prevented. Moreover, curcumin also attenuates DNA binding ability of NF-κB [18].

(TNF-α); macrophage inflammatory protein-1α; C-reactive protein (CRP); monocyte chemoattractant protein (MCP-1); the chemokines interleukin (IL)-1β, IL-2, IL-6, IL-8, and IL-12; chemokine receptor CXCR-4; and so on. Further, curcumin binds to lipoxygenase (5-LOX) and cyclooxygenase-2 (COX-2) to inhibit the activity in vivo and suppress inducible nitric oxide synthase (iNOS) through the inhibition of NF-κB [42]. Besides, curcumin also supports downregulation of the epidermal growth factor receptor and the activity of extracellular signal–regulated kinase 1/2 (also called mitogen-activated protein kinase [MAPK]) and c-Jun N-terminal kinase (JNK), as well as to inhibit the phosphatidylinositol-3 kinase/AKT pathway in a variety of cancer cells [19]. They can also completely suppress the activity of intercellular signaling protein kinase C, protamine kinase, autophosphorylation-activated protein kinase, and pp60c-src tyrosine kinase; these may be other plausible ways by which curcumin inhibits cytokine production. Curcumin could downregulate ROS-mediated inflammation through the nuclear factor (erythroid-derived 2)-related factor 2 (Nrf2) pathway. Reactive oxygen species (ROS) modulates the expression of the NF-κB and TNF-α pathway and has a central role in the inflammatory response [20]. Curcumin decreases the TNF-α-induced expression of intercellular adhesion molecule-1 and vascular cell adhesion molecule-1 in endometriotic stromal cells; curcumin displays antiinflammatory activity against adipose liver steatosis (e.g., in livers of obese mice) through phosphorylation of the signal transducer and activator of transcription 3 (STAT3), as well as via downregulation of suppressor of cytokine signaling 3 and sterol regulatory element–binding protein-1c; curcumin attenuates gene expression of mitochondrial DNA, nuclear respiratory factor 1, and mitochondrial transcription factor A, which are also involved in inflammation; curcumin exerts antiinflammation through epigenetic activities: modulation of histone deacetylases and histone acetyltransferases and inhibition of DNA methyltransferases [21].

14.2.4 *Salvia miltiorrhiza* (Danshen)

Danshen belongs to the family Labiatae and is derived from the dried roots of *Salvia miltiorrhiza*, tan shen, or red sage. The two major bioactives of danshen are danshensu and tanshinone IIA. (Fig. 14.6). Danshen is an antiinflammatory agent known to be effective in in the management of acute myocardial infraction.

14.2.4.1 *Pharmacokinetics and bioavailability*

Mostly, danshen are administered in the form of decoctions or extracted powder; there are other active moieties involved with them, so the evaluation of their pharmacokinetics cannot stand as accurate, and a possible idea could be drawn about their pharmacokinetics. The two major bioactives of danshen, namely, danshensu and tanshinone IIA, were noticed to absorb rapidly after oral administration in

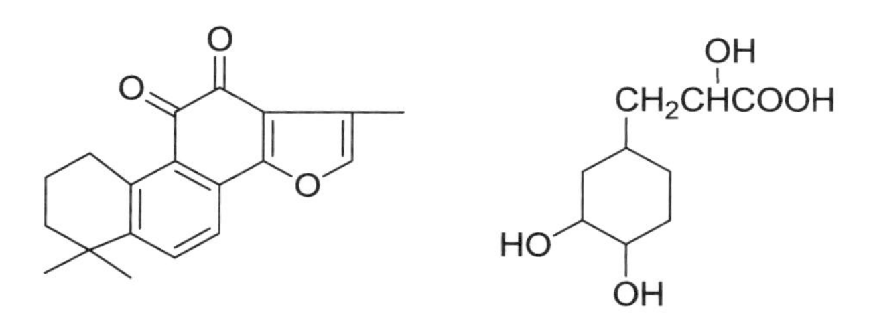

Fig. 14.6 Main chemical constituents of *Salvia miltiorrhiza* (danshen).

various animal studies. Danshensu is studied to be quickly absorbed in human subjects too. The pharmacokinetics of danshensu evaluated the plasma concentration profile by simple high-performance liquid chromatography after its intravenous administration to rabbits, which reflected the half-life of the drug to be 32 min, steady-state volume of distribution to be 149 mL/kg, mean residence time to be 48 min, and total clearance to be 3.13 mL/min/kg. Another bioactive compound of danshen, lithospermic acid B, is studied for its hydrolytic kinetics by reverse-phase high-performance liquid chromatography. The degradation compounds identified from lithospermic acid B were protocatechuic aldehyde, danshensu, lithospermic acid, and their isomers. Kim et al. also studied the pharmacokinetic of lithospermic acid B utilizing Caco-2 cell monolayer system of rats to estimate the rate of passage of lithospermic acid. Low permeability was observed through the system, which was indicative of the low bioavailability because of poor absorption and substantial metabolism. This low bioavailability was also confirmed in another study, which indicated poor absorption and extensive metabolism along with wide distribution of this compound. All the studied bioactives studied for their pharmacokinetic activities were administered after purification [44].

The knowledge about the bioavailability of danshen or any other bioactive highlights the basic amount or dose that could possibly be available for the treatment of various diseases and disorders such as cardiovascular, cerebrovascular, and cognitive diseases. Generally the bioavailability of danshen decoctions on oral uptake denotes the absorption, metabolism, and excretion of the ingested quantity. Numerous animal studies on danshen have provided ample reports on the uptake, absorption, metabolism, and excretion. The identification of their metabolites in vivo and pharmacokinetic properties has been reflected by analytical methods such as spectroscopic, spectrometric, and chromatographic. In this study, danshen decoctions were orally administered into miniature pigs, and ten phenolic compounds out of fifty compounds under investigation were studied for bioavailability under uptake, absorption, metabolism, circulation, and excretion. It was observed that before being excreted in urine, these active phenolic compounds underwent metabolic transformation in the liver and colon inclusive of glucuronidation, hydrolytic reactions, methylation, sulfation, hydrogenation, glycine conjugation, methylation, and decarboxylation. In addition, 87% of the urine metabolite fraction constituted of caffeic acid, danshensu, and protocatechuic aldehyde, which probably denoted their most significant components, and their concentration or circulation was enormously escalated by intestinal metabolism representing potential bioactivity. Moreover, fourfold increase in the monophenolic fraction occurred due to their colonic polyphenolic metabolism [45].

14.2.4.2 Pharmacodynamics

The pharmacodynamics study of danshen can be specifically found. A study suggests that about 12.5–100 μmolL^{-1} of ethanol extract of danshen was used to treat RAW 264.7 cells for 24 h, which showcased that the extract subdued the antiinflammatory cytokines such as IL-4, TGF-β, IL-10, and IL-1Ra and proinflammatory cytokines like NO, IL-6, TNF-α, and IL-1β. The expressions of IL-1β (at 6 and 12 h), platelet-activating factor (at 3 and 12 h), spleen (at 3 and 12 h), the pathological scores of thymus (at all-time points), and soluble IL-2 receptor (at 3 and 6 h) were remarkably lower in *S. miltiorrhiza*-treated group when compared with control because of which the acute pancreatitis severity, mortality rates in rats were reduced and pathological changes were observed in spleen, thymus, and small intestine. Their major bioactive component, tanshinone IIA, was reported to reduce ICAM-1 and VCAM-1 expression, which consequently reduced TNF-α-induced neutrophil adhesion to BMVECs in dose-dependent manner when administered in a dose range of 5–20 μg·mL^{-1} for 24 h. In brain microvascular endothelial cells, this component regulates TNF-α-induced expression of VCAM-1 and ICAM-1 via inhibition of ROS generation and NF-κB activation [22]. Other components of danshen have also reported to possess antiinflammatory activities. A neolignan of *S. miltiorrhiza* Bunge named ailanthoidol has been demonstrated to subdue NO generation, prostaglandin E2, cyclooxygenase-2, and inducible NOS with injected dose of 20 μmolL^{-1}. They also found to inhibit inflammatory cytokines, namely, IL-6 and IL-1β, in RAW 264.7 cells [23]. Cryptotanshinone, another component of Danshen that is a quinoid terpene, was reported effective in RAW264.7 cells when treated in dose range of 2.5–10 μmolL^{-1} by inhibiting the phosphorylation of mitogen-activated protein kinase (MAPKs) inclusive of p38MAPK, ERK1/2, and JNK, which are responsible for the regulation of proinflammatory mediator secretion. They also shown to fully remove LPS-triggered nuclear factor-KB activation when observed through western blot and immunofluorescence analysis [24].

14.2.5 Echinacea purpurea (Echinacea)

Echinacea, also known as *E. purpurea*, is grouped under Aster family (Asteraceae or Compositae). It is perennial prairie wild flowers from North America. The main bioactive component of *E. purpurea* is echinacoside (Fig. 14.7).

14.2.5.1 Pharmacokinetics and bioavailability

The pharmacokinetics of echinacea alkylamide was studied after ingestion of it via different formulations, namely, tablet, liquid, and

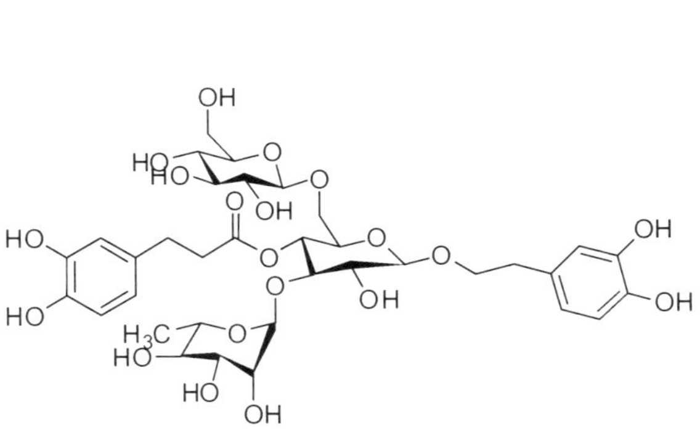

Echinacoside

Fig. 14.7 Main chemical constituents of *Echinacea purpurea* (echinacea).

lozenges. It was observed that the absorption of alkylamides was fast with time taken for attaining maximum serum alkylamide concentration being dependent upon the formulation typically about 20–45 min. In addition, tablet formulation took longer time than liquid preparation because the tablets had to disintegrate before the absorption process. Another study showcased that the absorption of echinacea alkylamide depends upon the food intake as high-fat breakfast investigated to prolong the time to attain maximum plasma concentration [46]. The pharmacokinetics of echinacoside, a water-soluble glycoside, present in traces in *E. purpurea* and is a secondary metabolite of *E. angustifolia* and *E. pallida* has also been investigated. Their absorption is extremely rapid Tmax of 15.0 min after intragastric administration in rats (100 mg/kg). The serum concentration was very low Cmax of 612.2 ± 320.4 ng/mL, elimination T1/2 was 74.4 min, and after 6-h postdosing, the concentration decreased to 36.3 ng/mL.The serum concentration–time curves for intragastric and intravenous administration were fitted to a one-compartment model and a two-compartment model, respectively. The distribution and elimination of echinacoside were extremely fast in rats about 12.4 min and 41.0 min, respectively, and after intravenous administration (5 mg/kg), and the mean concentrations decreased from 15,598.8 ng/mL in 2 min to 43.6 ng/mL in 4 h [47].

There are very few studies reported on the bioavailability of echinacea among which an in vitro study with a model of intestinal barrier demonstrated greater permeability in 12 echinacea alkylamides than caffeic acid derivatives. Similarly, in vivo studies also denoted that 8 echinacea alkylamides possessed certain bioavailability when compared with caffeic acid after oral intake.

14.2.5.2 Pharmacodynamics

Echinacea is currently being used in inflammation-related diseases such as infectious rhinosinusitis and pharyngitis. It is because of inflammation that is triggered by substances produced by macrophages that result in tissue swelling, which is indicative of capillary leakage and vasodilation and leukocyte infiltration. These substances are referred to as nitric oxides, peroxide; toxic oxygen radicals; interleukins (IL)-1, IL-6, Il-8, and IL-12; platelet-activating factor; leukotrienes; and tumor necrosis factor alpha (TNF-α). Besides, upper airway inflammation is triggered by various kinins and C-reactive protein, whereas pharyngeal and nasal inflammations are regulated by neural pathways that are basically parasympathetic. Echinacea seem to activate macrophages and control in vitro cytokine secretions. Their alkamide-rich extracts exert antiinflammatory actions by inhibition of 5-lipoxygenase and cyclooxygenase. Rininger's laboratory observed the reduction of prostaglandin E2 production and arachidonic acid metabolism by numerous *E. purpurea* products. The antiinflammatory effects of these preparations were also studied in paw edema of rats as well as their response to hyaluronidase, serotonin, formalin, and trypsin. Despite these encouraging discoveries, any randomized, blinded trials that have demonstrated clinically antiinflammatory properties are still to be reported [25].

14.2.6 *Allium sativum* L. (garlic)

One of the oldest cultivated herbal plant is garlic (*Allium sativum* L.), which was used for treatment of wound in World War II. This spice contains various compounds such as allicin, ajoene; thiacremonone; S-allylmercaptocysteine; proteins; diallyl sulfide; inorganic elements like iron, copper, and selenium; polyphenols; and amino acids (Fig. 14.8) [26]. Garlic and its bioactive compounds were used for cardiovascular diseases, atherosclerosis, arrhythmia, thrombosis, hyperlipidemia, diabetes, and hypertension in ancient period. They also possess antiinflammatory, antimicrobial, antioxidant, antineoplastic, and cardioprotective properties [70].

14.2.6.1 Pharmacokinetics and bioavailability

The pharmacokinetics of garlic is reflected when human urinary samples after garlic intake showcased *N*-acetylcysteine S-conjugates (mercapturic acids). Besides the urinary concentration also detected sulfur compounds, namely, diallyldisulfide, diallylsulfide, and dimethyldisulfide, when one volunteer ingested commercial capsules of

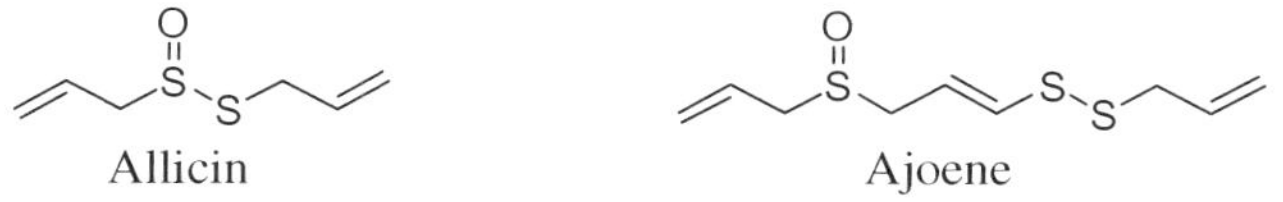

Allicin

Ajoene

Fig. 14.8 Main chemical constituents of *Allium sativum* L. (Garlic).

garlic. Another study portrayed that a single dose of garlic (200 mg) to six subjects found *N*-acetyl-S-allyl-L-cysteine (allylmercapturic acid) in urinary excretion [48].

Allicin is the major active component of garlic. In a study on allicin, availability or decomposition in the body during digestion is not affected by gastrointestinal pH, which is because of the inactivation of enzyme alliinase, which mainly functions to convert alliin into allicin at $\leq$ pH 3. Besides, allicin is not detected in urine or blood from 1 to 24 h after 25 g of oral ingestion of allicin or raw garlic (about 90 mg allicin). This can be very well overcome with the interconversion to their metabolites. S-Allylcysteine, one of their major metabolite after oral administration, has been reported to rapidly absorb and provide 100% bioavailability. Similarly, garlic extract with S-allylcysteine and S-allylmercaptocysteine has been studied for their bioavailability, where S-allylcysteine seemed to be easily detected in the kidney, liver, and plasma and is 103% bioavailable in mice, 87.2% in dogs, and 98.2% in rats. Hence, these metabolites form potent component for extensive studies in vivo or in vitro [49].

14.2.6.2 Pharmacodynamics

The pharmacodynamics study of the active constituents of garlic, namely, sulfur-containing compounds (S-allylmercaptocysteine, diallyl sulfide, ajoene) and thiacremonone and its oil, have been mainly studied. Arreola et al. studied the immunomodulatory and antiinflammatory effects of garlic and found that the various sulfur-containing bioactives of this spice, namely, S-allylmercaptocysteine, diallyl sulfide, and ajoene are responsible for the suppression of inflammatory factor NF-κB, and compounds like *Z*- and *E*-ajoene downregulated nitric oxide, tumor necrosis factor alpha (TNF-α), prostaglandin E2, interleukin-6, and Il-1β, whereas alliin also possessed antiinflammatory activity. It was reported that garlic decreased the activity of protease in chondrocyte-like cells, which are responsible for matrix degradation and inhibited the effect of matrix metalloproteinase and the tissue inhibitor of metalloproteinase-1 that are responsible for causing inflammation because of the presence of diallyl disulfide compounds. The inflammation-inducing cytokines and nitric oxide were reduced more effectively by fresh raw garlic than heated garlic. Garlic decreased the development of IL-6 and TNF-α and also brought reduction in IL-12 by increasing their dose [26]. Another study reflected that in IL-3-dependent murine pro-B-cells Ba/F3, its ethyl acetate soluble fraction was potent in inhibiting the NF-κB activation, COX-2 expression, and inducible nitric oxide synthase via Toll-like receptor-dependent pathway. In addition, the chloroform extract of garlic attenuated TNF-α-induced VCAM-1 expression through NF-κB-dependent pathway in human umbilical vein endothelial cells (HUVEC), which consequently decrease the adhesiveness of monocytes on endothelial cells. Thiacremonone, the bioactive compound of garlic, was investigated in paws of Sprague–Dawley rats and was demonstrated to inhibit 12-O-tetradecanoylphorbol-13-acetate-induced ear edema in ICR mice, carrageenan, and *Mycobacterium butyricum*-induced inflammatory and arthritic response. Moreover, garlic oil is reported to be efficient in suppressing 1-chloro-2,4-dinitrobenzene-induced contact hypersensitivity measured by ear swelling [27]. The prominent metabolite of garlic, namely, S-allylcysteine (SAC), has biological targets that exert antiinflammatory activities in brain, which is denoted in Fig. 14.9 [43].

14.2.7 Zingiber officinale roscoe (ginger)

Ginger (*Zingiber officinale* Roscoe) belongs to the family Zingiberaceae. *Z. officinale* is indigenous to tropical Asia, probably to southern China or India. The characteristic odor and flavor of ginger root are due to bioactive components of ginger such as zingerone, shogaols, gingerols, and volatile oils. Gingerols are a group of volatile phenolic compounds that are pungent in nature. 6-Gingerol is the major compound of the rhizome responsible for the pungency, while other gingerols, such as 4-, 8-, 10- and 12-gingerols, are present in lesser concentrations. Apart from that, there are 6-, 8- and 10-shogaols also present in ginger (Fig. 14.10). Ginger is firmly entrenched in culinary and the preferred choice of medical practitioners from time immemorial. This herb plays an important role in Ayurvedic, Chinese, Arabic, and African traditional medicines to treat headaches, nausea, colds, arthritis, rheumatism, muscular discomfort, and inflammation.

14.2.7.1 Pharmacokinetics and bioavailability

The pharmacokinetics of ginger has been highlighted through various studies, and it is tolerated well in humans. However, no steady-state pharmacokinetics of ginger is documented. The pharmacokinetic study by Mukkavilli et al. showed that ginger phenolics were detected only at subtherapeutic plasma concentration in mouse and human and major portion of these phenolics were existed as phase II metabolites in plasma (glucuronide conjugates). There was no inhibition or induction of clearance pathways with plasma subjection of phenolic ginger from day 1 to day 7 and their both free and conjugated form get accumulated in all tissues. Ginger is found to be most potent because of the reversible conversion of conjugated form of phenolics of ginger to free forms with the help of β-glucuronidase [50]. Another study explained that the pharmacokinetics of ginger extract by investigating their single dose that was raised from 100 mg to 2 g (standardized to 5% of total gingerols) was administered orally in 27

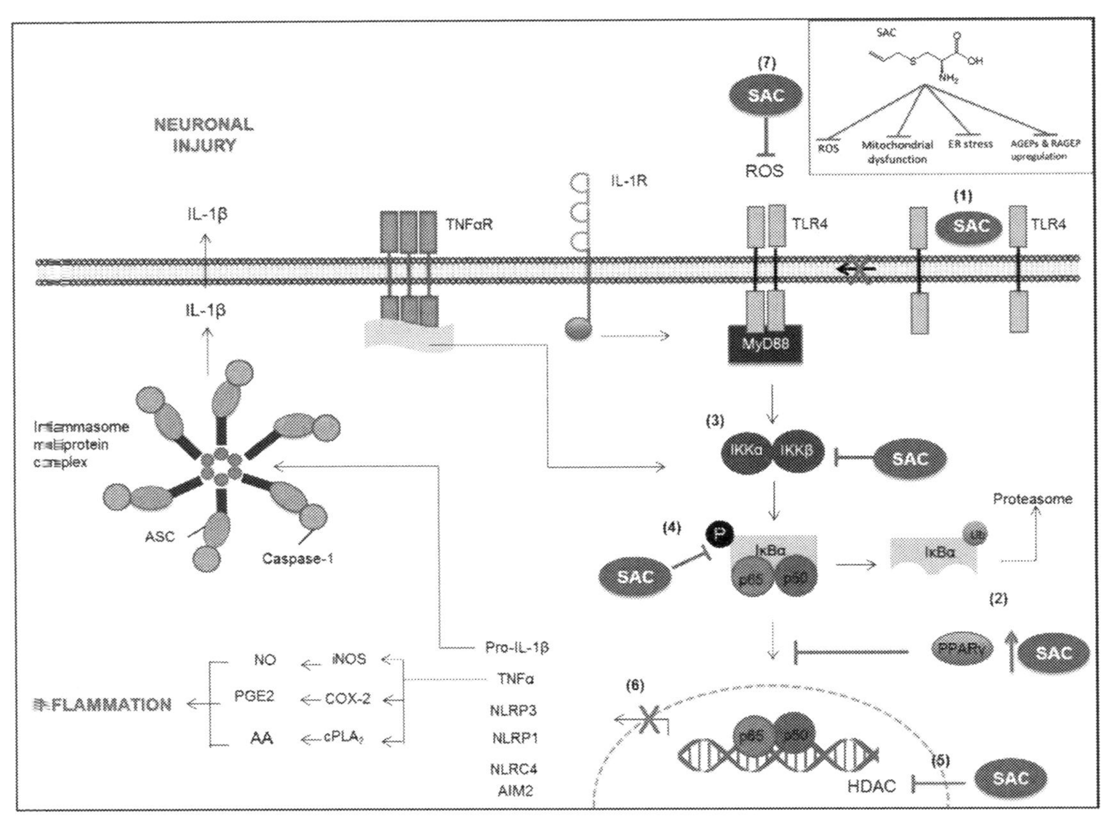

Fig. 14.9 Schematic representation of the many major biological targets for S-allylcysteine (SAC) in the brain (right upper corner) and the different cellular and molecular mechanisms by which SAC might exert antiinflammatory effects. SAC reduces reactive oxygen species (ROS) and oxidative stress, mitochondrial dysfunction, endoplasmic reticulum (ER) stress, and receptor for advanced glycation end product (RAGEP) upregulation. SAC also exerts antiinflammatory actions through the following mechanisms: (1) inhibition of the Toll-like receptor 4 (TLR4) at the extracellular level, reducing NF-κB pathway signaling; (2) increased expression of peroxisome proliferator–activated receptor g (PPARg) at the cytoplasmic level, inhibiting NF-κB nuclear translocation; (3) inhibition of the IkkB kinase activity, blocking IkBa phosphorylation and further release of NF-κB; (4) MAP kinase-mediated inhibition of IkBa phosphorylation, avoiding NF-κB nuclear translocation; (5) inhibition of histone deacetylase (HDAC), reducing NF-κB recognition at DNA sequence; (6) decreased levels of several proinflammatory cytokines, also limiting Pro-IL-1β-dependent IL-1β synthesis through inflammasome activation; and (7) inhibition of TLR4 activation by ROS scavenging [28].

healthy subjects to learn the pharmacokinetics of their active components, namely, 6-, 8-, and 10-gingerols and 6 shogaols. The active components of ginger were shown to quickly absorb at 2g dose with t1/2 from 75 to 120 min and Tmax 55 to 65.6 min. There was no free 6 shogaols or 6-, 8-, and 10-gingerols found in plasma; however, only glucuronides were observed as metabolites. Generally the conjugates of shogaol and gingerol were completely eliminated from plasma in

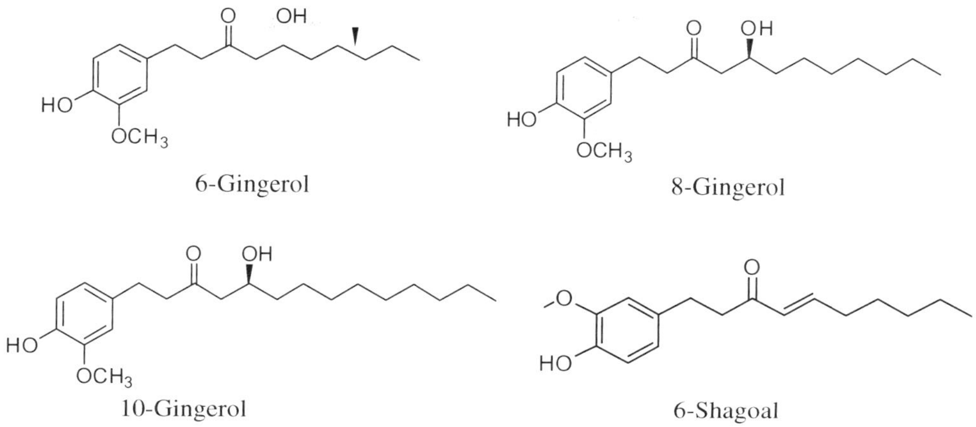

Fig. 14.10 Main chemical constituents of *Zingiber officinale* Roscoe (ginger).

4 h, and only conjugates of 6 shogaols were detected at dose below 1.0 g. For 6-gingerols the areas under the curve (AUC) at doses 250 and 500 mg were found to be 2.8 and 5.3 μg min/ml, respectively; Cmax at lower doses of 100 and 250 mg were 0.3 and 0.4 μg/mL, respectively; their sulfates were found at doses of 1.0–2.0 g. The sulfates of other active components of ginger, namely, 8- and 10-gingerol and 6-shogoal, were not detected even at high doses; however, their glucuronides were detected at low range other than 2 doses of ginger. Similar observations corresponding to these studies were observed in rats where after 6-gingerol administration, no free 6-gingerol was found in urine and bile but glucuronide was observed and the conjugation of 6-gingerol was found to be due to 2B7, UGT1A1, and 1A3 [51].

14.2.7.2 Pharmacodynamics

Ginger are potent candidate in exhibiting the antiinflammatory activities. Its active components, namely, gingerol and shogaol, are reported to inhibit leukotriene biosynthesis and prostaglandin via repression of prostaglandin synthetase or 5-lipoxygenase and also inhibit proinflammatory cytokines such as IL-1, IL-8, and TNF-α. Shogaols have demonstrated downregulate inflammatory COX-2 and iNOS expression in macrophages in another study. Besides the exorbitant production of NO, TNF-α, PGE, and IL-1β is successfully inhibited by hexane fraction of ginger extract. The activation of TNF-α and NF-κB is the primary cause for various inflammatory diseases such as myocardial infraction, arthritis, and cancer, and Habib et al. demonstrated that ginger extract decreased the excessive production of these expressions (Fig. 14.11) [29]. Ginger is also capable of inhibiting PGE2 production by inhibiting the LPS-induced COX-2 expression as investigated from the study of Lantz et al. In addition, 6-shogoal can be

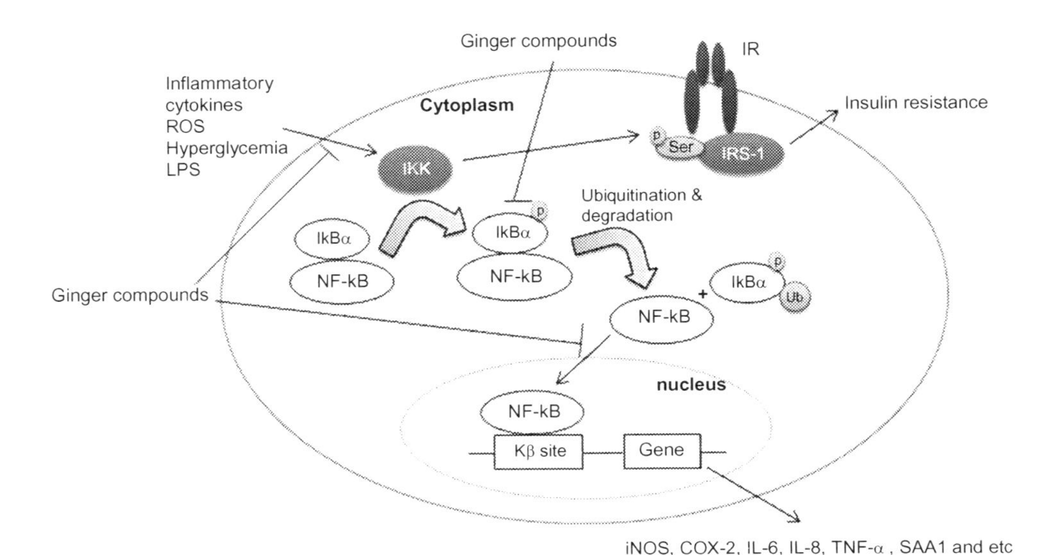

Fig. 14.11 Effect of ginger constituent on NF-κB signaling pathway [29].

efficiently used as a curative agent in gout patients because of its antioxidant and antiinflammatory activities. However, more trials, both in vitro and in vivo, need to be studied to expose the pharmacodynamics of ginger [30].

14.2.8 *Ginkgo biloba* L. (Ginkgo biloba)

Ginkgo biloba L. is a deciduous ornamental tree that belongs to the family Ginkgoaceae and is known since time immemorial. This antiinflammatory agent has been used for the cerebrovascular disorders, peripheral circulatory insufficiency, geriatric complaints, and Alzheimer's disease. Gingko biloba extract is available in the market in the form of tablets, capsule, tinctures, liquid extract, gel, sublingual spray, and tea. The relative percentage of this extract varies in context of flavonoids and terpenoids. Commercially a 40 mg extract contains standard 24% flavonoids (quercetin, isorhamnetin glycosides, and kaempferol) and 6% terpenes (ginkgolides and bilobalides) (Fig. 14.12). The fruits and leaves of this tree contain another category of constituents, namely, ginkgolic acids and related alkyl phenols. However, these components are linked to adverse reactions such as allergies and toxic effects. Therefore, according to the monograph of Commission E of German and WHO, the maximum safe concentration of ginkgolic acid is 5 μg/g. This constituent is reported to ameliorate cardiovascular blood flow by antagonizing platelet-activating factor, which is an endogenous mediator of inflammation that is developed by different inflammatory cells. Besides, bilobalide, a terpenoid fraction of ginkgo

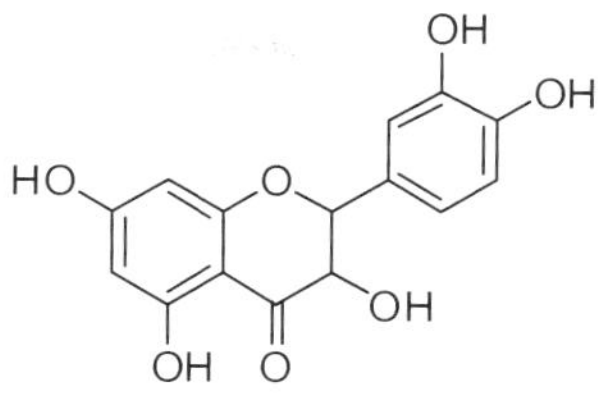

Quercetin

Fig. 14.12 Main chemical constituents of *Ginkgo biloba* L. (Ginkgo biloba).

biloba extract, exhibits antiischemic properties. Numerous clinical trials have been conducted with standardized aqueous acetone for their efficacy and evaluated for their pharmacokinetic characteristics in human and rats.

14.2.8.1 Pharmacokinetics and bioavailability

There are many studies that reflect the pharmacokinetics of the bioactive compounds of gingko biloba. In rats the radiolabeled EGb 761 standardized extract represents two compartmental pharmacokinetic models with a terminal half-life of around 4.5h [52]. Greater than 60% of the radioactivity was absorbed in stomach and small intestines, and more portions were detected in eyes and neuronal and glandular tissues. After oral administration of 30, 55, and 100mg/kg ginkgo extract EGb 761, the pharmacokinetics of bilobalide and ginkgolides A and B were found to be dose linear with half-lives of 2.2, 1.7, and 2.0h, respectively, and clearance ranged from 24.2 to 37.6mL/min/kg. In addition, through HPLC with diode array detection (DAD), another study demonstrated the pharmacokinetic parameters of various metabolites after oral administration of EGb 761 extract in rats and humans. About 7 metabolites were detected in urine samples of rat, which were 30% less than the actual flavonoids administered [53]. However, in humans, the metabolites were in the form substituted benzoic acids in free or conjugated form with no phenyl alkyl acids detected, which represented more extensive metabolism in them [53].

Among them one the study analyzed the pharmacokinetics of ginkgolides A and B and bilobalide by liquid chromatography/atmospheric pressure chemical ionization–mass spectrometry (LC/APCI-ITMS) in healthy volunteers after oral administration of 160mg extract consisting of 24% flavonol glycosides and 6% terpene lactones in free or phospholipid complex form. The Cmax of bilobalides and total ginkgolides were two to three times more than phospholipid complex formulation, and mean elimination half-life of individual terpene was in between 120 and 180min [54]. Another study reflected that the half-lives of ginkgolides A and B and bilobalide were 4.5, 10.6, and 3.2h, respectively. The pharmacokinetics is influenced by the type of extract, the formulation, and the dosage form. Pharmaceutically equivalent products with respect to dosage form and content of markers are not necessarily bioequivalent. The pharmacokinetic parameters were compared between two products in US market: Ginkgold® tablets with Egb 761 extract (reference) and ginkgo biloba capsules (test) from Centrum Herbals with a commercial dry extract as active pharmaceutical ingredients. This study was conducted on 12 healthy subjects and was single-dose, open, and crossover design. The total AUC for reference product was found to be 3 times greater for ginkgolide B, 1.5 times greater for ginkgolide A, and 1.2 times higher for bilobalide, and there were also obvious differences between Cmax. Intake on the reference product resulted in higher plasma concentrations, even though the reference product had lower ginkgolide and bilobalide contents [55]. In another striking study, the effect of increasing the doses (50, 100, and 300 mg) of ginkgo biloba, LI 1370, a commercial extract in the form of film-coated tablets, was demonstrated in 2 healthy subjects. It was observed that the flavonoids were available in plasma and urine in the form of glucuronic acid conjugates [56]. The peak plasma concentration of quercetin, kaempferol, flavonol aglycones, and isorhamnetin were observed for a period of 24 h after acid hydrolysis and measured [57]. The peak plasma concentrations of the total flavonol concentrations were reached within 2–3 h, which were proportional to the applied dose, and the half-life of total flavonols was 2–4 h and elimination was complete after 24 h. As is known from quercetin glycosides, absorption is strongly dependent on the sugar moiety attached to the aglycone. Based on the glycoside, maximum plasma concentrations of flavonol conjugates may be reached within 30 min or only after several hours, yielding completely different pharmacokinetic profiles.

A study on healthy persons after oral and intravenous administration of the extract of ginkgo biloba extract EGb 761 that contained 24% flavonoids and 6% terpenoids reflected their bioavailability. The absolute bioavailabilities of ginkgolides A and B were greater than 80%, and ginkgolide C indicated very low bioavailability after the oral intake of 80 mg of EGb 761. However, oral intake of 120 mg of EGb761 resulted in 70% bioavailability of bilobalide [58]. Another bioavailability study demonstrated that the mean bioavailabilities of ginkgolide A, ginkgolide B, and bilobalide were 80, 88, and 79%, respectively [71]. The bioavailability of ginkgolides from BioGinkgo (27% flavonoids and 7% terpenoids) was studied in rabbits by Li and Wong, which were compared with commercialized standard extracts (24% flavonoids and 6% terpenoids). This study suggested that higher concentrations of ginkgolides were maintained for prolonged residence time in BioGinkgo because of higher contents of terpenoids and enrichment of the extract with ginkgolide B with longer half-life than ginkgolide A.

14.2.8.2 Pharmacodynamics

G. biloba is reported to possess antiinflammatory effect by upregulating NF-κB factor and downregulating nitric oxide (NO) and PGE2 production inclusive of mRNA expression of iNOS, COX-2, and proinflammatory cytokinins (IL-6,IL-1β, and TNF-α) [72]. Their leaf extract efficiently suppresses the activation of macrophages and downregulates inflammatory stress markers like p53-phospho-serine-15 and p53; inflammatory expressions like iNOS, COX-2, TNF-α, and T-cell numbers like CD4 + and CD25-/FOX p3 were also reduced when investigated in the colon of mice depicting that they are efficient against chronic infections [59]. *G. biloba* extract was found to be effective in helping rats recover from colitis and inhibited mRNA expression of TNF-α, IL-6, and NF-κBp65 [73]. An atherosclerosis rat model after administration of 100 mg/kg per/day of *G. biloba* extract under 8 weeks of examination exhibited upregulation of inflammatory markers and downregulation of the mRNA expression of TNF-α and IL-1β when compared with control group [33].

14.2.9 Panax ginseng Meyer (ginseng)

Ginseng, a perennial herbaceous and half-shaped plant, is known as *Panax ginseng* Meyer. This herbal medicine is from the family Araliaceae with traditional importance in East Asian regions from the time of yore. Their active constituents are ginsenosides, alkaloids, phenolics, amino acids, flavonols (quercetin), polypeptides, and vitamins; among them, ginsenosides are considered biologically active component (Fig. 14.13). Ginsenosides are catalogued under dammarane-type

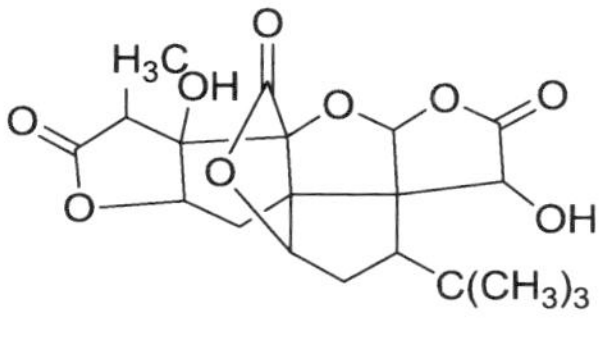

Ginkgolide A

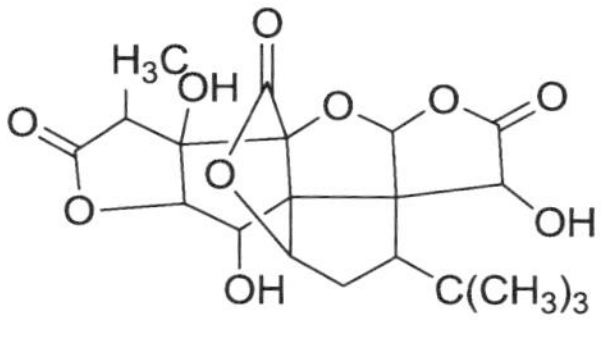

Ginkgolide B

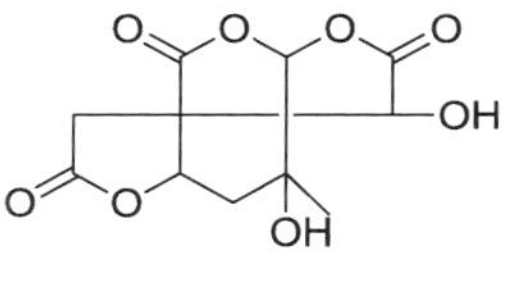

Bilobalide

Fig. 14.13 Main chemical constituents of *Panax ginseng* Meyer (ginseng).

triterpene saponins that consist of 17 carbons in a 4-ring structure named dammarane skeleton with varied sugar components like rhamnose, xylose, glucose, and arabinose that are attached to C-3 and C-20. The two major groups of ginsenosides are protopanaxatriol (PPT) and protopanaxadiol (PPD) where ginsenosides Rb1, Rb2, Rb3, Rd., and Rc are catalogued under PPD and ginsenosides Rg1 and Re come under PPT. In PPD group, Rg3 is a metabolite of Rb1, Rb2, Rb3, Rc, and Rd. and Rg3 is metabolized to Rh2, whereas in PPT group, Rh1 is metabolite of Re and Rg1 and Rg2 is a metabolite of Re. Ginsenosides of PPD type contain sugar moieties attached to β-OH at position C-3/C-20; however, sugar moieties are attached to α-OH at C-6 and β-OH at C-20 position in PPT-type ginsenoside [61]. They have been efficient against cardiovascular risks, hypertension, Alzheimer's, cognition, and diabetes (type 2), regulate glucose metabolism, boost immunity, and possess antiinflammatory activities [62].

14.2.9.1 Pharmacokinetics and bioavailability

The most active ingredient of ginseng is ginsenosides with a short half-life of about 0.2–18 h and low bioavailability. They are poorly absorbed from the gastrointestinal tract because it was undetected in plasma concentration as demonstrated when red ginseng was administered in healthy volunteers. However, after oral administration of ginsenosides in mice, rapid absorption was found of nearly 30% in 1 h. The bioactive components of ginseng such as Re, R1, Rg1, Rb1, and Rd. exhibited rapid GIT absorption with bioavailability of 7.06%. The in vitro studies conducted in models with intestinal bacterial and hepatic metabolism demonstrated that maximum ginsenosides undergone metabolic pathway. The metabolites of PPT were inclusive of Rh1 and PPT, whereas PPD metabolites included K, Rg3, Rh2's aglycone, and Rh2. Under mild acidic conditions, PPT ginsenoside is hydrolyzable to Rh1, and two degradation products of PPT ginsenosides, Rh1 and F1, reach in the systematic circulation in addition to compound K from stepwise deglycosylation of PPD ginsenosides. An intravenous dose in diabetic rat exhibited rapid and voluminous biliary excretion of about 43%–100%. Another study on Sprague–Dawley rats suggested that ginsenosides witnessed renal excretion engaging passive glomerular filtration instead of active tubular secretion, which was depicted by the CLR value of 624 mL/h/kg by glomerular filtration [62].

It is reported that ginsenosides possess extremely low bioavailability that is less than 10%, which limits its further applications. The poor aqueous solubility and low permeability are supposed to be the main causes of their low bioavailability. With an intravenous dose of 1 mg/kg of Rg3 or an oral dose of 10 mg/Kg of Rg3 that was administered to rats, the oral bioavailability of Rh3 was found to be 2.63%, whereas Rh2 was not been able to detect. However, the oral bioavailability of PPD

(36.8 ± 12.4) was greater than Rh2 and Rg3 ginsenosides. Moreover, when Rh2 was both orally and intravenously administered at a dose of 1 mg/kg and 0.1 mg/kg in native and micronized form, an improved result was monitored in micronized form of about 32%, which was greater than native form of about 16%. However, in rats, the oral bioavailability did not improve with 1 mg/kg of micronized Rh2 as it was only about 4.7% after oral administration. It was observed from a study that 50 mg/kg of ginsenoside in rats after oral administration resulted that the oral bioavailability was 4.35% for Rb1 and 18.40% for Rg1. PPT group of ginsenosides portrayed comparatively better bioavailability than PPD group [61].

14.2.9.2 Pharmacodynamics

In pharmacodynamics, it is very essential to select proper effective marker displaying the real-time effect after drug administration. Studies pertaining to the prevention of ginsenoside Rg1 and Rb1 on cardiovascular system linked to their stimulation of NO release in endothelial cells have been well documented. However, their effect on real-time channeling of NO release in myocardial ischemia is less. Therefore Zhan et al. investigated that ginsenosides Rg1 and Rb1 resulted about 40% upraise in NO levels in serum after 2 h of administration in isoproterenol (ISO)-induced myocardial ischemia rats when compared with control. These two moieties caused slow onset and significant increase of nitrate (NO_x^-) concentration from 0.5 to 5 h, which displayed their NO-inducing effect was drug concentration dependent as the drug concentration in serum lowered from 14,092.1 to 10.8 ng/mL for ginsenoside Rg1 and from 88,755.2 to 33,140.6 ng/mL for ginsenoside Rb1 at 0.17–5 h. In addition, a lag was observed between alteration of NO levels in serum and drug concentration, and Rg1 and Rb1 were found effective in producing real-time NO release in concentration-dependent manner in treated rats when their AUEC of serum NO was observed after ingestion [63].

14.2.10 Silybum marianum (L.) Gaertn. (Milk thistle)

Milk thistle is most prominently used in nontraditional therapies in Germany. It has numerous flavonolignans that are developed by a radical coupling of phenylpropanoid and flavonoids. The seed extracts of milk thistle contain mainly silybin about 50%–70% and their isomers, namely, silydianin, isosilybin, and silychristin. About 4%–6% of silymarin is contained in its fruit extract. Silymarin or silybin is beneficial for cirrhosis, hepatitis, and other gallbladder and liver conditions. Moreover the water-soluble derivative silibinin dihydrogensuccinate is useful for *Amanita phalloides* mushroom poisoning when administered intravenously. It is evident that silybin and its derivative have

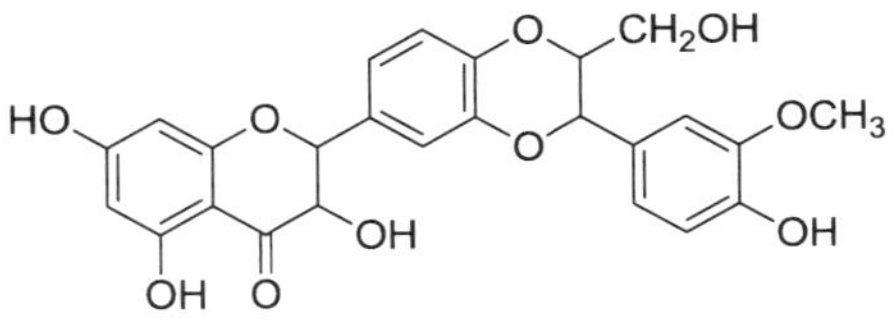

Fig. 14.14 Main chemical constituents of *Silybum marianum* (L.) Gaertn. (Milk Thistle).

P-glycoprotein modulating activities (Fig. 14.14). Silymarin is also currently investigated for anticarcinogenic, antioxidant, and antiinflammatory activities.

14.2.10.1 Pharmacokinetics and bioavailability

The pharmacokinetics of milk thistle are studied in vitro and in vivo. In rat models a comparative pharmacokinetic parameters were studied taking 200 mg/kg of silybin phosphatidylcholine (IdB1016) and silybin. It was observed that conjugated form of silybin in plasma is about 94% cumulative. The biliary excretion account for 3.7% and urinary excretion is 3.3% for IdB 1016, and for silybin administration, the biliary excretion is 0.001% and urinary excretion is 0.032% [64]. These observations were similar in another study with 200 ng/kg dose that accounted for cumulative biliary excretion and urinary excretion of 13% for of IdB1016 and 2% for silymarin and relative bioavailability of silybin 10 times greater than silymarin [65].

Weyhenmeyer et al. studied the pharmacokinetics of silybin diastereomers on 6 healthy male volunteers after different single doses (102, 153, 203, and 254 mg) and found area under the curve (AUC) linear to the given dose. Nearly 10% of silybin was present as conjugates in plasma, and only 5% of the dose was excreted in urine in approximately 6 h. Another study displayed the pharmacokinetics of silymarin in herbal products by comparing with standard. For achieving a desirable plasma concentration, it is recommended that about 70% or 80% of silymarin should be released in vitro within 15 or 30 min, respectively [66]. Silybin demonstrated greater lipophilicity and ameliorated penetration across membranes. After administration of about 80 mg silybin equivalent dose on healthy patients, the free and conjugated silybin concentration elevated at 2.4 and 3.8 h, respectively, and half-lives were found to be 1.6 and 3.4 h, respectively. This highlighted the fact that silybin is extensively converted to conjugated derivatives that are retained in circulation at relatively large concentration. The administration of 80 mg of silybin resulted in an increase of Cmax and AUC by two- and threefold respectively, when the formulations of hard and soft gelatin capsule were compared and evaluated for silybin pharmacokinetics. This study indicated that silybin has superior absorption quality from the oily medium [34]. Silybin concentration in bile after oral administration of both formulations of silymarin and silipide equivalent to 120 mg silybin by patients requiring T-tube biliary drainage demonstrated that they excreted primarily through biliary route, which is important with respect to its hepatoprotective activity. The amounts of free and conjugated form of silybin recovered in bile were 11% and 3%, respectively, after 48 h, and silydianin and silychristin were found to be 1% of the total dose. Moreover, AUC of total silybin was 40-fold more than free silybin when 120 mg of silybin was orally administered in patients with extrahepatic biliary obstruction reflected impaired excretion of the conjugates in bile [35].

Ghosh reported in a study that the bioavailability of silymarin is about 23%–47%, which indicated a low bioavailability profile due to poor solubility. A study on the comparison of biliary excretion of silybin in silipide and silymarin showcased that the former had bioavailability 4.2 times higher than the latter taken on an average. Also, as the bioavailability of silybin is less, the same silybin–phosphatidylcholine complex termed as silipide (IdB 1016) was developed by Gatti and Perucca [71].

14.2.10.2 Pharmacodynamics

The antiinflammatory mechanism by which silybin is reported to act is by the inhibition of synthesis of leukotriene B4 (IC50 15 μmoL/L) without any effect on the prostaglandin E2 formation at concentration up to 100 μmoL/L in isolated Kupffer cells of rats [72]. Silymarin showcased pathway-dependent inhibition to exert antiinflammatory effect as it was found to inhibit NF-κB DNA binding activity and its dependent gene expression induced by okadaic acid; however, no NF-κB activation induced by TNF-β was observed when it was treated on HEPG2 hepatoma cell line. Moreover, silymarin inhibited leukotriene synthesis via 5-lipoxygenase pathway and formation of prostaglandins through which it exerts its pharmacological activities [73].

14.3 Conclusion

This chapter deals with different natural products that possess antiinflammatory activities. Through this chapter a brief outlook on the pharmacokinetics, bioavailability, and pharmacodynamics that could aid in a better rational dosage regimen. This attempt could considerably contribute to more accurate and extensive scientific assessments based on quality, efficacy, and safety for establishing these herbal products as antiinflammatory drugs. Moreover, comprehensive studies relating to the bioavailability, pharmacokinetics, and pharmacodynamics should be initiated for taking full advantage of the therapeutic potential of these medicinal plants.

References

[1] N. Kishore, P. Kumar, K. Shanker, A.K. Verma, Human disorders associated with inflammation and the evolving role of natural products to overcome, Eur. J. Med. Chem. 179 (2019) 272–309.

[2] S. Sauer, A. Plauth, Health-beneficial nutraceuticals—myth or reality? Appl. Microbiol. Biotechnol. 101 (2017) 951–961.

[3] M.S.A. Khan, I. Ahmad, in: M.S.A. Khan, I. Ahmad, D. Chattopadhyay (Eds.), Herbal medicine: Current trends and future prospects, new look to phytomedicine, Elsevier, Academic Press, United States, 2019, pp. 3–13.

[4] X. Zhou, Y. Li, X. Chen, Computational identification of bioactive natural products by structure activity relationship, J. Mol. Graph. Model. 29 (2010) 38–45.

[5] N. Bernstein, M. Akram, M. Daniyal, H. Koltai, M. Fridlender, J. Gorelick, Chapter four - Antiinflammatory potential of medicinal plants: a source for therapeutic secondary metabolites, Adv. Agron. 150 (2018) 131–183.

[6] W.W. Chao, B.F. Lin, Bioactivities of major constituents isolated from Angelica sinensis (Danggui), Chinas Med. 6 (2011) 29.

[7] S. Sakai, H. Ochiai, K. Nakajima, K. Terasawa, Inhibitory effect of ferulic acid on macrophage inflammatory protein-2 production in a murine macrophage cell line, RAW264.7, Cytokine 9 (1997) 242–248.

[8] S.M. Jung, H.R. Schumacher, H. Kim, M. Kim, S.H. Lee, F. Pessler, Reduction of urate crystal induced inflammation by root extracts from traditional oriental medicinal plants: elevation of prostaglandin D2 levels, Arthritis Res. Ther. 9 (2007) R64–R72.

[9] R.H. Fu, H.J. Hran, C.L. Chu, C.M. Huang, S.P. Liu, Y.C. Wang, Y.H. Lin, W.C. Shyu, S.Z. Lin, Lipopolysaccharide stimulated activation of murine DC2.4 cells is attenuated by n-butylidenephthalide through suppression of the NF-κB pathway, Biotechnol. Lett. 33 (2011) 903–910.

[10] Z.Q. He, J.A. Findlay, Constituents of Astragalus membranaceus, J. Nat. Prod. 54 (1991) 810–815.

[11] J.H. Kim, M.H. Kim, G. Yang, Y. Huh, S.H. Kim, W.M. Yang, Effects of topical application of Astragalus membranaceus on allergic dermatitis, Immunopharmacol. Immunotoxicol. 35 (2013) 151–156.

[12] M. Ryu, E.H. Kim, M. Chun, S. Kang, B. Shim, Y.B.B. Yu, G. Jeong, J.S. Lee, Astragali radix elicits anti-inflammation via activation of MKP-1, concomitant with attenuation of p38 and Erk, J. Ethnopharmacol. 115 (2008) 184–193.

[13] S.S.S. Boyanapalli, A.T. Kong, "Curcumin, the king of spices": epigenetic regulatory mechanisms in the prevention of cancer, neurological, and inflammatory diseases, Curr. Pharmacol. Rep. 1 (2015) 129–139.

[14] S. Chun-Yan, M. Qian-Liang, R. Khalid, H. Ting, Q. Lu-Ping, Salvia miltiorrhiza, traditional medicinal uses, chemistry, and pharmacology, Chin. J. Nat. Med. 13 (2015) 163–182.

[15] J. Qiangqiang, Z. Ruyuan, T. Yimiao, C. Beibei, L. Rui, L. Lin, W. Lili, C. Yiwen, Z. Dandan, M. Fangfang, G. Sihua, Z. Dongwei, Salvia miltiorrhiza in diabetes: a review of its pharmacology, phytochemistry and safety, Phytomedicine 58 (2019) 152871.

[16] X. Li, X. Zheng, Z. Liu, Q. Xu, H. Tang, J. Feng, S. Yang, C.T. Vong, H. Gao, Y. Wang, Cryptotanshinone from salvia miltiorrhiza Bunge (Danshen) inhibited inflammatory responses via TLR4/MyD88 signaling pathway, Chinas Med. 2 (2020) 15–20.

[17] A. Saedisomeolia, M.M. Arzati, M. Abdolahi, M. Sedighiyan, A. Rangel, G. Muench, M. Zarezadeh, A. Jafarieh, N.M. Honarvar, Mechanisms of action of ginger in nuclear factor-kappaB signaling pathways in diabetes, J. Her. Med. 16 (2019) 100239.
[18] N.S. Mashhadi, R. Ghiasvand, G. Askari, M. Hariri, L. Darvishi, M.R. Mofid, Anti-oxidative and anti-inflammatory effects of ginger in health and physical activity: review of current evidence, Int. J. Prev. Med. 4 (2013) S36–S42.
[19] V.S. Kotakadi, Y. Jin, A.B. Hofseth, Ginkgo biloba extract EGb 761 has anti-inflammatory properties and ameliorates colitis in mice by driving effector T cell apoptosis, Carcinogenesis 29 (2008) 1799–1806.
[20] T. Belwal, L. Giri, A. Bahukhandi, M. Tariq, P. Kewlani, I.D. Bhatt, R.S. Rawal, Ginkgo biloba, in: S.M. Nabavi, A.S. Silva (Eds.), Nonvitamin and Nonmineral Nutritional Supplements, Academic Press, United States, 2018, pp. 241–250.
[21] S. Zhan, W. Guo, Q. Shao, X. Fan, Z. Li, Y. Cheng, A pharmacokinetic and pharmacodynamics study of drug-drug, interaction between ginsenoside Rg1,ginsenoside Rb1and schizandrin after intravenous administration to rats, J. Ethnopharmacol. 152 (2014) 333–339.
[22] P. Gua, A. Wusimana, S. Wanga, Y. Zhanga, Z. Liu, Y. Hu, J. Liu, D. Wang, Polyethylenimine-coated PLGA nanoparticles-encapsulated Angelica sinensis polysaccharide as an adjuvant to enhance immune responses, Carbohydr. Polym. 223 (2019) 115128.
[23] C.R. Huang, G.J. Wang, X.L. Wu, H. Li, H.T. Xie, H. Lv, J.G. Sun, Absorption enhancement study of astragaloside IV based on its transport mechanism in caco-2 cells, Eur. J. Drug Metab. Pharmacokinet. 31 (2006) 5–10.
[24] W.D. Zhang, C. Zhang, R.H. Liu, H.L. Li, J.T. Zhang, C. Mao, Preclinical pharmacokinetics and tissue distribution of a natural cardioprotective agent astragaloside IV in rats and dogs, Life Sci. 79 (2006) 808–815.
[25] R.A. Sharma, W.P. Steward, A.J. Gescher, Pharmacokinetics and pharmacodynamics of curcumin, Adv. Exp. Med. Biol. 595 (2007) 453–470.
[26] C.F. Moriano, E.G. Burgos, M.P. Gómez-Serranillos, Chapter 2—Curcumin: Current evidence of its therapeutic potential as a lead candidate for anti-inflammatory drugs-an overview, in: G. Brahmachari (Ed.), Discovery and Development of Anti-Inflammatory Agents from Natural Products, Natural Product Drug Discovery, Elsevier, United States, 2019, pp. 7–59.
[27] F. Toselli, A. Matthias, E.M.J. Gillam, Echinacea metabolism and drug interactions: the case for standardization of a complementary medicine, Life Sci. 85 (2009) 97–106.
[28] C. Jia, H. Shi, X. Wu, Y. Li, J. Chen, P. Tu, Determination of echinacoside in rat serum by reversed-phase high-performance liquid chromatography with ultraviolet detection and its application to pharmacokinetics and bioavailability, J. Chromatogr. B 844 (2006) 308–313.
[29] R. Mukkavilli, C. Yang, R.S. Tanwar, A. Ghareeb, L. Luthra, R. Aneja, Absorption, metabolic stability, and pharmacokinetics of ginger phytochemicals, Molecules 22 (2017) 553.
[30] Y. Yu, S. Zick, X. Li, P. Zou, B. Wright, D. Sun, Examination of the pharmacokinetics of active ingredients of ginger in humans, AAPS J. 13 (2011) 417.
[31] H.J. Won, H. Kim, T. Park, H. Kim, K. Jo, H. Jeon, S.J. Ha, J.M. Hyun, A. Jeong, J.S. Kim, Y.J. Park, Y.H. Eo, J. Lee, Nonclinical pharmacokinetic behavior of ginsenosides, J. Ginseng Res. 43 (2019) 354–360.
[32] H.M.N. Thimarul, C.W.Y. June, G. Manish, A.N.K. Muhammad, C.M. Long, A review of Panax ginseng as an herbal medicine, Arch. Pharm. Pract. 7 (2016) S61–S65.
[33] H.U. Schulz, M. Schürer, G. Krumbiegel, W. Wächter, R. Weyhenmeyer, G. Seidel, The solubility and bioequivalence of silymarin preparations, Arzneimittelforschung 45 (1995) 61–64.

[34] J.S. Jurenka (ASCP), Anti-inflammatory properties of curcumin, a major constituent of Curcuma longa: A review of preclinical and clinical research, Altern. Med. Rev. 14 (2009) 141–152.
[35] S.M.B. Asdaq, M.N. Inamdar, Pharmacodynamic and pharmacokinetic interactions of propranolol with garlic (Allium sativum) in rats, Evid. Complement. Altern. Med. 2011 (2011) 824042.
[36] W.W. Chao, Y.H. Kuo, W.C. Li, B.F. Lin, The production of nitric oxide and prostaglandin E2 in peritoneal macrophages is inhibited by Andrographis paniculata, Angelica sinensis and Morus alba ethyl acetate fractions, J. Ethnopharmacol. 122 (2009) 68–75.
[37] B.B. Aggarwal, K.B. Harikumar, Potential therapeutic effects of curcumin, the anti-inflammatory agent, against neurodegenerative, cardiovascular, pulmonary, metabolic, autoimmune and neoplastic diseases, Int. J. Biochem. Cell Biol. 41 (2009) 40–59.
[38] G. Kumar, S. Mittal, K. Sak, H.S. Tuli, Molecular mechanisms underlying chemopreventive potential of curcumin: current challenges and future perspectives, Life Sci. 148 (2016) 313–328.
[39] B. Sung, S. Prasad, V.R. Yadav, B.B. Aggarwal, Cancer cell signaling pathways targeted by spice-derived nutraceuticals, Nutr. Cancer 64 (2012) 173–197.
[40] A. Jeffry, J.M. Andrew, V.B. Alan, Curcumin as a clinically-promising anti-cancer agent: pharmacokinetics and drug interactions, Expert Opin. Drug Metab. Toxicol. 9 (2017) 953–972.
[41] B. Barrett, Medicinal properties of Echinacea: a critical review, Phytomedicine 10 (2003) 66–86.
[42] C.W. Tsai, H.W. Chen, L.Y. Sheen, C.K. Lii, Garlic: health benefits and actions, Biomedicine 2 (2012) 17–29.
[43] A.L. Colín-Gonzalez, S.F. Ali, I. Túnez, A. Santamaría, On the antioxidant, neuroprotective and anti-inflammatory properties of S-allyl cysteine: an update, Neurochem. Int. 89 (2015) 83–91.
[44] N. Kumar, A. Rai, N.D. Reddy, P.V. Raj, P. Jain, P. Deshpande, G. Mathew, N.G. Kutty, N. Udupa, C.M. Rao, Silymarin liposomes improves oral bioavailability of silybin besides targeting hepatocytes, and immune cells, Pharmacol. Rep. 66 (2014) 788–798.
[45] A.D. Kshirsagar, D.K. Ingawale, P. Ashok, N. Vyawahare, Silymarin: A comprehensive review, Pharmacog. Rev. 3 (2009) 126–134.
[46] J.B. Zhu, X.Y. Wan, Y.P. Zhu, X.L. Ma, Y.W. Zheng, T.B. Zhang, Effect of astragaloside IV on the embryo-fetal development of Sprague-Dawley rats and New Zealand white rabbits, J. Appl. Toxicol. 29 (2009) 381–385.
[47] Q. Zhang, L.L. Zhu, G.G. Chen, Y.U. Du, Pharmacokinetics of astragaloside iv in beagle dogs, Eur. J. Drug Metab. 32 (2007) 75–79.
[48] T.O. Cheng, Review cardiovascular effects of Danshen, Int. J. Cardiol. Heart Vasc. 12 (2009) 9–22.
[49] H.M. Hügel, N. Jackson, Danshen diversity defeating dementia, Bioorg. Med. Chem. Lett. 24 (2014) 708–716.
[50] P.A.C.M. De Smet, J.R.B. Brouwers, Pharmacokinetic evaluation of herbal remedies, Clin. Pharm. 32 (1997) 427–436.
[51] M.S. Rahman, Allicin and other functional active components in garlic: health benefits and bioavailability, Int. J. Food Prop. 10 (2007) 245–268.
[52] J. Kleijnen, P. Knipschild, Ginkgo biloba for cerebral insufficiency, Br. J. Clin. Pharmacol. 34 (1992) 352–358.
[53] P.G. Pietta, C. Gardana, P.L. Mauri, Identification of flavonoid metabolites after oral administration to rats of a Ginkgo biloba extract, J. Chromatogr. B 673 (1995) 75–80.

[54] J. Li, D. Li, J. Hu, Y. Bi, Simultaneous determination of ginkgolides a, B, C and bilobalide by LC-MS/MS and its application to a pharmacokinetic study in rats, Biomed. Chromatogr. 29 (2015) 1907–1912.
[55] S. Kressmann, W.E. Müller, H.H. Blume, Pharmaceutical quality of different Ginkgo biloba brands, J. Pharm. Pharmacol. 54 (2002) 661–669.
[56] D.G. Waston, E.J. Oliveira, Solid-phase extraction and gas chromatography-mass spectrometry determination of kaempferol and quercetin in human urine after consumption of Ginkgo biloba tables, J. Chromatogr. B 723 (1999) 203–210.
[57] M. Nieder, Pharmakokinetic der Ginkgo-flavonole im plasma, Münch Med Wochenschr 133 (1991) 61–62.
[58] K.H. Kimbel, Ginkgo biloba, Lancet 340 (1992) 1474.
[59] P. Morazzoni, M.J. Magistretti, C. Giachetti, G. Zanolo, Comparative bioavailability of Silipide, a new flavanolignan complex in rats, Eur. J. Drug Metab. Pharmacokinet. 17 (1992) 39–44.
[60] P. Morazzoni, A. Montalbetti, S. Malandrino, G. Pifferi, Comparative pharmacokinetics of silipide and silymarin in rats, Eur. J. Drug Metab. Pharmacokinet. 18 (1993) 289–297.
[61] D. Savio, P.C. Harrasser, G. Basso, Softgel capsule technology as an enhancer device for the absorption of natural principles in humans. A bioavailability cross-over randomised study on silybin, Arzneimittelforschung 48 (1998) 1104–1106.
[62] R. Schandalik, E. Perucca, Pharmacokinetics of silybin following oral administration of silipide in patients with extrahepatic biliary obstruction, Drugs Exp. Clin. Res. 20 (1994) 37–42.
[63] G. Gatti, E. Perucca, Plasma concentrations of free and conjugated silybin after oral intake of a silybin-phosphatidylcholine complex (silipide) in healthy volunteers, Int. J. Clin. Pharmacol. Ther. 32 (1994) 614–617.
[64] G.D. Lin, R.W. Li, Natural products targeting inflammation processes and multiple mediators, in: S.C. Mandal, V. Mandal, T. Konishi (Eds.), Nat. Prod. Drug Discovery, Int. Appr. Elsevier, United States, 2018, pp. 277–308.
[65] W.W. Chao, Y.H. Hong, M.L. Chen, B.F. Lin, Inhibitory effects of Angelica sinensis ethyl acetate extract and major compounds on NF-B trans-activation activity and LPS-induced inflammation, J. Ethnopharmacol. 129 (2010) 244–249.
[66] S. Li, Y. Sun, J. Huang, B. Wang, Y. Gong, Y. Fang, Y. Liu, S. Wang, Y. Guo, H. Wang, Z. Xu, Y. Guo, Anti-tumor effects and mechanisms of Astragalus membranaceus (AM) and its specific immunopotentiation: status and prospect, J. Ethnopharmacol. 258 (2020) 112797.
[67] S. Sakai, H. Kawamata, T. Kogure, N. Mantani, K. Terasawa, M. Umatake, H. Ochiai, Inhibitory effect of ferulic acid and isoferulic acid on the production of macrophage inflammatory protein-2 in response to respiratory syncytial virus infection in RAW264.7 cells, Mediators Inflamm. 8 (1999) 173–175.
[68] L. Xiaoxia, L. Qu, Y. Dong, L. Han, E. Liu, S. Fang, Y. Zhang, T. Wang, A review of recent research progress on the astragalus genus, Molecules 19 (2014) 18850–18880.
[69] S. Pareek, M. Dixit, S. Govil, I. Jadhav, D. Shrivastava, M. Vahedi, P.S. Bisen, Garlic and its role in arthritis management, in: R.R. Watson, V.R. Preedy (Eds.), Bioactive Food as Dietary Interventions for Arthritis and Related Inflammatory Diseases, second ed., Academic Press, United States, 2019, pp. 245–252.
[70] J. Adiwidjaja, A.J. McLachlan, A.V. Boddy, Curcumin as a clinically-promising anti-cancer agent: pharmacokinetics and drug interactions, Expert Opin. Drug Metab. Toxicol. 13 (2017) 953–972.
[71] J.B. Fourtillan, A.M. Brisson, J. Girault, I. Ingrand, J.P. Decourt, K. Drieu, P. Jouenne, A. Biber, Pharmacokinetic properties of Bilobalide and Ginkgolides A and B in healthy subjects after intravenous and oral administration of *Ginkgo biloba* extract (EGb 761), Therapie 50 (1995) 137–144.

[72] M.A. Mir, R.S. Albaradie, Immunomodulation of inflammatory markers in activated macrophages by leaf extracts of *Gingko biloba*, Adv. Neuroimmune Biol. 6 (2015) 9–17.
[73] Y.H. Zhou, J.P. Yu, Y.F. Liu, M.X.J. TengMing, P. Lv, P. An, S.Q. Liu, H.G. Yu, Effects of *Ginkgo biloba* extract on inflammatory mediators (SOD, MDA, TNF-α, NF-κBp65, IL-6) in TNBS-induced colitis in rats, Mediators Inflamm. 10 (2006) 1–9.

15

Supplements and diets for antiinflammation

A. Thahira Banu and Janeline Lunghar
School of Sciences, Department of Home Science, The Gandhigram Rural Institute—Deemed to be University, Gandhigram, Dindigul, Tamil Nadu, India

15.1 Introduction

Inflammation is a significant cause of global disease in today's scenario. When the immune system overreaches and begins attacking healthy body tissues, this results in an autoimmune disorder. Inflammatory effects also are linked to celiac and irritable bowel disease (IBD). Inflammation of the arteries causes heart disease, swelling in the joints causes arthritis, swelling of the colon causes ulcerative colitis, asthma creates inflamed in airways, inflammation related to diabetes affects insulin resistance, and so on. By and large, inflammation in the body is a severe threat to your health. The inflammatory response is triggered by trauma, extreme temperatures, radiations, and infections like viruses, bacteria, fungi, parasites, chemical toxins, irritants, and immune reactions (hypersensitivity). When one of these causative agents injured the body tissues, the damaged cells release chemicals that cause blood vessels to leak fluid into the tissue; therefore it generates inflammation and attracts white blood cells that "eat" germs and dead cells, resulting in damage.

15.2 Types of inflammatory

Inflammatory diseases are a group of clinical disorders that are characterized by abnormal inflammatory responses (i.e., chronic inflammation) as a major hallmark. Chronic inflammation is tightly linked to the pathogenesis of inflammatory diseases that it becomes difficult to pinpoint both the cause and effect of these disorders. For example, inflammation is caused by obesity, whereas chronic inflammation can lead to obesity-associated diabetes partly because of insulin resistance [1]. Similar feedback mechanisms are also evident for other inflammatory disorders.

Inflammation and Natural Products. https://doi.org/10.1016/B978-0-12-819218-4.00007-9
© 2021 Elsevier Inc. All rights reserved.

15.2.1 Acute inflammation

Acute inflammation is a short-term inflammation and therefore considered as a part of innate immunity, the first line of host defense against foreign invaders and danger molecules. Humanity has known the typical symptoms of inflammation for hundreds of years, which include redness, pain, swelling, and heat [2]. However, emerging literature suggests that inflammation operates as a much-sophisticated system than ever thought at the molecular level. The entire cause of inflammation comes with many different processes involved in its initiation, regulation, and resolution. There is a wide range of inflammations that have been identified, with many different forms initiated by numerous stimuli and governed by various regulatory mechanisms. As a result, we are currently far from being able to fully comprehend the consequence of inflammation in human health and diseases.

15.2.2 Chronic inflammation

Chronic inflammations have been subject to extensive studies over the decades not only for the growing burdens of the associated pathological conditions in modern societies but also for the underlying mechanisms that remain mostly unresolved. A chronic inflammation is believed to originate if the elimination of the triggering stimulus fails to happen, such as any persistent infection or chronic cellular injury [3, 4]. Nonetheless, the initial trigger for a vast majority of the chronic inflammatory conditions has not been well defined yet, making the understanding of their pathological processes much more complicated due to the lack of any microbial infection or tissue damages.

Chronic inflammation is a prolonged and persistent conditions and not just a primary cause of diseases, but it is the major driver of the pathogenesis. In chronic inflammatory conditions the significant damages done to the host are mediated by the host inflammatory response itself, not by the foreign invaders such as pathogens. The chronic inflammatory conditions in this modern era of medical science cannot be overrated because they are linked with various diseases like atherosclerosis, obesity, type 2 diabetes, asthma, inflammatory bowel diseases, neurodegenerative diseases, rheumatoid arthritis, and cancer.

Chronic inflammation evolved tightly with the pathogenesis of inflammatory diseases. For example, inflammation is caused by obesity, whereas chronic inflammation can lead to obesity-associated diabetes partly because of insulin resistance [1]. Similar feedback mechanisms are also evident for other inflammatory disorders. A typical another example is rheumatoid arthritis, a long-term inflammation-associated disorder. In rheumatoid arthritis the synovium, the lining of the joint, undergoes chronic inflammation by the infiltration of macrophages and lymphocytes and the activation of synoviocytes, synovial cells, which produce synovial fluids. Natural products are given due importance in the treatment of diseases owing to the negligible side effects, easy availability, and affordability when compared with synthetic drugs. Natural resources from both land and ocean have tapped for their therapeutic potential, and a presentation of the review of the supplements used for antiinflammation follows.

15.3 Antiinflammatory supplements

15.3.1 Omega-3 fatty acid

There has been a substantial amount of research on supplements of omega-3s, particularly the types of omega-3s found in seafood and fish oil and heart disease. There is some evidence that omega-3s of the types found in seafood and fish oil may be modestly helpful in relieving symptoms in rheumatoid arthritis. Polyunsaturated fatty acids (PUFAs) of the omega-3 series are essential nutrients since they are not produced by humans [5] and whose dietary intake, with food and supplements, is associated with several health benefits [6, 7]. Omega-3 PUFAs, primarily found in dietary fish oils [8], are also derived from plants [9] and are substrates able to reduce or limit inflammation in critical illness [10]. The underlying molecular mechanisms responsible for omega-3 PUFAs' biological effects are mediated by the production of proresolving mediators, which have been proposed to modulate and likely resolve inflammatory responses [11]. Lipid mediators synthesized from omega-6 and omega-3 fatty acids are known not only as antiinflammatory molecules but also to have a key role in inducing effective resolution of inflammation. One of the most effective antiinflammatory agents is omega-3 polyunsaturated fatty acids. Countries that have the highest fish consumption also have a lower incidence of neurodegenerative disease and depression. The biological basis for the effectiveness of fish oil in treating arthritis has been well documented with many positive clinical studies when compared with traditional pharmaceutical antiinflammatory agents [12–18].

15.3.2 Turmeric

Turmeric is a popular spice frequently used in Indian foods and curry. Curcumin (1,7-bis[4-hydroxy-3-methoxyphenyl]-1,6-heptadiene-3,5-dione) is the most active constituent of turmeric curcuminoids obtained from the rhizome *Curcuma longa* [19]. Curcumin is classified as a polyphenol compound that gives turmeric its bright yellow color. Besides being a popular dietary supplement, it is used as a food coloring agent. Curcumin holds a high place in ayurvedic medicine as a "cleanser of the body," and today, science has documented several diseased conditions that can be healed by the active ingredients of turmeric [20, 21]. Curcumin has the potential to act against diabetes, asthma, allergies, arthritis, atherosclerosis, neurodegenerative diseases, and other chronic illnesses like cancer since it possesses therapeutic properties [22, 23]. Turmeric, an active compound known as curcumin, is one of the effective antiinflammatory components. Turmeric health benefits prove valuable in an antiinflammatory diet [24]. The journal *Oncogene* published the results of a study that evaluated several antiinflammatory compounds. It is found that aspirin (Bayer) and ibuprofen (Advil and Motrin) are least potent, while curcumin is among the most potent antiinflammatory and antiproliferative agents in the world [25]. Due to its high antiinflammatory properties, turmeric is highly effective at helping people manage rheumatoid arthritis (RA). A recent study out of Japan evaluated its relationship with interleukin (IL)-6, the inflammatory cytokine known to be involved in the RA process, and discovered that curcumin "significantly reduced" these inflammatory markers [26].

15.3.3 Ginger

Ginger (*Zingiber officinale* Roscoe) is a medicinal plant containing active compound gingerols, which are described by their pungency [27]. The most abundant pungent component of ginger is 6-gingerol and claimed as antiinflammatory. Ginger root *Z. officinale* is used for thousands of years in the Far East. It is an important cooking flavor food and herbal medicine around the world, the ginger family (Zingiberaceae). The *Z. officinale* rhizome contains numerous chemical compounds, including [6] gingerol, α-zingiberene, gingerone, camphene, neral, camphene, and a host of other chemical constituents [28]. Gingerols are the major components of ginger [29]. It has a strong antioxidant action and antitumor and antiinflammatory properties and prevents the generation of free radicals, considered as a safe herbal medicine without side effects [27, 30–32].

Ginger is used fresh or dried or in supplement form. The extract of ginger helps in immune modulators, which reduces inflammation. Ayurvedic medicine has proved ginger's ability to boost the immune system before it records the history. It believes that because ginger is so effective at warming the body, it can help break down the accumulation of toxins in your organs and lower inflammation. It's also known to cleanse the lymphatic system, our body's sewage system. Ginger treats for inflammation in allergic and asthmatic disorders [33].

15.3.4 Probiotic

Probiotics are healthy bacteria that normally live in our GI tract (intestines). Our intestines typically contain trillions of these helpful bacteria. They help maintain healthy GI and immune function. They make mucin that decreases the ability of other bacteria to "stick" to the gut; they make vitamin K, foster IgA maturation, and make the intestine less leaky and more acidic, protecting from disease-causing bacteria (http://herbs-supplements.osu.edu/).

Inflammatory bowel disease (IBD) is a broad term that describes chronic inflammation of the intestine and colon and include Crohn's disease (CD), and ulcerative colitis (UC), a caused by a poor immune response to host intestinal microbiota in genetically susceptible subjects. Study reported that several fermented dairy products contain lactic acid bacteria (LAB) and bifidobacteria, some of which characterized as probiotics that can modify the gut microbiota and may be beneficial for the treatment and the prevention of IBD. CD is a chronic inflammatory condition of the gastrointestinal tract driven by abnormal T-cell responses to the intestinal microbiota [34]. Medical therapy includes a combination of aminosalicylates, and immunomodulators as maintenance for remission and corticosteroids for severe symptomatic patients [35, 36]. Nevertheless, the importance of the intestinal microbiota in the etiology of mucosal inflammation provides a rationale for therapeutic strategies using probiotics and prebiotics in patients with CD [37].

Numerous studies, in IBD patients, reported the influence of the intestinal bacteria and the development of UC (ulcerative colitis) [38]. Modulation of the intestinal microbiota can be performed either by probiotics [39]. The modification of the intestinal microbiota through direct supplementation with protective bacteria could play a protective role in the inflammatory process [40]. Another inflammatory known as pouchitis is a common troublesome condition in surgical patients with ileal pouch-anal anastomosis (IPAA) [41]. It is a nonspecific idiopathic inflammation of the ileal reservoir [42]. The daily administration of 500 mL of a fermented milk product (Cultura) containing live *L. acidophilus* (La-5) and *B. lactis* (Bb-12) for 4 weeks increased the number of desirable bacteria in the UC/IPAA patients and significantly remained increased for 1 week after the intervention. A similar report found that there is good evidence for the usefulness of probiotics in preventing an initial attack of pouchitis (VSL#3) and in preventing further relapse of pouchitis after the induction of remission with antibiotics. Probiotics can be recommended to patients with pouchitis of mild activity or as maintenance therapy for those in remission [43].

15.3.5 White willow bark

Bark from the white willow tree is one of the oldest herbal remedies for pain and inflammation, dating back to ancient Egyptian, Roman, Greek, and Indian civilizations, as an analgesic and antipyretic agent. Because of the gastric side effects of aspirin, there has been a revival in the use of white willow bark for the treatment of inflammatory syndromes. The mechanism of action of white willow bark is similar to that of aspirin, which is a nonselective inhibitor of COX-1 and COX-2, used to block inflammatory prostaglandins [44]. Various randomized, placebo-controlled studies comparing white willow bark with nonsteroidal agents have shown efficacy comparable with these agents and aspirin. Salicin from white willow bark is converted to salicylic acid by the liver and considered to have fewer side effects than aspirin. The usual dose of white willow bark is 240 mg/day [45–48].

15.3.6 Green tea

Green tea has long recognized to have cardiovascular and cancer preventative characteristics due to its antioxidant properties. Its use in the treatment of arthritic disease as an antiinflammatory agent is recognized more recently. The constituents of green tea are polyphenolic compounds called catechins, and epigallocatechin-3 gallate is the abundant catechin in green tea. Epigallocatechin-3-gallate inhibits IL-1-induced proteoglycan release and type 2 collagen degradation in cartilage explants [49]. In human in vitro models, it also suppresses IL-1β and attenuates activation of the transcription factor NF-κB. Green tea also inhibits the aggrecanases, which degrade cartilage. Green tea research now demonstrates both antiinflammatory and chondroprotective effects. Additionally, green tea research includes the "Asian paradox," which theorizes that increased green tea consumption in Asia may lead to significant cardiovascular, neuroprotective, and cancer prevention properties. The usual recommendation is 3–4 cups of tea a day. Green tea extract has a typical dosage of 300–400 mg [50].

15.3.7 Capsaicin

Capsaicin (chili pepper) (*Capsicum annuum*) is a small spreading shrub that is originally cultivated in the tropical regions of the Americas but is now grown throughout the world, including the United States. The small red fruit commonly used to accentuate chili owes its stinging pungency to the chemical capsaicin, which was isolated by chemists more than a century ago and constituted approximately 12% of the chili pepper. Capsaicin potently activates transient receptor potential vanilloid 1, which is a main receptor underlying nociception. It also inhibits NF-κB, thus producing an antiinflammatory effect. There are topical capsaicin formulations now available to treat postherpetic neuralgia. Other uses are studied for peripheral neuropathies and chronic musculoskeletal pain [51–53].

15.3.8 Seaweeds

Seaweeds have caused an emerging interest in the biomedical area, mainly due to their contents of bioactive substances, which show great potential as antiinflammatory, antimicrobial, antiviral, and antitumoral drugs. Among these compounds, polyphenols [54], polysaccharides [55], meroterpenoids [56], and terpenoids [57] are considered as promising bioactive molecules in the search for potential therapeutic drugs.

Fucoidans, a group of sulfated polysaccharides purified from brown algae, possess a variety of therapeutic effects, namely, in cancer and inflammatory conditions [58]. Other substances biosynthesized by algae with an impact on human health include carotenoids and natural pigments and are used as antioxidant compounds reducing the incidence of many diseases [59].

Chemical and biological investigations indicated that the main substances synthesized by brown algae with antiinflammatory potentials include sulfated polysaccharides, phlorotannins, and carotenoids. These compounds are a potent blocker and used experimentally to prevent inflammatory damage after ischemic events. Fucoidans inhibit phospholipase A2, an essential enzyme in the inflammatory cascade [60], and appear to inhibit the functions of macrophages, a predominant source of proinflammatory factors [61, 62]. Kim et al. reported that ethanolic extract of the brown algae, *Ishige okamurae*, was effective in inhibiting the production of inflammatory mediators, such as TNF-α, IL-1β, IL-6, and PG-E2 in Raw 264.7 macrophage cells, and by inactivation of NF-χB transcription factor in macrophages stimulated by lipopolysaccharide [63]. Myers et al. showed, in an open-label combined phase I and phase II pilot-scale study in osteoarthritis of knee, that formulation, containing a blend of extracts from three different species of brown algae, when taken orally by patients over 12 weeks decreased the symptoms of osteoarthritis in a dose-dependent manner [64]. Investigations conducted by Sugiura et al. demonstrated that MeOH/$CHCl_3$ extract from *E. arborea* inhibited inflammatory mediators (histamine and eicosanoids: LTB4 and PGD2) released from RBL cells and that phlorotannins and methanol/chloroform extract inhibited activities of enzymes (phospholipase A2, cyclooxygenase A2, and lipoxygenase) involved in eicosanoid synthesis in the arachidonate cascade. In addition to the polar components (fucoidans and phlorotannins), brown algae also produced nonpolar components, such as carotenoids, with antiinflammatory potential. Recently, fucoxanthin, one of the most abundant carotenoids isolated from brown algae, exerts antiinflammatory effect via inhibitory effect of nitric oxide production in lipopolysaccharide-induced Raw 264.7 macrophage cells [65]. Japanese, in their recommended dietary allowance, indicates 13 g per day consumption.

15.3.9 Brahmi

Bacopa monnieri (Brahmi) is an herb that effectively suppressed experimentally induced inflammatory reaction by inhibiting the prostaglandin synthesis and partly by stabilizing lysosomal membranes and without causing gastric irritation [66]. The ethanol extract of the whole plant of *B. monnieri* produced significant writhing inhibition in acetic acid-induced writhing in mice at the oral dose of 250 and 500 mg/kg.

15.3.10 Ocimum sanctum *L.*

Ocimum sanctum L. (Tulsi) is one such medicinal plant having numerous medicinal properties [67, 68]. The medicinal uses of Tulsi are well documented and extensively used in the traditional Indian systems of medicine for treating various diseases either alone or in combination with other herbal plants [69]. Tulsi has been used for thousands of years for its diverse healing properties and regarded in Ayurveda as the "elixir of life" that promotes longevity [70].

A methanol extract (500 mg/kg) and an aqueous suspension of *O. sanctum* are known to have analgesic, antipyretic, and antiinflammatory effects in acute and chronic inflammation in rats. The fixed oil and linolenic acid possess significant antiinflammatory activity against prostaglandin E2, leukotriene, and arachidonic acid by virtue of their capacity to block both the lipoxygenase and cyclooxygenase pathways of arachidonic acid metabolism [71].

Experimental studies of Tulsi have shown to inhibit acute and chronic inflammation in rats. This test conducted by carrageenan-induced paw edema, croton oil-induced granuloma, and exudates at a dose of 500 mg/kg, BW/day. The oils processed from fresh leaves and seeds of *O. sanctum* have revealed antiinflammatory effects on experimental animals induced by carrageenan, histamine, serotonin, and prostaglandin E2, according to some studies. These experimental rats administered with essential oil (200 mg/kg, BW, and fixed oil (0.1 mL/kg, bw) before the injection of phlogistic agents and compared with standard drug flurbiprofen. It noted that Tulsi extracts could significantly reduce the edema when compared with the saline-treated control [72].

15.4 Role of diet in body inflammation

Diet plays a vital role in our health and has been dramatically affected by these changes. A diet emphasizes minimally processed whole foods to prevent and reduce inflammation in the body. Some of the foods like added salt, refined sugar, hydrogenated fats, trans fats, and highly refined grains have increased the prevalence of low-grade, persistent inflammation in our bodies. One of the most researched examples of eating is the traditional Mediterranean diet, which is a dietary pattern inspired by some countries of the Mediterranean basin. People that more closely eat Mediterranean-like food have consistently lower levels of inflammation compared with other less healthy ways of eating [73, 74]. The Mediterranean diet has been extensively studied and is protective against many chronic health conditions, including cardiovascular disease, type 2 diabetes mellitus, Parkinson's and Alzheimer's disease, and some cancers [75, 76]. The Mediterranean diet is just one example of traditional food and happens to be the most researched classic diet pattern in the world. Many conventional diets are healthier than trendy modern diets because they centered on eating whole and unprocessed foods.

Researchers have proved that intake of refined grains and pure sugar is positively associated with the risk of developing inflammatory-mediated diseases, including diabetes mellitus and cardiovascular disease. Not only that, there seemed to be an increased level in inflammatory protein concentrations, which relate to the formation of chronic inflammatory conditions.

Many studies indicate that a diet high in fruits and vegetables is useful for decreasing inflammation. Colorful fruits and vegetables contain a diversity of antiinflammatory bioactive compounds. A bioactive compound is a general term for numerous substances naturally found within plant foods that provide multiple health benefits to humans when consumed. Although all fruits and vegetables are beneficial, green leafy vegetables, cabbage, onions, berries, cherries, pomegranate, and citrus fruits for their antiinflammatory properties. Most importantly, herbs and spices have strong antiinflammatory properties. Some of the powerful antiinflammatory herbs and spices include turmeric, ginger, garlic, cinnamon, rosemary, basil, and thyme [77].

Another healthy food to include for reducing inflammation is omega-3 and omega-6 fatty acid found in fishes like salmon, sardine, mackerel, and flaxseeds from plant sources. When our bodies break down omega-6 fatty acids and omega-3 fatty acids, the end products are eicosanoids. Eicosanoids are molecules in our cells that signal reactions in the body and exert control over many bodily systems, mainly related to inflammation or immunity. They also act as messengers in the central nervous system. Depending on the fat intake of a person's diet (omega-3's vs omega-6's), the signal for the production of antiinflammatory prostaglandins will vary. Omega-6's are proinflammatory, and omega-3's are antiinflammatory. At the end of the day, what we eat can affect totally in inflammation, and specific diets are more likely to lower the pain and other symptoms of disease. Estimated that 60% of chronic illnesses are prevented by a healthy diet [78]. People who eat foods that are consistent with an antiinflammatory diet are less likely to have health problems.

Additionally, omega-3 polyunsaturated fatty acids (PUFAs), eicosapentaenoic acid (EPA), and docosahexaenoic acid (DHA) are much more potent antiinflammatory agents than their precursor alpha-linolenic acid (ALA). ALA does convert into EPA and then to DHA, but less than 1% of the original amount of ALA is converted to the physiologically active DHA and EPA [79]. For this reason, flax oil (which is rich in ALA) is not as effective as EPA and DHA for inflammation. Fish oil contains preformed EPA and DHA (around 18% and 12%, respectively). Typically, plant sources of omega-3 included ALA, though there are now vegan supplements derived from algae that contain DHA.

When cooking, extra virgin olive oil is an excellent choice as it has shown to lower systolic blood pressure, fasting glucose, c-reactive protein (CRP), and LDL levels [80, 81]. It is also resistant to trans-fatty acid conversion with heat. There is currently an increase in consumption in using coconut oil. As a medium-chain fatty acid, it does not behave like most long-chain saturated fatty acids. It does not cause increased cholesterol and fat accumulation; however, fewer research papers are available on coconut oil usage in diets and its antiinflammatory effects.

Fatty acids found in nature are more balanced than the fats we typically consume in our typical diets. Chia seed benefits, for example, offer both omega-3 and omega-6, which are consumed in balance with one another [82]. Chia is an antioxidant and antiinflammatory powerhouse, containing essential fatty acids alpha-linolenic and linoleic acid; mucin; strontium; vitamins A, B, E, and D; and minerals including sulfur, iron, iodine, magnesium, manganese, niacin, and thiamine. Chia seeds' ability to reverse inflammation, regulate cholesterol, and lower blood pressure makes it extremely beneficial to consume for heart health.

Furthermore, diets high in surface help to decrease inflammation. A good goal is about 30 g/day, ideally from a diet rich in whole grains, fruits, and vegetables. Grain consumption has an inverse relationship with CRP, with more excellent protection seen at a total fiber level above 22 g/day [83].

Studies are indicating that some foods can reduce and some can increase the inflammation, and the following are the list of foods taken liberally to reduce the inflammation:

1. Foods rich in omega-3 fatty acids, namely, fishes (herring, mackerel, sardine, salmon), flaxseeds, and walnuts;
2. Foods high in fiber whole grains, pulses, and green leafy vegetables;

3. Foods rich in antioxidants, namely, dark leafy greens, capsicum, colored vegetables, onions, garlic ginger, citrus fruits, and green tea;
4. Spices and herbs like turmeric, black pepper, cloves, cinnamon, nutmeg, basil, thyme, oregano, sage, rosemary, and lemongrass.

Foods to be consumed in a restricted manner that otherwise can increase the inflammation are as follows:

1. Foods rich in trans fat and omega-6 fatty acid like baked foods, red meat, fried fish, and meat;
2. Partially hydrogenated oils;
3. Preserved foods and crackers;
4. Foods high in simple carbohydrates and with a high glycemic load;
5. Foods that cause allergy and intolerance like dairy, wheat, egg, and artificial flavor colors.

According to various interventional studies, there are foods associated with decreasing inflammatory markers in humans. Bogani et al. reported that consuming extra virgin olive oil in one meal of the day found to decreased TXB2 (thromboxane B2) and LTB4 (leukotriene B4) in comparison with standard oil or nonvirgin olive oil [84]. Wood et al. also reported that consuming 10 days of tomato juice reduced neutrophil airway influx in asthmatics [85]. Similarly, drinking red wine with a duration of 4 weeks reduces CRP (C-reactive protein) and fibrinogen. Similar studies observed by Faintuch et al. and Bahorun et al. reported that intake of flaxseed for 2 weeks and black tea for 12 weeks decreased the CRP, fibronectin, and serum amyloid A in obese subjects. Sweet cherries are not only high in antioxidants but also possess antiinflammatory properties [86, 87]. Intake of fresh cherries for a month is regularly reported to decrease CRP and CCL5-chemokine (c-c motif) ligand 5 in healthy adults [88].

15.5 Nutrient effects on chronic inflammation

Carbohydrates and glycemic index diets with relatively high glycemic index (GI) and glycemic load (GL) have been associated with an elevated risk of coronary heart disease, stroke, and type 2 diabetes mellitus, particularly among overweight individuals [89]. Dietary GI, the low propensity of carbohydrate in an individual's diet to increase blood glucose level, and dietary GL, the product of dietary GI and quantity of sugar, have shown inconsistent relationships to high-sensitivity C-reactive protein (HS-CRP) in observational studies. In the Harvard Women's Health Study, blood levels of HS-CRP showed a small but progressive increase across quintiles of dietary GI, with a highly significant difference between first and fifth quintiles, even with multivariate statistical adjustment. There was no significant association between HS-CRP and GL. Similar findings were reported in a smaller study from the Netherlands. In the Dutch study, each 10-unit increase in dietary GI associated with a 29% increase in HS-CRP. A study from Tufts University compared the metabolic response with weight loss in healthy, overweight individuals consuming diets of low and high GL. The only significant difference between the two groups was a greater decrease in HS-CRP among the low GL dieters [90]. In contrast a prospective study done at the University of Massachusetts found no relationship between GI or GL and HS-CRP among a population with a relatively high mean GI and GL [91]. The relationship between carbohydrate quality and inflammation may only be measurable with relatively low GI diets.

The fiber content of the food may influence the relationship between carbohydrate quality and systemic inflammation. A recent review of seven clinical trials done on relationship between weight loss and HS-CRP reported that six of the trials reported significant reduction on HS-CRP concentrations (25%–54%) with increased fiber consumption (3.3–7.8 g/MJ) [92]. In the survey report the consumption of flavonoids is inversely related to HS-CRP levels. Among the flavonoid compounds investigated, quercetin, kaempferol, malvidin, peonidin, daidzein, and genistein each had inverse associations with serum HS-CRP concentration, even after adjustment for total fruit and vegetable consumption [92]. Based upon studies in vitro, numerous mechanisms for antiinflammatory effects of flavonoids have described, many of which are derived from enzyme inhibition [93]. Concerning consumption between fruits and vegetables, numerous studies have shown an inverse correlation and serum levels of inflammatory markers [94]. This effect has been seen not only with adults but also with adolescents and confined to Western populations [95]. Many researchers also investigated the effects of fish oil consumption on reducing inflammation, including most of the studies that reported that fish oil is antiinflammatory and decreases a wide range of immune cell responses [96].

15.6 Conclusion

Antiinflammatory supplements like probiotics, omega-3 fatty acid, and correct diet pattern are the key roles in reducing chronic and low-grade inflammation. Many researchers have proved on the consumption of good fat such as ω-3 PUFAs, MUFAs, a reasonable amount of fiber, and flavonoids, and by lowering carbohydrate food, it can decrease the levels of inflammatory markers such as serum and c-reactive protein. In contrast, unhealthy fats like saturated fat, trans fat, and high GI carbohydrates are associated with an increased level of inflammation.

References

[1] G.S. Hotamisligil, Inflammation and metabolic disorders, Nature 444 (2006) 860–867.

[2] R. Medzhitov, Origin and physiological roles of inflammation, Nature 454 (2008) 428–435.

[3] V. Kumar, M. Carey, S.L. Robbins, Robbins Basic Pathology, seventh ed., Saunders, New York, 2003.

[4] G. Majno, I. Joris, Cells, Tissues, and Diseases, Oxford University Press, New York, 2004.

[5] M.G. Duvall, B.D. Levy, DHA- and EPA-derived resolvins, protectins, and maresins in airway inflammation, Eur. J. Pharmacol. 785 (2016) 144–155.

[6] A. Molfino, G. Gioia, F. Fanelli, M. Muscaritoli, The role for dietary omega-3 fatty acids supplementation in older adults, Nutrients 6 (2014) 4058–4072.

[7] A. Molfino, M. Amabile, M. Monti, S. Arcieri, F. Rossi Fanelli, M. Muscaritoli, The role of docosahexaenoic acid (DHA) in the control of obesity and metabolic derangements in breast cancer, Int. J. Mol. Sci. 17 (2016) 505.

[8] P.C. Calder, Omega-3 polyunsaturated fatty acids and inflammatory processes: nutrition or pharmacology? Br. J. Clin. Pharmacol. 75 (2013) 645–662.

[9] E.J. Baker, E.A. Miles, G.C. Burdge, P. Yaqoob, P.C. Calder, Metabolism and functional effects of plant-derived omega-3 fatty acids in humans, Prog. Lipid Res. 64 (2016) 30–56.

[10] W. Manzanares, P.L. Langlois, G. Hardy, Intravenous lipid emulsions in the critically ill, Curr. Opin. Crit. Care 22 (2016) 308–315.

[11] C.N. Serhan, S. Hong, K. Gronert, S.P. Colgan, P.R. Devchand, G. Mirick, R. Moussignac, Resolvins: a family of bioactive products of omega-3 fatty acid transformation circuits initiated by aspirin treatment that counter proinflammation signals, J. Exp. Med. 196 (2002) 1025–1037.

[12] J.E. Bernstein, D.R. Bickers, M.V. Dahl, J.Y. Roshal, Treatment of chronic postherpetic neuralgia with topical capsaicin, J. Am. Acad. Dermatol. 17 (1987) 93–96.

[13] C.L. Curtis, J.L. Harwood, C.M. Dent, B. Caterson, Biological basis for the benefit of nutraceutical supplementation in arthritis, Drug Discov. Today 9 (2004) 165–172.

[14] C.L. Curtis, S.G. Rees, C.B. Little, C.R. Flannery, C.E. Hughes, C. Wilson, C.M. Dent, I.G. Otterness, J.L. Harwood, B. Caterson, Pathologic indicators of degradation and inflammation in human osteoarthritic cartilage are abrogated by exposure to N-3 fatty acids, Arthritis Rheum. 46 (2002) 1544–1553.

[15] C.L. Curtis, C.E. Hughes, C.R. Flannery, C.B. Little, J.L. Harwood, B. Caterson, N-3 fatty acids specifically modulate catabolic factors involved in articular cartilage degradation, J. Biol. Chem. 275 (2000) 721–724.

[16] M.L. Daviglus, J. Stamler, A.J. Orencia, A.R. Dyer, K. Liu, P. Greenland, M.K. Walsh, D. Morris, R.B. Shekelle, Fish consumption and the 30-year risk of fatal myocardial infarction, N. Engl. J. Med. 336 (1997) 1046–1053.

[17] T.M. Haqqi, D.D. Anthony, S. Gupta, N. Ahmad, M.S. Lee, G.K. Kumar, H. Mukhtar, Prevention of collagen-induced arthritis in mice by a polyphenolic fraction from green tea, Proc. Natl. Acad. Sci. 96 (1999) 4524–4529.

[18] Q. Peng, Z. Wei, B.H. Lau, Pycnogenol inhibits tumor necrosis factor-alpha-induced nuclear factor kappa B activation and adhesion molecule expression in human vascular endothelial cells, Cell. Mol. Life Sci. 57 (2000) 834–841.

[19] B.T. Kurien, R.H. Scofield, Oral administration of heat-solubilized curcumin for potentially increasing curcumin bioavailability in experimental animals, Int. J. Cancer 125 (2009) 1992–1993.

[20] S. Mishra, K. Palanivelu, The effect of curcumin (turmeric) on Alzheimer's disease: an overview, Ann. Indian Acad. Neurol. 11 (2008) 13–19.

[21] H.B. Nair, B. Sung, V.R. Yadav, R. Kannappan, M.M. Chaturvedi, B.B. Aggarwal, Delivery of antiinflammatory nutraceuticals by nanoparticles for the prevention and treatment of cancer, Biochem. Pharmacol. 80 (2010) 1833–1843.
[22] S.K. Sandur, H. Ichikawa, M.K. Pandey, A.B. Kunnumakkara, B. Sung, G. Sethi, B.B. Aggarwal, Role of pro-oxidants and antioxidants in the anti-inflammatory and apoptotic effects of curcumin (diferuloylmethane), Free Radic. Biol. Med. 43 (2007) 568–580.
[23] C.-L. Lin, J.-K. Lin, Curcumin: a potential cancer chemopreventive agent through suppressing NF-κB signalling, J. Cancer Mol. 4 (2008) 11–16.
[24] D.S. Kim, S. Park, J. Kim, Curcuminoids from *Curcuma longa* L. (Zingiberaceae) that protect PC12 rat pheochromocytoma and normal human umbilical vein endothelial cells from βa(1–42) insult, Neurosci. Lett. 303 (2001) 57–61.
[25] Y. Takada, F.R. Khuri, B.B. Aggarwal, Protein farnesyltransferase inhibitor (SCH 66336) abolishes NF-κb activation induced by various carcinogens and inflammatory stimuli leading to suppression of NF-κb-regulated gene expression and up-regulation of apoptosis, J. Biol. Chem. 279 (2004) 26287–26299.
[26] M. Wada, H. Nagasawa, K. Kurita, S. Koyama, S. Arawaka, T. Kawanami, K. Tajima, M. Daimon, T. Kato, Cerebral small vessel disease and C-reactive protein: results of a cross-sectional study in community-based Japanese elderly, J. Neurol. Sci. 264 (2008) 43–49.
[27] B.H. Ali, G. Blunden, M.O. Tanira, A. Nemmar, Some phytochemical, pharmacological and toxicological properties of ginger (*Zingiber officinale* Roscoe): a review of recent research, Food Chem. Toxicol. 46 (2008) 409–420.
[28] J.A. Ojewole, Analgesic, anti-inflammatory and hypoglycaemic effects of ethanol extract of *Zingiber officinale* (Roscoe) rhizomes (Zingiberaceae) in mice and rats, Phytother. Res. 20 (2006) 764–772.
[29] A. Kato, Y. Higuchi, H. Goto, H. Kizu, T. Okamoto, N. Asano, J. Hollinshead, R.J. Nash, I. Adachi, Inhibitory effects of *Zingiber officinale* Roscoe derived components on aldose reductase activity in vitro and in vivo, J. Agric. Food Chem. 54 (2006) 6640–6644.
[30] K. Chun, K. Park, J. Lee, M. Kang, Y. Surh, Inhibition of mouse skin tumor promotion by anti-inflammatory diarylheptanoids derived from *Alpinia oxyphylla* Miquel (Zingiberaceae), Oncol. Res. Featur. Preclin. Clin. Cancer Therapeut. 13 (2002) 37–45.
[31] C. Shen, K. Hong, S.W. Kim, Effects of ginger (*Zingiber officinale* Rosc.) on decreasing the production of inflammatory mediators in sow Osteoarthrotic cartilage explants, J. Med. Food 6 (2003) 323–328.
[32] S.K. Verma, M. Singh, P. Jain, A. Bordia, Protective effect of ginger, *Zingiber officinale* Rosc on experimental atherosclerosis in rabbits, Indian J. Exp. Biol. 42 (2004) 736–738.
[33] T.T. Wada, Y. Araki, K. Sato, Y. Aizaki, K. Yokota, Y.T. Kim, H. Oda, R. Kurokawa, T. Mimura, Aberrant histone acetylation contributes to elevated interleukin-6 production in rheumatoid arthritis synovial fibroblasts, Biochem. Biophys. Res. Commun. 444 (2014) 682–686.
[34] R.B. Sartor, Microbial influences in inflammatory bowel diseases, Gastroenterology 134 (2008) 577–594.
[35] J.A. Katz, Treatment of inflammatory bowel disease with corticosteroids, Gastroenterol. Clin. N. Am. 33 (2004) 171–189.
[36] S. Friedman, General principles of medical therapy of inflammatory bowel disease, Gastroenterol. Clin. N. Am. 33 (2004) 191–208.
[37] J.L. Benjamin, C.R. Hedin, A. Koutsoumpas, S.C. Ng, N.E. McCarthy, A.L. Hart, M.A. Kamm, J.D. Sanderson, S.C. Knight, A. Forbes, A.J. Stagg, K. Whelan, J.O. Lindsay, Randomised, double-blind, placebo-controlled trial of fructo-oligosaccharides in active Crohn's disease, Gut 60 (2011) 923–929.

[38] H. Ishikawa, I. Akedo, Y. Umesaki, R. Tanaka, A. Imaoka, T. Otani, Randomized controlled trial of the effect of bifidobacteria-fermented milk on ulcerative colitis, J. Am. Coll. Nutr. 22 (2003) 56–63.
[39] A. Tursi, G. Brandimarte, A. Papa, A. Giglio, W. Elisei, G.M. Giorgetti, G. Forti, S. Morini, C. Hassan, M.A. Pistoia, M.E. Modeo, S. Rodino, T. D'Amico, L. Sebkova, N. Sacca, E. Di Giulio, F. Luzza, M. Imeneo, T. Larussa, A. Gasbarrini, Treatment of relapsing mild-to-moderate ulcerative colitis with the probiotic VSL#3 as adjunctive to a standard pharmaceutical treatment: a double-blind, randomized, placebo-controlled study, Am. J. Gastroenterol. 105 (2010) 2218–2227.
[40] K. Kato, S. Mizuno, Y. Umesaki, Y. Ishii, M. Sugitani, A. Imaoka, M. Otsuka, O. Hasunuma, R. Kurihara, A. Iwasaki, Y. Arakawa, Randomized placebo-controlled trial assessing the effect of bifidobacteria-fermented milk on active ulcerative colitis, Aliment. Pharm. Ther. 20 (2004) 1133–1141.
[41] K.O. Laake, A. Bjørneklett, G. Aamodt, L. Aabakken, M. Jacobsen, A. Bakka, M.H. Vatn, Outcome of four weeks' intervention with probiotics on symptoms and endoscopic appearance after surgical reconstruction with a J-configurated ileal-pouch-anal-anastomosis in ulcerative colitis, Scand. J. Gastroenterol. 40 (2005) 43–51.
[42] P. Gionchetti, F. Rizzello, C. Morselli, G. Poggioli, R. Tambasco, C. Calabrese, P. Brigidi, B. Vitali, G. Straforini, M. Campieri, High-dose probiotics for the treatment of active pouchitis, Dis. Colon Rectum 50 (2007) 2075–2084.
[43] World Gastroenterology Organisation, Global Guidelines, 2011.
[44] O.P. Gulati, Pycnogenol® in metabolic syndrome and related disorders, Phytother. Res. 29 (2015) 949–968.
[45] B. Schmid, I. Kótter, L. Heide, Pharmacokinetics of salicin after oral administration of a standardised willow bark extract, Eur. J. Clin. Pharmacol. 57 (2001) 387–391.
[46] J.J. Gagnier, M.W. Van Tulder, B.M. Berman, C. Bombardier, Herbal medicine for low back pain, Cochrane Database Syst. Rev. (2006), https://doi.org/10.1002/14651858.cd004504.pub3.
[47] R.J. Ko, Adulterants in Asian patent medicines, N. Engl. J. Med. 339 (1998) 847.
[48] S. Chrubasik, Treatment of low back pain with a herbal or synthetic antirheumatic: a randomized controlled study. Willow bark extract for low back pain, Rheumatology 40 (2001) 1388–1393.
[49] L.B. Tijburg, T. Mattern, J.D. Folts, U.M. Weisgerber, M.B. Katan, Tea flavonoids and cardiovascular diseases: a review, Crit. Rev. Food Sci. Nutr. 37 (1997) 771–785.
[50] S. Ghosh, M.J. May, E.B. Kopp, NF-κb and rel proteins: evolutionarily conserved mediators of immune responses, Annu. Rev. Immunol. 16 (1998) 225–260.
[51] M.J. Caterina, D. Julius, The Vanilloid receptor: a molecular gateway to the pain pathway, Annu. Rev. Neurosci. 24 (2001) 487–517.
[52] J.M. Chung, K.H. Lee, H. Yuichi, W.D. Willis, Effects of capsaicin applied to a peripheral nerve on the responses of primate spinothalamic tract cells, Brain Res. 329 (1985) 27–38.
[53] F. Andersohn, S. Suissa, E. Garbe, Use of first- and second-generation cyclooxygenase-2-selective nonsteroidal antiinflammatory drugs and risk of acute myocardial infarction, Circulation 113 (2006) 1950–1957.
[54] M.A. Naqvi, K.W. Glombitza, Phlorotannins, brown algal polyphenols, Prog. Phycol. Res. 4 (1986) 129–147.
[55] M.J. Kwon, T.J. Nam, A polysaccharide of the marine alga *Capsosiphon fulvescens* induces apoptosis in AGS gastric cancer cells via an IGF-IR-mediated PI3K/ Akt pathway, Cell Biol. Int. 31 (2007) 768–775.
[56] R. Valls, B. Banaigs, L. Piovetti, A. Archavlis, J. Artaud, Linear diterpene with antimitotic activity from the Brown alga *Bifurcaria bifurcata*, Phytochemistry 34 (1993) 1585–1588.

[57] G. Culioli, A. Ortalo-Magné, M. Daoudi, H. Thomas-Guyon, R. Valls, L. Piovetti, Trihydroxylated linear diterpenes from the brown alga *Bifurcaria bifurcata*, Phytochemistry 65 (2004) 2063–2069.

[58] S. Gupta, N. Abu-Ghannam, Bioactive potential and possible health effects of edible brown seaweeds, Trends Food Sci. Technol. 22 (2011) 315–326.

[59] A. Cantrell, D. McGarvey, T. George Truscott, F. Rancan, F. Böhm, Singlet oxygen quenching by dietary carotenoids in a model membrane environment, Arch. Biochem. Biophys. 412 (2003) 47–54.

[60] L.S. Ritter, J.G. Copeland, P.F. McDonagh, Fucoidin reduces coronary microvascular leukocyte accumulation early in reperfusion, Ann. Thorac. Surg. 66 (1998) 2063–2071.

[61] B. Lomonte, Y. Angulo, L. Calderón, An overview of lysine-49 phospholipase A2 myotoxins from crotalid snake venoms and their structural determinants of myotoxic action, Toxicon 42 (2003) 885–901.

[62] I. Wijesekara, N.Y. Yoon, S. Kim, Phlorotannins from *Ecklonia cava* (Phaeophyceae): biological activities and potential health benefits, Biofactors 36 (2010) 408–414.

[63] H.T. Kim, K.P. Kim, T. Uchiki, S.P. Gygi, A.L. Goldberg, S5a promotes protein degradation by blocking synthesis of nondegradable forked ubiquitin chains, EMBO J. 28 (2009) 1867–1877.

[64] S. Myers, J. O'Connor, J.H. Fitton, L. Brooks, M. Rolfe, P. Connellan, H. Wohlmuth, P.A. Cheras, C. Morris, A combined phase I and II open label study on the effects of a seaweed extract nutrient complex on osteoarthritis, Biologics 4 (2010) 33–44.

[65] Y. Sugiura, K. Matsuda, Y. Yamada, M. Nishikawa, K. Shioya, H. Katsuzaki, K. Imai, H. Amano, Anti-allergic Phlorotannins from the edible Brown alga, Eisenia Arborea, Food Sci. Technol. Res. 13 (2007) 54–60.

[66] M. Saraf, S. Prabhakar, P. Pandhi, A. Anand, *Bacopa monniera* ameliorates amnesic effects of diazepam qualifying behavioral-molecular partitioning, Neuroscience 155 (2008) 476–484.

[67] A. Sharma, N. Kumar, D. Kumar, V. Kumari, S. Saraswati, K. Chandel, A review paper on antimicrobial activity of medicinal plant Tulsi (*Ocimum* spp.) and pudina (*Mentha* spp.), Int. J. Curr. Res. 5 (2013) 487–489.

[68] G. Cragg, D. Newman, Natural product drug discovery in the next millennium, Pharm. Biol. 39 (2001) 8–17.

[69] G.V. Satyavati, M.K. Raina, M. Sharma, Medicinal Plants of India, vol. 1, Indian Council of Medical Research, Medicinal Plants of India, New Delhi, 2008.

[70] S.K. Gupta, J. Prakash, S. Srivastava, Validation of traditional claim of Tulsi, *Ocimum sanctum* Linn as a medicinal plant, Indian J. Exp. Biol. 40 (2002) 765–773.

[71] S. Singh, D.K. Majumdar, Evaluation of anti-inflammatory activity of fatty acids of *Ocimum sanctum* fixed oil, Indian J. Exp. Biol. 35 (1997) 380–383.

[72] S. Singh, Comparative evaluation of anti-inflammatory potential of fixed oil of different species of *Ocimum* and its possible mechanism of action, Indian J. Exp. Biol. 36 (1998) 1028–1031.

[73] E. Neale, M. Batterham, L. Tapsell, Consumption of a healthy dietary pattern results in significant reductions in C-reactive protein levels in adults: a meta-analysis, Nutr. Res. 36 (2016) 391–401.

[74] M. Ruiz-Canela, I. Zazpe, N. Shivappa, J.R. Hébert, A. Sánchez-Tainta, D. Corella, J. Salas-Salvadó, M. Fitó, R.M. Lamuela-Raventós, J. Rekondo, J. Fernández-Crehuet, M. Fiol, J.M. Santos-Lozano, L. Serra-Majem, X. Pinto, J.A. Martínez, E. Ros, R. Estruch, M.A. Martínez-González, Dietary inflammatory index and anthropometric measures of obesity in a population sample at high cardiovascular risk from the PREDIMED (PREvención con DIeta MEDiterránea) trial, Br. J. Nutr. 113 (2015) 984–995.

[75] F. Sofi, F. Cesari, R. Abbate, G.F. Gensini, A. Casini, Adherence to Mediterranean diet and health status: meta-analysis, BMJ 337 (2008) a1344.
[76] A. Trichopoulou, M.A. Martínez-González, T.Y. Tong, N.G. Forouhi, S. Khandelwal, D. Prabhakaran, D. Mozaffarian, M. De Lorgeril, Definitions and potential health benefits of the Mediterranean diet: views from experts around the world, BMC Med. 12 (2014) 112.
[77] Natural Medicines Database, Anti Inflammatory Diet, 2015, Available at: https://naturalmedicines.therapeuticresearch.com/databases/food,-herbs-supplements/professional.aspx?productid=1294#background. Naturalmedicines.therapeuticresearch.com.
[78] W.C. Willett, The Mediterranean diet: science and practice, Public Health Nutr. 9 (2006) 105–110.
[79] R.J. Pawlosky, J.R. Hibbeln, J.A. Novotny, N. Salem, Physiological compartmental analysis of α-linolenic acid metabolism in adult humans, J. Lipid Res. 42 (2001) 1257–1265.
[80] R. Estruch, M.A. Martínez-González, D. Corella, J. Salas-Salvadó, V. Ruiz-Gutiérrez, M.I. Covas, M. Fiol, E. Gómez-Gracia, M.C. López-Sabater, E. Vinyoles, F. Arós, M. Conde, C. Lahoz, J. Lapetra, G. Sáez, E. Ros, PREDIMED study investigators, effects of a Mediterranean-style diet on cardiovascular risk factors: a randomized trial, Ann. Intern. Med. 145 (2006) 1–11.
[81] J.S. Perona, R. Cabellomoruno, V. Ruizgutierrez, The role of virgin olive oil components in the modulation of endothelial function, J. Nutr. Biochem. 17 (2006) 429–445.
[82] N. Mohd Ali, S.K. Yeap, W.Y. Ho, B.K. Beh, S.W. Tan, S.G. Tan, The promising future of Chia, *Salvia hispanica* L., J. Biomed. Biotechnol. 2012 (2012) 1–9.
[83] J.A. Griffith, Y. Ma, L. Chasan-Taber, B.C. Olendzki, D.E. Chiriboga, E.J. Stanek, P.A. Merriam, I.S. Ockene, Association between dietary glycemic index, glycemic load, and high-sensitivity C-reactive protein, Nutrition 24 (2008) 401–406.
[84] P. Bogani, C. Galli, M. Villa, F. Visioli, Postprandial anti-inflammatory and antioxidant effects of extra virgin olive oil, Atherosclerosis 190 (2007) 181–186.
[85] L.G. Wood, M.L. Garg, H. Powell, P.G. Gibson, Lycopene-rich treatments modify noneosinophilic airway inflammation in asthma: proof of concept, Free Radic. Res. 42 (2008) 94–102.
[86] J. Faintuch, L.M. Horie, H.V. Barbeiro, D.F. Barbeiro, F.G. Soriano, R.K. Ishida, I. Cecconello, Systemic inflammation in morbidly obese subjects: response to oral supplementation with alpha-linolenic acid, Obes. Surg. 17 (2007) 341–347.
[87] T. Bahorun, A. Luximon-Ramma, T.K. Gunness, D. Sookar, S. Bhoyroo, R. Jugessur, D. Reebye, K. Googoolye, A. Crozier, O.I. Aruoma, Black tea reduces uric acid and C-reactive protein levels in humans susceptible to cardiovascular diseases, Toxicology 278 (2010) 68–74.
[88] D.S. Kelley, R. Rasooly, R.A. Jacob, A.A. Kader, B.E. Mackey, Consumption of Bing sweet cherries lowers circulating concentrations of inflammation markers in healthy men and women, J. Nutr. 136 (2006) 981–986.
[89] E.B. Levitan, N.R. Cook, M.J. Stampfer, P.M. Ridker, K.M. Rexrode, J.E. Buring, J.E. Manson, S. Liu, Dietary glycemic index, dietary glycemic load, blood lipids, and C-reactive protein, Metabolism 57 (2008) 437–443.
[90] H. Du, D.L. Van der A, M.M. Van Bakel, C.J. Van der Kallen, E.E. Blaak, M.M. Van Greevenbroek, E.H. Jansen, G. Nijpels, C.D. Stehouwer, J.M. Dekker, E.J. Feskens, Glycemic index and glycemic load in relation to food and nutrient intake and metabolic risk factors in a Dutch population, Am. J. Clin. Nutr. 87 (2008) 655–661.
[91] A.G. Pittas, S.B. Roberts, S.K. Das, C.H. Gilhooly, E. Saltzman, J. Golden, P.C. Stark, A.S. Greenberg, The effects of the dietary glycemic load on type 2 diabetes risk factors during weight loss, Obesity 14 (2006) 2200–2209.

[92] C.J. North, C.S. Venter, J.C. Jerling, The effects of dietary fibre on C-reactive protein, an inflammation marker predicting cardiovascular disease, Eur. J. Clin. Nutr. 63 (2009) 921–933.
[93] O.K. Chun, S. Chung, K.J. Claycombe, W.O. Song, Serum C-reactive protein concentrations are inversely associated with dietary flavonoid intake in U.S. adults, J. Nutr. 138 (2008) 753–760.
[94] J. González-Gallego, M.V. García-Mediavilla, S. Sánchez-Campos, M.J. Tuñón, Anti-inflammatory, immunomodulatory, and prebiotic properties of dietary flavonoids polyphenols, in: Prevention and Treatment of Human Disease, Elsevier, 2018, pp. 327–345.
[95] A. Nanri, M.A. Moore, S. Kono, Impact of C-reactive protein on disease risk and its relation to dietary factors, Asian Pac. J. Cancer Prev. 8 (2007) 167–177.
[96] A. Esmaillzadeh, M. Kimiagar, Y. Mehrabi, L. Azadbakht, F.B. Hu, W.C. Willett, Dietary patterns and markers of systemic inflammation among Iranian women, J. Nutr. 137 (2007) 992–998.

16

Values of natural products to future antiinflammatory pharmaceutical discovery

Augustine Amalraj and Sreeraj Gopi
R&D Centre, Aurea Biolabs (P) Ltd, Kolenchery, Cochin, Kerala, India

16.1 Introduction

Inflammation is a biological function that is generated with mechanical tissue disruption or from the reactions of physicochemical or biological agents in the body. These are described as contaminants, particularly toxic compounds, burns, or shock, and they may be microorganisms such as bacteria, fungi, and viruses or even hypersensitivity [1]. Inflammation lies at the heart of numerous prevalent diseases, including osteoarthritis (OA), rheumatoid arthritis (RA), diabetes, atherosclerosis, allergy, infections, and cancer. Multiple signaling pathways form a linkage of proinflammatory, immunomodulatory, and proresolving cascades, which state, by their interaction, the physiological and pathophysiological characteristics of inflammation [2]. It becomes more and more evident that for complex diseases such as inflammation, an interference with multiple targets is superior to pointing a particular key factor concerning drug efficiency, side effects, and adverse compensatory mechanisms [3]. It can be defined as acute and chronic based on the types of response and the capability of the action to eliminate the foreign agent or damaged tissues. Acute inflammation is very rapid with the link of plasma protein, fluid, and neutrophil migration whereas chronic inflammation is a lengthy development that is related to vascular proliferation, macrophages, fibrosis, and tissue destruction [4].

Acute inflammation happens within minutes with tissue injury, which mechanically involves the movement of plasma proteins, fluids, and neutrophils in the damaged region. These beneficial responses are part of the defense mechanism for repairing wounds and destroying microorganisms at the active sites [5–7]. Chronic inflammation is a lengthy process resulting in tissue necrosis, and it is due to the presence of macrophages and lymphocytes. It is described by

Inflammation and Natural Products. https://doi.org/10.1016/B978-0-12-819218-4.00009-2
© 2021 Elsevier Inc. All rights reserved.

the development of deteriorating sicknesses, including acquired immunodeficiency disorder, Alzheimer's disease, congestive heart failure, rheumatoid arthritis, central nervous system–neurodegenerative disorders, aging, asthma, atherosclerosis, multiple sclerosis, cancer, diabetes, gout, heart disease, inflammatory bowel disease, and infections (Fig. 16.1). It is also worried about the cause of muscle loss which is related with aging. Numerous tumorous situations are initiated from carcinomas complicated by the neoplastic transformation by chronic inflammation. A number of cancerous conditions such as colon carcinoma, chronic ulcerative colitis, esophageal adenocarcinoma, and liver, bladder and blood cancer because of disorders in mucosa-connected lymphoid tissue are predominantly considered their link with chronic inflammation [7, 8].

Inflammatory responses are important acts provided by the immune system during infection and tissue injury to maintain normal tissue homeostasis. Inflammation is a relatively complicated process at the molecular level, related to particular targets and the association of numerous proinflammatory expressions. Numerous health complications are related to prolonged inflammation, which affects almost all

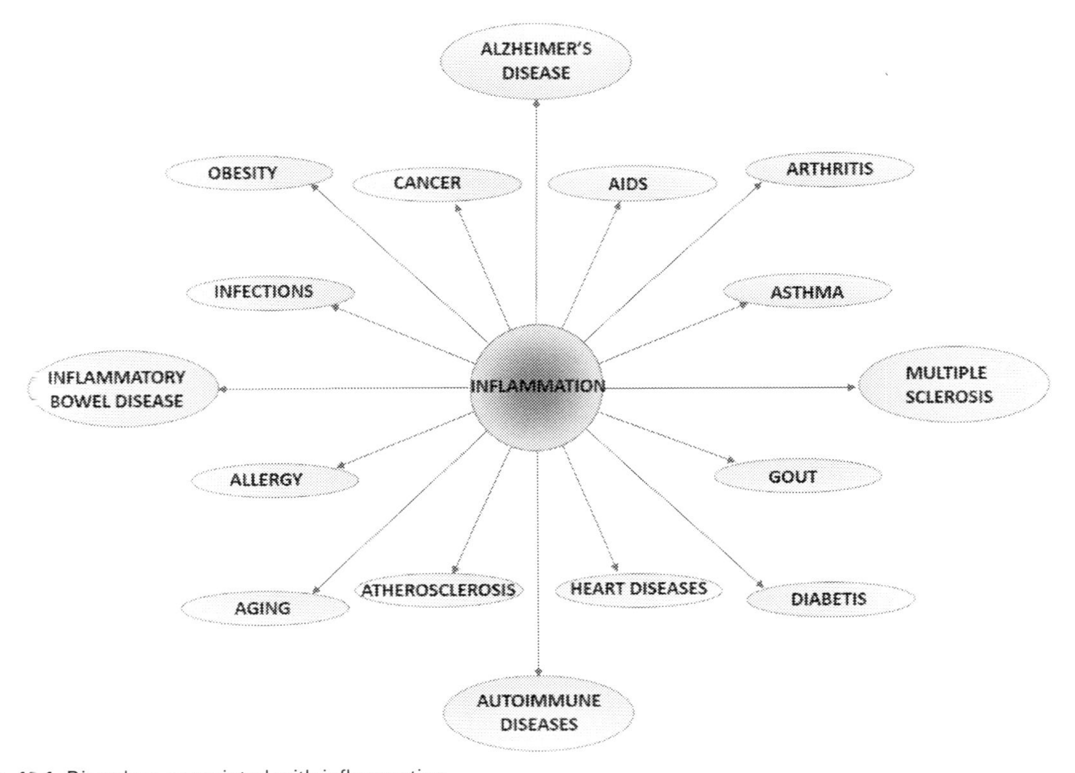

Fig. 16.1 Disorders associated with inflammation.

minor to major diseases [9]. Subsequently, the identification of proinflammatory cytokines in the past few decades has added significant value in inflammatory investigations. The production of proinflammatory mediators during inflammation is stimulated by macrophages, which comprise the tumor necrosis factor (TNF-α), various interleukins (ILs), prostaglandins (PGs), reactive oxygen species (ROS), and nitric oxide (NO). Chronic worsening diseases such as asthma, cancer, arthritis, and other related disorders are linked with the overproduction of these proinflammatory mediators [10]. The molecular and epidemiological studies showed that inflammation is meticulously connected with numerous infectious and noninfectious diseases [11].

16.2 Inflammatory mediators

The human body relates to numerous facets frequently termed heat, redness, pain, and edema that have long been strongly associated with inflammatory responses [12]. The release of numerous intermediaries, chemokines, and cytokines is encouraged by these inflammatory responses, which accomplish cellular penetration that spontaneously reconstructs tissue integrity. Conversely, persistent provocative stimuli or an unregulated mechanism of the determination point may tip to chronic inflammation [13]. Clinical confirmation recommends that several cytokine mediators regulate the antiinflammatory responses, including IL-4, IL-10, IL-13, NO, TNF-α, PGE2, and interferon (INF)-γ as a noncytokine mediator [14]. These factors play vital roles in the pathogenesis of different irregular metabolic illnesses during inflammation. The most effective approach to create an inflammatory response is to either block toll-like receptors (TLRs) or to downregulate proinflammatory gene expression by blocking nuclear factor kappa B (NF-κB) or c-Jun N-terminal kinases. Definite plant compounds, particularly polyphenols, have been shown, at least in vitro, to be capable of interacting with these receptors and transcription factors [15, 16].

16.3 Antiinflammatory therapy and response

Inflammation is a complex physiological process that is controlled by multiple signaling pathways. This necessitates the interaction of different cell types and modulates a wide spectrum of cellular responses, including immune cell maturation and function as well as tissue homeostasis [16]. Pharmacological approaches to conquer inflammation work on the agonists of the glucocorticoid receptor (glucocorticoids), interfering with eicosanoid biosynthesis [nonsteroidal antiinflammatory drugs (NSAIDs)] and the barrier of proinflammatory cytokine signaling such as biologics targeting TNF-α and IL-1 signaling [17].

To control inflammation and the equilibrium among an inflammatory and normal state, antiinflammatory cytokines such as IL-4, IL-10, and IL-13 as well as IFN-γ and transforming growth factor are released by macrophages [15]. The transcription factor NF-κB regulates the expression of different genes encoding proinflammatory cytokines, adhesion molecules, chemokines, growth factors, and chemo-attractants such as cyclooxygenase-2 (COX-2), monocyte chemoattractant protein-1, and inducible nitric oxide synthase (iNOS). Accordingly, through NF-κB the inflammatory response can be downregulated and the characteristic tissue reactions of inflammation disappear. Inflammation is regularly ameliorated by preventing iNOS and COX [16].

Inhibitors of eicosanoid biosynthesis control antiinflammatory designed multiple ligands because eicosanoids are dominant lipid mediators with pleiotropic, often contrasting, activities. They combine proinflammatory, immunomodulatory, proresolving, and homeostatic properties depending on the structure, concentration, and responsive tissue [18]. The exact interference with proinflammatory eicosanoid development without disturbing physiologically appropriate levels of eicosanoids is consequently considered key to avoid side effects. Strategies to achieve such a well-balanced modulation of the eicosanoid profile focus on the selective inhibition of terminal isoenzymes of eicosanoid biosynthesis and on multitarget inhibitors [19–21].

16.4 Nonsteroidal antiinflammatory drugs vs inflammation

To rectify inflammatory-caused disorders, it is mandatory to regulate the creation of several chemical mediators secreted during the entire process. A multiplicity of drugs is accessible to standardize the inflammatory situations and numerous more are under development. Still, existing drugs for inflammation are not effective in all circumstances and also report side effects [22, 23]. The most common drugs worldwide utilized for inflammation and associated diseases are a class of NSAIDs [24, 25]. The frequently used NSAIDs are ibuprofen [26], flurbiprofen [27], indomethacin [28], aspirin [29], diclofenac [30], and ketoprofen [31]. They belong to an analog of different classes of compounds such as heterocyclic, carboxylic, propionic, and phenyl acetic acid derivatives. The stated drug comprises organic acid that prevents the entry of arachidonic acid to the enzyme on the active site and inhibits the COX pathway [7].

Even with the outstanding antiinflammatory action of the NSAIDs, their healing practice is limited because of the association with severe side effects in the gastrointestinal pathway [32–34]. The acidic nature of NSAIDs affects the mucosal irritation, which leads to abdominal damage by preventing mucosal defensive PGE production [35, 36].

The terms ulceration, dyspepsia, abdominal pain, bleeding, diarrhea, and gastrointestinal irritation are categorized under gastrointestinal toxicity that may be caused due to the extended use of NSAIDs [37–39]. It has also been observed in studies that the use of NSAIDs reduces the risk of colorectal cancer [40, 41]. NSAID administration in about 50% of patients leads to mucosal damage in the small intestine. The injury to the mucosal membrane by using several NSAIDs has been well defined with the capability to prevent PG synthesis [42, 43]. These drugs have a multiplicity of toxic and unwanted effects, including interference with hemostasis, gastric erosion, and adverse effects on the liver and kidneys [44]. To resolve the problem, research toward plant-based antiinflammatory metabolites is needed to develop potential candidates. Therefore, the plant-based formulations have received extensive recognition by medical experts to treat several human diseases. To enhance human health and the value of life, the current necessity is to develop cheaper, safer, and more potent medications for inflammation.

16.5 Biodiversity of plant natural products

The natural compounds drug development shielded significant aids to the new way in drug development with their ethnopharmacological confirmation [45, 46]. Because of their extensive distribution, chemical multiplicity, and scientifically recognized biological promises, natural products have induced improved awareness as an extraordinary source for medications that might be beneficial to treat numerous human ailments [47]. Based on the several biological actions of medicinal plants and the chemical diversity also depend on their specialized metabolism pathway by which they derived. The abundant class of secondary metabolites has been derived by plants due to the molecular origin of heredity and plant evolution [48]. Therefore, a particular species of plant is able to develop exact metabolites that create distinctiveness with other plant species. These plant metabolites provide security to the plants to form environmental interactions and also have a wide range of considerable pharmacological properties. The world is customary for the use of medicinal plants to treat almost all type of human disorders since ancient time. The recognition goes to plants for the innovation of novel medicines as well as a key source of drug development [49, 50].

16.6 Medicinal plants as gift of nature

Since ancient times, medicinal plants have been utilized to treat a range of diseases. Medicinal plants and their active components contribute to most of the clinically used drugs to treat infectious diseases as well as cancers of different origins. However, the healing importance of the bioactive molecules is expected to develop sustained innovation for the global prevalence of microbial infections, cardiovascular diseases, and different cancers. Even though traditional medicines have been utilized along with the rapid growth of effective management against inflammation, the objectionable side effects limit the use of antiinflammatory drugs. The studies of natural compounds in association with ethnobotanical and pharmacological evidence are significant contributions for additional successful of these traditional components. [51]. The natural compounds have extensively been used for antiinflammatory purposes. Investigations of phytochemicals and ethnopharmacologicals has played a key role in the identification, isolation, characterization, and research on the mechanism of action of a diversity of natural active components.

Extraordinary levels of phytochemicals in vegetables, fruits, legumes, spices, and other edible herbs have shown substantial antiinflammatory potential. These phytochemicals are deliberated as natural strategic supporters for the improvement of human health. Terpenoids and phenolics have displayed greater antiinflammatory action compared to other classes of secondary metabolites [52]. Several plant-based natural products have the ability to substantially inhibit proinflammatory enzymes and mediators, as has been shown by several in vitro and in vivo investigations [53]. Phytochemicals have great multifunctionality to prevent cancer by interfering in different levels via altering the mechanistic pathway associated with inflammatory action [54–56]. There are several potential phytochemicals capable of preventing COX-2, iNOS, 5-lipoxygenase (5-LOX), TNF-α, and NF-1β, which are involved in inflammation-related cancer.

The flavonoid-rich fractions from the fruit peels of *Citrus limetta* displayed antimalarial activity by the inhibition of the proinflammatory mediators involved such as IFN-γ, TNF-α, and IL-6 in malarial pathogenesis [57, 58]. *Withania somnifera*, also known as Ashwagandha, the Indian winter cherry or Indian ginseng, is one of the most prevalent herbs used to stimulate human health. Several reports have linked the health benefits of Ashwagandha to its antiinflammatory, antistress, antioxidant, adaptogenic, and antidepressant properties [59–61] due to the ability of the bioactive components of Ashwagandha to inhibit NF-κB activation. The bioactive components of Ashwagandha were shown to hamper TNF and lipopolysaccharide (LPS)-induced NF-κB activation by conquering the degradation of the phosphorylated nuclear factor of the kappa light polypeptide gene enhancer in the B-cell inhibitor-α. Moreover, the bioactive components of Ashwagandha can block the TNF-induced activation of NF-κB by inhibiting the inhibitor of κB kinase β (IKKβ) kinase action through a thioalkylation-sensitive redox mechanism upon targeting Cys179 in the catalytic site of IKKβ

[62–69]. The extracts of the leaves of *Leucosidea sericea* showed substantial inhibition of IL-8 and TNF-α in human U937 cells [70]. The extracts of the roots of *Ocimum sanctum* demonstrated considerable antiinflammatory, analgesic, and antipyretic potential without any side effects [71]. *Bacopa monnieri* is a traditional Ayurveda medicinal herb recognized for its efficacy in relieving acute pain and inflammation, as related to the selective inhibition of the COX-2 enzyme and the consequent reduction in COX-2-mediated prostanoid mediators. *B. monnieri* downregulated NO and TNF-α in stimulated RAW 246.7 macrophages and IFN-γ in stimulated human blood cells. Moreover, in human blood cells, IL-10 was slightly elevated, indicating polarization toward a regulatory T cell phenotype, which justified the clinical evaluation of *B. monnieri* to handle diseases involving chronic systemic and brain inflammation driven by the innate immune system [72].

16.7 Herbal remedies in traditional medication for inflammation

A report from the World Health Organization (WHO) specified that about 75% of the world's population depend on herbal medicine for a healthy life [73, 74]. The plants *Artemisia herba-alba* and *Magnolia officinalis* are used for the treatment of inflammatory disorders in folklore medicine. These plants displayed their antiinflammatory effects by the stimulation and inhibition of IL-4 and IL-12 production while additionally decreasing the level of NO [75]. The flavone chrysoeriol is found in a the medicinal herb, *Lonicera japonica* Thunb., which is commonly used to treat inflammatory diseases. The chrysoeriol-ameliorated 12-*O*-tetradecanoylphorbol-13-acetate induced ear edema in mice and inhibited NF-κB and signal transducers and activators of transcription 3 (STAT3) pathways, which provides the pharmacological groundwork for developing chrysoeriol as a novel antiinflammatory agent [76]. The herbal infusion of three plants, namely, *Buddleia scordioides, Chamaemelum nobile,* and *Litsea glaucescens,* was found to exhibit significant antiinflammatory potential in Mexican folklore via regulating the COX-2, TNF-α, NF-κB, and IL-8 cytokines [77]. Liquiritin is a kind of flavonoid isolated from the root of *Glycyrrhiza uralensis* that attenuated rheumatoid arthritis via reducing inflammation, suppressing angiogenesis, and inhibiting the mitogen-activated protein kinase signaling pathway [78]. The methanolic extracts from the green and red species of the genus Brassica significantly inhibited LPS-induced NO production in a dose-dependent manner via COX-2 and iNOS suppression [79]. The Chinese skullcap root extract suppressed abnormal androgen events in prostate tissue and inhibited the development of benign prostatic hyperplasia by targeting inflammation and apoptosis-related markers [80]. The leaves of the herb *Smallanthus sonchifolius* were recently used in traditional therapy for their capacity to inhibit PGE2, TNF-α, and NO due to the presence of chlorogenic acid [81]. A nutrient mixture made with the germinated seeds of green gram and horse gram with herbs such as turmeric and fenugreek ameliorated CCl_4-induced liver cirrhosis due to their antiinflammatory and antioxidant activities [82]. The herb *Clitoria ternatea* L., commonly known as the butterfly pea, showed its antiinflammatory action in RAW 264.7 macrophage cells by suppressing LPS-induced inflammation due to the presence of polyphenols [83]. *Emblica officinalis* treatment attenuated aluminum-induced proinflammatory responses and endoplasmic reticulum stress through its antiinflammatory, antioxidant, and antiapoptotic activities, which suggest that *E. officinalis* could be a good therapeutic lead to manage Alzheimer's disease due to its neuroprotective efficacy [84]. *Cyperus rotundus* L. extract exhibited topical antiinflammatory activity in models of irritative dermatitis and skin hyperproliferation due to the antiinflammatory and antiproliferative effects [85].

16.8 Ayurvedic formulation for inflammation

The concept of polyherbalism is peculiar to Ayurveda, although it is difficult to explain in terms of modern parameters. Historically, the Ayurvedic literature "Sarangdhar Samhita" highlighted the concept of synergism behind polyherbal formulations. These formulations have a wide therapeutic range, fewer side effects, ecofriendliness as well as being cheaper and readily available. A polyherbal formulation (Vati), a mixture of *Boswellia serrata*, *Commiphora wightii*, *Hemidesmus indicus*, *Aloe barbadensis*, *W. somnifera*, *Zingiber officinale*, *Berberis aristata*, and *Cucurma longa*, was recorded to inhibit LPS-induced TNF-α production by immature dendritic cells [86].

An Ayurvedic formulation derived from the composition of three plants (*Glycyrrhiza glabra*, *O. sanctum*, and *Phyllanthus emblica*) is termed CIM-Candy. This standardized herbal formulation displayed significant humoral immune response to enhance the hemagglutination antibody titer response at a dose of 10 and 100 mg/kg body weight in Swiss albino mice [87]. The essential oil from the *Houttuynia cordata* displayed good inhibition of the release of LPS-induced PGE2 from the mouse peritoneal macrophages with an IC50 value of 44.8 mg/mL and COX-2 inhibition with an IC50 value of 30.9 mg/mL in a dose-dependent manner without affecting COX-1 production [88]. The Chinese herbal prescription JieZe-1 (JZ-1) is composed of *Phellodendri Chinensis* Cortex, Ginkgo Semen, *Solanum nigrum* L., *Taraxacum mongolicum* Hand., Taraxaci Herba, *Thlaspi arvense* Linn. (Herba Patriniae), *Dictamnus dasycarpus* Turcz. (Dictamni Cortex), *Smilax glabra* Roxb.

(Smilacis Glabrae Rhizoma), *Paeonia suffruticosa* Andr. (Moutan Cortex), *Mentha haplocalyx* Briq. (Menthae Haplocalycis Herba), and *Borneolum Syntheticum*. It significantly attenuated the increased cytokines IFN-α, IFN-β, IL-6, and TNF-α by herpes simplex virus-2 infection [89]. A herbal formulation was formulated by the ethanol extract of five major herbs, *M. haplocalyx* Briq., *Magnolia biondii* Pamp, *Xanthium sibiricum* Patr., *Asarum sieboldii* Miq., and *Caryophyllus aromaticus* L., and used against inflammatory diseases. The formulation suppressed severe lung inflammation in mice, which seemed to be associated with the formulation activating nuclear factor erythroid 2-related factor 2, suggesting that this therapeutic effect can be an alternative to conventional herbal remedies [90]. Treatment with a mixture of three herbs such as *Astragalus membranaceus* (Fisch.) Bunge., *Rehmanniae Radix* Libosch, and *Epimedium brevicornu* Maxim ameliorated airway remodeling and alleviated asthmatic features in a chronic asthma model due to the antiinflammatory and antioxidant effects [91]. Fangxiao Formula is an herbal medicine that is formulated by mixtures of 11 Chinese medicinal ingredients such as *Radix Adenophorae, Radix Glehniae, Radix Codonopsis, Radix Ophiopogonis, Cortex Lycii Radicis, Radices Paeoniae Alba, Concha Haliotidis, Poria cocos, Atractylodes macrocephala, Pinellia ternata*, and *Pericarpium Citri Reticulatae*. It alleviated airway inflammation and remodeling in rats with asthma via the suppression of the transforming growth factor-β/Smad3 signaling pathway [92]. A Chinese medicine formula containing mixture of the herbs *Panax ginseng* C.A. Mey., *Polygala tenuifolia* Wild., *Acorus tatarinowii* Schott., and *P. cocos* (Schw.) Wolf. ameliorated depression-like behaviors in chronic unpredictable mild stressed mice by regulating the gut microbiota-inflammation-stress system [93].

16.9 Natural products—A promising antiinflammatory pharmaceutical drug discovery

Traditional Ayurveda medicines and natural phytonutrients could be used as a better alternative strategy for the effective treatment of inflammation and other biological disorders. In recent decades, the popularity of herbal drugs has grown tremendously due to potentially better safety, availability, and economic benefits. The WHO also encourages developing countries such as India to use herbal drugs for future pharmaceutical discoveries and as an alternative to modern medications [94, 95]. Thus, there is an increasing need to investigate the potential of standardized herbal formulations for use as complementary treatment. It has been estimated that about 80% of the people from developing countries depend upon drugs from natural origin. However, quite a number of plant metabolites have been used as precursors in synthesis to improve their biological potential [96]. The literature reports on bioactive compounds confirmed the antiinflammatory potential of a variety of metabolites. The potent bioactive entities isolated from the plants, namely curcuminoids [97], piperine [98], β-caryophyllene [99, 100], boswellic acid [101], gingerol, shogaol [102], rosmarinic acid, and carnosic acid [103, 104], have proved their efficacy and potency against inflammation and inflammatory mediators. They are available as a nutraceutical delivery system and are shown in Fig. 16.2.

16.9.1 Turmeric—Curcuminoids

Natural products have been used in traditional medicine for thousands of years, and have shown promise as a source of components for the development of new drugs. Turmeric (*C. longa* Linn) is a member of the Zingiberaceae family and is cultivated in tropical and subtropical regions around the world. It originates from India, Southeast Asia, and Indonesia. Turmeric powder is used extensively

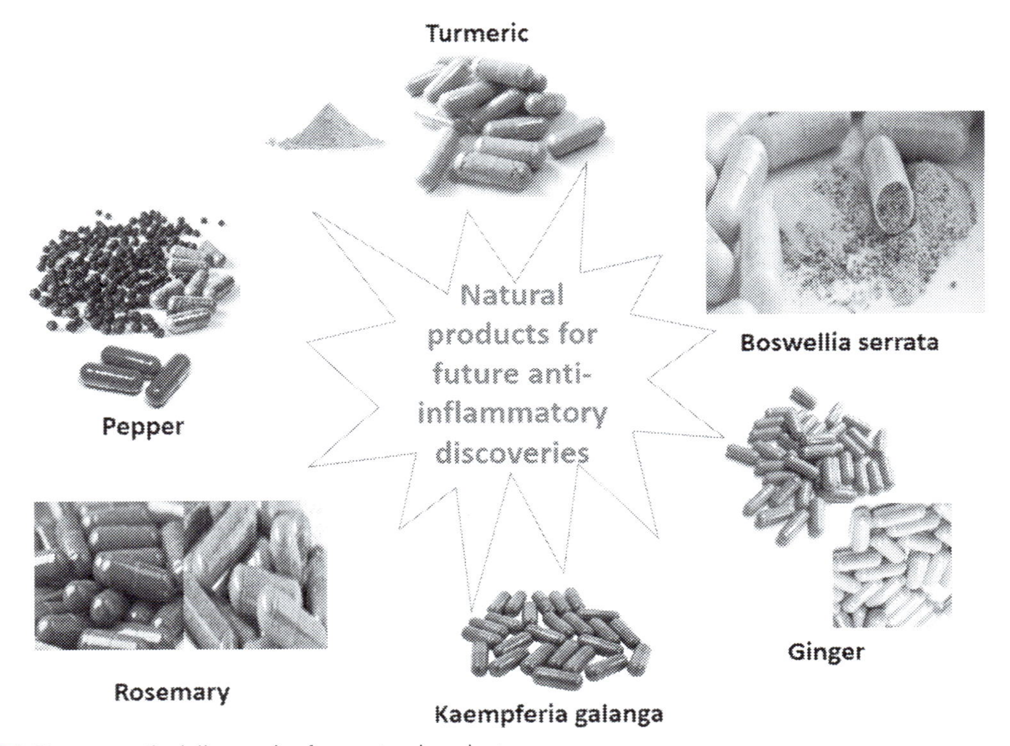

Fig. 16.2 Pharmaceutical discoveries from natural products.

as a coloring and flavoring agent in curries and mustards. Turmeric has been used in India to maintain oral hygiene. It has traditionally been used for medical purposes for many centuries in countries such as India and China to treat jaundice and other liver ailments. Turmeric is one of the most popular medicinal herbs, with a wide range of pharmacological activities such as antiinflammatory, antioxidant, antiprotozoal, antivenom, antimicrobial, antimalarial, antiproliferative, antiangiogenic, antitumor, and antiaging properties. It has also been used to treat ulcers, parasitic infections, various skin diseases, antiimmune diseases, and the symptoms of colds and flus [97]. The pharmacological activity of turmeric has been attributed mainly to curcuminoids consisting of curcumin (72%–78%), demethoxycurcumin (12%–18%), and bisdemethoxycurcumin (3%–8%) with a purity of $\geq$ 95%. Chemically, curcumin is a diferuloyl methane molecule [(1*E*,6*E*)-1,7-bis (4-hydroxy-3-methoxyphenyl)-1, 6-heptadiene-3,5-dion] containing two ferulic acid residues joined by a methylene bridge. In recent decades, curcuminoid has drawn great attention for its broad spectrum of therapeutic actions, including antiinflammatory, antioxidant, anticancer, antimicrobial, wound healing, hepatoprotective, and potential ability in preventing neurodegenerative diseases [105]. The WHO stated the acceptable daily intake of curcuminoids as a food additive in the range of 0–3 mg/kg. Curcuminoids and turmeric products have been characterized as safe by the US Food and Drug Administration (FDA). The average intake of turmeric in the Indian diet is approximately 2–2.5 g for a 60 kg individual, which corresponds to a daily intake of approximately 60–100 mg of curcumin. Curcuminoids have achieved potential therapeutic interest to cure immune-related, metabolic diseases, and cancer due to a vast number of biological targets and virtually no side effects [97, 105].

Numerous studies have demonstrated the antiinflammatory, antioxidant, antimicrobial, anticarcinogenic, and proapoptotic properties of curcumin. Curcumin has also been shown to exhibit therapeutic effects against arthritis and various neurological, pulmonary, hematological, and cardiovascular ailments. Curcuminoids exhibit pleiotropic actions, interacting with numerous molecular targets involved in inflammation, including TNF-α, IL-1β, IL-6, IL-8, NF-κB, and COX-2 [97, 105, 106]. Besides, it has a superior safety profile determined by a clinical study that as high as 8 g/day of dosage would not induce any observable adverse effects. This safe profile has been reflected by the continuous increase of preparations based on curcuminoid marketed as a food ingredient or constituent of dietary supplements. However, the functional applications of curcuminoid have been seriously limited by its very low systemic bioavailability, attributable to poor absorption, fast metabolic alterations, and rapid elimination. The available evidence indicates that only minute amounts of curcuminoid reach the circulation after high-dose oral administration in animals and humans. The majority of the orally administered curcuminoid is excreted in the feces and the urine, with very little being detected in the blood plasma. Curcuminoid has very low solubility in aqueous media due to inter- and intrahydrogen bonding. Higher solubility was observed in an alkaline solution when dissolved curcuminoid was quickly degraded into vanillin, ferulic acid, and feruloyl methane. Other environmental factors such as ultraviolet (UV) irradiation and heating also contribute to decomposition of this yellowish polyphenolic compound. These dramatically affect the absorption and bioavailability of this active molecule with a consequent unsatisfactory pharmacokinetic profile and reduced efficacy [97, 105, 106]. These are the very big challenges for future antiinflammatory pharmaceutical discoveries.

To increase the water solubility, stability, bioavailability, and potential applications, different methods have been proposed and investigated. Several strategies such as nanoparticles, liposomes, solid dispersions, solid lipid nanoparticles, microemulsions, and complexation with phospholipids and cyclodextrins have been developed to improve the bioavailability of curcumin/curcuminoid [107–111]. A bioavailable form of curcuminoid–"Cureit"–was developed based on the recreation of the complete natural turmeric matrix (CNTM) with active curcuminoids (~ 50%) by a polar-nonpolar-sandwich (PNS) technology to preserve functional properties, improve the stability of compounds, enhance health benefits, control the release of bioactive compounds at the desired time and at a specific target, and increase the bioavailability of the bioactive compounds [105].

Recently, in vitro and in vivo studies indicated that hydrogenated curcumin such as tetrahydro curcuminoids, hexahydrocurcuminoids, and octahydrocurcuminoids exhibit antiinflammatory, antioxidant, antitumor, antidiabetic, and antihepatotoxicity properties. On this basis, these molecules can be potential therapeutic agents for the prevention and/or treatment of various malignant diseases such as arthritis, Alzheimer's disease, allergies, and other inflammatory illnesses. A high bioavailable form of hydrogenated curcuminoids was prepared by encapsulation with β-cyclodextrin, which reduced inflammation in RA patients. The antiinflammatory action of the bioavailable form of hydrogenated curcuminoids clearly indicated that their oral administration registered a superior percentage of improvement in patients with RA and strongly favored safe and effective applications for the management of inflammatory action [112–114].

16.9.2 *Boswellia serrata*

Traditionally, *B. serrata* extract is used in Indian Ayurvedic medicine to treat inflammatory diseases [101]. Boswellic acids (BA), the main active ingredients of *B. serrata* extracts, have potent antiinflammatory properties and represent promising therapeutic agents with no serious, long-term, or irreversible adverse effects for the treatment of many inflammatory disorders such as RA, OA, chronic colitis, ulcerative colitis, bronchial asthma, and inflammatory bowel disease [115, 116]. The pharmacological effects of the *B. serrata* extracts were attributed to the pentacyclic triterpenic boswellic acids, especially to 3-*O*-acetyl-11-keto-β-boswellic acid (AKBA). Boswellic acids were proposed to act as inhibitors of 5-LOX, the NF-κB pathway, human leukocyte elastase, cathepsin G, and microsomal PGE2 synthase (mPGES)-1 in various pharmacological effects [95, 117].

BA was able to inhibit the IL-1β and TLR4-mediated induction of several inflammatory mediators from OA synovial explant tissue [116]. The pharmacological effects of *B. serrata* gum resin extracts (BSE) were mainly attributed to 11-keto-β-boswellic acid (KBA) and AKBA [101]. The antioxidant and antiarthritic activities of BSE in collagen-induced arthritis were effective in bringing significant changes in articular elastase. BSE markedly inhibited the clinical signs of joint swelling, significantly decreased the free radical load, and modulated inflammatory mediators in arthritic rats [115]. The protective role of BSE was mediated via its antioxidant effect through the suppression of lipid peroxidation and boosting the antioxidant defense system. It was demonstrated that the AKBA was able to block inflammatory reactions in both acute and chronic inflammation models [118–120]. In another study, it was found that AKBA inhibits the activation of NF-κB in an in vivo mice model [121]. BA reduces cartilage loss, osteophyte formation, and synovitis in a mice model study and provides evidence to implicate the inhibition of both IL-1β and TLR signaling as potential mechanisms mediating this protective effect [116]. BSE has been claimed to decrease glycosaminoglycan degradation, which helps to keep the cartilage in good condition. This might be responsible for the recovery of patients with OA and might stop the progression of this condition. BSE is effective on the production of antibodies and cell-mediated immunity while also inhibiting human leukocyte elastase. This could be of help in autoimmune disorders such as rheumatoid arthritis [122, 123]. Moreover, the study encountered statistically significant improvement in the efficacy variables in patients with knee OA treated with BSE [124].

16.9.3 *Kaempferia galanga*

The *Kaempferia galanga* rhizome is rich in essential oils and is used traditionally for the treatment of indigestion, cold, pectoral and abdominal pain, and headaches as well as an expectorant, diuretic, and carminative [125]. *K. galanga* extracts are also used for the treatment of swelling [126]; cytotoxic [127], antihypertensive [128], and hypolipidemic effects; [125] and antiinflammatory effects [129, 130]. *K. galanga* has been used as an alternative medicine for antirheumatic activities through its antiinflammatory activity. Alcoholic and petroleum ether extracts of *K. galanga* were tested against adjuvant-induced chronic inflammation in rats. The study suggested that the extracts of *K. galanga* effectively suppress the progression of acute and chronic inflammation in rats [131]. The antinociceptive and antiinflammatory activities of the aqueous extracts of *K. galanga* using female Balb/c mice and Sprague-Dawley rats were also tested. The extract produced a significant antiinflammatory activity when assessed using a carrageenan-induced paw-edema test [132]. Hence, *K. galanga* can be a potential therapeutic agent to treat inflammation.

16.9.4 Pepper

Black pepper (*Piper nigrum*), which is widely used in seasoning, contains bioactive ingredients in its oleoresin fraction, such as essential oils and the alkaloid piperine. Various antiinflammatory effects of substances extracted from black pepper are known for many therapeutic applications in modern medicine to treat different diseases [133]. In particular, some ethanolic and hexane extracts of black pepper have exposed significant antiinflammatory activity in mice and rats, using different dosage protocols [134]. Piperine can be considered the main ingredient of black pepper, and it revealed antiinflammatory activity in the IL-1β activated fibroblast-like synoviocytes [135], inhibiting LPS-stimulated endotoxins [136]. Further, piperine might be viewed as a potent immunomodulator, inhibiting airway inflammation in a murine model of asthma by the enhanced expression of transforming growth factor-beta gene in the lungs [137]. Piperine was also detected to reduce the production of IL-6, matrix metalloproteinase-13, and PGE at the concentration range of 10–100 mg/mL [135]. In another study, piperine was coadministered with curcumin from *C. longa* to suppress a high fat diet-induced inflammation in C57BL/6 mice and for the prevention of metabolic syndrome [138]. Apart from that, piperine antiinflammatory potential was investigated at colorectal sites, inhibiting free fatty acid-induced TLR4-mediated inflammation and acetic acid-induced ulcerative colitis in mice [139]. Finally, this compound was evaluated in a carrageenan-induced inflammation assay in mice to assess the analgesic and antiinflammatory activities of piperine activities at an oral dose of 6 mg/kg/day [140]. Several studies have been performed on the mechanism of piperine, reporting that it acts by reducing the expression of intercellular adhesion molecule 1, TNF-α, iNOS, and NF-κB; inhibiting cytochrome P450 enzymes21 and *p*-glycoprotein activities; and preventing endoplasmic reticulum stress [141–144].

β-Caryophyllene (trans-(1*R*,9*S*)-4,11,11-trimethyl-8-methylene bicycle [7.2.0] undec-4-ene) [145] seems to be an excellent candidate as it is readily available as a natural compound found in the essential oils of many species and food plants, particularly in black pepper. It inhibited inflammation and tissue damage in models of colitis and nephrotoxicity in a cannabinoid receptor type 2 (CB2) receptor-dependent manner without side effects [95, 99, 100]. By incorporating β-caryophyllene in the formulation, the subjects with active OA would experience a reduction in pain due to the cannabinoid action [146]. Moreover, β-Caryophyllene is a potent phyto-cannabinoid that is reported to have antiinflammatory activity. β-Caryophyllene is the first natural CB2 receptor agonist that could orally attenuate thermal hyperalgesia, mechanical allodynia, and reduced inflammatory responses in different animal models [100, 147]. β-Caryophyllene significantly decreased arthritis, which was evident with the arthritis index, paw volume, and maintenance of biochemical parameters. β-Caryophyllene suppressed the NF-κB activity, joint inflammation, and destruction in adjuvant arthritic rats. Bone structures were recalcified upon treatment with the β-caryophyllene dose-dependently. The histopathology and radiology also revealed the control in inflammation with β-caryophyllene. β-Caryophyllene has prominent antiarthritic activity, which may be attributed to its antiinflammatory activity [147]. The antiinflammatory properties of β-caryophyllene could be beneficial for the prevention and management of inflammation-related diseases. A natural herbal formulation (Acujoint) is prepared with four herbs—the bioavailable form of curcumin, Cureit; AKBA from *B. serrata*; β-caryophyllene from *P. nigrum*; and the aqueous extract of *K. galanga* in a fixed ratio based on a complete natural matrix as health supplements with the concept of Ayurveda. Acujoint is a safe and effective natural formulation that has promising therapeutic utility in the treatment of osteoarthritis [95].

16.9.5 Ginger

Z. officinale Roscoe, the commonly used herb known as ginger, belongs to the family Zingiberacae, which has been cultivated and used for medicinal purposes since ancient times. It was an important ingredient in herbal medicines for catarrh and rheumatism. The plant has a number of chemical components responsible for its medicinal properties, such as zingerone, gingerol, shogoals, yakuchinone A, diarylheptanoid, and 12-dehydrogingerdione. There is scientific evidence regarding the various pharmacological activities of ginger powder, extract, or its bioactive components that show effective antiinflammatory, antiarthritic, antidiabetic, antibacterial, antifungal, and anticancer properties. The ginger constituents are reported to significantly inhibit inflammation by (a) regulating the arachidonic acid pathway and inhibiting PGE2 and COX2, inhibits iNOS mRNA. (b) By inhibiting NF-κB pathway and inhibiting the production of TNF-α, IL-1β, IL-6, and IL-12, the active component of ginger modulate the inflammation. Abundant clinical trials demonstrate compelling evidence for the potential use of Zingiber as a treatment for a variety of immunomodulatory, antiinflammatory, and allergic disorders such as acute respiratory distress syndrome, asthma, inflammatory conditions, and associated pain such as rheumatism and musculoskeletal disorders, during which varying degrees of relief from pain and swelling complaints are achieved with ginger extract [51].

Lipid peroxides and activated macrophages play a crucial role in arthritis and other inflammatory diseases. Experimental studies have shown that ginger constituents inhibit the inflammation process by inhibiting the arachidonic acid metabolism, a key pathway [148]. Both in vitro and in vivo animal models have shown that ginger and its constituents inhibit both cyclooxygenase and LOX [149], and also act as an inhibitor of leukotriene synthesis [150]. The antiinflammatory property of ginger and its bioactive component zingerone has been evaluated on LPS-induced inflammation in mice [151]. Ginger and zingerone suppressed LPS-induced NF-κB activities in cells in a dose-dependent manner, with maximal inhibition at 100 mg/mL. The production of LPS-induced proinflammatory cytokines was significantly reduced by dietary ginger and zingerone. The study suggested that ginger and zingerone act as broad-spectrum antiinflammatory agents via suppressing the activation of NF-κB, the production of IL-1β, and the infiltration of inflammatory cells.

The antiinflammatory activity of 6-, 8-, 10-gingerol and 6-shogaol isolated from the ginger rhizome has been studied in an LPS-induced in vitro system [152]. Crude organic extracts of ginger that predominantly contain the above compounds were capable of inhibiting LPS-induced PGE2 production and were less effective in inhibiting TNF-α. Ginger extracts or standard compounds, particularly gingerols, inhibited LPS-induced COX-2 expression while shogaol-containing extracts did not affect the same, thus suggesting that compounds found in ginger are capable of inhibiting PGE2 production. The antiinflammatory/antioxidant activity of ginger has been shown in rat adjuvant-induced arthritis [153]. Ginger significantly suppressed the incidence and severity of arthritis by increasing/decreasing the production of the antiinflammatory/proinflammatory cytokines, respectively, and activating the antioxidant defense system. The effect of hydroalcoholic extracts of ginger rhizomes on the classical models of rat paw and skin edema has been reported [154]. The carrageenan or serotonin-induced rat paw edema was inhibited significantly by the administration of alcoholic ginger extract. The antiedematogenic activity seems to be related, at least partially, to an antagonism of the serotonin receptor. In the backdrop of the traditional use of ginger and its extracts in antiinflammatory remedies, the antiinflammatory effect of ginger essential oils was assessed in Lewis rats with streptococcal cell wall (SCW)-induced arthritis [155]. Ginger essential oils prevented chronic joint inflammation, suggesting that ginger's antiinflammatory properties are attributable to the combined effects of both the pungent-tasting gingerols as well as its aromatic essential oils. A few clinical studies, mostly randomized controlled trials, investigating the beneficial effect of ginger in ameliorating arthritic knee pain in patients with symptomatic osteoarthritis and rheumatoid arthritis have been reported [148, 156–158]. The antiinflammatory effect of ginger extract was observed in 56 osteoarthritic patients as indicated by reduced arthritic knee pain, although the effectiveness was less than that of ibuprofen [59]. A moderate but statistically significant reduction in knee pain through the ginger extract was reported in subsequent clinical studies [157, 158]. These trials found that the pain level of the participants in the intervention group were significantly lower than those in the placebo group. In addition, the decreased use of nonsteroidal antiinflammatory drugs and analgesics was also observed [148]. The results of the clinical studies conducted so far have been encouraging. Considering the broad spectrum of antiinflammatory actions of ginger and its safety record, this herbal product is likely to be a valuable dietary supplement in the treatment of inflammatory disorders [159].

16.9.6 Rosemary

Rosemary, *Rosmarinus officinalis* L. is also another promising medicinal plant known for various biological activities, predominantly its antiinflammatory, antioxidant, antineurodegenerative, antitumor, antidepressant, and hepatoprotective effects. These are attributed to the extracts of rosemary that contain several fractions of biologically active compounds, particularly carnosic acid, carnosol, rosmarinic acid, caffeic acid, and ursolic acid [103, 160]. Rosemary extract produced a stronger inhibition of proinflammatory cytokine secretion, reducing the TNF-α, IL-1β, and IL-6 secretions in two cell models of THP-1 human macrophages activated with LPS and with human oxidized low-density lipoproteins, simulating a general inflammatory response and an atherosclerotic-related inflammation, respectively [161]. Rosemary extract and its bioactive components inhibited inflammatory responses stimulated by LPS through the NO and proinflammatory factor as well as TNF-α inhibition [162]. Morphine withdrawal syndrome was attenuated with ethanolic and aqueous extracts of rosemary [103, 163].

The carnosic acid is an abietane diterpenoid that also has various important pharmacological effects, including antiinflammatory, antioxidation, antitumor, antibacterial, and antifungal activities as well as gastroprotection [160]. Kuo et al. [164] demonstrated that the inhibitory effects of carnosic acid on LPS-induced NO and TNF-α production are related to the suppression of iNOS and COX-2 expression, resulting from the inhibition of NF-κB signaling. Carnosol decreased LPS-induced iNOS mRNA and downregulated the inhibitor NF-κB kinase activity in the mouse macrophage RAW 264.7 cell line [165]. Moreover, these components suppressed the formation of proinflammatory leukotrienes in intact human polymorphonuclear leukocytes and inhibited the activity of 5-LOX [166].

However, these may not be the only components solely responsible for the antiinflammatory activity of rosemary; in fact, rosmarinic acid has also been studied [167]. Rosmarinic acid, an ester of caffeic acid and 3,4-dihydroxyphenyllactic acid, has a broad range of applications and a variety of biological activities, including antioxidative, antiinflammatory, antiapoptotic, antitumor, antiallergic, and antiviral activities [103, 160, 168]. This pure compound was recently tested in different rat models of local (carrageenan-induced paw edema) and systemic inflammation (liver ischemia–reperfusion and thermal injury models), producing a notable activity on both. Rosmarinic acid administered at 25 mg/kg efficiently reduced paw edema at 6 h by more than 60%, exhibiting a dose-response effect, whereas the same amount was also able to reduce multiorgan dysfunction markers by modulating NF-κB and metalloproteinase-9 in the thermal injury model [167]. Rosmarinic acid exhibited significant neuroprotective effects during diabetic cerebral I/R injury, including attenuation of the blood-brain barrier breakdown, a decrease of infarct volume, alleviation of cerebral damage, reduction of high mobility group box protein 1—HMGB1 expression, and phosphorylated IκB-α and NF-κB protein expression in ischemic brain tissue, which pointed out the therapeutic potential for rosmarinic acid as a useful antiinflammatory lead compound in early diabetic cerebral I/R injury [103].

16.10 Major challenges in upgrading natural products

The efficient discoveries of natural products are still dependent on various factors, which leads to major challenges in natural product drug development. Instead of these major tasks, natural products still have a dynamic role in pharmaceutical science and drug discovery. These key tasks include the collection of prospective plants with an ethnic nature, rare species, shortage of plant samples, high cost, capricious composition of crude extracts, lack of reproducible outcomes, isolation and purification of bioactive components from crude extract, complications to replication at a large scale from natural sources to obtain vital drugs, and modification in the drug discovery tactics of combinatorial chemical properties [7, 169].

16.11 Future prospective and conclusions

Many natural products with a "folk medicinal" background. Including curcuminoids, piperine, β-caryophyllene, boswellic acid, gingerol, shogaol, rosmarinic acid. and carnosic acid. Surprisingly have no or marginal side effects, even when provided at a high oral dosage in animal and human studies. Hence, the enhancement of these natural products may lead to the development of new, more potent antiinflammatory agents from these screened metabolites from plant sources. Multiple natural matrices promise higher therapeutic efficacy and safety, especially for the treatment of complex disorders such as inflammation. These scientific reports convey the attention toward the discovery of antiinflammatory drugs from plant sources. This chapter summarizes the current status of research and findings of natural products with their mechanism of action. This chapter provides a platform to researchers who are working to upgrade antiinflammatory agents.

References

[1] N.T. Ashley, Z.M. Weil, R.J. Nelson, Inflammation: mechanisms, costs, and natural variation, Annu. Rev. Ecol. Evol. Syst. 43 (2012) 385–406.

[2] R. Medzhitov, Origin and physiological roles of inflammation, Nature 454 (2008) 428–435.

[3] R. Morphy, Z. Rankovic, Designed multiple ligands. An emerging drug discovery paradigm, J. Med. Chem. 48 (2005) 6523–6543.

[4] S. Riegsecker, D. Wiczynski, M.J. Kaplan, S. Ahmed, Potential benefits of green tea polyphenol EGCG in the prevention and treatment of vascular inflammation in rheumatoid arthritis, Life Sci. 93 (2013) 307–312.

[5] I. Striz, I. Trebichavsky, Calprotectin—a pleiotropic molecule in acute and chronic inflammation, Physiol. Res. 53 (2004) 245–253.

[6] B. de las Heras, S. Hortelano, Molecular basis of the anti-inflammatory effects of terpenoids, Inflamm. Allergy Drug Targets 8 (2009) 28–39.

[7] N. Kishore, P. Kumar, K. Shanker, A.K. Verma, Human disorders associated with inflammation and the evolving role of natural products to overcome, Eur. J. Med. Chem. 179 (2019) 272–309.

[8] N. Kumar, S. Drabu, S.C. Mondal, NSAID's and selectively COX-2 inhibitors as potential chemoprotective agents against cancer, Arab. J. Chem. 6 (2013) 1–23.

[9] A.U. Ahmed, An overview of inflammation: mechanism and consequences, Front. Biol. 6 (2011) 274–281.

[10] K.L. Rock, H. Kono, The inflammatory response to cell death, Annu. Rev. Pathol. Mech. Dis. 3 (2008) 99–126.

[11] P. Hunter, The inflammation theory of disease, EMBO Rep. 13 (2012) 968–970.
[12] M. Perretti, F. D'Acquisto, Annexin A1 and glucocorticoids as effectors of the resolution of inflammation, Nat. Rev. Immunol. 9 (2009) 62–70.
[13] F. D'Acquisto, F. Maione, M. Pederzoli-Ribeil, From IL-15 to IL-33: the never-ending list of new players in inflammation. Is it time to forget the humble aspirin and move ahead? Biochem. Pharmacol. 79 (2010) 525–534.
[14] M.S. Bahia, Y.K. Katare, O. Silakari, B. Vyas, P. Silakari, Inhibitors of microsomal prostaglandin E2 synthase-1 enzyme as emerging anti-inflammatory candidates, Med. Res. Rev. 34 (2014) 825–855.
[15] J.J. Turner, K.M. Foxwell, R. Kanji, C. Brenner, S. Wood, B.M. Foxwell, M. Feldmann, Investigation of nuclear factor-kappaB inhibitors and interleukin-10 as regulators of inflammatory signalling in human adipocytes, Clin. Exp. Immunol. 162 (2011) 487–493.
[16] A. Jungbauer, S. Medjakovic, Anti-inflammatory properties of culinary herbs and spices that ameliorate the effects of metabolic syndrome, Maturitas 71 (2012) 227–239.
[17] S.P. Weisberg, D. McCann, M. Desai, M. Rosenbaum, R.L. Leibel, A.W. Ferrante Jr., Obesity is associated with macrophage accumulation in adipose tissue, J. Clin. Invest. 112 (2003) 1796–1808.
[18] M.J. Stables, D.W. Gilroy, Old and new generation lipid mediators in acute inflammation and resolution, Prog. Lipid Res. 50 (2011) 35–51.
[19] F. Celotti, S. Laufer, Anti-inflammatory drugs: new multitarget compounds to face an old problem. The dual inhibition concept, Pharmacol. Res. 43 (2001) 429–436.
[20] A. Koeberle, O. Werz, Inhibitors of the microsomal prostaglandin E(2) synthase-1 as alternative to non steroidal anti-inflammatory drugs (NSAIDs)—a critical review, Curr. Med. Chem. 16 (2009) 4274–4296.
[21] A. Koeberle, O. Werz, Multi-target approach for natural products in inflammation, Drug Discov. Today 19 (2014) 1871–1882.
[22] J.L. Wallace, Prostaglandins, NSAIDs, and gastric mucosal protection: why doesn't the stomach digest itself? Physiol. Rev. 88 (2008) 1547–1565.
[23] K.V. Nemmani, S.V. Mali, N. Borhade, A.R. Pathan, M. Karwa, V. Pamidiboina, S.P. Senthilkumar, M. Gund, A.K. Jain, N.K. Mangu, N.P. Dubash, D.C. Desai, S. Sharma, A. Satyam, NO-NSAIDs, gastric-sparing nitric oxide-releasable prodrugs of non-steroidal anti-inflammatory drugs, Bioorg. Med. Chem. Lett. 19 (2009) 5297–5301.
[24] S.K. Suthar, M. Sharma, Recent developments in chimeric NSAIDs as safer anti-inflammatory agents, Med. Res. Rev. 35 (2015) 341–407.
[25] S. Jain, S. Tran, M.A. Gendy, K. Kashfi, P. Jurasz, C.A. Velazquez-Martinez, Nitric oxide release is not required to decrease the ulcerogenic profile of nonsteroidal anti-inflammatory drugs, J. Med. Chem. 55 (2012) 688–696.
[26] R. Bushra, N. Aslam, An overview of clinical pharmacology of ibuprofen, Oman Med. J. 25 (2010) 155–161.
[27] B. Chowdhury, M. Adak, S.K. Bose, Flurbiprofen, a unique non-steroidal anti-inflammatory drug with antimicrobial activity against Trichophyton, Microsporum and Epidermophyton species, Lett. Appl. Microbiol. 37 (2003) 158–161.
[28] S. Nalamachu, R. Wortmann, Role of indomethacin in acute pain and inflammation management: a review of the literature, Postgrad. Med. 126 (2014) 92–97.
[29] S. Wu, J. Han, A.A. Qureshi, Use of aspirin, nonsteroidal anti-inflammatory drugs, and acetaminophen (paracetamol), and risk of psoriasis and psoriatic arthritis: a cohort study, Acta Derm. Venereol. 95 (2015) 217–222.
[30] N. Ercan, M.O. Uludag, E.R. Agis, E. Demirel-Yilmaz, The anti-inflammatory effect of diclofenac is considerably augmented by topical capsaicinoids-containing patch in carrageenan-induced paw oedema of rat, Inflammopharmacology 21 (2013) 413–419.

[31] M.D. Sakeena, M.F. Yam, S.M. Elrashid, A.S. Munavvar, M.N. Azmin, Anti-inflammatory and analgesic effects of ketoprofen in palm oil esters nanoemulsion, J. Oleo Sci. 59 (2010) 667–671.
[32] T. Funatsu, K. Chono, T. Hirata, Y. Keto, A. Kimota, M. Sasamata, Mucosal acid causes gastric mucosal microcirculatory disturbance in nonsteroidal anti-inflammatory drug-treated rats, Eur. J. Pharmacol. 554 (2007) 53–59.
[33] R. Blackler, S. Syer, M. Bolla, E. Ongini, J.L. Wallace, Gastrointestinal-sparing effects of novel NSAIDs in rats with compromised mucosal defence, PLoS One 7 (2012) e35196.
[34] J. Li, Y. Kuang, J. Shi, Y. Gao, J. Zhou, B. Xu, The conjugation of nonsteroidal anti-inflammatory drugs (NSAID) to small peptides for generating multifunctional supramolecular nanofibers/hydrogels, Beilstein J. Org. Chem. 9 (2013) 908–917.
[35] H. Nishio, S. Terashima, M. Nakashima, E. Aihara, K. Takeuchi, Involvement of prostaglandin E receptor EP2 subtype and prostacyclin IP receptor in decreased acid response in damaged stomach, J. Physiol. Pharmacol. 58 (2007) 407–421.
[36] M. Gund, F.R. Khan, A. Khanna, V. Krishnakumar, Nicotinic acid conjugates of nonsteroidal anti-inflammatory drugs (NSAIDs) and their anti-inflammatory properties, Eur. J. Pharm. Sci. 49 (2013) 227–232.
[37] S. Fujimora, K. Gudis, Y. Takahashi, T. Seo, Y. Yamada, A. Ehara, T. Kobayashi, K. Mitsui, M. Yonezawa, S. Tanaka, A. Tatsuguchi, C. Sakamoto, Distribution of small intestinal mucosal injuries as a result of NSAID administration, Eur. J. Clin. Invest. 40 (2010) 504–510.
[38] Y.J. Huang, B.B. Zhang, N. Ma, M. Murata, A.Z. Tang, G.W. Huang, Nitrative and oxidative DNA damage as potential survival biomarkers for nasopharyngeal carcinoma, Med. Oncol. 28 (2011) 377–384.
[39] H.M. Amaro, R. Barros, A.C. Guedes, I. Sousa-Pinto, F.X. Malcata, Microalgal compounds modulate carcinogenesis in the gastrointestinal tract, Trends Biotechnol. 31 (2013) 92–98.
[40] I.H. Sahin, M.M. Hassan, C.R. Garrett, Impact of non-steroidal anti-inflammatory drugs on gastrointestinal cancers: current state-of the science, Cancer Lett. 345 (2014) 249–257.
[41] R. Francescone, V. Hou, S.I. Grivennikov, Cytokines, IBD, and colitis associated cancer, Inflamm. Bowel Dis. 21 (2015) 409–418.
[42] C. Musumba, D.M. Pritchard, M. Pirmohamed, Review article: cellular and molecular mechanisms of NSAID-induced peptic ulcers, Aliment. Pharmacol. Ther. 30 (2009) 517–531.
[43] J.L. Wallace, P.R. Devchand, Emerging roles for cyclooxygenase-2 in gastrointestinal mucosal defence, Br. J. Pharmacol. 145 (2005) 275–282.
[44] M.C. Hochberg, R.D. Altman, K.D. Brandt, B.M. Clark, P.A. Dieppe, M.R. Griffin, R.W. Moskowitz, T.J. Schnitzer, Guidelines for the medical management of osteoarthritis (part 1 and 2), Arthritis Rheumatol. 38 (1995) 1535–1546.
[45] M.C. Bonito, C. Cicala, M.C. Marcotullio, F. Maione, N. Mascolo, Biological activity of bicyclic and tricyclic diterpenoids from Salvia species of immediate pharmacological and pharmaceutical interest, Nat. Prod. Commun. 6 (2011) 1205–1215.
[46] F. Maione, C. Cicala, G. Musciacco, V. De Feo, A.G. Amat, A. Ialenti, N. Mascolo, Phenols, alkaloids and terpenes from medicinal plants with antihypertensive and vasorelaxant activities. A review of natural products as leads to potential therapeutic agents, Nat. Prod. Commun. 8 (2013) 539–544.
[47] A.N. Shikov, O.N. Pozharitskaya, A.S. Krishtopina, V.G. Makarov, Naphthoquinone pigments from sea urchins: chemistry and pharmacology, Phytochem. Rev. 17 (2018) 509–534.
[48] D.S. Fabricant, N.R. Farnsworth, The value of plant used in traditional medicine for drug discovery, Environ. Health Perspect. 109 (2001) 69–75.

[49] D.J. Newman, G.M. Cragg, Natural products as sources of new drugs from 1981 to 2014, J. Nat. Prod. 79 (2016) 629–661.
[50] T. Debnath, D.H. Kim, B.O. Lim, Natural products as a source of anti-inflammatory agents associated with inflammatory bowel disease, Molecules 18 (2013) 7253–7270.
[51] S. Tasneem, B. Liu, B. Li, M.I. Choudhary, W. Wang, Molecular pharmacology of inflammation: medicinal plants as anti-inflammatory agents, Pharmacol. Res. 139 (2019) 126–140.
[52] F. Zhu, B. Du, B. Xu, Anti-inflammatory effects of phytochemicals from fruits, vegetables, and food legumes: a review, Crit. Rev. Food Sci. Nutr. 58 (2018) 1260–1270.
[53] N. Muhammad, S.R. Lal, A. Adhikari, A. Wadood, H. Khan, A.Z. Khan, F. Maione, N. Mascolo, V.D. Feo, First evidence of the analgesic activity of govaniadine, an alkaloid isolated from *Corydalis govaniana* wall, Nat. Prod. Res. 29 (2015) 430–437.
[54] N.P. Yadav, R. Khatri, D.U. Bawankule, A. Pal, K. Shanker, P. Srivastava, A.K. Gupta, D. Chanda, Topical anti-inflammatory effects of *Ocimum basilicum* leaf extract in the phorbol-12, 13-dibutyrate model of mouse ear inflammation, Planta Med. 75 (2009) PA72.
[55] D.K. Singh, K. Shanker, M. Singh, Jyotshna, S. Luqman, Hepatoprotective effect of *Lawsonia inermis* L. in acetaminophen-induced oxidative stress in mice and its chemical profiling by Rp-hplc, Indian J. Pharm. 45 (2013) S184.
[56] S.V. Singh, A. Manhas, S.P. Singh, S. Mishra, N. Tiwari, P. Kumar, K. Shanker, K. Srivastava, K.V. Sashidhara, A. Pal, A phenolic glycoside from *Flacourtia indica* induces heme mediated oxidative stress in *Plasmodium falciparum* and attenuates malaria pathogenesis in mice, Phytomedicine 30 (2017) 1–9.
[57] B.S. Wang, G.J. Huang, Y.H. Lu, L.W. Chang, Anti-inflammatory effects of an aqueous extract of Welsh onion green leaves in mice, Food Chem. 138 (2013) 751–756.
[58] S. Mohanty, A.K. Maurya, Jyotshna, A. Saxena, K. Shanker, A. Pal, D.U. Bawankule, Flavonoids rich fraction of *Citrus limetta* fruit peels reduces proinflammatory cytokine production and attenuates malaria pathogenesis, Curr. Pharm. Biotechnol. 16 (2015) 544–552.
[59] B. Jayaprakasam, Y. Zhang, N.P. Seeram, M.G. Nair, Growth inhibition of human tumor cell lines by withanolides from *Withania somnifera* leaves, Life Sci. 74 (1) (2003) 125–132.
[60] N. Singh, M. Bhalla, P. de Jager, M. Gilca, An overview on ashwagandha: a rasayana (rejuvenator) of ayurveda, Afr. J. Tradit. Complement. Altern. Med. 8 (5 Suppl) (2011) 208–213.
[61] W. Vanden Berghe, L. Sabbe, M. Kaileh, G. Haegeman, K. Heyninck, Molecular insight in the multifunctional activities of Withaferin A, Biochem. Pharmacol. 84 (2012) 1282–1291.
[62] M. Kaileh, W. Vanden Berghe, A. Heyerick, J. Horion, J. Piette, C. Libert, D. De Keukeleire, T. Essawi, G. Haegeman, Withaferin a strongly elicits IkappaB kinase beta hyperphosphorylation concomitant with potent inhibition of its kinase activity, J. Biol. Chem. 282 (2007) 4253–4264.
[63] J.H. Oh, T.J. Lee, J.W. Park, T.K. Kwon, Withaferin A inhibits iNOS expression and nitric oxide production by Akt inactivation and down-regulating LPS-induced activity of NF-kappaB in RAW 264.7 cells, Eur. J. Pharmacol. 599 (2008) 11–17.
[64] R. Mohan, H.J. Hammers, P. Bargagna-Mohan, X.H. Zhan, C.J. Herbstritt, A. Ruiz, L. Zhang, A.D. Hanson, B.P. Conner, J. Rougas, V.S. Pribluda, Withaferin A is a potent inhibitor of angiogenesis, Angiogenesis 7 (2004) 115–122.
[65] J.H. Oh, T.K. Kwon, Withaferin A inhibits tumor necrosis factor-alpha-induced expression of cell adhesion molecules by inactivation of Akt and NF-kappaB in human pulmonary epithelial cells, Int. Immunopharmacol. 9 (2009) 614–619.

[66] F. Martorana, G. Guidotti, L. Brambilla, D. Rossi, Withaferin A inhibits nuclear factor-kappaB-dependent pro-inflammatory and stress response pathways in the astrocytes, Neural Plast. 2015 (2015) 381964.
[67] S.S. Jackson, C. Oberley, C.P. Hooper, K. Grindle, S. Wuerzberger-Davis, J. Wolff, K. McCool, L. Rui, S. Miyamoto, Withaferin A disrupts ubiquitin-based NEMO reorganization induced by canonical NF-kappaB signaling, Exp. Cell Res. 331 (2015) 58–72.
[68] K. Heyninck, M. Lahtela-Kakkonen, P. Van der Veken, G. Haegeman, W. Vanden Berghe, Withaferin A inhibits NF-kappaB activation by targeting cysteine 179 in IKKbeta, Biochem. Pharmacol. 91 (2014) 501–509.
[69] B. Hassannia, E. Logie, P. Vandenabeele, T.V. Berghe, W.V. Berghe, Withaferin A: from ayurvedic folk medicine to preclinical anti-cancer drug, Biochem. Pharmacol. 173 (2020) 113602.
[70] R. Sharma, N. Kishore, A. Hussein, N. Lall, The potential of *Leucosidea sericea* against *Propionibacterium acnes*, Phytochem. Lett. 7 (2014) 124–129.
[71] A. Kumar, K. Agarwal, A.K. Maurya, K. Shanker, U. Bushra, S. Tandon, D.U. Bawankule, Pharmacological and phytochemical evaluation of *Ocimum sanctum* root extracts for its anti-inflammatory, analgesic and antipyretic activities, Pharmacogn. Mag. 11 (2015) S217–S224.
[72] R. Williams, G. Münch, E. Gyengesi, L. Bennett, *Bacopa monnieri* (L.) exerts anti-inflammatory effects on cells of the innate immune system in vitro, Food Funct. 5 (2014) 517–520.
[73] L. Rubio, M.-J. Motilva, M.-P. Romero, Recent advances in biologically active compounds in herbs and spices: a review of the most effective antioxidant and anti-inflammatory active principles, Crit. Rev. Food Sci. Nutr. 53 (2013) 943–953.
[74] M.C. Recio, I. Andujar, J.L. Rios, Anti-inflammatory agents from plants: progress and potential, Curr. Med. Chem. 19 (2012) 2088–2103.
[75] D. Messaoudene, H. Belguendouz, M.L. Ahmedi, T. Benabdekader, F. Otmani, M. Terahi, C. Touil-boukoffa, Ex vivo effects of flavonoids extracted from *Artemisia herba alba* on cytokines and nitric oxide production in Algerian patients with Adamantiades-Behçet's disease, J. Inflamm. 8 (2011) 35.
[76] J.Y. Wu, Y.J. Chen, L. Bai, Y.X. Liu, X.Q. Fu, P.L. Zhu, J.K. Li, J.Y. Chou, C.L. Yin, Y.P. Wang, J.X. Bai, Y. Wu, Z.Z. Wu, Z.L. Yu, Chrysoeriol ameliorates TPA-induced acute skin inflammation in mice and inhibits NF-κB and STAT3 pathways, Phytomedicine 68 (2020) 153173.
[77] E. Herrera-Carrera, M.R. Moreno-Jimenez, N.E. Rocha-Guzman, J.A. Gallegos-Infante, J.O. Diaz-Rivas, C.I. Gamboa-Gomez, R.F. Gonzalez-Laredo, Phenolic composition of selected herbal infusions and their anti-inflammatory effect on a colonic model in vitro in HT-29 cells, Cogent Food Agric. 1 (2015) 1059033.
[78] K.F. Zhai, H. Duan, C.Y. Cui, Y.Y. Cao, J.L. Si, H.J. Yang, Y.C. Wang, W.G. Cao, G.Z. Gao, Z.J. Wei, Liquiritin from *Glycyrrhiza uralensis* attenuating rheumatoid arthritis via reducing inflammation, suppressing angiogenesis, and inhibiting MAPK signaling pathway, J. Agric. Food Chem. 67 (2019) 2856–2864.
[79] H.A. Jung, S. Karki, N.Y. Ehom, M.H. Yoon, E.J. Kim, J.S. Choi, Anti-diabetic and anti-inflammatory effects of green and red kohlrabi cultivars (*Brassica oleracea var. gongylodes*), Prev. Nutr. Food Sci. 19 (2014) 281–290.
[80] B.R. Jin, K.S. Chung, H.J. Kim, H.J. An, Chinese Skullcap (*Scutellaria baicalensis* Georgi) inhibits inflammation and proliferation on benign prostatic hyperplasia in rats, J. Ethnopharmacol. 235 (2019) 481–488.
[81] R.B. Oliveira, D.A. Chagas-Paula, A. Secatto, T.H. Gasparoto, L.H. Faccioli, A.P. Campanelli, F.B. Da Costa, Topical anti-inflammatory activity of yacon leaf extracts, Rev. Bras. Farm. 23 (2013) 497–505.
[82] S. Nithyananthan, P. Keerthana, S. Umadevi, S. Guha, I.H. Mir, J. Behera, C. Thirunavukkarasu, Nutrient mixture from germinated legumes: enhanced medicinal value with herbs attenuated liver cirrhosis, J. Food Biochem. 44 (2020) e13085.

[83] V. Nair, W.Y. Bang, E. Schreckinger, N. Andarwulan, L. Cisneros-Zevallos, Protective role of ternatin anthocyanins and quercetin glycosides from butterfly pea (*Clitoria ternatea*-Leguminosae) blue flower petals against lipopolysaccharide (LPS)-induced inflammation in macrophage cells, J. Agric. Food Chem. 63 (2015) 6355–6365.
[84] M. Dhivya Bharathi, A. Justin-Thenmozhi, T. Manivasagam, M. Ahmad Rather, C. Saravana Babu, M. Mohamed Essa, G.J. Guillemin, Amelioration of aluminum maltolate-induced inflammation and endoplasmic reticulum stress-mediated apoptosis by tannoid principles of *Emblica officinalis* in neuronal cellular model, Neurotox. Res. 35 (2019) 318–330.
[85] F.G. Rocha, M.M. Brandenburg, P.L. Pawloski, B.D.S. Soley, S.C.A. Costa, C.C. Meinerz, I.P. Baretta, M.F. Otuki, D.A. Cabrini, Preclinical study of the topical anti-inflammatory activity of *Cyperus rotundus* L. extract (Cyperaceae) in models of skin inflammation, J. Ethnopharmacol. 254 (2020) 112709.
[86] P. Joshi, G.S. Yadaw, S. Joshi, R.B. Semwal, D.K. Semwal, Antioxidant and anti-inflammatory activities of selected medicinal herbs and their polyherbal formulation, S. Afr. J. Bot. 130 (2020) 440–447.
[87] D.U. Bawankule, D. Man, A. Pal, K. Shanker, N.P. Yadav, S. Yadav, A.K. Srivastava, J. Agarwal, A.K. Shasany, M.P. Darokar, M.M. Gupta, S.P.S. Khanuja, Immunopotentiating effect of an ayurvedic preparation from medicinal plants, J. Health Sci. 55 (2009) 285–289.
[88] W. Li, P. Zhou, Y. Zhang, L. He, *Houttuynia cordata*, a novel and selective COX-2 inhibitor with anti-inflammatory activity, J. Ethnopharmacol. 133 (2011) 922–927.
[89] Q. Shao, T. Liu, W. Wang, Q. Duan, T. Liu, L. Xu, G. Huang, Z. Chen, The Chinese herbal prescription JZ-1 induces autophagy to protect against herpes simplex Virus-2 in human vaginal epithelial cells by inhibiting the PI3K/Akt/mTOR pathway, J. Ethnopharmacol. 254 (2020) 112611.
[90] J.-W. Han, K.H. Kim, M.-J. Kwun, J.-Y. Choi, S.-J. Kim, S.-I. Jeong, B.-J. Lee, K.-I. Kim, R. Won, J.H. Jung, H.J. Jung, M. Joo, Suppression of lung inflammation by the ethanol extract of Chung-Sang and the possible role of Nrf2, BMC Complement. Altern. Med. 19 (2019) 15.
[91] J. Cui, F. Xu, Z. Tang, W. Wang, L.l. Hu, C. Yan, Q. Luo, H. Gao, Y. Wei, J. Dong, Bu-Shen-Yi-Qi formula ameliorates airway remodeling in murine chronic asthma by modulating airway inflammation and oxidative stress in the lung, Biomed. Pharmacother. 112 (2019) 108694.
[92] Y. Ge, R. Cheng, S. Sun, S. Zhang, L. Li, J. Jiang, C. Yang, X. Xuan, J. Chen, Fangxiao formula alleviates airway inflammation and remodeling in rats with asthma via suppression of transforming growth factor-β/Smad3 signaling pathway, Biomed. Pharmacother. 119 (2019) 109429.
[93] C. Cao, M. Liu, S. Qu, R. Huang, M. Qi, Z. Zhu, J. Zheng, Z. Chen, Z. Wang, Z. Han, Y. Zhu, F. Huang, J. Duan, Chinese medicine formula Kai-Xin-San ameliorates depression-like behaviours in chronic unpredictable mild stressed mice by regulating gut microbiota-inflammation-stress system, J. Ethnopharmacol. 261 (2020) 113055.
[94] P. Ozorio, World Health Organization encourages traditional medicine in the third world, Dev. Dir. 2 (1979) 16.
[95] A. Amalraj, J. Jacob, K. Varma, A.B. Kunnumakkara, C. Divya, S. Gopi, Acujoint™, a highly efficient formulation with natural bioactive compounds, exerts potent anti-arthritis effects in human osteoarthritis—a pilot randomized double blind clinical study compared to combination of glucosamine and chondroitin, J. Herb. Med. 17–18 (2019) 100276.
[96] P. Avato, M. Argentieri, Plant biodiversity: phytochemicals and health, Phytochem. Rev. 17 (2018) 645–656.

[97] A. Amalraj, A. Pius, S. Gopi, S. Gopi, Biological activities of curcuminoids, other biomolecules from turmeric and their derivatives—a review, J. Tradit. Complement. Med. 7 (2016) 205–233.
[98] P. Shrivastava, K. Vaibhav, R. Tabassum, A. Khan, T. Ishrat, M.M. Khan, A. Ahmad, F. Islam, M.M. Safhi, F. Islam, Anti-apoptotic and anti-inflammatory effect of Piperine on 6-OHDA induced Parkinson's rat model, J. Nutr. Biochem. 24 (2013) 680–687.
[99] A. Bento, R. Marcon, R. Dutra, R. Claudino, M. Cola, D. Leite, J. Calixto, Caryophyllene inhibits dextran sulfate sodium-induced colitis in mice through CB2 receptor activation and PPARγ pathway, Am. J. Pathol. 178 (2011) 1153–1166.
[100] A.L. Klauke, I. Racz, B. Pradier, A. Markert, A.M. Zimmer, J. Gertsch, A. Zimmer, The cannabinoid CB2 receptor-selective phytocannabinoid beta-caryophyllene exerts analgesic effects in mouse models of inflammatory and neuropathic pain, Eur. Neuropsychopharmacol. 24 (2014) 608–620.
[101] K. Gerbeth, J. Hüsch, G. Fricker, O. Werz, M. Schubert-Zsilavecz, M. Abdel-Tawab, In vitro metabolism, permeation, and brain availability of six major boswellic acids from *Boswellia serrata* gum resins, Fitoterapia 84 (2013) 99–106.
[102] C.Y. Hsiang, H.M. Cheng, H.Y. Lo, C.C. Li, P.C. Chou, Y.C. Lee, T.Y. Ho, Ginger and zingerone ameliorate lipopolysaccharide-induced acute systemic inflammation in mice, assessed by nuclear factor-κB bioluminescent imaging, J. Agric. Food Chem. 63 (2015) 6051–6058.
[103] M.G. Rahbardar, B. Amin, S. Mehri, S.J. Mirnajafi-Zadeh, H. Hosseinzadeh, Rosmarinic acid attenuates development and existing pain in a rat model of neuropathic pain: an evidence of anti-oxidative and anti-inflammatory effects, Phytomedicine 40 (2018) 59–67.
[104] S.A. Farr, M.L. Niehoff, M.A. Ceddia, K.A. Herrlinger, B.J. Lewis, S. Feng, A. Welleford, D.A. Butterfield, J.E. Morley, Effect of botanical extracts containing carnosic acid or rosmarinic acid on learning and memory in SAMP8 mice, Physiol. Behav. 165 (2016) 328–338.
[105] A. Amalraj, S. Jude, K. Varma, J. Jacob, S. Gopi, O.S. Oluwafemi, S. Thomas, Preparation of a novel bioavailable curcuminoid formulation (Cureit™) using Polar-Nonpolar-Sandwich (PNS) technology and its characterization and applications, Mater. Sci. Eng. C Mater. Biol. Appl. 75 (2017) 359–367.
[106] A. Amalraj, K. Varma, J. Jacob, C. Divya, A.B. Kunnumakkara, S.J. Stohs, S. Gopi, A novel highly bioavailable curcumin formulation improves symptoms and diagnostic indicators in rheumatoid arthritis patients: a randomized, double-blind, placebo-controlled, two-dose, three-arm, and parallel-group study, J. Med. Food 20 (2017) 1022–1030.
[107] M.C. Bergonzi, R. Hamdouch, F. Mazzacuva, B. Isacchi, A.R. Bilia, Optimization, characterization and in vitro evaluation of curcumin microemulsions, LWT Food Sci. Technol. 59 (2014) 148–155.
[108] S. Chaurasia, R.R. Patel, P. Chaubey, N. Kumar, G. Khan, B. Mishra, Lipopolysaccharide based oral nanocarrier for the improvement of bioavailability and anticancer efficacy of curcumin, Carbohydr. Polym. 130 (2015) 9–17.
[109] N. Sanoj Rejinold, P.R. Sreerekha, K.P. Chennazhi, S.V. Nair, R. Jayakumar, Biocompatible, biodegradable and thermo-sensitive chitosan-g-poly (Nisopropylacrylamide) nanocarrier for curcumin drug delivery, Int. J. Biol. Macromol. 49 (2011) 161–172.
[110] C. Righeschi, M.C. Bergonzi, B. Isacchi, C. Bazzicalupi, P. Gratteri, A.R. Bilia, Enhanced curcumin permeability by SLN formulation: the PAMPA approach, LWT Food Sci. Technol. 66 (2016) 475–483.
[111] P.R. Sarika, N.R. James, P.R. Anil Kumar, D.K. Raj, Preparation, characterization and biological evaluation of curcumin loaded alginate aldehyde-gelatin nanogels, Mater. Sci. Eng. C 68 (2016) 251–257.

[112] P. Anand, S.G. Thomas, A.B. Kunnumakkara, C. Sundaram, K.B. Harikumar, B. Sung, S.T. Tharakan, K. Misra, I.K. Priyadarsini, K.N. Rajasekharan, B.B. Aggarwal, Biological activities of curcumin and its analogues (Congeners) made by man and Mother Nature, Biochem. Pharmacol. 76 (2008) 1590–1611.

[113] S. Gopi, J. Jacob, R. George, T.R. Sreeraj, A unique formulation of hydrogenated curcuminoids with higher bio-availability and the application in food matrices, J. Nutr. Food Sci. 6 (2016) 1–4.

[114] J. Jacob, A. Amalraj, K.K.J. Raj, C. Divya, A.B. Kunnumakkara, S. Gopi, A novel bioavailable hydrogenated curcuminoids formulation (CuroWhite™) improves symptoms and diagnostic indicators in rheumatoid arthritis patients—a randomized, double blind and placebo controlled study, J. Tradit. Complement. Med. 9 (2018) 346–352.

[115] S. Umar, K. Umar, A.H. Sarwar, A. Khan, N. Ahmad, S. Ahmad, C.K. Katiyar, S.A. Husain, H.A. Khan, *Boswellia serrata* extract attenuates inflammatory mediators and oxidative stress in collagen induced arthritis, Phytomedicine 21 (2014) 847–856.

[116] Q. Wang, X. Pan, H.H. Wong, C.A. Wagner, L.J. Lahey, W.H. Robinson, J. Sokolove, Oral and topical boswellic acid attenuates mouse osteoarthritis, Osteoarthr. Cartil. 22 (2014) 128–132.

[117] Y. Takada, H. Ichikawa, V. Badmaev, B.B. Aggarwal, Acetyl-11-ketobeta-boswellic acid potentiates apoptosis, inhibits invasion, and abolishes osteoclastogenesis by suppressing NF-kappa B and NF-kappa B-regulated gene expression, J. Immunol. 176 (2006) 3127–3140.

[118] B.B. Aggarwal, S. Prasad, S. Reuter, R. Kannappan, V.R. Yadev, B. Park, J.H. Kim, S.C. Gupta, K. Phromnoi, C. Sundaram, S. Prasad, M.M. Chaturvedi, B. Sung, Identification of novel anti-inflammatory agents from Ayurvedic medicine for prevention of chronic diseases: "reverse pharmacology" and "bedside to bench" approach, Curr. Drug Targets 12 (2011) 1595–1653.

[119] P.K. Kokkiripati, L.M. Bhakshu, S. Marri, K. Padmasree, A.T. Row, A.S. Raghavendra, S.D. Tetali, Gum resin of *Boswellia serrata* inhibited human mono-cytic (THP-1) cell activation and platelet aggregation, J. Ethnopharmacol. 137 (2011) 893–901.

[120] R.A. Mothana, Anti-inflammatory, antinociceptive and antioxidant activities of the endemic Soqotraen *Boswellia elongata* Balf. f. and *Jatropha unicostata* Balf. F. in different experimental models, Food Chem. Toxicol. 49 (2011) 2594–2599.

[121] C. Cuaz-Perolin, L. Billiet, E. Bauge, C. Copin, D. Scott-Algara, F. Genze, B. Buchele, T. Syrovets, T. Simmet, M. Rouis, Anti-inflammatory and anti-atherogenic effects of the NF-kappaB inhibitor acetyl-11-keto-beta-boswellic acid in LPS-challenged ApoE −/− mice, Arterioscler. Thromb. Vasc. Biol. 28 (2008) 272–277.

[122] H. Safayhi, B. Rall, E.R. Sailer, H.P. Ammon, Inhibition by boswellic acids of human leukocyte elastase, J. Pharmacol. Exp. Ther. 281 (1997) 460–463.

[123] P.R. Vuddanda, S. Singh, S. Velaga, Boswellic acid—medicinal use of an ancient herbal remedy, J. Herb. Med. 6 (2016) 163–170.

[124] N. Kimmatkar, V. Thawani, L. Hingorani, R. Khiyani, Efficacy and tolerability of *Boswellia serrata* extract in treatment of osteoarthritis of knee—a randomized double blind placebo controlled trial, Phytomedicine 10 (2003) 3–7.

[125] H. Riasari, R. Rachmaniar, Y. Febriani, Effectiveness of anti-inflammatory plaster from Kencur (*Kaempferia galanga* L.) rhizome ethanol extract, Int. J. Pharm. Sci. Res. 7 (2016) 1746–1749.

[126] A.M. Vittalrao, T. Shanbhag, M. Kumari, K.L. Bairy, S. Shenoy, Evaluation of anti-inflammatory and analgesic activities of alcoholic extract of *Kaempferia galanga* in rats, Indian J. Physiol. Pharmacol. 55 (2011) 13–24.

[127] M.G. Lequesne, The algofunctional indices for hip and knee osteoarthritis, J. Rheumatol. 24 (1997) 779–781.

[128] M.K. Jeengar, S.V.K. Rompicharla, S. Shrivastava, N. Chella, N.R. Shastri, V.G.M. Naidua, R. Sistla, Emu oil based nano-emulgel for topical delivery of curcumin for amelioration of rheumatoid arthritis, Int. J. Pharm. 506 (2016) 222–236.

[129] K. Coradini, R.B. Friedrich, F.N. Fonseca, M.S. Vencato, D.F. Andrade, C.M. Oliveira, A.P. Battistel, S.S. Guterres, M.I.U.M. da Rocha, A.R. Pohlmann, R.C. Beck, A novel approach to arthritis treatment based on resveratrol and curcumin coencapsulated in lipid-core nanocapsules: in vivo studies, Eur. J. Pharm. Sci. 78 (2015) 163–170.

[130] B. Chandran, A. Goel, A randomized, pilot study to assess the efficacy and safety of curcumin in patients with active rheumatoid arthritis, Phytother. Res. 26 (2012) 1719–1725.

[131] P.C. Jagadish, K.P. Latha, J. Mudgal, G.K. Nampurath, Extraction, characterization and evaluation of *Kaempferia galanga* L. (Zingiberaceae) rhizome extracts against acute and chronic inflammation in rats, J. Ethnopharmacol. 24 (2016) 434–439.

[132] M.R. Sulaiman, Z.A. Zakaria, I.A. Daud, F.N. Ng, Y.C. Ng, M.T. Hidayat, Antinociceptive and anti-inflammatory activities of the aqueous extract of *Kaempferia galanga* leaves in animal models, J. Nat. Med. 62 (2008) 221–227.

[133] S. Shityakov, E. Bigdelian, A.A. Hussein, M.B. Hussain, Y.C. Tripathi, M.C. Khan, M.A. Shariati, Phytochemical and pharmacological attributes of piperine: a bioactive ingredient of black pepper, Eur. J. Med. Chem. 176 (2019) 149–161.

[134] F. Tasleem, I. Azhar, S.N. Ali, S. Perveen, Z.A. Mahmood, Analgesic and anti-inflammatory activities of *Piper nigrum* L, Asian Pac. J. Trop. Med. 7 (2014) 461–468.

[135] J.S. Bang, H.M. Choi, B.J. Sur, S.J. Lim, J.Y. Kim, H.I. Yang, K.S. Kim, Anti-inflammatory and antiarthritic effects of piperine in human interleukin 1β-stimulated fibroblast-like synoviocytes and in rat arthritis models, Arthritis Res. Ther. 11 (2009) 49.

[136] G.S. Bae, M.S. Kim, W.S. Jung, S.W. Seo, S.W. Yun, S.G. Kim, S.J. Park, Inhibition of lipopolysaccharide-induced inflammatory responses by piperine, Eur. J. Pharmacol. 642 (2010) 154–162.

[137] S.H. Kim, Y.C. Lee, Piperine inhibits eosinophil infiltration and airway hyperresponsiveness by suppressing T cell activity and Th2 cytokine production in the ovalbumin-induced asthma model, J. Pharm. Pharmacol. 61 (2009) 353–359.

[138] T. Miyazawa, K. Nakagawa, S.H. Kim, M.J. Thomas, L. Paul, J.M. Zingg, G.G. Dolnikowski, S.B. Roberts, F. Kimura, T. Miyazawa, A. Azzi, M. Meydani, Curcumin and piperine supplementation of obese mice under caloric restriction modulates body fat and interleukin-1beta, Nutr. Metab. 15 (2018) 12.

[139] R.A. Gupta, M.N. Motiwala, N.G. Dumore, K.R. Danao, A.B. Ganjare, Effect of piperine on inhibition of FFA induced TLR4 mediated inflammation and amelioration of acetic acid induced ulcerative colitis in mice, J. Ethnopharmacol. 164 (2015) 239–246.

[140] A. Yasir, S. Ishtiaq, M. Jahangir, M. Ajaib, U. Salar, K.M. Khan, Biology-oriented synthesis (BIOS) of piperine derivatives and their comparative analgesic and anti-inflammatory activities, Med. Chem. 14 (2018) 269–280.

[141] K. Vaibhav, P. Shrivastava, H. Javed, A. Khan, M.E. Ahmed, R. Tabassum, M.M. Khan, G. Khuwaja, F. Islam, M.S. Siddiqui, M.M. Safhi, F. Islam, Piperine suppresses cerebral ischemia-reperfusion-induced inflammation through the repression of COX-2, NOS-2, and NF-κB in middle cerebral artery occlusion rat model, Mol. Cell. Biochem. 367 (2012) 73–84.

[142] D. Hu, Y. Wang, Z. Chen, Z. Ma, Q. You, X. Zhang, Q. Liang, H. Tan, C. Xiao, X. Tang, Y. Gao, The protective effect of piperine on dextran sulfate sodium induced inflammatory bowel disease and its relation with pregnane X receptor activation, J. Ethnopharmacol. 169 (2015) 109–123.

[143] H.G. Kim, E.H. Han, W.S. Jang, J.H. Choi, T. Khanal, B.H. Park, T.P. Tran, Y.C. Chung, H.G. Jeong, Piperine inhibits PMA-induced cyclooxygenase-2 expression through downregulating NF-κB, C/EBP and AP-1 signaling pathways in murine macrophages, Food Chem. Toxicol. 50 (2012) 2342–2348.

[144] M. Mohammadi, H. Najafi, Z.M. Yarijani, G. Vaezi, V. Hojati, Protective effect of piperine in ischemia-reperfusion induced acute kidney injury through inhibition of inflammation and oxidative stress, J. Tradit. Complement. Med. (2019), https://doi.org/10.1016/j.jtcme.2019.07.002.

[145] J. Gertsch, R.G. Pertwee, V. Di Marzo, Phytocannabinoids beyond the Cannabis plant-do they exist? Br. J. Pharmacol. 160 (2010) 523–529.

[146] R. Othman, H. Ibrahim, M.A. Mohd, K. Awang, A.U. Gilani, M.R. Mustafa, Vasorelaxant effects of ethyl cinnamate isolated from *Kaempferia galanga* on smooth muscles of the rat aorta, Planta Med. 68 (2002) 655–657.

[147] J.Y. Cho, H.Y. Kim, S. Kim, J.H.Y. Park, H.J. Lee, H.S. Chun, β-Caryophyllene attenuates dextran sulfate sodium-induced colitis in mice via modulation of gene expression associated mainly with colon inflammation, Toxicol. Rep. 2 (2015) 1039–1045.

[148] K. Srinivasan, Ginger rhizomes (*Zingiber officinale*): a spice with multiple health beneficial potentials, PharmaNutrition 5 (2017) 18–28.

[149] T. Mustafa, K.C. Srivastava, K.B. Jensen, Drug development report 9 pharmacology of ginger, *Zingiber officinale*, J. Drug Dev. 6 (1993) 25–39.

[150] F. Kiuchi, S. Iwakami, M. Shibuya, F. Hanaoka, U. Sankawa, Inhibition of prostaglandin and leukotriene biosynthesis by gingerols and diarylheptanoids, Chem. Pharm. Bull. (Tokyo) 40 (1992) 387–391.

[151] C.Y. Hsiang, H.M. Cheng, H.Y. Lo, C.C. Li, P.C. Chou, Y.C. Lee, T.Y. Ho, Ginger and zingerone ameliorate lipopolysaccharide-induced acute systemic inflammation in mice, assessed by nuclear factor-kb bioluminescent imaging, J. Agric. Food Chem. 63 (2015) 6051–6058.

[152] R.C. Lantz, G.J. Chen, M. Sarihan, A.M. Sólyom, S.D. Jolad, B.N. Timmermann, The effect of extracts from ginger rhizome on inflammatory mediator production, Phytomedicine 14 (2007) 123–128.

[153] G. Ramadan, M.A. Al-Kahtani, W.M. El-Sayed, Anti-inflammatory and antioxidant properties of *Curcuma longa* (turmeric) versus *Zingiber officinale* (ginger) rhizomes in rat adjuvant-induced arthritis, Inflammation 34 (2011) 291–301.

[154] S.C. Penna, M.V. Medeiros, F.S. Aimbire, H.C. Faria-Neto, J.A. Sertié, R.A. Lopes-Martins, Anti-inflammatory effect of the hydralcoholic extract of *Zingiber officinale* rhizomes on rat paw and skin edema, Phytomedicine 10 (2003) 381–385.

[155] J.L. Funk, J.B. Frye, J.N. Oyarzo, J. Chen, H. Zhang, B.N. Timmermann, Antiinflammatory effects of the essential oils of ginger (*Zingiber officinale* Roscoe) in experimental rheumatoid arthritis, PharmaNutrition 4 (2016) 123–131.

[156] H. Bliddal, A. Rosetzsky, P. Schlichting, M.S. Weidner, L.A. Andersen, H.H. Ibfelt, K. Christensen, O.N. Jensen, J. Barslev, A randomized, placebo-controlled, crossover study of ginger extracts and ibuprofen in osteoarthritis, Osteoarthr. Cartil. 8 (2000) 9–12.

[157] R.D. Altman, K.C. Marcussen, Effects of a ginger extract on knee pain in patients with osteoarthritis, Arthritis Rheum. 44 (2001) 2531–2538.

[158] I. Wigler, I. Grotto, D. Caspi, M. Yaron, The effects of Zintona EC (a ginger extract) on symptomatic gonarthritis, Osteoarthr. Cartil. 11 (2003) 783–789.

[159] R. Grzanna, L. Lindmark, C.G. Frondoza, Ginger—an herbal medicinal product with broad anti-inflammatory actions, J. Med. Food 8 (2005) 125–132.

[160] A. Amalraj, N.P. Sukumaran, A. Nair, S. Gopi, Preparation of a unique bioavailable bacoside formulation (Cognique®) using polar-nonpolar-sandwich (PNS) technology and its characterization, in vitro release study, and proposed mechanism of action, Regen. Eng. Transl. Med. (2020), https://doi.org/10.1007/s40883-020-00162-2.

[161] E. Arranz, L. Jaime, M.R. García-Risco, T. Fornari, G. Reglero, S. Santoyo, Antiinflammatory activity of rosemary extracts obtained by supercritical carbon dioxide enriched in carnosic acid and carnosol, Int. J. Food Sci. Technol. 50 (2015) 674–681.

[162] C.-H. Peng, J.-D. Su, C.-C. Chyau, T.-Y. Sung, S.-S. Ho, C.-C. Peng, R.Y. Peng, Supercritical fluid extracts of rosemary leaves exhibit potent antiinflammation and anti-tumor effects, Biosci. Biotechnol. Biochem. 71 (2007) 2223–2232.

[163] H. Hosseinzadeh, M. Nourbakhsh, Effect of *Rosmarinus officinalis* L. aerial parts extract on morphine withdrawal syndrome in mice, Phytother. Res. 17 (2003) 938–941.

[164] C.F. Kuo, J.D. Su, C.H. Chiu, C.C. Peng, C.H. Chang, T.Y. Sung, S.H. Huang, W.C. Lee, C.C. Chyau, Antiinflammatory effects of supercritical carbon dioxide extract and its isolated carnosic acid from *Rosmarinus officinalis* leaves, J. Agric. Food Chem. 59 (2011) 3674–3685.

[165] A.H. Lo, Y.C. Liang, S.Y. Lin-Shiau, C.T. Ho, J.K. Lin, Carnosol, an antioxidant in rosemary, suppresses inducible nitric oxide synthase through down-regulating nuclear factor-κB in mouse macrophages, Carcinogenesis 23 (2002) 983–991.

[166] D. Poeckel, C. Greiner, M. Verhoff, O. Rau, L. Tausch, C. Hörnig, D. Steinhilber, M. Schubert-Zsilavecz, O. Werz, Carnosic acid and carnosol potently inhibit human 5-lipoxygenase and suppress pro-inflammatory responses of stimulated human polymorphonuclear leukocytes, Biochem. Pharmacol. 76 (2008) 91–97.

[167] A.P. Sánchez-Camargo, M. Herrero, Rosemary (*Rosmarinus officinalis*) as a functional ingredient: recent scientific evidence, Curr. Opin. Food Sci. 14 (2017) 13–19.

[168] C.R. Wu, C.W. Tsai, S.W. Chang, C.Y. Lin, L.C. Huang, C.W. Tsai, Carnosic acid protects against 6-hydroxydopamine-induced neurotoxicity in in vivo and in vitro model of Parkinson's disease: involvement of antioxidative enzymes induction, Chem. Biol. Interact. 225 (2015) 40–46.

[169] B.B. Mishra, V.K. Tiwari, Natural products: an evolving role in future drug discovery, Eur. J. Med. Chem. 46 (2011) 4769–4807.

17

Identification of toxicology biomarker and evaluation of toxicity of natural products by metabolomic applications

N.S.K. Gowthaman[a], H.N. Lim[b], Sreeraj Gopi[c], and Augustine Amalraj[c]

[a]Materials Synthesis and Characterization Laboratory, Institute of Advanced Technology, Universiti Putra Malaysia, Serdang, Selangor, Malaysia, [b]Department of Chemistry, Faculty of Science, Universiti Putra Malaysia, Serdang, Selangor, Malaysia, [c]R&D Centre, Aurea Biolabs (P) Ltd, Kolenchery, Cochin, Kerala, India

17.1 Background

Natural product has a long history of use to prevent and treat diseases since the appearance of civilization, and the popularity is increasing for health care and chronic disease treatment. In ancient times, natural product was extremely significant to resist the attack of diseases. Even in this seemingly advanced era, natural product is an essential part as medications and prodrugs, directly or indirectly treating diseases. The resources of natural product are plants, animals, and minerals. However, safety issues of natural products have been raised, including the identification of plant materials and active principles, the method of preparation, dosing regimens, the potential to interact with other herbal remedies and conventional drugs, and assurances that herbal products are genuine and do not contain toxins or contaminants [1–5].

The metabolome is a set of small molecular mass organic compounds found in a given biological medium. Polymerized structures such as proteins and nucleic acids are excluded from the metabolome, but small peptides such as the tripeptide glutathione are included. Molecules that constitute the metabolome are called metabolites. For some scientists the concept of metabolite includes all the organic substances naturally occurring from the metabolism of a living organism and that do not directly come from gene expression. It should be

Inflammation and Natural Products. https://doi.org/10.1016/B978-0-12-819218-4.00015-8
© 2021 Elsevier Inc. All rights reserved.

stressed here that this definition could be applied as well to a microorganism, a human being, or a plant [6–10]. Two different kinds of metabolites can be distinguished based on their origin: endogenous and exogenous metabolites [4, 11, 12].

Endogenous metabolites could be classified as primary and secondary metabolites. Primary metabolite has a broad distribution in living species and is directly involved in essential life processes such as growth, development, and reproduction (e.g., amino acids and glycolysis intermediates). In controversy, secondary metabolites are species specific, have a restricted distribution, and are synthesized for a particular biological function, as alkaloids for plants or hormones for mammals [11]. Exogenous metabolites represent the biotransformation or metabolism products of exogenous compounds, resulting from phase I (modification of the original molecule to introduce a functional group) and/or phase II (conjugation) enzymatic conversion [12].

A systematic qualitative and quantitative analysis of metabolites in a given organism or biological sample that then quantificationally describes changes of endogenous metabolites before and after stimulations or disturbances is known as metabolomics [6]. The terms metabolomics and metabonomics can be interchanged. Initially, metabolomics refers to the measurement of the pool of cell metabolites [13, 14], whereas metabonomics describes "the quantitative measurement of the dynamic multiparametric metabolic response of living systems to pathophysiological stimuli or genetic modification" [4, 6, 15]. Nicholson's definition underlines the role of two major scientific disciplines used in metabonomics are analytical chemistry and biostatistics [16, 17]. By consistency, we use the term metabolomics in this chapter. Biochemists have long been doing metabolomics, just like the prose without knowing it, meaning that they suspected that patterns of biochemical substances could explain or describe interindividual variation [17].

Without knowing toxic compositions, concentrations, and mechanisms, the use of natural product may be unhealthy, harmful, and even lethal. Feasible and efficient steps need to be carried out to solve this problem before toxins cause physiological or pathological damages. Metabolomics appears ideal for this purpose. Metabolomics has minimal damage to the body due to little sample needed, which makes it convenient compared with proteomics and genomics [1, 10]. Metabolomics belongs to the "omics" techniques together with genomics, transcriptomics, and proteomics that are related to the genome (DNA), the transcriptome (RNA), and proteome (proteins), respectively (Fig. 17.1). The term metabolomics was coined on the basis of the genome and transcriptome. It appeared for the first time in a publication by Oliver in 1998 [17]. The metabolome reflects past events that include whole metabolism and the interaction with the environment, whereas the genome reflects the real and potential functional information of organism.

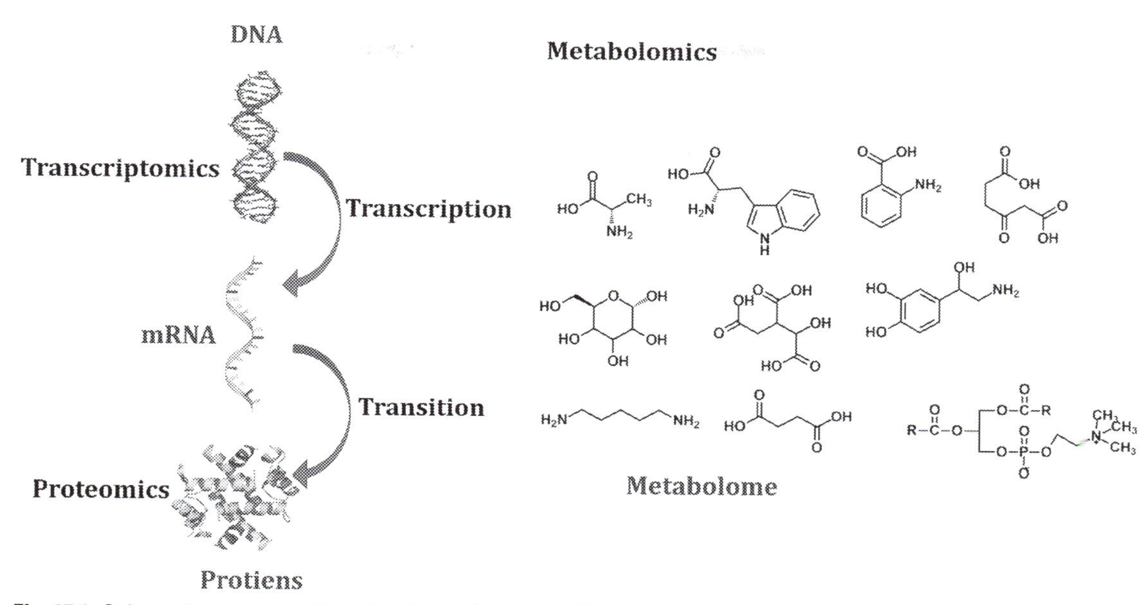

Fig. 17.1 Schematic representation of omics technologies. The flow of information starts from genes to metabolites running through transcripts and proteins. Copyrights @Elsevier (2011). Ref. J.K. Nicholson, J.C. Lindon, E. Holmes, 'Metabonomics': understanding the metabolic responses of living systems to pathophysiological stimuli via multivariate statistical analysis of biological NMR spectroscopic data, Xenobiotica 29 (1999) 1181–1189.

Metabolomics allows the quantitative measurement of large numbers of low molecular weight (<1-kDa) endogenous metabolites, including lipids, amino acids, peptides, nucleic acids, organic acids, vitamins, and carbohydrates, which play important roles in biological systems and represent attractive candidates to understand phenotypes [18–20]. Metabolomics is suitable for observing abnormal changes of endogenous metabolites before the appearance of physiological or pathological damages. As a systemic approach, metabolomics adopts a "top-down" strategy to reflect the function of organisms from terminal symptoms of metabolic network and understand metabolic changes of a complete system caused by interventions in a holistic context [21]. Since the metabolome represents the physiological or pathological status of organisms [8], metabolomics can be used in toxicity evaluation [9] and toxicological biomarker identification.

Biological markers or biomarkers are measurable internal indicators of molecular and/or cellular alterations that may appear in an organism after or during exposure to a toxicant and possible disease [22, 23]. This definition is used in environmental and occupational toxicology and is larger than that of the National Institutes of Health (NIH) that focuses on drug development and defines a biomarker as "a characteristic that is objectively measured and evaluated as an indicator of normal biologic processes, pathogenic processes, or pharmacological processes to a therapeutic intervention" [24].

Biomarkers are compounds or a set of compounds (metabolomic profile) that must be quantitatively, sensitively, specifically, and easily measurable on noninvasively collected biological media [25]. Biomarkers could be divided into several categories that include biomarkers of exposure, biomarkers of effect, and biomarker of susceptibility. A biomarker of exposure is an indication of the occurrence and extent of exposure. It depends on the chemical fate of the exposed toxicant in the body. The biomonitoring of exposure has been used for a long time in occupational settings, for example, for the determination of lead [26] or benzene metabolites [27] in blood or urine.

Biomarkers of effect indicate that exposure has resulted in an interaction between the toxicant and a biological target. Mutagenic and carcinogenic substances that possess electrophilic functions bind to macromolecules such as proteins, DNA, or lipids. Hemoglobin is often used in biomonitoring because of its long lifespan and ease of access. Oxidative stress perturbs the homeostasis of cell and leads to the production of specific substances such as 8-hydroxy-2′-deoxyguanosine or to an imbalance of glutathione pathway [28].

Biomarkers of susceptibility describe interindividual differences in response to toxicants from genetic causes or from nongenetic factors (age, liver disease, kidney disease, diet, and dietary supplementation). Polymorphisms of activating/detoxificating enzymes have been identified as key factors in the relationship between external (e.g., ambient air) and internal exposure (e.g., urinary excretion).

17.2 Metabolomic technology

Modern metabolomic technologies allow for qualitative and quantitative measurement of a vast number of metabolites in complex biological systems. Frequently used technologies in metabolomics are proton nuclear magnetic resonance (^{1}H NMR) spectroscopy, gas chromatography-mass spectrometry (GC-MS), capillary electrophoresis-MS, high-performance liquid chromatography (HPLC)-MS, ultra-performance liquid chromatography (UPLC)-MS, and LC-solid-phase extraction (SPE)-NMR. The different metabolomic techniques have been applied to different herbal medicine research [29]. ^{1}H NMR allows the rapid, high-throughput, and automated analysis of crude extracts and then quantitatively detects metabolites in many different groups [30, 31], as well as providing structural information including stereochemical details [32]. However, ^{1}H NMR fails to obtain valid data when the concentrations of metabolites in complex sample are lower [33]. Due to the enrichment effect of SPE, the sensitivity can be improved, and LC-SPE-NMR can solve the aforementioned problems. Compared with MS-based approaches, NMR is less sensitive, and its data are limited [30, 31, 34]. GC-MS is particularly suitable for the detection of thermally stable metabolites with volatile derivatives (volatile metabolites). GC-MS, with high resolution and sensitivity, is usually used for quality control and qualitative and quantitative determination of active components in natural product [35]. However, the application range of GC-MS is limited. Only volatile metabolites in sample can be directly analyzed, but most metabolites are nonvolatile that we might lose the information.

With continuous technical updates of LC-MS, the utilization is more frequent. LC has the ability to isolate different kinds of metabolites in a complex system. MS can provide structural information to help identify metabolites. LC-MS costs less and provides more details of large and submerged portions than NMR [34]. LC-MS allows the analysis of thermally labile nonvolatile metabolites [36]. The molecules that can be detected by LC-MS range from polar sugars and non-aromatic organic acids [37] to various lipids [38]. The ability of LC-MS to analyze various kinds of metabolites depends strongly on the ionization source and the chromatographic method that is used to separate a complex mixture of analytes [1]. UPLC from Waters Corporation, with sub-2-μm chromatographic particles and a fluid system capable of operating at pressures up to 15,000 psi, elevates better chromatographic resolution than traditional HPLC operated with larger particle [39]. MSE can obtain highly accurate parent ion and fragment ion information in one analytical run. MSE provides parallel alternating scans for acquisition at either low collision energy to obtain precursor ion information or ramping of high collision energy to obtain full-scan accurate mass fragment, precursor ion, and neutral loss information [40, 41]. UPLC-MSE is considered to be appropriate for metabolomic study, especially for large-scale untargeted metabolic profiling in complex biological sample.

17.3 Sample preparation

Sample preparation, including its source, storage, and extraction, has significant effects on the results of metabolomic analysis. Plasma, serum, urine, and tissue are usually biological samples in metabolomic analysis [42]. To decrease the changes of potential metabolites in metabolomic samples, biological samples usually can be restored in − 80°C. For ^{1}H NMR analysis the change of pH and ionic strength caused by the change of the chemical shift is the primary problem, and the addition of pH buffer during the sample extraction can solve the problem [43]. Compared with ^{1}H NMR the samples extracted for

MS-based metabolomics are more complicated. For LC-MS analysis, biological samples are complex and contain various endogenous and exogenous acidic, basic, and neutral compounds with high polarity. The samples usually require to be centrifuged and diluted with deionized water before metabolomic analysis [44]. For GC-MS analysis, most potential biomarkers in biological samples are high polar and nonvolatile; thus the samples must be derivatized before analysis [45].

Minimal sample preparation steps are supposed to perform on urinary samples to decrease the loss of potential biomarkers, as urine is a complex sample including various endogenous and exogenous acidic, basic, and neutral compounds with high polarity. Urine sample needs to be centrifuged at 13,000 rpm to remove solid and diluted with deionized water before metabolomic analysis. SPE can extract substance with special properties [4, 5, 10].

Plasma and serum samples usually can be restored in − 80°C for 24 months or in − 20°C for 1 month. Prior to analysis, plasma and serum samples need be thawed at room temperature. Acetonitrile or methanol is added to serum and vortex-mixed vigorously for several minutes. The mixture is then centrifuged at 13,000 rpm for minutes at 4°C. The supernatant was pipetted out and then analyzed directly or lyophilized.

Tissue is harvested after in situ cardioperfusion and then immediately washed with physiological saline and stored at − 80°C for the following metabolomic study. The sample is homogenized in acetonitrile in an ice bath. Samples were then vortex-mixed for minutes and put on ice in between. Following centrifugation (13,000 rpm, 4°C), the supernatant is removed and then lyophilized. The extract was resuspended before analysis [4, 5, 10].

17.4 Data analysis

Data analysis is the crucial one, since the data matrix generated in metabolomic study is generally large and complex. Data preprocessing is the first step of metabolomic data analysis. The main objective is to transform the data in such way that the samples in the dataset are more comparable to ease and improve the data analysis [46]. The preprocessing of ^{1}H NMR data usually includes baseline correction, alignment, binning, normalization, and scaling [47]. For MS, many software such as MetAlign, MZmine, and XCMS have been established to process the raw data [48]. Multivariate statistical methods are professionally approached for analyzing and maximizing information retrieval from complex metabolomic data. The multivariate statistical methods can be classified into unsupervised and supervised methods. Unsupervised methods consist of principal component analysis (PCA), hierarchical cluster analysis

(HCA), *k*-means, and statistical total correlation spectroscopy. Supervised methods such as partial least squares discriminant analysis (PLS-DA), orthogonal partial least squares discriminant analysis (OPLS-DA), quadratic discriminant analysis, and linear discriminant analysis can reveal the most important factors of variability characterizing the metabolomic datasets [49].

The commonly used software for metabolomic multivariate statistical analysis are Shimadzu Class-VP software and SIMCA-P software. The identification of metabolites and their pathway analysis are also essential components of metabolomic data analysis. The updating commercial software is crucial for identifying potential metabolites, while accurate mass, isotopic pattern, fragments information, and available biochemical databases are also necessary. Presently a number of metabolites databases such as Human Metabolome Database (HMDB), Kyoto Encyclopedia of Genes and Genomes (KEGG), Biochemical Genetic and Genomic (BiGG), ChemSpider, and PubChem are emerging and have been applied in the identification of metabolites and biomarkers. For metabolic pathway analysis, KEGG, Ingenuity Pathway Analysis, Cytoscape, and Reactome Pathway Database are commonly used databases and softwares and were listed in Table 17.1. Besides, MetaboAnalyst, a powerful web-based tool, provides plenty of analysis methods including data processing, normalization, multivariate statistical analysis, biomarker,

Table 17.1 Softwares and databases for pathway analysis and interactions in metabolomics.

Name	Website	Description
KEGG	www.kegg.jp	Free database resource. Helpful in the understanding of high-level functions and utilities of the biological system, especially large-scale molecular datasets generated by high-throughput experimental technologies
Ingenuity pathway analysis	www.ingenuity.com	Instinctive web-based applications for quick and accurate analysis and interpretation of biological meaning in metabolomics
Cytoscape	www.cytoscape.org	Free software for integrated models of biomolecular interaction networks and functional annotation
Reactome pathway database	www.reactome.org	Free online database of biopathways that can be used to browse pathways and submit data to analysis tools
BioCarta	www.biocarta.com	An open database that provides metabolite interaction and enrichment and pathway analysis
Biological networks	http://biologicalnetworks.net/	A software for biological pathway analysis, querying, and visualization of metabolic pathways and protein interaction networks

and pathway analysis. MetaboAnalyst allows users to process data online and accepts several data input types produced from NMR, LC-MS, or GC-MS, which make it popular in metabolomics. The flowchart of typical metabolomic experiment including sample preparation, metabolomic technology, data analysis, and pathway analysis is shown in Fig. 17.2.

A flowchart of a typical targeted and untargeted metabolomics experiment is shown in Fig. 17.3. As analytical instruments progress, data acquisition is not the challenge of metabolomic research. NMR and MS-based technology can output data automatically and directly, and the corresponding software can help to process and analyze data. Before analysis the evaluations of the method including stability, accuracy, precision, and reproducibility need to be done [10]. As described in Fig. 17.3, metabolomics can be divided into targeted metabolomics and untargeted metabolomics. Targeted metabolomics usually quantify the most abundant metabolites and use simpler statistical approach than untargeted metabolomics, while untargeted metabolomics usually use multivariate statistical analysis, such as PCA, PLS-DA, and OPLS-DA.

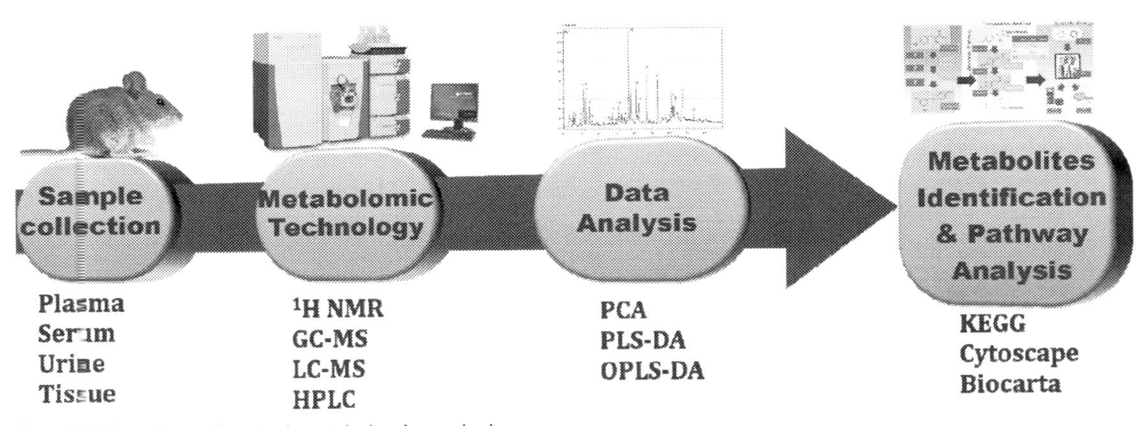

Fig. 17.2 Flowchart of typical metabolomic analysis.

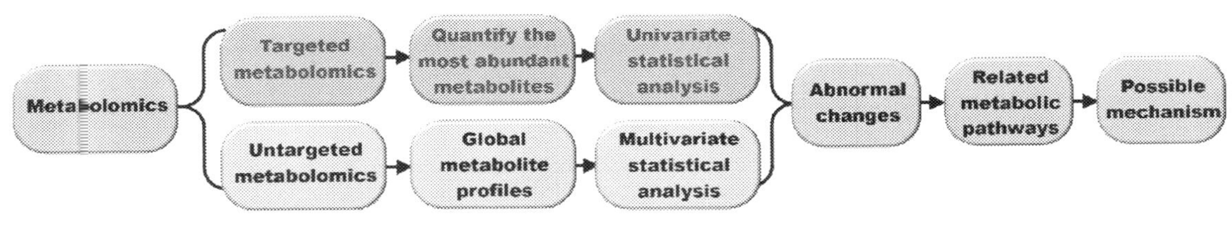

Fig. 17.3 Workflow of typical metabolomics.

17.5 Metabolomics in toxicity evaluation and biomarker identification of natural products

The natural products are in the category of pure compound and extract and compound prescription. Pure compound was isolated from crude extract of natural products, and its chemical structure was identified as one chemical component. Extracts contain more than one chemical compound in the sample, and crude extracts of natural products usually contain different chemical components. In this chapter the toxicity induced by pure compound, crude extract, and compound prescription were indicated in Tables 17.2–17.4, respectively.

17.5.1 Metabolomics in nephrotoxicity evaluation of natural products

Metabolomics is a useful tool to evaluate toxicity and identify toxicological biomarkers of bioactive compounds from natural products. *Aristolochic acid* is a mixture of structural-related nitrophenanthrene carboxylic acid derivatives existed in *Aristolochia, Bragantia, and Asarum* genera, such as *Aristolochiae* Fructus, *Stephaniae tetrandrae* Radix, and Asari Radix et Rhizoma [78]. Aristolochic acid is a toxicant that can cause a common and rapidly progressive interstitial nephropathy called aristolochic acid nephropathy, which can lead to end-stage renal disease [79]. The pathophysiology and underlying mechanisms of the aristolochic acid nephropathy have been studied using metabolomic approach by different analysis methods. Metabolomics was used to study the toxicology of aristolochic acid in rats. Significant changes of two metabolite biomarkers, kynurenic acid and hippuric acid, were detected in rat urine [50]. Hu et al. employed a GC-MS-based metabolomic technique to analyze urinary metabolites in aristolochic acid-treated rats. Eight metabolites were selected as potential metabolic biomarkers including methylsuccinic acid, nicotinamide, 3-hydroxyphenylacetic acid, citric acid, creatinine, uric acid, glycolic acid, and gluconic acid. The identified metabolites suggested that the pathways of energy metabolism, gut microbiota, and purine metabolism were associated with aristolochic acid-induced nephrotoxicity [80]. In another LC-MS-based urinary metabolomic study, the results suggested that the nephrotoxicity of aristolochic acid could be characterized via systemic disturbance of metabolic network including tricarboxylic acid cycle, gut microflora metabolism, amino acid metabolism, purine metabolism, and bile acid biosynthesis, which were partly consistent with the results of GC-MS-based metabolomic study [55].

^{1}H NMR spectroscopy-based metabolomics results showed that aristolochic acid caused a renal proximal tubular and papillary lesion and a slight hepatic impair. The renal proximal tubule lesion was the

Table 17.2 Summary of pure compound-induced toxicity.

Toxic compound	Toxicity	Analytical technique	Specimen types	Identified biomarkers
Aristolochic acid	Nephrotoxicity	LC-MS	Rat urine	L-Leucine, creatine, creatinine, D-serine, homocysteine, L-aspartyl, adenosine, 5-CH_3-tetrahydrofolate, L-arginine, valeric acid, caprylic acid, arachidonic acid, L-methionine, carbamoyl phosphate, hippuric acid, spermine, phenylacetylglycine, 5-L-glutamyl-taurine, cholic acid, 3-methyldioxyindole, citrate, indole-3-carboxylic acid, acontiate, aspartic acid, uridine, taurine, 2,4-dihydroxybenzoic acid, 4-hydroxynonenal, fumarate, glucose, *p*-cresol sulfate, uric acid, uracil, allantoin, indoxyl sulfate, *p*-cresol glucuronide, kynurenic acid, 2-heptanone, aconitate, β-lactose, serine, and citric acid
		LC-MS	Rat kidney tissue	15 phosphatidylcholines, 7 lysophosphatidylcholines, 6 triglycerides, 4 lysophosphatidylethanolamines, 3 phosphatidyletha-nolamines, and 1 ceramide
		LC-MS	Rat kidney plasma	Bile acids, LPC (20:5), LPC (20:4), LPC (20:4), and arachidonic acid
		GC-MS	Rat urine	Citrate, aconitate, isocitrate, succinate, *m*-hydroxyphenylpropionat *p*-cresol, *p*-hydroxyphenylacetate, cystine, and cysteine
		^{1}H NMR	Rat urine	Valine, ethanol, lactate, acetate, succinate, 2-oxoglutarate, citrat creatinine, TMAO (trimethylamine-*N*-oxide), glycine, allantoin, phenylalanine, leucine, succinate, α-oxoglutarate, glutamine, trimethylamine, creatine, choline, taurine, glucose, and hippurate
Ricin	Nephrotoxicity and pulmonary toxicity	^{1}H NMR	Rat urine	Isoleucine, leucine, valine, ethanol, lactate, alanine, acetate, succinate, 2-oxoglutarate, citrate, dimethylamine, trimethylamin dimethylglycine, creatine, creatinine, choline, phosphocholine, taurine, betaine, glycine, urea, allantoin, kynurenic acid, tyrosine phenylalanine, tryptophan, hippurate, kynurenine, trigonelline
Triptolide	Hepatotoxicity and male infertility	GC-MS	Mice serum and testicular tissue	Inosine, adenosine, oleic acid, stearic acid, octadecadienoic acid, indole-3-carboxylic acid, arachidonic acid, palmitic acid, palmitoleic acid, heptadecanoic acid, eicosapentaenoic acid, uric acid, glucose, pyruvic acid, citric acid, butanedioic acid, 3-hydroxybutyric acid, fumaric acid, and supraene
Aconitine, mesaconitine, and hypaconitine	Cardiotoxicity and central nervous system toxicity	^{1}H NMR	Rat urine	Lactic acid, alanine, 3-hydroxybutyric acid, valine, urea, glycerol, phosphoric acid, isoleucine, glycine, glyceric acid, serine, threonine, aspartic acid, malic acid, proline, 2,3,4-erythronic aci creatinine, phenylalanine, phosphoglyceride, glutamine, ornithin and citric acid, L-deoxyglucose, fructose, mannose, glucose, lysine, tyrosine, hexadecanoic acid, inositol, uric acid, tryptopha octadecanoic acid, and cholesterol
Yuanhuapine	Cytotoxicity and irritancy	LC-MS	Rat urine	Homocysteine, indole-3-carboxylic acid, tryptophyl-proline, 3-methyldioxyindole, 5-L-glutamyl-taurine, 2-oxoarginine, homo itrulline, arachidonic acid ethyl ester, C^{16} sphinganine, hippuric acid, D-glucuronic acid, L-phosphate, phenylacetylglycine, *p*-cres and L-prolyl-L-phenylalanine

Related metabolic pathways	Possible toxic mechanism	References
Homocysteine formation, the folate cycle, arachidonic acid biosynthesis, Krebs cycle, gut microflora metabolism, amino acid metabolism, purine metabolism, bile acid biosynthesis, fatty acid generation, and energy metabolism	Inhibition of eicosanoid synthesis from arachidonic acid by nonspecific blocking of the enzyme cyclooxygenase leading to vasoconstriction and reversible mild renal impairment in volume contracted states	[50–55]
Fatty acid, phospholipid, and glycerolipid metabolisms		[51]
Fatty acid and energy metabolisms		[56]
Fatty acid generation and energy and amino acid metabolisms		[52]
Ketone body, amino acid, fatty acid, tricarboxylic acid cycle, and energy metabolisms		[57, 58]
Energy metabolism, oxidative stress, and nitrogen, amino acid, and kynurenine metabolisms	Shifting the energy metabolism from an aerobic to an anaerobic state and causing membrane damage and provoked cell and organelles rupture	[59]
Purine metabolism, energy metabolism, mitochondrial beta-oxidation of long-chain saturated fatty acids, fatty acid oxidation, glycolysis, citric acid cycle, pyruvate metabolism, citric acid cycle, and ketone body metabolism	Abnormal fatty acid level in testicular tissue by the downregulation of the protein expression of PPARα and PPARβ and reduced concentrations of L-carnitine and its derivatives	[60]
Energy, fatty acid, amino acid, and purine metabolisms	Aconitine-induced arrhythmia is the disruption of intracellular Ca^{2+} homeostasis in the cardiomyocytes	[61]
Amino acid, lipid, carbohydrate, and fatty acid metabolisms		[62]

Table 17.3 Summary of extract-induced toxicity.

Toxic compound	Toxicity	Resources	Analytical technique	Specimen types
Cinnabar	Hepatotoxicity	More than about 96% mercuric sulfide (HgS)	^{1}H NMR	Rat urine
The mother root of *Aconitum carmichaelii*	Cardiotoxicity and central nervous system toxicity	Mother roots of *Aconitum carmichaelii*	LC-MS	Rat urine
The daughter root of *Aconitum carmichaelii*	Cardiotoxicity and central nervous system toxicity	Dried daughter or lateral roots of *Aconitum carmichaelii*	LC-MS	Mice blood
Euphorbia kansui	Inflammation, skin irritation, tumor promotion, and liver and kidney lesions	Dried roots of *Euphorbia kansui*	^{1}H NMR	Rat urine
Pinellia ternata	Tongue numbing, swelling, salivation, slurred speech, inflammatory reaction, and liver injury	Dried roots of *Pinellia ternata*	LC-MS	Ras serum

Identified biomarkers	Related metabolic pathways	Possible toxic mechanism	References
Citrate, succinate, 2- oxoglutarate, trimethyl-*N*-oxide, dimethylamine, dimethylglycine, creatine, taurine, phenylacetyglycine, hippurate, formate, acetate, acetoacetate	Energy metabolism, oxidative stress and amino acid metabolism	Mercury poisoning: the ability to form stable complex with sulfhydryl of protein and causing alteration in enzyme functions	[63]
5-Hydroxy-6-methoxy-indole glucuronide, 4,6-dihydroxyquinoline, 5-L-glutamyl-taurine, 7-dehydropregnenolone, glycitin, glucuronide, 3-methyldioxyindole, L-phenylalanyl-L-hydroxyproline, adenylsuccinic acid, 4-(2-amino-3-hydroxyphenyl)-2,4-dioxobutanoic acid, *N*-acetyl-9-*O*-lactoylneuraminic acid, palmitoyl glucuronide, 3-indole carboxylic acid, 3-oxohexadecanoic acid, 2-phenylethanol glucuronide, D-glucuronic acid L-phosphate, tryptophan, and phytosphingosine	Pentose and glucuronate interconversions; alanine, aspartate and glutamate, starch, and sucrose metabolisms; amino sugar and nucleotide sugar, purine, and tryptophan metabolisms; taurine and hypotaurine metabolism; fructose and mannose metabolism and fatty acid metabolism	Exciting the vagus nerve system, resulting in atrioventricular block and symptomatic bradycardia and causing arrhythmia through interfering with sodium channels	[64, [64]
Stearic acid, 5-tridecynoic acid, arachidonic acid, linoleic acid, eicosapentaenoic acid, LPC (22:6), LPC (16:0), PS (21:0), sphinganine, tryptophan, and dihydrosphingosine	Phospholipid metabolism, sphingolipid metabolism, saturated fatty acid oxidation, and unsaturated fatty acid peroxidation	Disruption of intracellular Ca^{2+} homeostasis in the cardiomyocytes	[64] [65]
Lactate, alanine, succinate, α-ketoglutaric acid, citrate, creatine, creatinine, trimethyl-*N*-oxide, taurine, betaine, glucose, glycine, phenylalanine, and hippurate	TCA cycle, anaerobic glycolysis, lipid metabolism, and amino acid metabolism	Unclear	[66]
L-Carnitine, L-acetyl-carnitine, tryptophan, LPC (18:2), PC (34:2), PC (36:5), PC (36:4)	Phospholipid, amino acid, and carnitine metabolisms		[67]

Continued

Table 17.3 Summary of extract-induced toxicity—cont'd

Toxic compound	Toxicity	Resources	Analytical technique	Specimen types
Rhizoma coptidis	Diarrhea	Dried roots of *Rhizoma coptidis*	^{1}H NMR and GC-MS	Rat serum
Xanthii Fructus	Hepatotoxicity	Fruits of *Xanthii* Fructus	LC-MS	Rat urine
Stephania tetrandra	Nephrotoxicity	Dried roots of *Stephania tetrandra*	^{1}H NMR	Rat urine
Aristolochia fangchi	Nephrotoxicity	Dried roots of *Aristolochia fangchi*	^{1}H NMR	Rat urine
Realgar	Nephrotoxicity	Ore crystal (more than 90% As4S4)	^{1}H NMR	Rat urine

main damage caused by aristolochic acid, and the renal toxicity was a progressive course with the accumulation of dosage by monitoring the toxicological processes from onset, development, and part recovery [57]. These disturbed metabolic pathways were indicated in Fig. 17.4. The declined arachidonic acid may be one of the major causes of induced acute renal failure. Phospholipase A2 (PLA2) hydrolyzes phospholipids into phosphatidylinositol and lecithin. During this process, arachidonic acid is produced, which is an important intermediate in the synthesis of prostaglandins (PGs). Aristolochic acid can greatly

Identified biomarkers	Related metabolic pathways	Possible toxic mechanism	References
Alanine, L-proline, glycine, isoindole, serine, Glutamate, L-ornithine, and D-glucose	Energy, amino acid, lipid, and fatty acid metabolisms	Disturbance in the normal gut microbiota	[68]
6-Hydroxy-5-methoxy-indole glucuronide, 4,6-dihydroxyquinoline, L-phenylalanyl-L-hydroxyproline, uridine, 3-methyldioxyindole, indoxyl sulfate, hippuric acid, sebacic acid, and arachidonic acid	Energy, amino acid, lipid, and fatty acid metabolisms	A disruption of oxidative phosphorylation that is an essential process for the cell's energy metabolism and transfer system	[69]
Citrate, 2-oxoglutarate, succinic acid, taurine, hippurate, glucose, *N*-acetyl glycoprotein, acetic acid, TMAO, creatine, and creatinine	Unsaturated fatty acid metabolism, energy acid metabolism, amino acid metabolism, tricarboxylic acid cycle, and glycolysis	The same toxic mechanisms as aristolochic acid	[70]
Citrate, 2-oxoglutarate, succinic acid, taurine, hippurate, glucose, *N*-acetyl glycoprotein, acetic acid, TMAO, creatine, and creatinine	Unsaturated fatty acid metabolism, energy acid metabolism, amino acid metabolism, tricarboxylic acid cycle, and glycolysis	The same toxic mechanisms as aristolochic acid	[70] [71]
Pyruvate, α-ketoglutarate, succinate, citrate, 3-D-hydroxybutyrate, acetoacetate, taurine, dimethylglycine, TMAO, hippurate, phenylacetylglycine, acetate, acetone, alanine, betaine, citric acid, dimethylamine, formic acid, glycine, hippuric acid, lactic acid, methylguanidine, pyruvic acid, succinic acid, trigonelline, and α-ketoglutaric acid	Energy, amino acid, lipid, fatty acid, choline, pyruvate and ketone body metabolisms, citric acid cycle, choline metabolism, betaine metabolism, folate, alanine, phenylalanine and metabolisms, protein catabolism, bile acid biosynthesis and niacin metabolism	Arsenic poisoning: carbohydrate depletion and inhibition of gluconeogenesis were important toxic mechanisms of arsenic in rats	[72] [73]

inhibit PLA2 and result in the downregulation of arachidonic acid, which decreased PGs (Fig. 17.4b)

Castor beans, the seeds of *Ricinus communis*, have been used as emetic and cathartic. The kernels of castor beans contain the highest concentration of ricin, which is a large and water-soluble glycoprotein as a principal toxin of castor beans. Long-term ricin treatment produced perturbations on energy metabolism, nitrogen metabolism, amino acid metabolism and kynurenine metabolism pathway and evoked oxidative stress. Fig. 17.5 indicated the interrelationship

Table 17.4 Summary of compound prescription-induced toxicity.

Toxic compound	Toxicity	Analytical technique	Specimen types	Identified biomarkers	Related metabolic pathways	Possible toxic mechanism	References
Bu-Fei-A-Jiao-Tang	Nephrotoxicity	HPLC and ^{1}H NMR	Mice urine	Creatine, glycine, creatinine, TMAO, valine, hippurate, DMG (dimethyl-amine), citrate, lactate, alanine, glucose, fumarate, and formate	Energy and amino acid metabolisms	The same toxic mechanisms as aristolochic acid	[74]
Niuhuang Jiedu Tablet	Nephrotoxicity	^{1}H NMR	Rat urine	Leucine, isoleucine, 3-hydroxybutyrate, lactate, alanine, acetate, pyruvate, 2-oxoglutarate, citrate, creatine, choline, taurine, TMAO, betaine, creatinine, phenylalanine, hippurate, and leucine	Energy, amino acid, lipid, and fatty acid metabolisms	The same toxic mechanisms as realgar	[75] [76]
Zhusha Anshen Wan	Hepatotoxicity	^{1}H NMR	Rat urine	Citrate, TMAO, succinic acid, 2-oxoglutarate, hippurate, taurine, creatinine, lactate, glycine, 3-hydroxybutyric acid, and acetate	Energy, lipid, choline, and amino acid metabolisms	The same toxic mechanisms as cinnabar	[77]

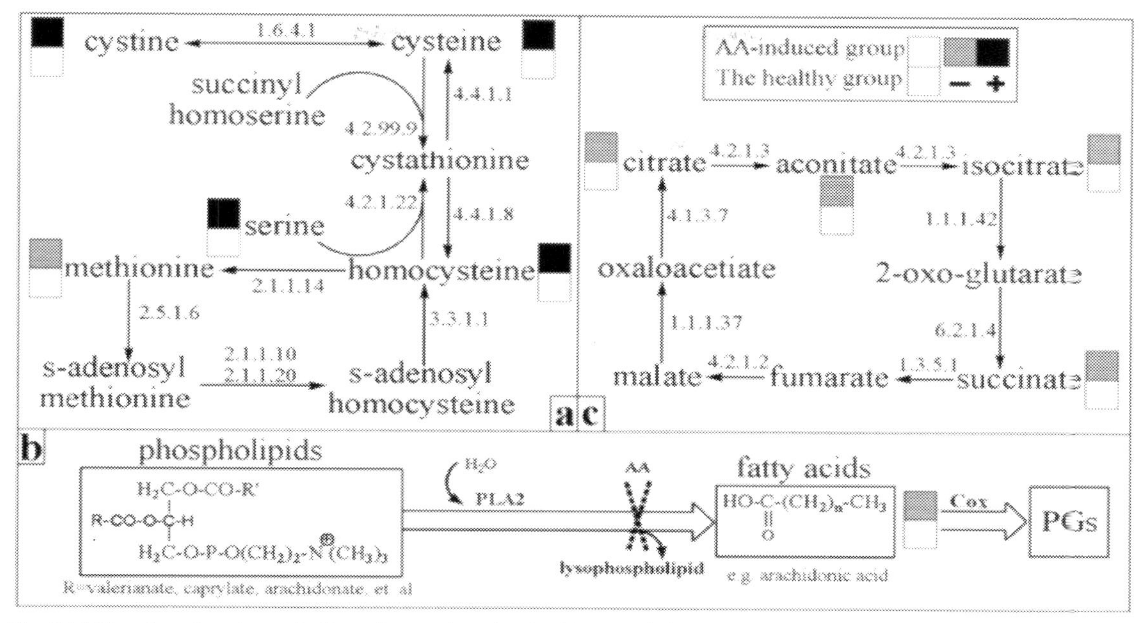

Fig. 17.4 The disordered metabolic pathways by aristolochic acid. (A) Aristolochic acid-induced metabolic disorders in amino acids. (B) Aristolochic acid-disturbed TCA cycle. (C) PLA2-related mechanisms of aristolochic acid nephropathy. Dark square indicated the upregulation of metabolites were observed in the urine of aristolochic acid nephropathy rat, whereas gray square showed the downregulation of metabolites. Copyright – Elsevier (2016). Ref. D-Q. Chen, H. Chen, L. Chen, D-D. Tang, H. Miao, Y-Y. Zhao, Metabolomic application in toxicity evaluation and toxicological biomarker identification of natural product, Chem. Biol. Interact. 252 (2016) 114–130.

of these identified metabolic pathways. These findings could well explain *R. communis*-induced nephrotoxicity and pulmonary toxicity and provide several potential biomarkers for diagnostics of these toxicities [59].

The roots of *Aristolochia fangchi* (AF) and the radix of *Stephania tetrandra* (ST) have been widely used for the treatment of rheumatic arthritis, but the inappropriate intake may cause lesion on kidney. During a slimming regimen in Belgium in the early 1990s, because of accidental replacement of ST by aristolochic acid-containing AF, about 100 cases of renal disease were reported. Some of the patients died and most of them required dialysis or kidney transplant [81]. The toxicity differences between AF and ST were investigated. Compared with control group the urinary citrate, 2-oxoglutarate, taurine, hippurate, TMAO, creatine and the plasma 3-D-hydroxybutyrate, acetone, *N*-acetyl-L-cysteine, and creatinine were all changed. The liver and kidney toxicity of ST were all more serious than the AF [70]. Metabolomics was carried out to study the changes of urinary metabolites after treatment of AF. Compared with control group, urinary

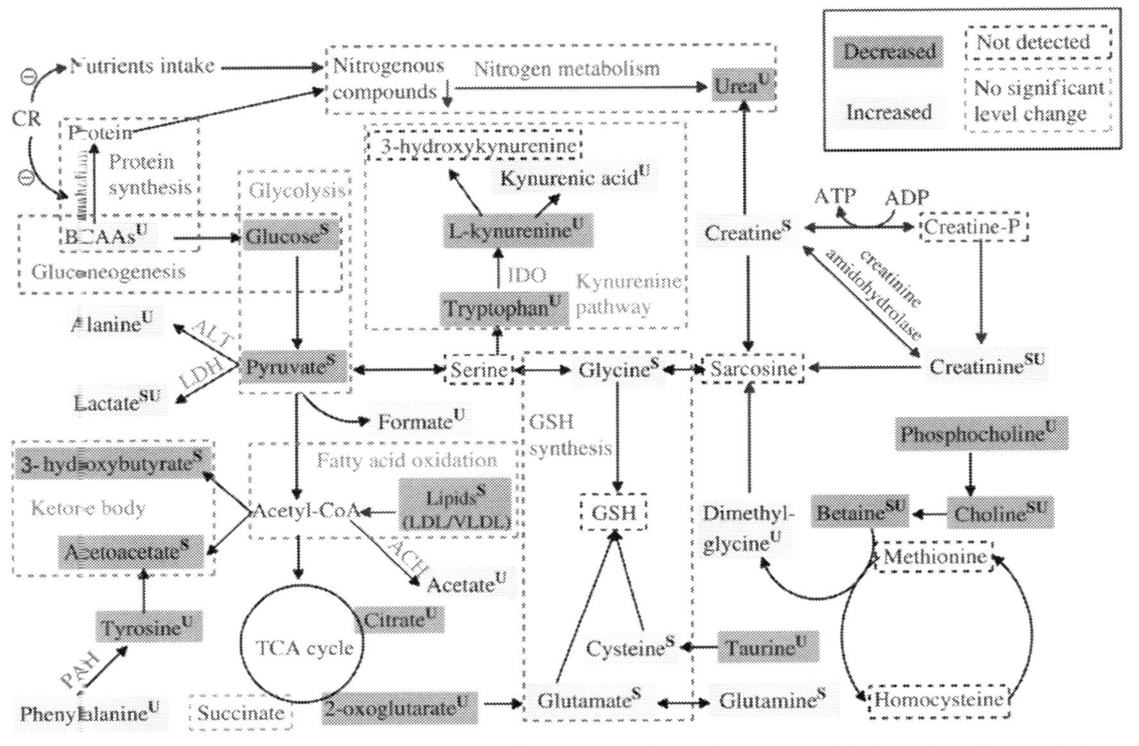

Fig. 17.5 Schematic diagram of the disturbed metabolic pathways. "–" indicated the inhibition effect. The words in blue (*light gray in print version*) were enzymes and inhibited pathways, and words in red (*gray in print version*) were enzymes and promoted pathway. Metabolites with superscript "U" indicated observed significant fluctuation in urine; "S" indicated observed significant fluctuation in serum; "SU" indicated observed significant fluctuation in serum and urine. Copyright – Elsevier (2016). Ref. D-Q. Chen, H. Chen, L. Chen, D-D. Tang, H. Miao, Y-Y. Zhao, Metabolomic application in toxicity evaluation and toxicological biomarker identification of natural product, Chem. Biol. Interact. 252 (2016) 114–130.

taurine was increased time-dependently, while citrate was decreased in AF-treated group during the whole administration period. Hippurate level was increased at weeks 2 and 6 (2 weeks after drug withdrawal) but decreased at week 4. The creatinine and creatine were increased at week 4 but decreased at week 6, while 2-oxoglutarate was decreased and TMAO was increased at weeks 4 and 6. High-dose AF can induce renal lesion, and its seriousness is related to the lasting of administration [71].

Realgar, an ore crystal containing more than 90% tetraarsenic tetrasulfide (As_4S_4), has been used in the treatment of syphilis, psoriasis, malaria, and parasitic infections for more than 1500 years [82], but it is toxic and exhibits severe adverse effects especially in long-term use. ^{1}H NMR-based metabolomics showed that the perturbations of energy metabolism, altered transmethylation and gut microflora environment, impairment of amino acid metabolism, oxidative injury, and slight liver and kidney injury were observed in realgar-treated rats [72]. Significant perturbations in amino acid metabolism, TCA cycle, choline metabolism, and porphyrin metabolism were found in urine of realgar-treated rats by employing UPLC-MS and ^{1}H NMR spectroscopy.

Bu-Fei-A-Jiao-Tang (BFAJT) is a common compound prescription used for some lung-related symptoms and contains *Fructus Aristolochiacontorta* (FA), an Aristolochiaceae herb. The toxicities of aristolochic acid standard sample, FA, and BFAJT were investigated. The different degrees of acute renal tubular injuries were observed in dosed group. Both metabolomics and pathological studies revealed that aristolochic acid standard sample, FA, and BFAJT were all nephrotoxicity. The compositions of compound prescription did not diminish the nephrotoxicity caused by aristolochic acid [74].

Niuhuang Jiedu Tablet (NJT) is an effective formula in China used to treat acute tonsillitis, pharyngitis, periodontitis, and mouth ulcer, which composes of Realgar (As_2S_2), Bovis Calculus Artificialis, Borneolum Synthcticum, Gypsum Fibrosum, Rhei Radix et Rhizoma (RR), Scutellariae Radix (SR), Platycodonis Radix (PR), and Glycyrrhizae Radix et Rhizoma (GR). Although NJT is effective in treatment, the arsenic contained in realgar was still a potential toxic element. ^{1}H NMR-based metabolomics showed that it was more secure and much less toxic for counterbalanced realgar in NJT. The effective material bases of toxicity alleviation to realgar were RR, SR, PR, and GR, which regulated energy metabolism, choline metabolism, amino acid metabolism, and gut flora disorder by realgar exposure [75, 76].

17.5.2 Metabolomics in hepatotoxicity evaluation of natural products

Triptolide, a bioactive diterpenoid compound isolated from *Tripterygium wilfordii*, exhibits diverse biological activities such as antiinflammatory, immunomodulatory, and antiproliferative activities [62]. However, the further clinical research and application of triptolide is confined by its severe toxicity on the liver, kidney, and reproductive systems [83]. Zhao et al. developed a LC-MS-based metabolomic method to investigate the hepatotoxicity of triptolide in mice. Mice were administered triptolide by gavage to establish the acute liver injury model. Metabolomic results showed that a total of thirty metabolites were significantly changed by triptolide treatment and the abundance of twenty-nine metabolites was correlated with toxicity. Pathway analysis indicated that the mechanism of triptolide-induced hepatotoxicity was related to alterations in multiple metabolic pathways, including glutathione metabolism, tricarboxylic acid cycle, purine metabolism, glycerophospholipid metabolism, taurine and hypotaurine metabolism, pantothenate and coenzyme A biosynthesis, pyrimidine metabolism, and amino acid metabolism [84]. Further a GC-MS-based serum metabolomics found that treatment with triptolide at either 0.6 or 2.4 mg/kg dose caused deviations in the metabolic pattern and resulted in perturbation of taurine, creatinine, free fatty acids, β-hydroxybutyrate, TCA cycle intermediates, and amino acids, which indicated the dysfunction of β-oxidation of free fatty acids and impairment of the mitochondria and confirmed the hepatic toxicity of triptolide [85]. Targeted metabolomics was employed to elucidate triptolide-induced toxicity mechanism.

Cinnabar has been used for 2000 years in China for sedative therapy, which contains more than about 96% mercuric sulfide (HgS) [86]. Cinnabar can induce toxicological effects at the dose of 2 g/kg, and serum metabolomics showed cinnabar-induced disturbance in energy metabolism, amino acid metabolism, and gut microflora environment as well as slight liver and kidney injury [63]. *Curcuma longa* and curcumin have the effective protection function to the cinnabar-induced liver and kidney injury through regulating energy metabolism, intestinal microflora, and amino acid metabolism [87].

Xanthii Fructus (XF) is frequently used for the treatment of sinusitis, headache, rheumatism, and skin itching. However, the inappropriate use of XF can cause different degrees of the damages in various organs, particularly in the liver. Previous study showed that after treatment with XF for 4 weeks, serum alanine transaminase, aspartate aminotransferase, and alkaline phosphatase in rat were significantly increased [88]. Urinary metabolomics showed that the metabolic characters in XF-treated rats were perturbed in a dose-dependent manner and 10 metabolites can be identified as potential toxicity biomarker [69]. ^{1}H NMR-based metabolomics demonstrated that the major hepatotoxicity constituents were atractyloside, carboxyatractyloside, and 40-desulphate-atractyloside and the hepatotoxicity of XF involved in mitochondrial inability, fatty acid metabolism, and some amino acid metabolism [89]. *Xanthii Fructus* is the mature fruit with involucres of *Xanthium sibirium* Patr. and widely used for the treatment of sinusitis, headache, rheumatism, and skin itching [90]. Xue et al. performed an integrated metabolomic study using ^{1}H NMR combined with multivariate statistical analysis to elucidate the hepatotoxicity of *Xanthii Fructus*. When rats were treated with *Xanthii Fructus* at 30.0 g/kg, the hepatotoxicity was reflected in the changes observed in serum biochemical profiles and by the histopathological examination of the liver. The results demonstrated that atractyloside, carboxyatractyloside, and 40-desulphate-atractyloside were the major hepatotoxicity constituents in *Xanthii Fructus*. Moreover the hepatotoxicity of *Xanthii Fructus* mainly associated with mitochondrial inability, fatty acid metabolism, and some amino acid metabolism [91]. The urinary metabolic perturbations associated with toxicity induced by Xanthii Fructus were also studied using UPLC-MS. The results showed that the metabolic characters in Xanthii Fructus-treated rats were perturbed in a dose-dependent manner and 10 metabolites including 6-hydroxy-5-methoxyindole glucuronide/5-hydroxy-6-methoxyindole glucuronide, 4,6-dihydroxyquinoline, 3-methyldioxyindole, phenylalanine, indoxyl sulfate, hippuric acid, uridine, L-phenylalanyl-L-hydroxyproline, sebacic acid, and arachidonic acid were preliminarily identified as potential toxicity biomarkers [92].

Zhusha Anshen Wan (ZSASW) is composed of cinnabar (cinnabaris), Coptidis Rhizoma (*Coptis chinensis* French.), Angelicae Sinensis Radix (*Angelica sinensis* (Oliv.) Diels), uncooked Rehmanniae Radix (*Rehmannia glutinosa* Libosch.), honey fried Glycyrrhizae Radix et Rhizoma (*Glycyrrhiza uralensis* Fisch.), which is widely used for sedative therapy. Cinnabar is the chief component of ZSASW and possesses the toxicities. ZSASW had the effects of protecting from the toxicity induced by cinnabar alone, and metabolomics indicated that ZSASW may be more secure and less toxic than cinnabar alone for clinic treatment. Serum citrate, TMAO, succinate, 2-oxoglutarate, hippurate, formate, taurine, creatine, creatinine, lactate, glycine, 3-hydroxybutyric acid, and acetate were identified as biomarkers [77].

17.5.3 Metabolomics in cardiotoxicity evaluation of natural products

Herbal medicines derived from *Aconitum* species, including Aconiti Kusnezoffii Radix, Aconiti Radix, and Aconiti Lateralis Preparata Radix, have a long history of clinical use. These herbs have been shown to exhibit biological effects on various diseases, including rheumatic fever, painful joints, bronchial asthma, gastroenteritis, collapse, syncope, diarrhea, edema, and tumors. Modern research revealed that *Aconitum* herbs and *Aconitum* alkaloids are not only active ingredients but also toxic components [93]. Aconitine, mesaconitine, and hypaconitine are the main *Aconitum* alkaloids derived from Aconiti lateralis Radix praeparata, the lateral root of *Aconitum carmichaelii* Debx. These alkaloids have analgesic, antipyretic, and local anesthetic activities and have beneficial effects against rheumatosis and rheumatoid arthritis. But the strong toxicity and the narrow margin between therapeutic and toxic doses limited clinical application of the *Aconitum* alkaloids. Sun et al. investigated the metabolic changes in rats caused by aconitine, mesaconitine, and hypaconitine using ^{1}H NMR and GC-MS. Compared with control group the results revealed larger deviations in the aconitine and mesaconitine groups and smaller deviations in hypaconitine group, illustrating the different toxicity mechanisms of these alkaloids. Metabolomic analysis indicated that most of the metabolic biomarkers were related to tricarboxylic acid cycle [61].

The potential cardiotoxicity of Aconiti Radix (the mother roots of *A. carmichaelii* Debx) was frequently reported because of its narrow therapeutic window. A metabolomic method was performed to characterize the potential mechanisms of Aconiti Radix-induced cardiotoxicity by UPLC/Q-TOF-MS. Seventeen biomarkers were identified in urinary samples, which were associated with pentose and glucuronate interconversions and alanine, aspartate, and glutamate metabolism [94]. Meanwhile the levels of the identified toxicity biomarkers were modulated to the normal ranges by Glycyrrhizae Radix, Paeoniae Alba Radix, and Zingiberis Rhizoma. The results indicated that these three compatible herbal medicines could be the effective detoxifying substances against the toxicity of Aconiti Radix [95].

The daughter root of *A. carmichaelii* is beneficial for the treatment of rheumatism and painful joints, which has a potential cardiotoxicity with a relatively narrow margin of clinical safety. UPLC-MS-based metabolomics investigated comprehensive metabolic characters and biomarkers in rat plasma following dosing with the mother roots of *A. carmichaelii*, the daughter roots, and their processed products. Both the mother and daughter roots of *A. carmichaelii* could lead to serious cardiotoxicity, and these toxicities reveal a time and dose dependency. The activities of mother and daughter roots were related to sphingolipid metabolism, glycerophospholipid metabolism, aminoacyl-tRNA biosynthesis, and tryptophan metabolism, while their toxicities were responsible for abnormal sphingolipid metabolism, aminoacyl-tRNA biosynthesis, and tryptophan metabolism. Processing could detoxify by regulating sphingolipid metabolism and glycerophospholipid metabolism [64]. After treated with the daughter roots of *A. carmichaelii*, the changes of lipids in mice heart were investigated, and 14 lipid metabolites were identified as potential biomarkers, which were primarily involved in phospholipid metabolism, sphingolipid metabolism, saturated fatty acid oxidation, and unsaturated fatty acid peroxidation [65].

17.5.4 Metabolomics in other toxicity evaluation of natural products

In addition to the aforementioned hepatotoxicity, nephrotoxicity, and cardiotoxicity, reproduction toxicity of triptolide is also the main obstacle for its clinical applications. Ma et al. developed a GC-MS-based metabolomic approach to evaluate the mechanism of triptolide-induced reproductive toxicity in male mice and identify potential biomarkers for the early detection of spermatogenesis dysfunction. The results indicated that the testicular toxicity of triptolide may be caused by abnormal lipid and energy metabolism in the testis via downregulation of peroxisome proliferator-activated receptor mediation [60].

Cinnabar, a traditional mineral medicine containing more than 96% mercuric sulfide, has been used as a sedative and soporific for more than 2000 years. It was reported that cinnabar can impact central nervous system and cause neurotoxicity through blood-brain barrier [96]. Wei et al. investigated the neurotoxicity of cinnabar in rats by ^{1}H NMR-based metabolomics combined with multivariate pattern recognition. The metabolite variations induced by cinnabar were characterized by increased levels of glutamate, glutamine, myoinositol, and choline, as well as decreased levels of γ-amino-n-butyrate, taurine, *N*-acetylaspartate, and *N*-acetylaspartylglutamate in tissue extracts of the cerebellum and cerebrum. The results indicated that cinnabar induced glutamate excitotoxicity, neuronal cell loss, osmotic state changes, membrane fluidity disruption, and oxidative injury in the brain [97].

Kansui Radix, the dried root of *Euphorbia kansui* T. N. Liou ex T. P. Wang, was widely used for treatment of edema, ascites, and asthma [98]. The clinical application of Kansui Radix is greatly restricted since it can induce toxic symptoms such as stomachache, diarrhea, dehydration, and respiratory failure. The metabolites responsible for the toxicity of Kansui Radix were evaluated by ^{1}H NMR-based metabolomics. The toxicity of Kansui Radix accumulated with dosing time and persisted even when treatment was stopped. The metabolomic results revealed that the levels of alanine, lactate, taurine, betaine, hippurate, phenylalanine, and glucose were increased, while the levels of succinate, citrate, glycine, creatine, and creatinine were decreased. The corresponding biochemical pathway alterations included inhibited tricarboxylic acid cycle, increased anaerobic glycolysis, and perturbed amino acid metabolism [66].

Coptidis Rhizoma has been used as a heat-clearing and detoxifying agent in China for 2000 years. Coptidis Rhizoma is relatively safe in normal dosage, but an extensive dosage can cause side effects such as diarrhea. A combination of ^{1}H NMR-based and GC-MS-based metabolomic approach was applied to discover the endogenous metabolites related to the diarrhea induced by Coptidis Rhizoma. In the study, 12 marker metabolites from ^{1}H NMR and eight from GC-MS were identified; among those metabolites, hippurate, acetate, alanine, glycine, and glutamate were likely to break the balance of gut microbiota, whereas lactate and 2-ketoisovalerate were associated with energy metabolism [68].

Natural products are generally used in the form of prescriptions (the combination of several different herbal medicines). The bioactive constituents and fundamental mechanisms of most natural product prescriptions remain unclear due to the complex components of remedies. Metabolomics could provide a holistic view and deeper insight into the efficacy and toxicity of natural product prescriptions. It might also be a promising approach to investigate the detoxification of natural medicines and reasonable combination of prescriptions. Niuhuang Jiedu Tablet, composed of Realgar, Bovis Calculus Artificialis, Borneolum Synthcticum, Gypsum Fibrosum, Rhei Radix et Rhizoma, Scutellariae Radix, Platycodonis Radix, and Glycyrrhizae Radix et Rhizoma, is an effective natural product prescription used for treatment of acute tonsillitis, pharyngitis, periodontitis, and mouth ulcer [76]. In the prescription, significant level of realgar is a potentially toxic element. Xu et al. proposed a ^{1}H NMR-based metabolomic approach to investigate the toxicity of realgar after being counterbalanced by other herbal medicines in Niuhuang Jiedu Tablet. The results showed that it was more secure and much less toxic for counterbalanced realgar in Niuhuang Jiedu Tablet [75].

Zhusha Anshen Wan, composed of cinnabar, Coptidis Rhizoma, Angelicae Sinensis Radix, Rehmanniae Radix, and Glycyrrhizae Radix et Rhizoma, is a widely used natural product prescription for sedative therapy. Cinnabar is the chief component of Zhusha Anshen Wan and possesses certain toxicity. A metabolomic analysis suggested that Zhusha Anshen Wan may be more secure and much less toxic than cinnabar alone, and the four combined herbal medicines of Zhusha Anshen Wan had the effects of protecting from the toxicity induced by cinnabar alone [77].

17.6 Concluding remarks and perspectives

In recent years, metabolomic analysis has increased markedly in efficacy, quality control, action of mechanism, and active components discovery of natural products. The components and mechanisms of most natural product are still obscure. Meanwhile the toxicity of natural products has attracted a wide range of concerns and aroused many toxicity studies on natural product-based medicines. Unfortunately, there is no standard and objective basis for natural product toxicity evaluation and no uniform standard for safety assessment up to now, which seriously hinders the development of natural product toxicology research. To assure the safeness of such products, a deeper knowledge is strongly required. Animal experiment, cell experiment, and systems biology approach are common in toxicological research. As a systemic approach, metabolomics focuses on the analysis of global metabolites and their functions in the biological system, and hence it is a suitable choice to fit the holistic concept of multicomponent theory, which characterized with endpoint amplification of the global biological and functional status. It allows quantitative measurement of large numbers of low molecular endogenous metabolites involved in metabolic pathways and thus reflects fundamental metabolism status of body.

Among many approaches in metabolomics, LC-MS-based metabolomics has been frequently applied to toxicity study of pure compounds, extracts, and compound prescriptions and is a good tool to evaluate toxicity of natural product systematically, find potential biomarkers, and explore the mechanisms of toxicities. However, metabolomic application on toxicology of natural product still has great challenges. The result of metabolomics is hard to establish relationships with other experimental result. Moreover the present approaches are far from meeting the ultimate goal of metabolomics that detects the whole metabolites in cells, tissues, or organs. The promising analytical approaches are required to be developed to handle problems in toxicological research on natural product.

References

[1] M. Commisso, P. Strazzer, K. Toffali, M. Stocchero, F. Guzzo, Untargeted metabolomics: an emerging approach to determine the composition of herbal products, Comput. Struct. Biotechnol. J. 4 (2013), e201301007.

[2] D.G. Cox, J. Oh, A. Keasling, K.L. Colson, M.T. Hamann, The utility of metabolomics in natural product and biomarker characterization, Biochim. Biophys. Acta 1840 (2014) 3460–3474.

[3] J.G. Bundy, E.M. Lenz, N.J. Bailey, C.L. Gavaghan, C. Svendsen, D. Spurgeon, P.K. Hankard, D. Osborn, J.M. Weeks, S.A. Trauger, P. Speir, I. Sanders, J.C. Lindon, J.K. Nicholson, H. Tang, Metabonomic assessment of toxicity of 4-fluoroaniline, 3,5-difluoroaniline and 2-fluoro-4-methylaniline to the earthworm *Eisenia veneta* (rosa): identification of new endogenous biomarkers, Environ. Toxcol. Res. 21 (2002) 1966–1972.

[4] A. Roux, D. Lison, C. Junot, J.-F. Heilier, Applications of liquid chromatography coupled to mass spectrometry-based metabolomics in clinical chemistry and toxicology: a review, Clin. Biochem. 44 (2011) 119–135.

[5] L. Duan, L. Guo, L. Wang, Q. Yin, C.-M. Zhang, Y.-G. Zheng, E.-H. Liu, Application of metabolomics in toxicity evaluation of traditional Chinese medicines, Chin. Med. 13 (2018) 60.

[6] J.K. Nicholson, J.C. Lindon, E. Holmes, 'Metabonomics': understanding the metabolic responses of living systems to pathophysiological stimuli via multivariate statistical analysis of biological NMR spectroscopic data, Xenobiotica 29 (1999) 1181–1189.

[7] J.R. Idle, F.J. Gonzalez, Metabolomics, Cell Metab. 6 (2009) 348–351.

[8] S.K. Yan, R.H. Liu, H.Z. Jin, X.R. Liu, J. Ye, L. Shan, W.D. Zhang, "omics" in pharmaceutical research: overview, applications, challenges, and future perspectives, Chin. J. Nat. Med. 13 (2015) 3–21.

[9] D.G. Robertson, Metabonomics in toxicology: a review, Toxicol. Sci. 85 (2005) 809–822.

[10] D.-Q. Chen, H. Chen, L. Chen, D.-D. Tang, H. Miao, Y.-Y. Zhao, Metabolomic application in toxicity evaluation and toxicological biomarker identification of natural product, Chem. Biol. Interact. 252 (2016) 114–130.

[11] R.B. Herbert, The Biosynthesis of Secondary Metabolites, second ed., Chapman and Hall, 1989.
[12] L. Shargel, A. Yu, Applied Biopharmaceutics and Pharmacokinetics, fourth ed., McGraw-Hill, 1999.
[13] D.G. Robertson, P.B. Watkins, M.D. Reily, Metabolomics in toxicology: preclinical and clinical applications, Toxicol. Sci. 120 (S1) (2011) S146–S170.
[14] O. Fiehn, Metabolomics—the link between genotypes and phenotypes, Plant Mol. Biol. 48 (2002) 155–171.
[15] R.D. Beger, J. Sun, L.K. Schnackenberg, Metabolomics approaches for discovering biomarkers of drug-induced hepatotoxicity and nephrotoxicity, Toxicol. Appl. Pharmacol. 243 (2010) 154–166.
[16] J.C. Lindon, J.K. Nicholson, E. Holmes, H. Antti, M.E. Bollard, H. Keun, O. Beckonert, T.M. Ebbels, M.D. Reily, D. Robertson, G.J. Stevens, P. Luke, A.P. Breau, G.H. Cantor, R.H. Bible, U. Niederhauser, H. Senn, G. Schlotterbeck, U.G. Sidelmann, S.M. Laursen, A. Tymiak, B.D. Car, L. Lehman-McKeeman, J.-M. Colet, A. Loukaci, C. Thomas, Contemporary issues in toxicology the role of metabonomics in toxicology and its evaluation by the COMET project, Toxicol. Appl. Pharmacol. 187 (2003) 137–146.
[17] S.G. Oliver, M.K. Winson, D.B. Kell, F. Baganz, Systematic functional analysis of the yeast genome, Trends Biotechnol. 16 (1998) 373–378.
[18] P.L. Wood, Mass spectrometry strategies for clinical metabolomics and lipidomics in psychiatry, neurology, and neuro-oncology, Neuropsychopharmacology 39 (2014) 24–33.
[19] H. Gu, J. Du, F.C. Neto, P.A. Carroll, S.J. Turner, E.G. Chiorean, R.N. Eisenman, D. Raftery, Metabolomics method to comprehensively analyze amino acids in different domains, Analyst 140 (2015) 2726–2734.
[20] L.Y. Liu, H.J. Zhang, L.Y. Luo, J.B. Pu, W.Q. Liang, C.Q. Zhu, Y.-P. Li, P.-R. Wang, Y.-Y. Zhang, C.-Y. Yang, Z.-J. Zhang, Blood and urinary metabolomic evidence validating traditional Chinese medicine diagnostic classification of major depressive disorder, Chin. Med. 13 (2018) 53.
[21] E. Holmes, R.L. Loo, J. Stamler, M. Bictash, I.K. Yap, Q. Chan, T. Ebbels, M.D. Iorio, I.J. Brown, K.A. Veselkov, M.L. Daviglus, H. Kesteloot, H. Ueshima, L. Zhao, J.K. Nicholson, P. Elliott, Human metabolic phenotype diversity and its association with diet and blood pressure, Nature 453 (2008) 396–400.
[22] D.A. Bennett, M.D. Waters, Applying biomarker research, Environ. Health Perspect. 108 (2000) 907–910.
[23] D.J. Paustenbach, The practice of exposure assessment, in: A.W. Hayes (Ed.), Principles and Methods of Toxicology, Taylor and Francis, London, 2001.
[24] Biomarkers Definitions Working Group, Biomarkers and surrogate endpoints: preferred definitions and conceptual framework, Clin. Pharmacol. Ther. 69 (2001) 89–95.
[25] J.A. Timbrell, Biomarkers in toxicology, Toxicology 129 (1998) 1–12.
[26] R.A. Kehoe, F. Thamann, J. Cholak, Lead absorption and excretion in relation to the diagnosis of lead poisoning, J. Ind. Hyg. Toxicol. 15 (1933) 320–340.
[27] S.J. Pearce, H.H. Schrenk, W.P. Yant, United States Bureau of Mines, 1936.
[28] J. Angerer, U. Ewers, M. Wilhelm, Human biomonitoring: state of the art, Int. J. Hyg. Environ. Health 210 (2007) 201–228.
[29] L.F. Shyur, C.P. Liu, S.C. Chien, Handbook of Plant Metabolomics, Wiley-VCH Verlag GmbH & Co. KGaA, Weinheim, 2013, pp. 155–174.
[30] H.K. Kim, Y.H. Choi, R. Verpoorte, NMR-based metabolomic analysis of plants, Nat. Protoc. 5 (2010) 536–549.
[31] H.K. Kim, Y.H. Choi, R. Verpoorte, NMR-based plant metabolomics: where do we stand, where do we go? Trends Biotechnol. 29 (2011) 267–275.

[32] C. Seger, S. Sturm, Analytical aspects of plant metabolite profiling platforms: current standings and future aims, J. Proteome Res. 6 (2007) 480–497.

[33] J. Meiler, M. Will, Genius: a genetic algorithm for automated structure elucidation from ^{13}C NMR spectra, J. Am. Chem. Soc. 124 (2002) 1868–1870.

[34] L.W. Sumner, P. Mendes, R.A. Dixon, Plant metabolomics: large-scale phytochemistry in the functional genomics era, Phytochemistry 62 (2003) 817–836.

[35] P. Liu, S.L. Liu, D.Z. Tian, P. Wang, The applications and obstacles of metabonomics in traditional Chinese medicine, Evid. Based Complement. Altern. Med. 2012 (2012) 945824.

[36] A.D. Hegeman, Plant metabolomics-meeting the analytical challenges of comprehensive metabolite analysis, Brief. Funct. Genom. 9 (2010) 139–148.

[37] D.S. Wishart, Metabolomics: applications to food science and nutrition research, Trends Food Sci. Tech. 19 (2008) 482–493.

[38] J. Hummel, S. Sequ, Y. Li, S. Irgang, J. Jueppner, P. Giavalisco, Ultra performance liquid chromatography and high resolution mass spectrometry for the analysis of plant lipids, Front. Plant Sci. 2 (2011) 54.

[39] H. Miao, H. Chen, X. Zhang, L. Yin, D.Q. Chen, X.L. Cheng, X. Bai, F. Wei, Urinary metabolomics on the biochemical profiles in diet-induced hyperlipidemia rat using ultraperformance liquid chromatography coupled with quadrupole time-of-flight SYNAPT high-definition mass spectrometry, J. Anal. Methods Chem. 2014 (2014) 184162.

[40] Y.Y. Zhao, X.L. Cheng, F. Wei, X. Bai, X.J. Tan, R.C. Lin, Q. Mei, Intrarenal metabolomic investigation of chronic kidney disease and its TGF-β1 mechanism in induced-adenine rats using UPLC Q-TOF/HSMS/MS^E, J. Proteome Res. 12 (2013) 692–703.

[41] Y.Y. Zhao, L. Zhang, Y.L. Feng, X.L. Cheng, X. Bai, F. Wei, R.C. Lin, UPLC-Q-TOF/HSMS/MSE-based metabonomics for adenine-induced changes in metabolic profiles of rat faeces and intervention effects of ergosta-4,6,8(14),22-tetraen-3-one, Chem. Biol. Interact. 301 (2013) 31–38.

[42] D. Vuckovic, Current trends and challenges in sample preparation for global metabolomics using liquid chromatography-mass spectrometry, Anal. Bioanal. Chem. 403 (2012) 1523–1548.

[43] R.B. Gil, R. Lehmann, P. Schmitt-Kopplin, S.S. Heinzmann, ^{1}H NMR-based metabolite profiling workflow to reduce inter-sample chemical shift variations in urine samples for improved biomarker discovery, Anal. Bioanal. Chem. 408 (2016) 4683–4691.

[44] D.G. Sitnikov, C.S. Monnin, D. Vuckovic, Systematic assessment of seven solvent and solid-phase extraction methods for metabolomics analysis of human plasma by LC-MS, Sci. Rep. 6 (2016) 38885.

[45] K. Tong, Z.L. Li, X. Sun, S. Yan, M.J. Jiang, M.S. Deng, J. Chen, J. Li, M. Tian, Metabolomics approach reveals annual metabolic variation in roots of *Cyathula officinalis* Kuan based on gas chromatography-mass spectrum, Chin. Med. 12 (2017) 12.

[46] I. Martínez-Arranz, R. Mayo, M. Pérez-Cormenzana, I. Mincholé, L. Salazar, C. Alonso, J.M. Mato, Enhancing metabolomics research through data mining, J. Proteome 127 (2015) 275–288.

[47] A. Smolinska, L. Blanchet, L.M. Buydens, S.S. Wijmenga, NMR and pattern recognition methods in metabolomics: from data acquisition to biomarker discovery: a review, Anal. Chim. Acta 750 (2012) 82–97.

[48] M. Eliasson, S. Rännar, R. Madsen, M.A. Donten, E. Marsden-Edwards, T. Moritz, J.P. Shockcor, E. Johansson, J. Trygg, Strategy for optimizing LC-MS data processing in metabolomics: a design of experiments approach, Anal. Chem. 84 (2012) 6869–6876.

[49] P.S. Gromski, H. Muhamadali, D.I. Ellis, Y. Xu, E. Correa, M.L. Turner, R. Goodacre, A tutorial review: metabolomics and partial least squares-discriminant analysis—a marriage of convenience or a shotgun wedding, Anal. Chim. Acta 879 (2015) 10–23.
[50] W. Chan, Z.W. Cai, Aristolochic acid induced changes in the metabolic profile of rat urine, J. Pharm. Biomed. Anal. 46 (2008) 757–762.
[51] M.J. Chen, M.M. Su, L.P. Zhao, J. Jiang, P. Liu, J.Y. Cheng, Y.J. Lai, Y.M. Liu, W. Jia, Metabonomic study of aristolochic acid-induced nephrotoxicity in rats, J. Proteome Res. 5 (2006) 995–1002.
[52] Y. Ni, M.M. Su, Y.P. Qiu, M.J. Chen, Y.M. Liu, A.H. Zhao, W. Jia, Metabolic profiling using combined GC–MS and LC–MS provides a systems understanding of aristolochic acid-induced nephrotoxicity in rat, FEBS Lett. 581 (2007) 707–711.
[53] W. Chan, K.C. Lee, N. Liu, R.N.S. Wong, H.W. Liu, Z.W. Cai, Liquid chromatography/mass spectrometry for metabonomics investigation of the biochemical effects induced by aristolochic acid in rats: the use of information-dependent acquisition for biomarker identification, Rapid Commun. Mass Spectrom. 22 (2008) 873–880.
[54] Y.Y. Zhao, H.L. Wang, X.L. Cheng, F. Wei, X. Bai, R.C. Lin, N.D. Vaziri, Metabolomics analysis reveals the association between lipid abnormalities and oxidative stress, inflammation, fibrosis and Nrf2 dysfunction in aristolochic acid-induced nephropathy, Sci. Rep. 5 (2015) 12936.
[55] Y.Y. Zhao, D.D. Tang, H. Chen, J.R. Mao, X. Bai, X.H. Cheng, X.Y. Xiao, Urinary metabolomics and biomarkers of aristolochic acid nephrotoxicity by UPLC-QTOF/HDMS, Bioanalysis 7 (2015) 685–700.
[56] S.H. Lin, W. Chan, J.H. Li, Z.W. Cai, Liquid chromatography/mass spectrometry for investigating the biochemical effects induced by aristolochic acid in rats: the plasma metabolome, Rapid Commun. Mass Spectrom. 24 (2010) 1312–1318.
[57] X.Y. Zhang, H.F. Wu, P.Q. Liao, X.J. Li, J.Z. Ni, F.K. Pei, NMR-based metabonomic study on the subacute toxicity of aristolochic acid in rats, Food Chem. Toxicol. 44 (2006) 1006–1014.
[58] X. Liu, Y. Xiao, H.C. Gao, J. Ren, D.H. Lin, NMR-based metabonomic study on the subacute toxicity of aristolochic acid in rats, J. Chin. Univ. 31 (2010) 927–932.
[59] P.P. Guo, J.S. Wang, G. Dong, D.D. Wei, M.H. Li, M.H. Yang, L.Y. Kong, NMR-based metabolomics approach to study the chronic toxicity of crude ricin from castor bean kernels on rats, Mol. BioSyst. 10 (2014) 2426–2440.
[60] B. Ma, H. Qi, J. Li, H. Xu, B. Chi, J.W. Zhu, L.S. Yu, G.H. An, Q. Zhang, Triptolide disrupts fatty acids and peroxisome proliferator-activated receptor (PPAR) levels in male mice testes followed by testicular injury: a GC–MS based metabolomics study, Toxicology 336 (2015) 84–95.
[61] B. Sun, L. Li, S.M. Wu, Q. Zhang, H.J. Li, H.B. Chen, F.M. Li, F.T. Dong, X.Z. Yan, Metabolomic analysis of biofluids from rats treated with aconitum alkaloids using nuclear magnetic resonance and gas chromatography/time-of-flight mass spectrometry, Anal. Biochem. 395 (2009) 125–133.
[62] Y.Y. Chen, J.A. Duan, J.M. Guo, E.X. Shang, Y.P. Tang, Y.F. Qian, W.W. Tao, P. Liu, Yuanhuapine-induced intestinal and hepatotoxicity were correlated with disturbance of amino acids, lipids, carbohydrate metabolism and gut microflora function: a rat urine metabonomic study, J. Chromatogr. B 1026 (2015) 183–192.
[63] L. Wei, P. Liao, H. Wu, X. Li, F. Pei, W. Li, Y. Wu, Toxicological effects of cinnabar in rats by NMR-based metabolic profiling of urine and serum, Toxicol. Appl. Pharm. 227 (2008) 417–429.
[64] X.J. Wang, H.Y. Wang, A.H. Zhang, X. Lu, H. Sun, H. Dong, P. Wang, Metabolomics study on the toxicity of aconite root and its processed products using ultraperformance liquid-chromatography/electrospray-ionization synapt high-definition mass spectrometry coupled with pattern recognition approach and ingenuity pathways analysis, J. Proteome Res. 11 (2012) 1284–1301.

[65] Y.M. Cai, Y. Gao, G.G. Tan, S. Wu, X. Dong, Z.Y. Lou, Z.Y. Zhu, Y.F. Chai, Myocardial lipidomics profiling delineate the toxicity of traditional Chinese medicine *Aconiti Lateralis* radix praeparata, J. Ethnopharmacol. 147 (2013) 349–356.

[66] B.W. Tang, J.J. Ding, F.H. Wu, L. Chen, Y.X. Yang, F.Y. Song, ^{1}H NMR-based metabonomics study of the urinary biochemical changes in Kansui treated rat, J. Ethnopharmacol. 141 (2012) 134–142.

[67] Z.H. Zhang, Y.Y. Zhao, X.L. Cheng, Z. Dai, C. Zhou, X. Bai, R.C. Lin, General toxicity of *Pinellia ternata* (Thunb.) Berit. in rat: a metabonomic method for profiling of serum metabolic changes, J. Ethnopharmacol. 149 (2013) 303–310.

[68] Y.T. Zhou, Q.F. Liao, M.N. Lin, X.J. Deng, P.T. Zhang, M.C. Yao, L. Zhang, Z.Y. Xie, Combination of ^{1}H NMR- and GC-MS-based metabonomics to study on the toxicity of Coptidis rhizome in rats, PLoS ONE 9 (2014), e88281.

[69] F. Lu, M. Cao, B. Wu, X.Z. Li, H.Y. Liu, D.Z. Chen, S.M. Liu, Urinary metabonomics study on toxicity biomarker discovery in rats treated with Xanthii Fructus, J. Ethnopharmacol. 149 (2013) 311–320.

[70] Q. Liang, C. Ni, X.Z. Yan, M. Xie, Y.X. Zhang, Q. Zhang, M.J. Yang, S.Q. Peng, Y.Z. Zhang, Comparative study on metabonomics and on liver and kidney toxicity of Aristolochia fangchi and Stephania tetrandra, China J. Chin. Mater. Med. 35 (2010) 2882–2888.

[71] Q. Liang, C. Ni, M. Xie, Q. Zhang, Y.X. Zhang, X.Z. Yan, M.J. Yang, S.Q. Peng, Y.Z. Zhang, Nephrotoxicity study of *Aristolochia fangchi* in rats by metabonomics, J. Chin. Integrat. Med. 7 (2009) 746–752.

[72] L. Wei, P.Q. Liao, H.F. Wu, X.J. Li, F.K. Pei, W.S. Li, Y.J. Wu, Metabolic profiling studies on the toxicological effects of realgar in rats by ^{1}H NMR spectroscopy, Toxicol. Appl. Pharmacol. 234 (2009) 314–325.

[73] Y. Huang, Y. Tian, G. Li, Y. Li, X. Yin, C. Peng, F. Xu, Z. Zhang, Discovery of safety biomarkers for realgar in rat urine using UFLC-IT-TOF/MS and ^{1}H NMR based metabolomics, Anal. Bioanal. Chem. 405 (2013) 4811–4822.

[74] D.M. Tsai, J.J. Kang, S.S. Lee, S.Y. Wang, I.L. Tsai, G.Y. Chen, H.W. Liao, L. Wei-Chu, C.H. Kuo, Y.J. Tseng, Metabolomic analysis of complex chinese remedies: examples of induced nephrotoxicity in the mouse from a series of remedies containing aristolochic acid, Evid. Based Complement. Altern. Med. 2013 (2013) 263757.

[75] W.F. Xu, H.F. Wang, G. Chen, W. Li, R.W. Xiang, Y.H. Pei, ^{1}H NMR-based metabonomics study on the toxicity alleviation effect of other traditional Chinese medicines in Niuhuang Jiedu tablet to realgar (As_2S_2), J. Ethnopharmacol. 148 (2013) 88–98.

[76] W.F. Xu, H.F. Wang, G. Chen, W. Li, R.W. Xiang, X.L. Zhang, Y.H. Pei, A metabolic profiling analysis of the acute toxicological effects of the realgar (As_2S_2) combined with other herbs in Niuhuang Jiedu tablet using ^{1}H NMR spectroscopy, J. Ethnopharmacol. 153 (2014) 771–781.

[77] H.F. Wang, J. Bai, G. Chen, W. Li, R.W. Xiang, G.Y. Su, Y.H. Pei, A metabolic profiling analysis of the acute hepatotoxicity and nephrotoxicity of Zhusha Anshen Wan compared with cinnabar in rats using ^{1}H NMR spectroscopy, J. Ethnopharmacol. 146 (2013) 572–580.

[78] J. Michl, M.J. Ingrouille, M.S. Simmonds, M. Heinrich, Naturally occurring aristolochic acid analogues and their toxicities, Nat. Prod. Rep. 31 (2014) 676–693.

[79] Y.Y. Zhao, R.C. Lin, Metabolomics in nephrotoxicity (Chapter 3), Adv. Clin. Chem. 65 (2014) 69–89.

[80] X. Hu, J. Shen, X. Pu, N. Zheng, Z. Deng, Z. Zhang, H. Li, Urinary time- or dose-dependent metabolic biomarkers of aristolochic acid-induced nephrotoxicity in rats, Toxicol. Sci. 156 (2017) 123–132.

[81] J.L. Vanherweghem, M. Depierreux, C. Tielemans, D. Abramowicz, M. Dratwa, M. Jadoul, C. Richard, D. Vandervelde, D. Verbeelen, R. Vanhaelenfastre, M. Vanhaelen, Rapidly progressive interstitial renal fibrosis in young women: association with slimming regimen including Chinese herbs, Lancet 341 (1993) 387–391.

[82] J. Wu, Y. Shao, J. Liu, G. Chen, P.C. Ho, The medicinal use of realgar (As_4S_4) and its recent development as an anticancer agent, J. Ethnopharmacol. 135 (2011) 595–602.

[83] U. Mengs, C.D. Stotzem, Renal toxicity of aristolochic acid in rats as an example of nephrotoxicity testing in routine toxicology, Arch. Toxicol. 67 (1993) 307–311.

[84] J. Zhao, C. Xie, X. Mu, K.W. Krausz, D.P. Patel, X. Shi, X. Gao, Q. Wang, F.J. Gonzalez, Metabolic alterations in triptolide-induced acute hepatotoxicity, Biomed. Chromatogr. 32 (2018), e4299.

[85] J.Y. Aa, F. Shao, G.J. Wang, Q. Huang, W.B. Zha, B. Yan, T. Zheng, L.S. Liu, B. Cao, J. Shi, M.J. Li, C.Y. Zhao, X.W. Wang, Z.M. Wu, Gas chromatography time-of-flight mass spectrometry based metabolomic approach to evaluating toxicity of triptolide, Metabolomics 7 (2011) 217–225.

[86] T. Efferth, P.C. Li, V.S. Konkimalla, B. Kaina, From traditional Chinese medicine to rational cancer therapy, Trends Mol. Med. 13 (2007) 353–361.

[87] H.F. Wang, G.Y. Su, G. Chen, J. Bai, Y.H. Pei, ^{1}H NMR-based metabonomics of the protective effect of *Curcuma longa* and curcumin on cinnabar-induced hepatotoxicity and nephrotoxicity in rats, J. Funct. Foods 17 (2015) 459–467.

[88] B. Wu, M. Cao, S.M. Liu, Y. Jin, Y. Wang, Experiment study on hepatotoxicity induced by the water extract of cocklebur fruit in rats [J], Advers. Drug React. J. 12 (2010) 381–386.

[89] L.M. Xue, Q.Y. Zhang, P. Han, Y.P. Jiang, R.D. Yan, Y. Wang, K. Rahman, M. Jia, T. Han, L.P. Qin, Hepatotoxic constituents and toxicological mechanism of *Xanthium strumarium* L. fruits, J. Ethnopharmacol. 152 (2014) 272–282.

[90] T.H. Lee, J.J. Choi, D.H. Kim, K.R. Lee, M. Son, M. Jin, Gastroprokinetic effects of DA-9701, a new prokinetic agent formulated with Pharbitis semen and Corydalis tuber, Phytomedicine 15 (2008) 836–843.

[91] C. Ma, K.S. Bi, M. Zhang, D. Su, X.X. Fan, W. Ji, C. Wang, X.H. Chen, Metabonomic study of biochemical changes in the urine of morning glory seed treated rat, J. Pharm. Biomed. Anal. 53 (2010) 559–566.

[92] C. Ma, K.S. Bi, M. Zhang, D. Su, X.X. Fan, W. Ji, C. Wang, X.H. Chen, Toxicology effects of morning glory seed in rat: a metabonomic method for profiling of urine metabolic changes, J. Ethnopharmacol. 130 (2010) 134–142.

[93] E. Nyirimigabo, Y. Xu, Y. Li, Y. Wang, K. Agyemang, Y. Zhang, A review on phytochemistry, pharmacology and toxicology studies of *A. conitum*, J. Pharm. Pharmacol. 67 (2015) 1–19.

[94] H. Dong, A.H. Zhang, H. Sun, H.Y. Wang, X. Lu, M. Wang, B. Ni, X.J. Wang, Ingenuity pathways analysis of urine metabolomics phenotypes toxicity of Chuanwu in Wistar rats by UPLC-Q-TOF-HDMS coupled with pattern recognition methods†, Mol. BioSyst. 8 (2012) 1206–1221.

[95] H. Dong, G.L. Yan, Y. Han, H. Sun, A.H. Zhang, X.N. Li, X.J. Wang, UPLC-Q-TOF/MS-based metabolomic studies on the toxicity mechanisms of traditional Chinese medicine Chuanwu and the detoxification mechanisms of Gancao, Baishao, and Ganjiang, Chin. J. Nat. Med. 13 (2015) 687–698.

[96] J. Liu, J.Z. Shi, L.M. Yu, R.A. Goyer, M.P. Waalkes, Mercury in traditional medicines: is cinnabar toxicologically similar to common Mercurials? Exp. Biol. Med. 233 (2008) 810–817.

[97] L. Wei, R. Xue, P. Zhang, Y. Wu, X. Li, F. Pei, ^{1}H NMR-based metabolomics and neurotoxicity study of cerebrum and cerebellum in rats treated with cinnabar, a traditional Chinese medicine, OMICS 19 (2015) 490–498.

[98] J. Shen, J. Kai, Y. Tang, L. Zhang, S. Su, J.A. Duan, The chemical and biological properties of *Euphorbia kansui*, Am. J. Chin. Med. 44 (2016) 253–273.

Index

Note: Page numbers followed by *f* indicate figures and *t* indicate tables.

B

Printed in the United States
By Bookmasters